TRANSPORTATION INFRASTRUCTURE:
ROADS, HIGHWAYS, BRIDGES, AIRPORTS AND MASS TRANSIT

TRAFFIC INDUCED ENVIRONMENTAL VIBRATIONS AND CONTROLS

THEORY AND APPLICATION

TRANSPORTATION INFRASTRUCTURE: ROADS, HIGHWAYS, BRIDGES, AIRPORTS AND MASS TRANSIT

Additional books in this series can be found on Nova's website under the Series tab.

Additional e-books in this series can be found on Nova's website under the e-book tab.

U.S. TRANSIT, TRANSPORTATION AND INFRASTRUCTURE: CONSIDERATIONS AND DEVELOPMENTS

Additional books in this series can be found on Nova's website under the Series tab.

Additional e-books in this series can be found on Nova's website under the e-book tab.

TRANSPORTATION INFRASTRUCTURE:
ROADS, HIGHWAYS, BRIDGES, AIRPORTS AND MASS TRANSIT

TRAFFIC INDUCED ENVIRONMENTAL VIBRATIONS AND CONTROLS

THEORY AND APPLICATION

H. XIA
AND
R. CALÇADA
EDITORS

nova publishers
New York

Library of Congress Cataloging-in-Publication Data

Traffic induced environmental vibrations and controls : theory and application / editors, H. Xia (School of Civil Engineering, Beijing Jiaotong University, R. Calgada, Faculty of Engineering, University of Porto, Portugal, Beijing, China).
pages cm
Includes bibliographical references and index.
ISBN: 978-1-62618-240-0 (hardcover)
1. Traffic noise. 2. Noise control. 3. Railroads--Equipment and supplies--Vibration. 4. Motor vehicles--Vibration. 5. Soils--Vibration. 6. Roads--Environmental aspects. 7. Transportation, Automotive--Environmental aspects. I. Xia, H. (He), editor of compilation. II. Calgada, Rui, editor of compilation. III. Xia, H. (He). Problem of traffic-induced environmental vibrations and controls, an overview. [2011]
TD893.6.T7.T67 2013
620.3--dc23

2013011831

Published by Nova Science Publishers, Inc. † New York

CONTENTS

PREFACE

With the rapid development of modern industry and continuous expansion of big cities, the problem of vibration influence on the living and working environments has caused more concern of the people. The traffic-induced vibration, as one of the serious environmental pollutions, is one of the fundamental problems to be solved in planning urban traffic systems. How to predict, evaluate and mitigate the traffic-induced environmental vibration has become a worldwide research subject. This book contains the fundamental theory as well as numerical calculations and field experiments of many practical engineering projects. It contains many case studies concerning ground, underground and elevated rail transit systems.

In Chapter 1, a general review is made on the problem of traffic-induced environmental vibrations and the investigations on this subject. The characteristics of traffic-induced ground vibrations and their propagations from the aspects of duration, intensity distribution, vehicle load and speed are introduced. The vibration influences on the daily life of people, the operation of high-tech devices and the safety of buildings, the vibration properties of ground soil and buildings adjacent to underground, ground and elevated railway systems are summarized. The reduction and isolation measures against traffic-induced vibrations and the corresponding principles for planning and designing traffic systems are introduced.

Chapter 2 presents the railway traffic induced vibrations, comprising the prediction models, its experimental validation and its application on the analysis of mitigation measures. A 2.5D approach is adopted in developing the numerical model to consider the very large domain for solution of this problem. Moreover, the ability of the model for dealing efficiently with the train-track-ground dynamic interaction is highlighted. After the description of the numerical model main characteristics, its ability for dealing with real situations is evaluated through the comparison between prediction and experimental results for a case study developed by the authors. The experimental activities performed in that case study are presented, being the measurements of the vibrations induced by the train passage compared with the predicted results. Finally, after the experimental validation of the numerical model, its potentialities are highlighted namely in the analysis of the efficiency of ballast mat solutions on the abatement of vibrations induced by traffic.

In Chapter 3, the basic theory and analysis approaches in traffic-induced environmental vibrations are presented, mainly including vibration source, wave propagation, dynamic characteristics of soil, analytical model of train-track-ground, Green functions of ground displacements, and so on. By application of the Betti-Rayleigh Dynamic Reciprocal Theorem, the reciprocal property of Green function and the solution of soil/ground vibration subjected to moving loads are derived. Case studies show for receivers near the source, the R-wave contributes to ground vibration much more than the P-wave and the S-wave; while for receivers far from the source, the contribution of the P-wave is more obvious. For the source of moving constant-load, the ground vibration is featured of typical low frequency; while for moving harmonic-load, its frequency range is controlled by the Rayleigh's wave velocity of the top layer soil.

In Chapter 4, an efficient and accurate solution for the simulation of the dynamic behavior of very long structures, i.e., railways, roads or tunnel, is proposed by using 2.5D techniques in the context of the finite elements method, and adopting perfect matched layers (PML's) for the treatment of the boundaries due to the truncation of the finite elements mesh. Since the resource to 2.5D FEM-PML approach is not very usual, the equations of the 2.5D PML are derived, highlighting the compatibility with the 2.5D FEM. After that brief presentation of the model, a range of validation examples are depicted, showing the accuracy of the model. By last, a parametric study is presented in order to evaluate the influence of some tunnel and ground properties in the vibrations assessed at the ground surface.

Chapter 5 compares the ground vibration induced by trains moving on bridges using finite element analyses and field experiments. The finite element meshes included bridge superstructures, bridge foundations, moving trains, soils, and absorbing boundaries. The comparison indicates that the predicted numerical and measured experimental results are in good agreement, and the train-induced vibration will be large at the dominant frequencies determined by the train speed and the distance between two carriage centers. Different from the behavior of trains moving on embankments, the train-bridge-induced ground vibrations are not greatly influenced by rail irregularities, and only in the high frequency range over 15 Hz, the rail irregularity effect may not be ignored. However, those high frequency vibrations will be damped fast at a far field, such as 50 m from the railway.

In Chapter 6, the modeling and numerical simulation of human induced vertical vibrations of pedestrian bridges are reviewed. In particular, the load models of Bachmann and Seiler/Hüttner are discussed and compared to others. The simplified formula to assess the maximum accelerations of bridges with and without TMD's are presented. The measurements of human induced vibrations were performed on an aluminum bridge, which was excited in the second harmonic of the load function. Furthermore, a steel bridge being excited by the first harmonic of the load is investigated. The results obtained by a numerical simulation with a finite element analysis are in close agreement with the measured values. However, some models in literature are found to possess very large values of the Fourier coefficient of the second harmonic. In cases where the second harmonic of the load function is predominant, a realistic modeling of its size is important. Recommendations are given for the second Fourier term of the load, in order to obtain results which are in accordance with the measurements on real bridges. The measurements also indicate that higher Fourier coefficients may be significant for the response of the bridge at several measuring points.

In Chapter 7, the simulation of train loads, the propagation of metro-induced ground-borne vibration and the prediction of environmental vibration from underground trains are studied by combination of theoretic analysis, numerical simulation, in situ tests and empirical estimations. The simulation of train loads is achieved based on the analytical vehicle-track coupling system, and the propagation of metro-induced ground vibration in low frequency is studied both by an experiment in the Laboratory of track vibration abatement and by a 3-D track-tunnel-soil FE model simulation. As an extension research, a prediction method of environmental vibration from metro trains is studied by the coupled periodic FE-BE model which offers full modeling flexibility on the tunnel and free field compared to traditional 3-D FE model. Based on the research, two case-studies of vibration effect prediction from metro trains are presented, one for a laboratory equipped with high-precision transmission electron microscopy nearby a metro line in Beijing, another for a historic monument close to a metro line in Chengdu.

In Chapter 8, a three-dimensional numerical approach to simulate the ground vibration around high-speed railway viaducts is developed, in which the dynamic interactions between the train and the bridge as well as between the foundation and the ground are considered. To simplify the problem, the entire interaction system is divided into two subsystems: train-bridge interaction and foundation-ground interaction. In the stage of the train-bridge interaction subsystem, the dynamic responses of viaducts are calculated using a developed computer program to obtain dynamic reaction forces at the pier bottoms. Then, applying those reaction forces as input excitation to the foundation-ground interaction subsystem, the ground vibration around the viaducts is simulated using a general-purpose program. The numerical results are compared with experimental values to confirm the validity of the established numerical approach. Then, based on the dynamic characteristics of the bridge vibration, countermeasures to reduce the excessive vibration of both the bridge and the ground are designed and numerically evaluated.

In Chapter 9, the propagation properties for vibration waves and corresponding isolation measures are investigated in analytical, numerical and field measurement. Integral equations governing Rayleigh wave scattering are derived based on the Green's solution of Lamb problem. The integral equations, which are used to evaluate the complicated Rayleigh wave field generated by irregular scatter embedded in an elastic half-space solid, can be solved accurately and efficiently with an iteration technique. The passive isolation effectiveness of ground vibration by the piles and in-filled trench for screening Rayleigh wave is further studied in detail. In addition, the behavior of WIB isolation is conducted both by TLM and in-situ measurement. As case studies, the vibration isolation performance of these barriers are analyzed and explained.

In Chapter 10, the ground vibration and the building vibration caused by railway trains are studied by theoretical analysis, numerical simulation and field tests. Based on vehicle dynamics, track dynamics and the Green's functions of subsoil, the train-track-subsoil dynamic interaction model is established, in which the influence of track structure and wave motion in the three-dimensional layered-soil on ground vibrations are taken into account. The characteristics of ground vibrations are analyzed by case studies and experimental results. Combining with the train-track-subsoil model, the roadbed-subsoil-building dynamic interaction model is proposed by finite element method. Some laws of building vibrations are also summarized by results of field experiment. Based on the calculated results, the prediction formulae are proposed and validated by experimental results, which can be used to predict the ground vibration and building vibration.

In Chapter 11, a three-dimensional interaction model of coupled train-bridge-ground-building vibration system is established to investigate the environmental vibration caused by elevated rail transit, which is composed of the train-track-bridge submodel and the pier-foundation-soil-building submodel. The two submodels interact with each other through the contact forces on the pier tops of the viaduct. A field experiment was carried out at the elevated trial section of the Beijing Metro Line 5, where the ladder track is installed, to analyze the ground vibration induced by urban rail transit. The vibration responses were measured at rails, sleepers, bridge deck as well as the surrounding ground surface when the train running on the ladder track and the common slab track. A prediction formula for ground vibration induced by trains running on elevated bridges is established, based on the statistical regression analysis of the calculation results and verified by measured data.

In Chapter 12, the design and functional information of some typical elevated railway stations are introduced, and the structural characteristics and vibration problems of elevated railway stations are summarized. The dynamic interaction system of elevated railway stations and high-speed trains under the loads of train running, braking or starting are analyzed based on the vehicle model and the wheel-rail interaction assumptions considering the longitudinal motion. As a case study, the Wuhan Railway Station is analyzed with the cases of the high-speed train and ordinary speed train passing or braking in the station building, the effect of train speed, types and loading positions are analyzed and concluded.

The book also contains many case studies concerning practical engineering applications with real trains, tracks, bridges, tunnels and elevated railway stations. Many field experiments on the ground vibrations and building vibrations near the urban rail transit lines and high-speed railways in China, Portugal, Japan, Taiwan and Germany.

The book is intended to be a reference for solving traffic-induced environmental vibration problem that has to be considered in the planning and design of urban rail transit systems and high-speed railways. Moreover, results of recent scientific research are presented which are of interest for researchers active in this research field.

H. Xia
R. Calçada
December 15, 2012

In: Traffic Induced Environmental Vibrations and Controls ISBN: 978-1-62618-240-0
Editors: H. Xia and R. Calçada

Chapter 1

PROBLEM OF TRAFFIC-INDUCED ENVIRONMENTAL VIBRATIONS AND CONTROLS: AN OVERVIEW

He Xia*, Huile Li and Kunpeng Cui
School of Civil Engineering, Beijing Jiaotong University, Beijing, China

ABSTRACT

In this chapter, a general review is made on the problem of traffic-induced environmental vibrations and the investigations on this subject. The characteristics of traffic-induced ground vibrations and their propagations from the aspects of duration, intensity distribution, vehicle load and speed are introduced. The vibration influences on the daily life of people, the operation of high-tech devices and the safety of buildings, the vibration properties of ground soil and buildings adjacent to underground, ground and elevated railway systems are summarized. The reduction and isolation measures against traffic-induced vibrations and the corresponding principles for planning and designing traffic systems are introduced.

1. INTRODUCTION

As one of the main environmental pollutions, the vibration of environment that seemed to have been tolerated in the past is today increasingly being considered as a nuisance. With the rapid development of modern industry and the continuous expansion of big cities, the influence of vibration on the living and working environments of the people has caused to the attentions of metropolitan administrators, designers of traffic system and some experts of environmental problem in China and abroad.

According to the statistical analysis [1], traffic-induced vibrations occupy a high percentage in total public complaints on environmental vibrations, which is 14.2% in Japan, second only to those from industries and construction sites (see Figure 1). In fact, the factories may be built or removed away from the residential areas, and the vibration of

* E-mail address: hxia88@163.com.

construction sites occurs only during the period of construction. However, traffic-induced vibration often occurs around the busiest areas of the city, day and night, and thus has long-term influence on the residents or people working there, which can hardly be avoided.

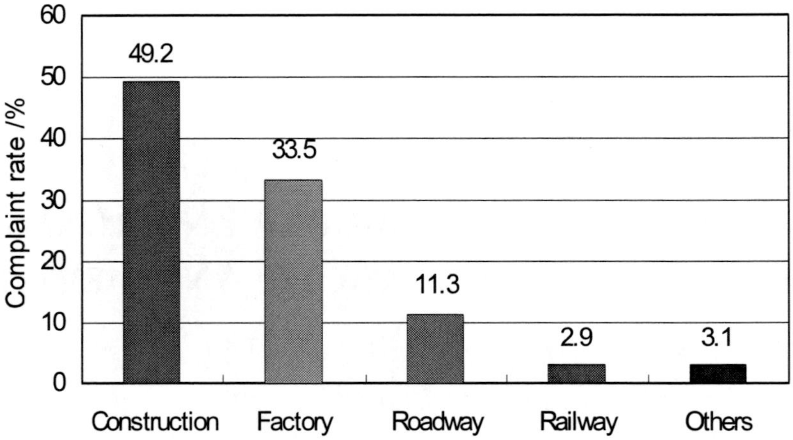

Figure 1. Public complaint on environmental vibrations (Japan).

In recent years, considerations on the environmental influence of traffic-induced vibrations become more important in designing and planning traffic systems. In the past, the residential blocks and buildings were relatively sparse, while at present, with the rapid growth of modern cities, the multi-level roads, metro lines and urban railways are forming a spatial cross traffic system, extending and intruding from underground, ground and space into the crowded residential areas, commercial centers and even cultural and high-tech research zones. In Tokyo and some other big cities, for instance, there are some elevated roads of 5 to 7 levels, with the minimum distances to adjacent buildings only several meters, or even inside the buildings (see Figure 2).

(a)

(b)

(c)

(d)

Figure 2. (a) Multi-level viaducts in Tokyo, (b) Elevated road close to a building in Sendai, (c) Monorail through a building in Chongqing and (d) Elevated rail through a building in Beijing.

On the other hand, with the raise of people's living level, the demand on the environment quality becomes higher. All these factors make the influence of traffic-induced vibrations become more intense, and have brought urgent requirement on study of this problem. The generation mechanism of traffic-induced vibrations, their propagation properties in the soil, the pollution to the environment, injuries to human health, and the countermeasures against this problem are all under study [2–21].

2. Traffic Induced Environmental Vibrations

2.1. Generation of Traffic-induced Vibrations

The vibration pollution refers to the vibrations of ground, buildings, traffic vehicles and so on induced by human activities, which influences the routine life and the physical and psychological health of the residents living near the vibration sources, the working conditions of the people, the safety of buildings and the normal operation of precise instruments.

The routing of a traffic line is based on the predicted passenger flows, so it is inevitable to go through crowded residential zones.

As shown in Figure 3, when a vehicle travels on a traffic line, it has an impact on the supporting structure and excites the structure to vibrate. The vibration is then transmitted from the structure to the ground, through the subgrade, the piers and foundations of a bridge, or the lining and invert foundation of a tunnel, and further spreads around, which causes the nearby buildings to vibrate, and further induces secondary vibrations and noises of the indoor objects.

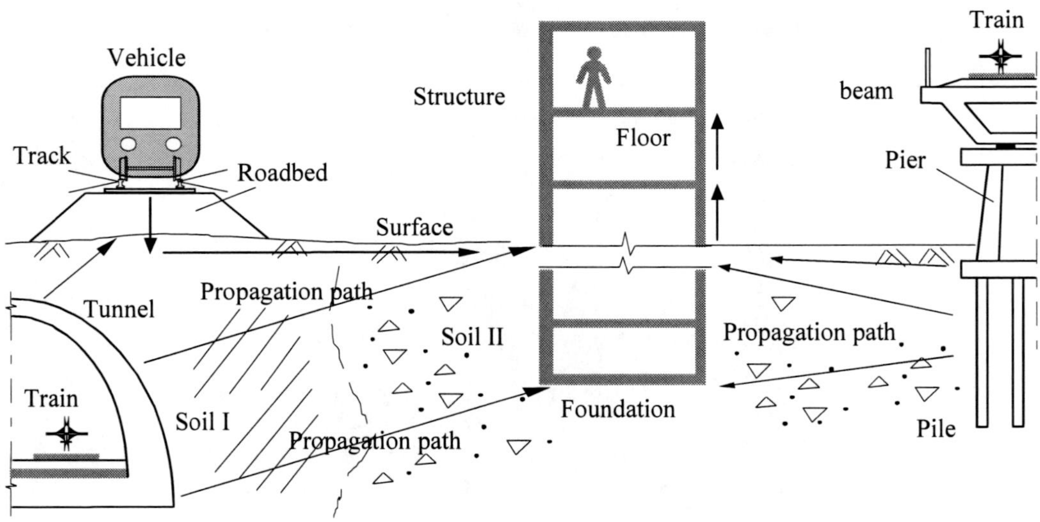

Figure 3. Generation of traffic-induced environmental vibrations.

2.2. Influence of Vibrations

2.2.1. Influence of Vibrations on Routine Living and Working of People

Generally, environmental vibration may not directly hurt a man's body, however, it may disturb his normal living, cause him to feel uncomfortable or upset, or even influence his sleeping, working and studying.

According to the investigation on 600 families in Japan [2], the highest complaint rate on traffic-induced vibration is its influence on human's sleeping, which is about 45% of the total, while the second is the mental injury to human and then the damage to buildings, which are 21% and 20%, respectively, as shown in Figure 4.

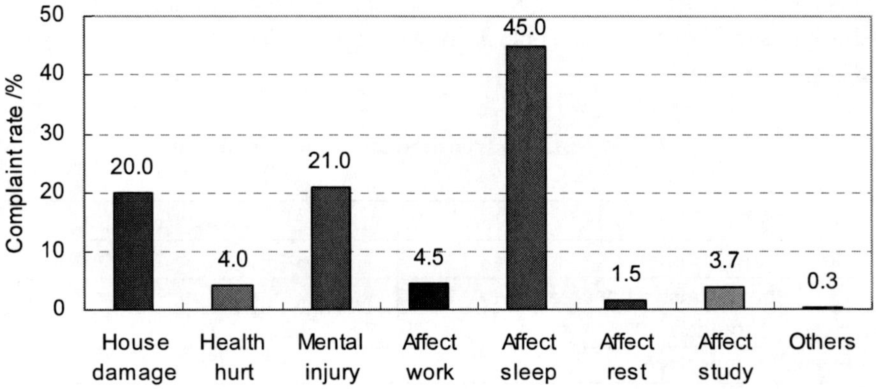

Figure 4. Public complaints on influence of traffic-induced vibrations (Japan).

An experiment shows that the higher the vibration level, the greater the influence on a man's sleeping. A vibration up to 60 dB has little effect on a normal man's sleeping, but it may affect a man sensitive to vibration or in unhealthy status. A 65 dB vibration has slight effect on a sleeper, a 69 dB may awake a light sleeper, a 74 dB may awake all the men but the sound sleeper, and a 79 dB will wake up all sound sleepers.

The upset feeling of a man is in accordance with his sensitivity to vibration, because the vibration sensing organs spread over a man's body. It has been proved that vibration may induce resonance of the inside organs of a man, thus even a slight vibration may cause upset of the man, and even harm his health.

Regarding influences on a man's working, vibration may affect his vision and hand action, which leads to decline of operating accuracy and speed, and even disoperation. In addition, vibration may obstruct a man from mind focus. Especially in the environment where vibration and noise exit simultaneously, the brain is interfered, thus it is difficult for a man to focus on judging, thinking, calculating and operating, thus leading to deficiency of his work.

2.2.2. Influence of Vibrations on Human's Health

Vibration influences on human's health involve various aspects of a man, such as blood circulatory system, respiratory system, digestive system, nervous system, auditory system, and visual system. An intense vibration may injure the bones, muscles, joints and ligaments. In the digestive system, it will increase stomach and intestinal peristalsis, decrease gastric secretion and digestion, and disorder liver detoxify metabolic. In the nervous system, it will cause sympathetic tone, tendon reflection diminish or disappear, fingers trembling, or insomnia. The influence of vibrations on human's health depends on the following factors:

(1) Frequency

Vibrations with different frequencies may induce different feeling and pathologies of the human. The perceptible frequencies of human beings are from 1 Hz to 1000 Hz, with the most sensitive range of 1−80 Hz in which concentrate the resonant frequencies of the internal organs. The human body consists of elastic tissue such as muscles, so its reaction to vibration is similar to that of a complicated mechanical system, where some predominant frequencies can be identified. Resonance can be expected when the vibration frequency is close to that of an organ, which brings various kinds of damage. The natural frequencies of different organs are also associated with the body posture. For example, a standing man is sensitive to 4−8 Hz most and a lying man to 1−2 Hz. Table 1 shows the typical natural frequencies of different organs of a standing man.

Table 1. Natural frequencies of man's organs

Organ system	Frequency /Hz
Body − lower vertebra	4 − 6
Body above waist − upper vertebra	10 − 14
Head − shoulder	20 − 30
Eyeball − inner eye	60 − 90
Lower jaw − cranial bones	100 − 200

(2) Amplitude

For a certain frequency, higher vibration amplitudes may induce greater effects on human's organs. Herein, the amplitude is a generalized one, including the amplitudes of displacement, velocity, as well as acceleration. The perceptive level of a man on vibration is estimated by acceleration or velocity, and acceleration is more popular. The higher the acceleration is, the greater the shock is and the more harmful to human health is. The influence of vibration amplitude on human body varies individually: for a certain amplitude, some people can stand, while some others may have a strong response.

(3) Exposure period

According to the duration time, vibrations can be divided into steady vibration, intermittent vibration and shock vibration. The steady vibration is induced by continuous excitation, such as machine operation. The intermittent vibration refers to that periodically takes place, such as the vibration induced by trains or trucks. The shock vibration is induced by impulsive forces, such as pile hammers and forging machines.

For all kinds of vibrations, the longer the exposure period is, the more serious the effect is. For those persons exposing to vibration environment, some of their organs may be injured by vibrations with different frequencies and exposure periods.

Drivers are persons working under vibration environment. An investigation on 370 tractor drivers reports 71%, 52% and 8% of them have pathological changes in bone joints, chest and lumbar vertebras, and as high as 40% of them in both chest and lumbar vertebras. The longer they worked, the higher the ratios of pathological changes, and a very high ratio of

80% is reported for the drivers who worked more than 10 years. Another investigation on 117 truck drivers and 95 bank staffs reveals that the ratios of pathological change in their vertebra are 43% and 20%, respectively, and the ages of truck drivers who have pathological changes are much younger.

In addition, influences of vibration on human health are related with the environment conditions such as temperature and air pollution, and individual physique such as age, body constitution, nutrition, smoking or not, and sensitivity to vibration.

2.2.3. Influence of Vibrations on Buildings

Traffic-induced vibrations may affect the structural safety of the ancient/old buildings or the working and daily life of the people inside the buildings. In Czech, some ancient masonry buildings close to busy roads were cracked owing to the vibrations induced by the passing vehicles (see Figure 5 left). With the growth of the cracks, some old churches and houses were damaged or even collapsed in Prague, Hustopece and hrusov areas [3].

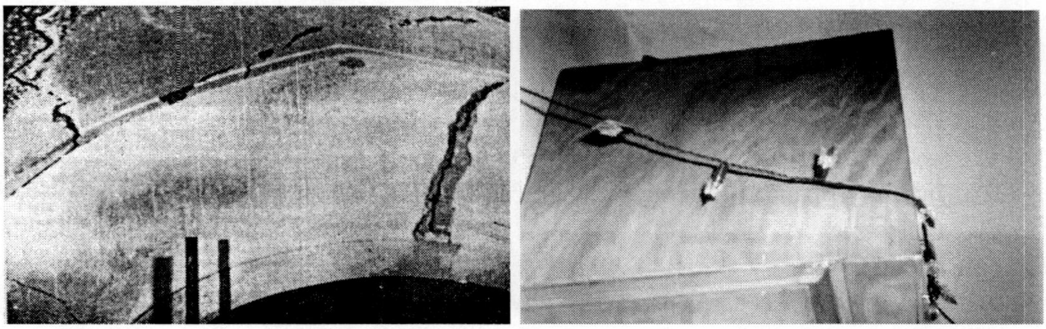

Figure 5. Cracks of an old church (left) and furniture shift (right) caused by environmental vibrations.

In many big cities such as Paris, Brussels and Tokyo, people can often feel vibration of the floors when they are at home or stay in hotels near rail traffics when the metro trains or urban trains are running. There have been also some reports on the public complaints against this problem in China.

In Beijing, a family living in a 5-storey building at 150 m to a railway line reflected that they could feel vibration and secondary noise in their rooms when a train passed, especially at night. Moreover, after a period, the furniture inside the room shifted a bit due to the vibration, which can be seen in Figure 5 (right) [4].

2.2.4. Influence of Vibrations on Precise Instruments

Environmental vibrations may influence the normal operation of precise instruments such as laser devices, electronic microscopes and electronic scales, and manufacture of high-precision products such as semiconductor integrated cores and photo-plate making, which may result in inaccurate readings, decrease the precisions of instruments, shorten their service life or even interrupt their work. Illustrated in Figure 6 is a comparison between the images under a microscope a 150 000 multiple in conditions with an ambient vibration and without vibration [1], from which an obvious difference can be observed.

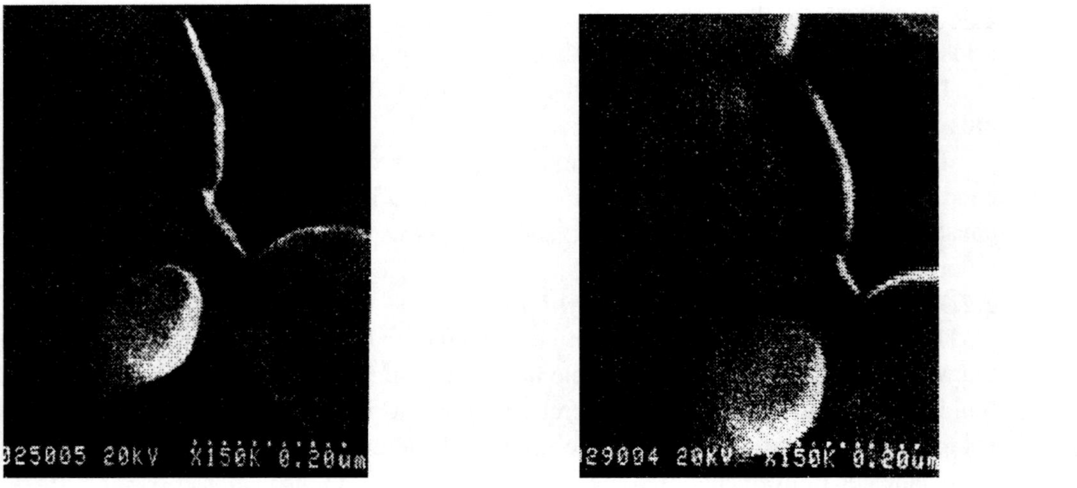

Figure 6. Disturbance of vibration on images in an electronic microscope.

There are other instances of vibration disturbance on precise instrument operation. During the construction of the No. 4 Metro Line in Beijing, the anxiety arose about the possible effect of train-induced vibration on the precise instruments in the nearby labs of Peking University (see Figure 7 left). Despite in this area, the bottom of metro tunnel is 17 m below the ground surface, and the horizontal distances to the concerned buildings are from 39 m to 70 m. In this case, the requirement of the instruments on the environmental vibration is 0.5-100 μ*g* for the frequency 0−10 Hz, and 2 μ*g* for the frequency ≤1 Hz. A special investigation was carried out, and this problem has been successfully solved with some countermeasures.

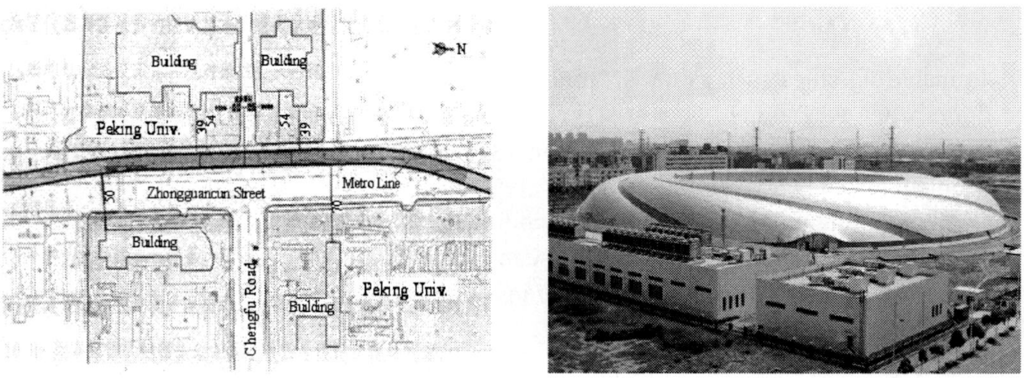

Figure 7. Influence of metro trains on precise instruments in nearby buildings.

Similar concern occurred in the Shanghai Light-source Project that contains sophisticated equipment, such as electron linear accelerator, beam line station and electronic storage ring, which requires the vertical stability of the track to be less than 1 μm. After investigation, a conch-like house (Figure 7 right) with vibration-isolation foundation was built to ensure the stability of the scientific facilities against disturbance of metro and ground traffics around it.

2.3. Research on Traffic-induced Environmental Vibrations

In recent years, new challenges have risen for research of traffic-induced vibration and its influences on the environment, and drawn much attention of scholars in China and abroad.

Japan is a country with serious vibration pollution. In its Basic Law of Pollution Harms and Policies, vibration is stipulated as one of the seven main pollution harms, and efficient measures shall be taken to limit the vibration level. In the Law of Limiting Vibration, traffic-induced vibration is specially mentioned. Yasutoshi [5], Takemiya and Kojima [6] studied the generation mechanism and propagation laws of traffic-induced vibrations, and proposed a method for predicting vibration levels of surrounding environments.

Facing the strong public complaints, the UK Railway organized the experiment on ground vibration induced by trains. Specialists studied the influences of train speeds, exciting frequencies and track structure parameters, based on which the BS7385 Standard for evaluation of environmental vibrations was issued. Krylov [7] established an analysis model to study the influences of track, train and subsoil on ground vibration. To verify the effect of critical speed for high-speed trains, he conducted a joint-experiment with SNCF on the high-speed railway between Courtalain and Tours in France, with the TGV train running at a speed higher than 500 km/h. In this experiment, vibration with dramatically high amplitude and propagation effect of vibration wave were observed in the track and the ground soil.

In the United States, Wilson et al. [8] proposed that vibration induced by railway trains could be mitigated through improving the structure type of track and optimizing the bogie structure to lower the wheel/rail contact force. Kurzweil [9] investigated the propagation path and attenuation characteristics of metro train-induced vibration waves in different strata, and the secondary vibration and noise in the surrounding buildings.

In planning the metro line passing the residential areas, business districts, churches and museums in Zurich downtown, Zach and Rutishauser studied the vibration frequency and acceleration characteristics of the train and the tunnel structure. They proposed the measures via improving track structure to mitigate the metro train-induced vibrations.

In many European countries, the construction of high-speed railway network has brought notice of people on traffic-induced vibrations, and several investigations have been carried out to find the solution. The high-speed railways connecting Paris, Brussels, Amsterdam and Cologne cross some streets and cultural monuments. It was predicted several ancient buildings nearby might be injured by the train-induced vibration, or their operation functions be disturbed. This has been concerned in the early stage of the construction.

In Belgium, several high-speed railways reforming from existing railways often traverse towns, and some of their tracks are closely against the foundation of existing buildings (Figure 8 left). To study the transmission laws of ground vibration induced by high-speed trains, Degrande and Lombaert [10] carried out a filed test on the high-speed railway between Paris and Brussels. They also performed an impact-loading test with a truck and numerical simulation to study the properties of the vibration sources and the transfer relationships between the vibration source and the receivers (see Figure 8 right).

At the Antwerp station, the high-speed railway runs beneath an ancient viaduct with a history of more than 200 years. To prevent the viaduct from the vibration disturbance of the trains, millions of Euros were invested by the Belgium government to reinforce it by building an integrated PC support structure (see Figure 9).

Figure 8. High-speed railway through a town (left) and impact test using a truck (right).

Figure 9. Protection of an ancient viaduct against disturbance of train-induced vibration (Belgium).

In France, Netherlands, Germany and some other countries, efforts have also been bestowed on this problem [11-14]. Many field tests on high-speed train induced environmental vibrations have been carried out, such as those on Brussels-Paris in Belgium (200−314 km/h), Amsterdam-Utrent in the Netherlands (40−160 km/h) and ICE in Germany (100−300 km/h). The propagating and attenuating properties of traffic-induced vibrations were systematically studied by statistical analysis based on the measured data.

In Sweden, the high-speed railway from Göteborg to Malmö traverses the area with soft soil, especially in the western coast, where the Rayleigh wave speed is only 45 m/s. In 1998, the Royal Institute of Technology carried out a field experiment in Ledsgård [15]. When the X2000 train ran at 180 km/h, exceeding the Rayleigh wave speed, the ground vibration was enlarged by more than 10 times, and when the train ran at 200 km/h, the maximum track displacement was as high as 14 mm, much exceeded the limit for running safety of train. Such intensive vibration may not only affect the running safety of the train, but also accelerate the accumulative damage of the track and the subgrade.

There are various vibration isolation and reduction measures for traffic-induced vibrations. They are implemented in two aspects: (1) vibration source control, such as using heavy rails, isolation mats and floating tracks; (2) propagation path control, such as adopting ground barriers, open trenches, filling ditches, row-piles, and wave impeding blocks.

Source control is one of the main measures to mitigate ground vibration induced by traffics. Researchers developed various kinds of vibration isolation or reduction tracks, and achieved good effect. Floating tracks, for example, were installed to reduce the vibration of trams in some cities in the US, Germany, France and some other countries. In Brussels, the CDM Company designed a “calm track” composed of elastic sleepers and embedded rails (Figure 10 left), which can supply sufficient damping to mitigate the impact of wheels on the ground, thus the tram can run on the road calmly.

The Basle Concert Hall in Switzerland was built in 1876, which is one of the top 10 best concert halls in the world. However, its service was severely affected by the secondary noise, which once reached 46 dB(A), caused by the indoor acceleration when nearby trams passed. To isolate the vibration, the Gerb Company installed 1240 mm-thick floating slabs with steel springs under the 170 m-long track in front of the hall (Figure 10 right), which reduced the secondary indoor noise by more than 20 dB(A).

Figure 10. The “calm track” on the street in Belgium (left) and the isolation devices under the track in front of the Basle Concert Hall in Switzerland (right).

For the control of propagation path, Kattis et al. [16] studied the isolation system using discontinuous barriers. With the homogenization theory of composite material, they proposed an equivalent approximation method between discontinuous and continuous barriers. Takemiya [17], Peplow and Kaynia [18] studied the WIB isolation system by numerical simulation and experimental analysis. Adam analyzed the isolation effect of open trenches and filling ditches using a coupled method of finite element and boundary element.

In china, there are more and more cities where the rail transit systems are in operation or under construction. In construction of these systems, influences of traffic-induced vibrations on the environments were predicted, such as that of Metro Line 4 on nearby universities and scientific institutes, and Metro Line 6 on the ancient buildings near the Forbidden City in Beijing, the Beijing-Shanghai high-speed railway on the Huqiu Inclined Tower in Suzhou. Researchers studied traffic-induced vibrations via theoretical analyses, numerical simulations and field tests, and acquired successful experiences. Therein, a typical case is the sensitive section of Beijing Metro Line 13, which passes underground between two residence buildings, with the gap between the tunnel edge and the building foundation only 1.7 m. After installing the floating slab tracks, the vibration has been reduced by 25 dB. Now, the residents feel no vibration when trains pass. In the National Grand Theater, the isolation technique was applied for the recording room, gaining a 40 dB transmission loss of vibration, efficiently eliminated the influence of the nearby metro trains. The environmental vibration assessment has become one of the necessary procedures in planning a new railway system.

In Taiwan, the possible influence of the Taipei-Kaohsiung high-speed railway on the Science and Technology Park in southern Taiwan brought the attention of researchers. Yang [19] and Ju [20] systematically studied the high-speed train induced vibrations and the corresponding countermeasures to reduce such influence.

In 2003, the first international symposium on environmental vibration was held by Zhejiang University in China. Afterwards, the second to fifth symposiums were respectively organized by Okayama University (2005), National Taiwan University (2007), Beijing Jiaotong University (2009) and Southwest Jiaotong University (2011), which promoted the research on the environmental vibration.

3. CHARACTERISTICS OF TRAFFIC INDUCED VIBRATIONS

3.1. Duration

Long duration is the first characteristic of traffic-induced vibrations. In a traffic system, there are always continuous flows of trains or automobiles, year after year, every day in a week, or even day and night.

For high-speed railways, the vibrations induced by trains are intermittent, while the time of each train's passage is relatively short. The standard vibration level of the Shinkansen train induced ground vibration is of trapezoid form, with the lasting duration being 7 s for a 16-car train at a speed of 200 km/h (Figure 11 left); while for a 12-car train at 240 km/h, the duration may be shortened to 5.4 s. However, owing to its huge traffic volume, such vibration takes place in every 3–6 min, thus the influence is considerable.

For ordinary double track railways, the interval between trains can be as short as 5−10 min, and the lasting period, according to the speed and the total length of the train, can be as long as 2−4 min (Figure 11 right). Therefore, the influence time of train induced environmental vibrations can be 1/5−1/3 of the total time.

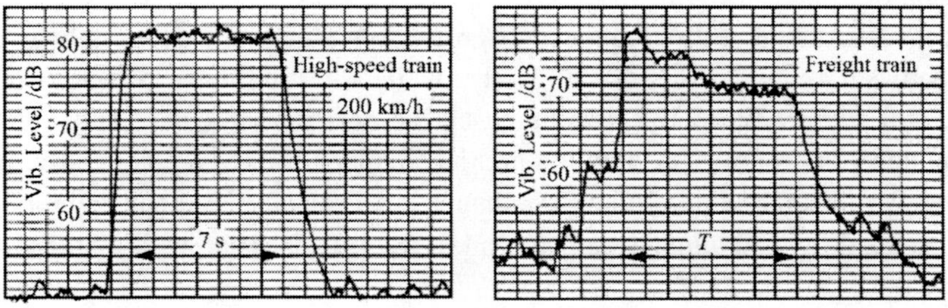

Figure 11. Duration characteristics of high-speed trains (left) and freight trains (right).

For metro systems, the normal operation time is longer than 18 hours, while the departure interval is as short as 2 minutes. According to the investigation, there can be 25−30 or more pairs of trains in two directions per hour in a metro line. For each train passage, the induced ground vibration will last about 10−15 seconds. Therefore, the influence duration of metro train induced vibration will occupy 15−20% of the total operation time of the metro system.

3.2. Circulation Times

The second characteristic of traffic-induced environmental vibration is its numerous circulation times. Illustrated in Figure 12 are the effective circulation times of different vibration sources for structures with 30 year's service period. It shows the railway and roadway, as the vibration sources, have the most influence in circulation times.

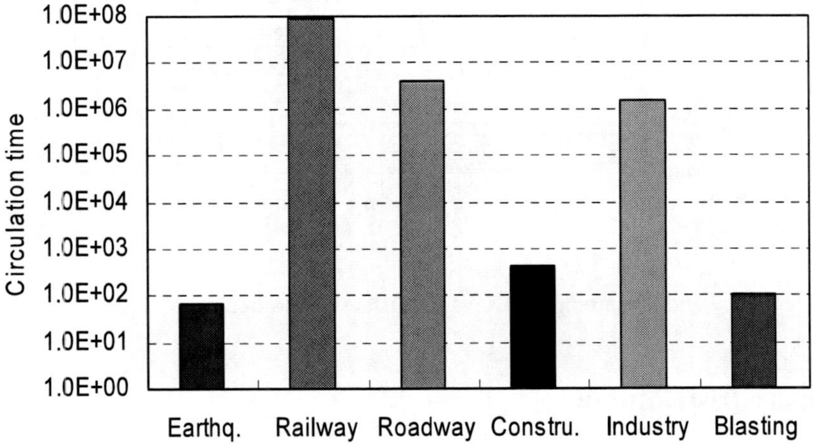

Figure 12. Effective circulation times of different vibration sources.

3.3. Amplitude and Frequency

Another characteristic of traffic-induced environmental vibration is micro amplitude and low frequency, which is completely different from earthquake, construction, industry and other vibration sources (Figure 13).

It is investigated that the frequency of ground vibration induced by railway and roadway ranges between several Hertz to 30 Hertz, while the amplitude is very small, with the dynamic soil strain around 10^{-5}, remaining elastic deformation stage. Thus, the research on traffic-induced environmental vibration and its countermeasures are generally focused in the micro scope with ground amplitude less than 1 mm, velocity less than 1 mm/s, acceleration less than several μg, and dynamic soil strain less than 10^{-4}, as shown in Figure 14.

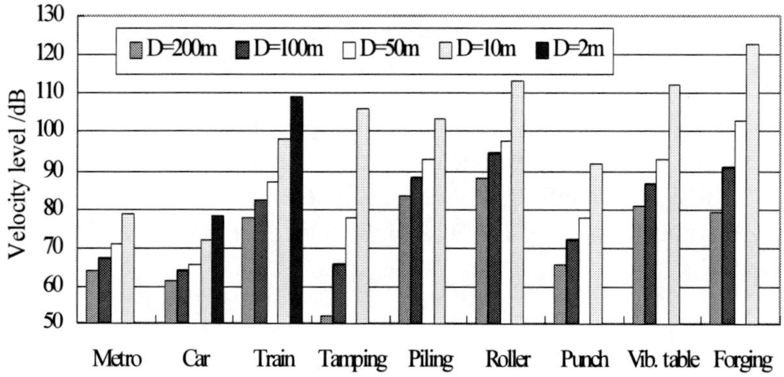

Figure 13. Ground velocity levels induced by different vibration sources (D is distance to track).

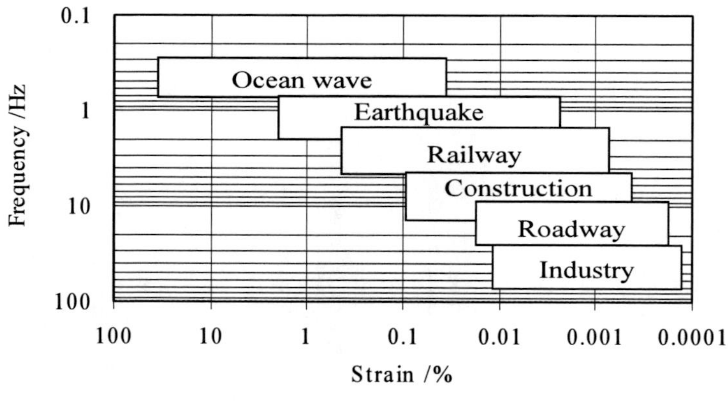

Figure 14. Frequency and strain ranges of soils for different vibration sources [2].

3.4. Intensity Distribution

3.4.1. Distribution of Vibration Intensity versus Depth

The most commonly used distribution curves of vibration attenuating versus the depth under ground were proposed by Richart and Woods in 1970 [21], which gave a group of horizontal and vertical vibrations induced by vertical harmonic loads versus the depth in soils with different Poisson ratios, as seen in Figure 15. It shows that the vibration close to the ground surface is the maximum; it attenuates rapidly with the increase of the depth, and becomes very small when the depth is bigger than the vibration wavelength.

Some other experiments [17] pointed the approximation of the curves in Figure 15. They showed the acceleration at 2 m underground was 20%−50% that at the surface, and further decreased to 10%−30% at 4 m. In general, the vibration intensity decreased with the depth, while at frequency around 24 Hz, there occurred some resonance zone at certain depth, where the vibration was amplified. As the soils are layered, vibrations with different frequencies attenuate in different extents in different layers. Therefore, depth is a complicated factor to be considered in metro design.

3.4.2. Distribution of Vibration Intensity vs. Distance to Source

Traffic-induced ground vibration propagates in the soil in three forms: S (shear) wave, P (compression) wave and R (Rayleigh, surface) wave. Owing to the spread of energy and the absorption of soils, the vibration intensity attenuates during its all-around propagation.

Many investigations believe a general law that train-induced ground vibrations attenuate versus the distance to the track. Some analyses show that for the point source, the vibration attenuates versus distance in the following pattern:

$$A(r) = A_0 \cdot r^{-n} \cdot \exp(-\frac{2\pi \cdot f \cdot \beta_c \cdot r}{c}) \tag{1}$$

where: A_0 is vibration amplitude near the source; $A(r)$ is amplitude at distance r to the source; f is excitation frequency; β_c is inner damping coefficient of the soil; c is propagation speed; n is a factor, $n = 0.5$ for surface wave and $n = 0.2$ for body wave in infinite half-space.

For the line vibration source, the attenuation can be expressed as the following form:

$$A(r) = \frac{1}{2} A_0 \cdot K_0 \cdot r^{-k_z} \cdot \exp(-\alpha_z \cdot r) V^{3/4} \tag{2}$$

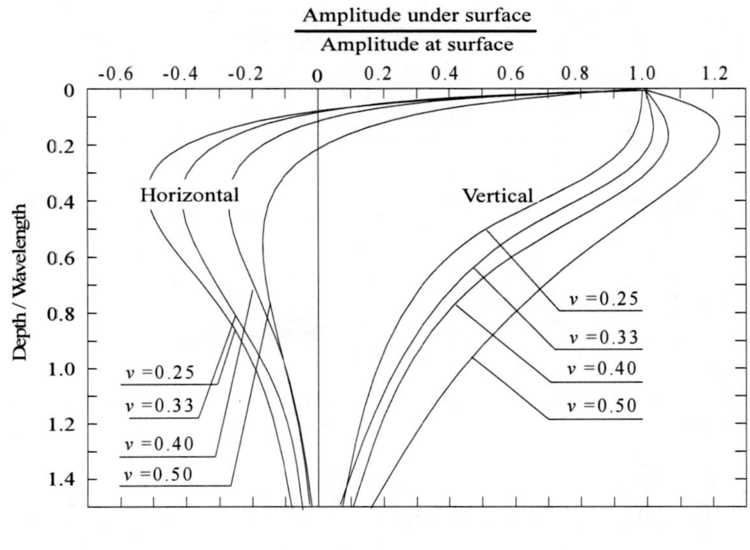

Figure 15. Vibration attenuation under ground surface.

where: A_0 is vibration amplitude near the source; $A(r)$ is amplitude at distance r to the source; V is train speed; k_z is comprehensive attenuation factor, $k_z = 0.75$ for vertical vibration and $k_z = 0.3$ for horizontal. K_0 and α_z are the amplitude factor and the energy absorption factor for different soils, which can be found in [4].

As for the area vibration sources, the attenuation pattern can be expressed as:

$$A(r) = A_0 \cdot \left[\frac{r_0}{r} \cdot \xi + \sqrt{\frac{r_0}{r}} \cdot (1-\xi) \right] \cdot \exp[-\alpha_0 \cdot f_0 \cdot (r - r_0)] \tag{3}$$

where: A_0 is vibration amplitude near the source; $A(r)$ is amplitude at distance r to the source; f_0 is the excitation frequency; r_0 is the radius of the area source; ξ is the geometrical attenuation factor, and α_0 is energy absorption factor of soils, which can be found in [22].

The characteristics of vibration attenuation depend on the types of vibration sources and the properties of soil. Some filed measurements reveal that generally the bigger the viscous coefficient of the soil is, the faster the vibration attenuates; while the higher the density of the soil is, the slower the vibration attenuates.

Shown in Figure 16 (top) are the investigated environmental vibrations in several cities [23]. The results show rather high levels of railway-induced ground vibrations: the vibrations measured at about 30 m away from the tracks in most of the places are as high as 80 dB, while they attenuate rapidly with distance to the track.

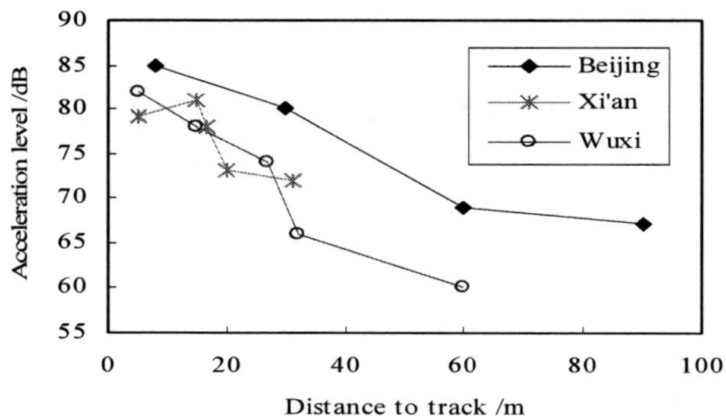

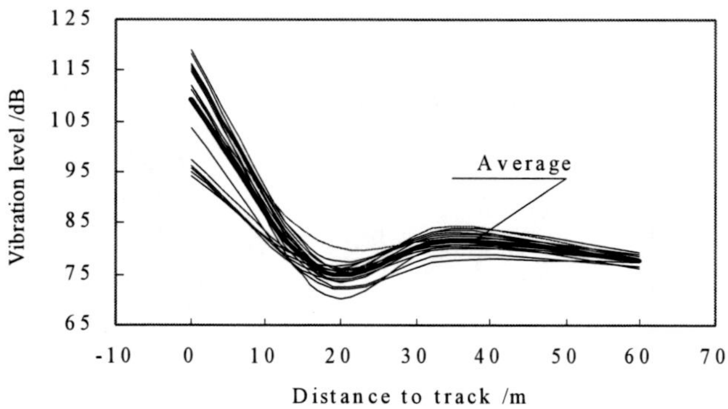

Figure 16. Vibration attenuation vs. distance measured in several cities (top) and in Shenyang (bottom).

Many investigations support the above laws, while there are also some reports showing different patterns. In the distribution curves of vibrations versus distance from a railway site in Shenyang [24], there exists a zone where the vibrations are magnified, see Figure 16 (bottom). This law is in accordance with that from the theoretical analysis [23].

In the field measurement on high-speed train induced ground vibrations, Degrande et al. [10] found the similar phenomenon that at the train speed of 256–314 km/h, there occurred also a magnifying zone in the area of 20–40 m to the HSR line, as shown in Figure 17 (left).

The measurement at a Shinkansen bridge in Japan shows the ground vibrations attenuate in different patterns with their frequency component [25]: for f>30 Hz, they monotonously decrease with the distance to the railway, while for f=10–20 Hz, there occurred several zones where the ground accelerations were magnified, which can be seen in Figure 17 (right).

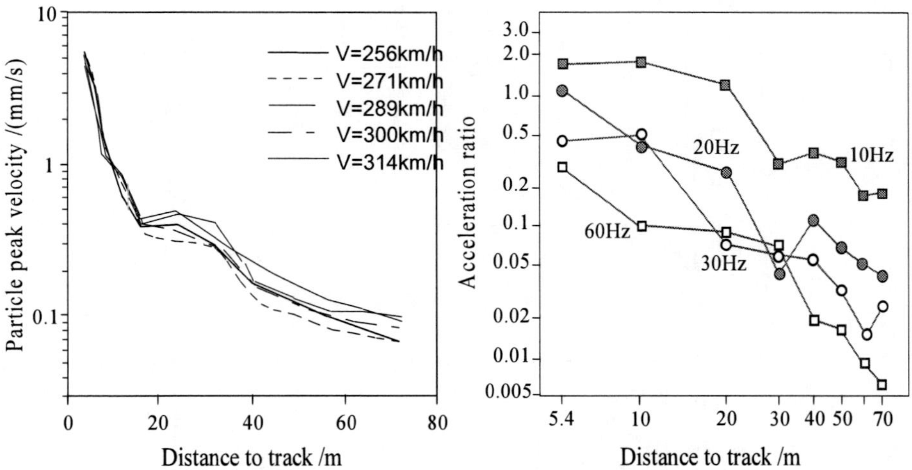

Figure 17. Distribution of ground vibrations vs. distance to railway lines measured at HSR in Belgium (left) and Shinkansen in Japan (right).

Careful observation on Figure 16 (top) may reveal the same law. This phenomenon thus seems to be universal, which can be explained by the repeated reflection of the vibration waves in the soft layer between the bedrock (hard layer) and the ground surface (Figure 18).

The appearance of the magnifying zone is related to the depth of bedrock, the properties of soil and the transmission speed of vibration waves, while Equations (1) and (2) cannot describe such a law.

For road traffic induced vibrations, the vibration level also attenuates with the distance from the road. Generally, a distance of a lane width may show the attenuation of 3–4 dB, thus at the place of 50 m away form the road, the vibration may be 40–50 dB lower than that at the road center, as shown in Figure 19 (top).

Owing to the filtering function of the soil, the higher components of vibration attenuate fast during its propagation. Shown in Figure 19 (bottom) is distribution of measured frequencies of train-induced ground vibrations versus distance to track [2], which shows that the dominant frequency is 30–50 Hz at distance ≤ 25 m, while lowers to 10–30 Hz at longer distance.

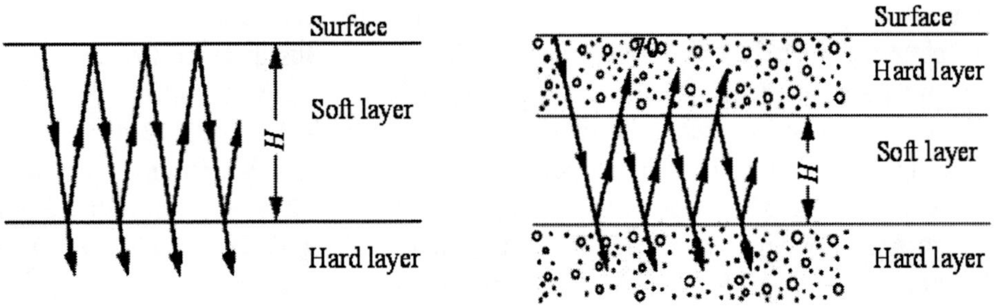

Figure 18. Repeated reflection of vibration waves in soil layers.

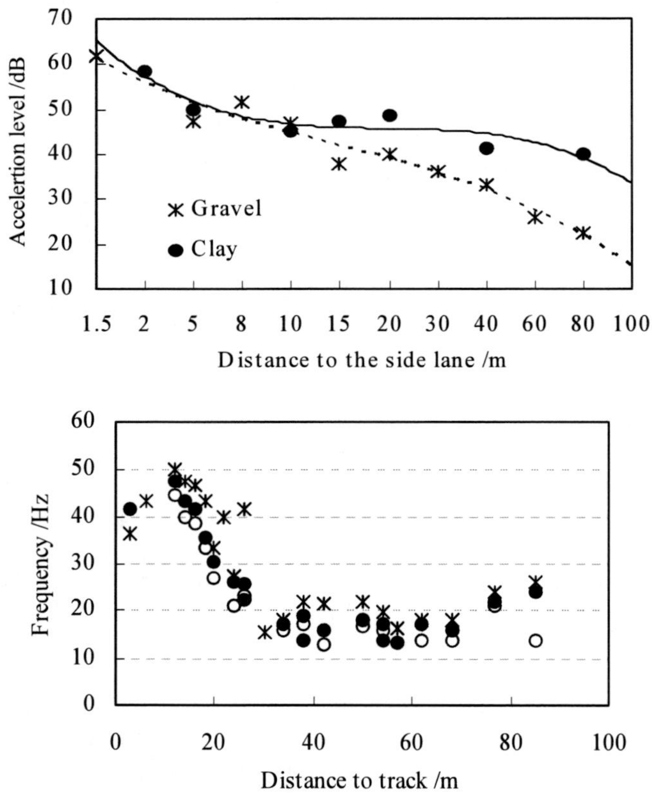

Figure 19. Distributions of ground acceleration vs. distance to the side lane on roadway (top) and frequency versus distance to track (bottom).

3.5. Vehicle Speed

Generally, running train induced ground vibration increases with the train speed. Based on the measurements at three fields in Spain, Volberg [26] proposed the relationship of vibration level with train speed (see Figure 20 top) as:

$$L_v = 64 + 20\lg(V/40) \ \text{(dB)} \tag{4}$$

where: L_v is ground velocity level with the reference of 5×10^{-8} m/s; V is train speed (km/h).

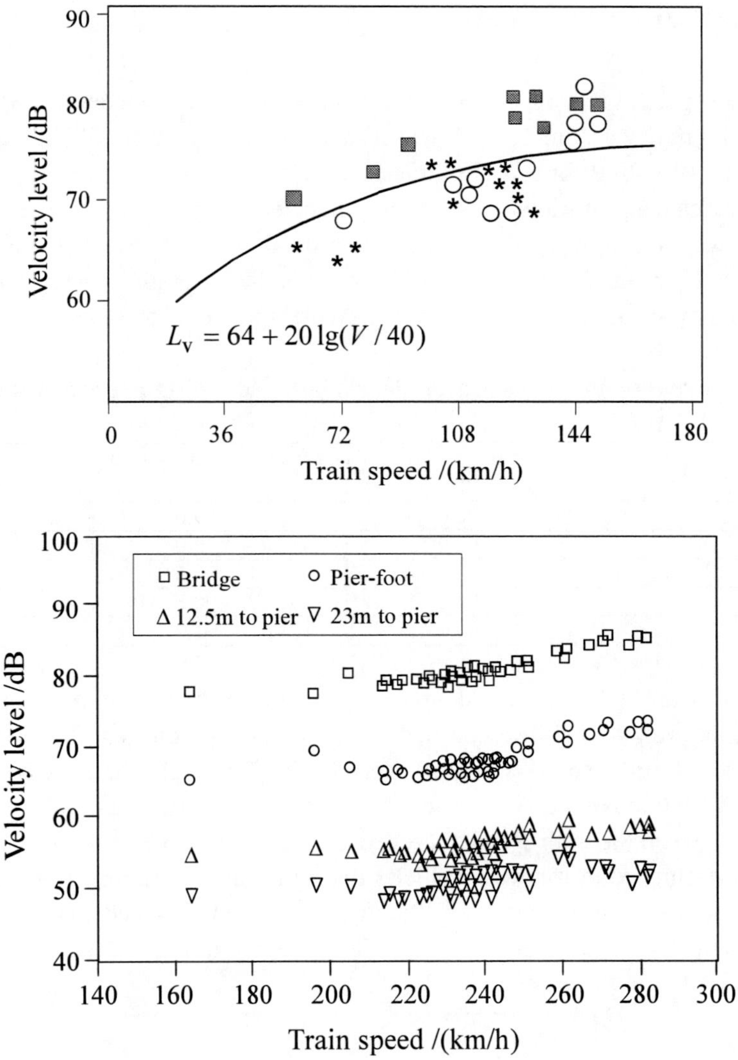

Figure 20. Ground vibration versus train speed measured in Spain (top) and in Japan (bottom).

Yoshioka [27] measured the ground vibration near a Shinkansen bridge. The result shows that in the speed range of 160−280 km/h, the accelerations show a difference of 5−8 dB (see Figure 20 bottom), and can be expressed as:

$$L_v = 10n\lg(V/V_0) + L_{v0} \text{ (dB)} \tag{5}$$

where: V_0 is the reference train speed (km/h); L_{v0} is the average vibration level at V_0; n is a coefficient with a choice range of 1.5−3.5.

The intense and influence range of metro system induced vibrations are related to the speed of trains: the higher the train speed, the greater the vibration, and the broader the influence range. According to investigations in Beijing and Tianjin, the vertical acceleration at the tunnel lining can be estimated by [28]:

$$L_V = 67 + 20\lg(V/65) + \Delta L \text{ (dB)} \tag{6}$$

where: V is train speed (km/h); ΔL is correction value for wheel-rail conditions: ΔL = 0 dB for CWR tracks with newly ground smooth wheels and rails, 5−11 dB for CWR tracks with worn wheels and rails, 10−22 dB for short rail tracks and worn wheels and rails.

The China Academy of Railway Sciences conducted a high-speed train experiment on the Test Loop Railway in Beijing, and measured the ground vibration, as shown in Table 2. It can be found that at 22 m and 30 m to the track, the ground acceleration increased with the train speed. When the train speed was 205 km/h, the acceleration level at 30 m reached 91 dB.

Table 2. Measured ground acceleration levels induced by high-speed trains (Unit: dB)

Train speed /(km/h)	Distance to the track center /m		
	22	30	60
170	91.6	89.6	86.0
180	92.0	90.0	86.0
190	93.5	90.6	85.8
200	94.0	90.8	86.2
205	94.0	91.0	86.5

Investigations are also performed in theory on the influence of running train-induced ground vibrations. Krylov [7] deduced that the ground vibration may increase sharply with further increase of train speed when it exceeds the Rayleigh wave velocity. Later, the Geotechnical Institue in Norway confirmed this deduction in a field test: they observed severe ground vibration when the train speed exceeded the Rayleigh wave velocity.

The speed influence of roadway vehicles on the ground vibration is greater than that of railway trains. Some measurements show that for every 10 km/h speed change of vehicles, the ground acceleration can be influenced by 2−3 dB, as shown in Figure 21.

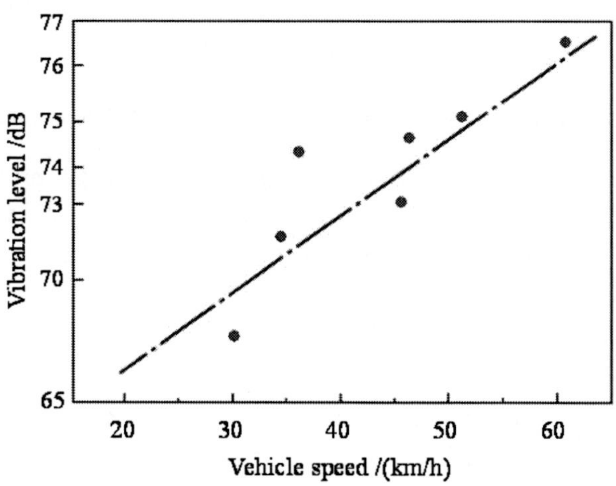

Figure 21. Relationship of ground vibration intensity versus vehicle speed (Japan).

3.6. Vehicle Load

It is believed the load of vehicle has big influences on the ground vibration: generally the heavier the load, the greater the impact of vehicle on the ground. The measured results by Xia et al. [29] show that freight trains with heavier loads induce 2−10 dB higher ground vibrations than passenger trains, and the nearer the site to the track, the bigger the difference, as shown in Figure 22 (left). During the experiment of heavy-load trains on the Test Loop Railway by the China Academy of Railway Sciences, the running of the train with 25 t axle-loads excited much stronger ground vibrations than the ordinary trains, which incurred intensive complaints from the nearby residents. According to an investigation in Japan, large vehicles may induce 10 dB bigger ground vibration than small vehicles, as shown in Figure 22 (right).

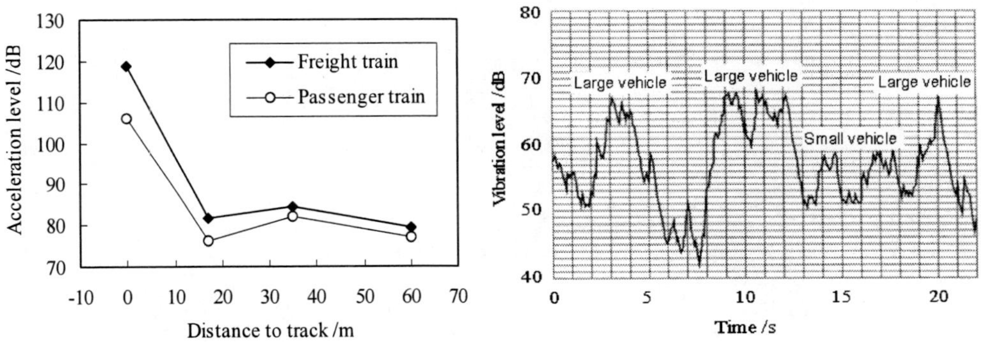

Figure 22. Ground vibrations induced by different loads of trains (left) and automobiles (right).

4. Vibration Characteristics of Several Railway Systems

4.1. Elevated Railway System

In addition to the general laws mentioned above, ground vibrations induced by elevated railway system have the following characteristics:

(1) Bridge is a line-shaped, elevated structure supported by piers at certain spacing on the ground. In the elevated railway system, the train loads and the vibration are transmitted to the subsoil through pier foundations at discrete positions, therefore, the vibration impact on the environment has the feature of array point source.

(2) The train loads are transmitted to the pier through the beam and the bearings, thus the pier-top force equals to the sum of reaction forces of the two beams on this pier. Therefore, the quasi-static loading by moving train gravity composes a dominant component in the pier-top force curve, especially at low train speed. With the increase of train speed, more and more impact from dynamic loads adds on the quasi-static load curve, as shown in Figure 23.

(3) For a bridge, the positions of these point vibration sources are fixed. However, when the train runs on the bridge, it arrives at different piers with a certain time interval, forming a phase gap between any two sources at adjacent piers. Moreover, the distances from different piers to the receiver point P are different, which forms another type of phase gaps among the

ground vibrations from different pier foundations to the receiver point. In analysis of ground vibration caused by elevated railway, both the two phase gaps should be taken into account, as shown in Figure 24.

(4) The structural forms of bridges, including the type, length, weight and stiffness of spans, the type and stiffness of bearings, the type, height, weight and stiffness of piers, the type, size and weight of foundations, as well as the structure of tracks, have influences on the environmental vibration.

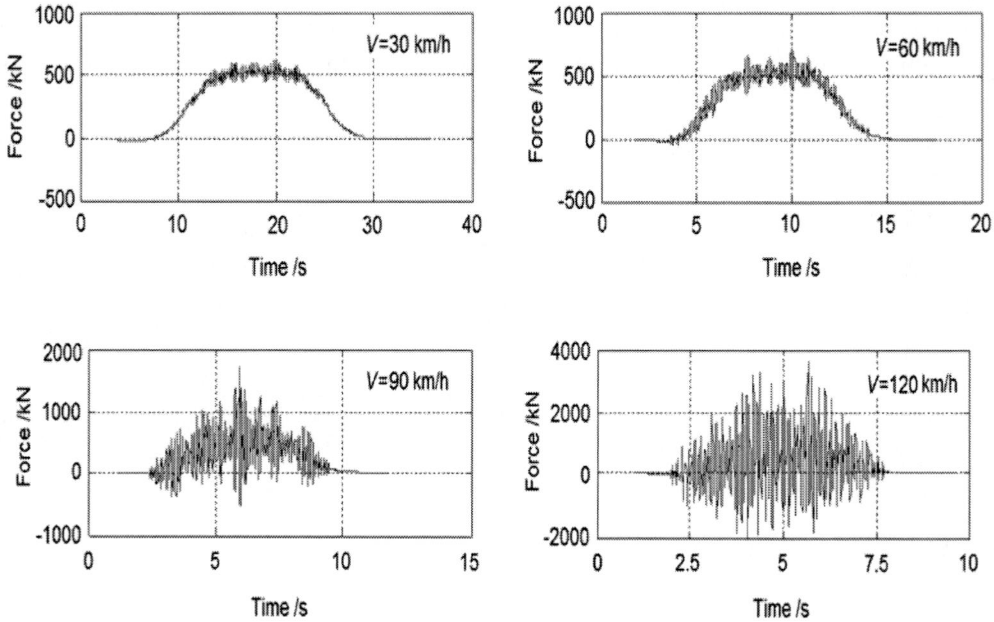

Figure 23. Pier-top force time-histories under different train speeds.

Shown in Figure 25 (left) is the comparison of ground vibrations measured on bridges with different spans. Generally, there is a decreasing trend in ground vibration with bridge span. In a statistical analysis [1], a quantitative formula $L_v=-0.36L$ (dB) was proposed to predict the ground vibration near bridges, where L is the span length (m). That is, for every 10m increase of the bridge span, the ground vibration decreases by 3.6 dB.

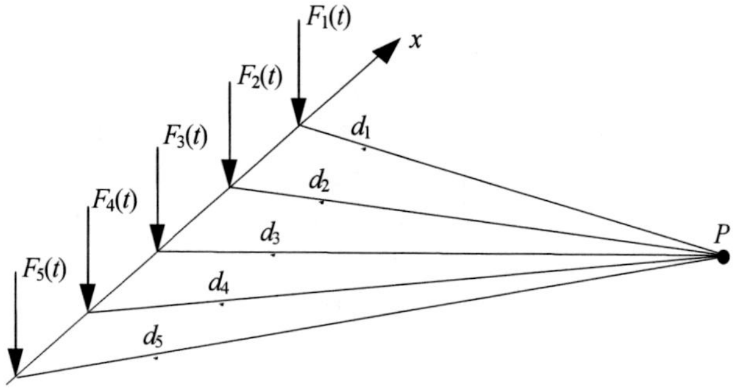

Figure 24. Dual space effects caused by ground vibration from elevated bridge.

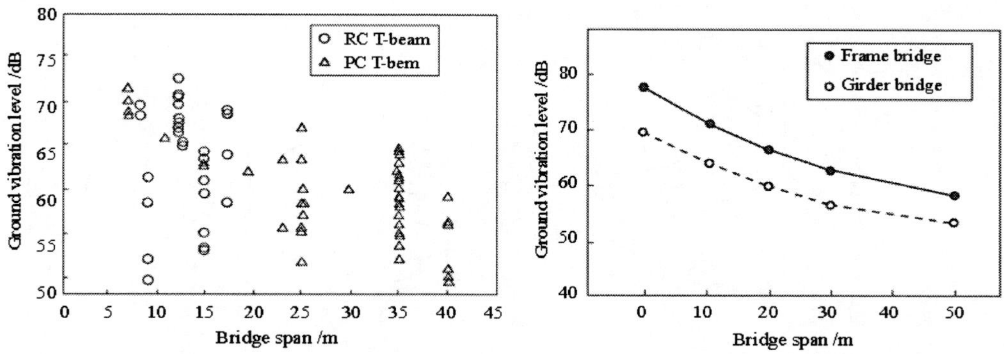

Figure 25. Influence of bridge spans (left) and structural forms (right) on ground vibrations.

Compared in Figure 25 (right) are the vertical vibration levels at ground near two types of bridges on the Tokaido Shinkansen and Sanyo Shinkansen in Japan. It shows that the ground vibration by the rigid frame bridge is about 5−8 dB larger than that by the girder bridge.

The larger the beam mass, the smaller the ground vibration. Reference [1] proposed a quantitative index to predict the ground vibration, $\Delta L_{\mathrm{v}}=-0.0789W$ (dB), where W is the concrete volume (m^3/m). This shows that for every 10 m^3 increase of concrete used in the beam per unit length, the ground vibration decreases by 0.8 dB.

The type of bridge foundation has a big influence on ground vibration. For pile foundation, vibration caused by trains can transmit to deep soil stratum, thus the displacement, velocity and acceleration of ground vibration are all smaller than those with flat foundation, and attenuate faster versus the distance from the track.

The vibration transmitted from bridge foundation is affected by the depth (pile foundation) and supporting area (flat foundation) of the foundation. Generally, the deeper the foundation and the larger the supporting area is, the smaller the ground vibration is. For flat foundation, a quantitative index for predicting the influence of foundation bearing area on the ground vibration is $\Delta L_{\mathrm{v}}=-0.0206A$ (dB), where A is the bottom area (m^2) [1]. It means for every 10 m^2 more bottom area, the ground vibration decreases by 0.2 dB.

(5) Comparison of vibrations at different parts of the elevated bridge. Illustrated in Figure 26 are the measured vibration levels of a track-beam-pier-ground system on the Shinkansen [27], where D is the distance of the observation point to the pier. It clearly shows the distribution of high-speed train induced vibrations at different part of the system. The acceleration is as high as 90 dB at the track, and attenuated by about 20 dB at the pier top, and the law of the ground vibration attenuation versus the distance to the pier is obvious. For different measuring points, the accelerations show peak values at the frequencies of 12−30 Hz.

(6) Compared to ground railways, the elevated system can reduce the environmental vibration level, showing 5−10 dB lower at a place of 30 m to the line [30].

In addition, the bridge structure of elevated railway system should be carefully designed to avoid the resonance between the running train and the structure to reduce the intensity of vibration source.

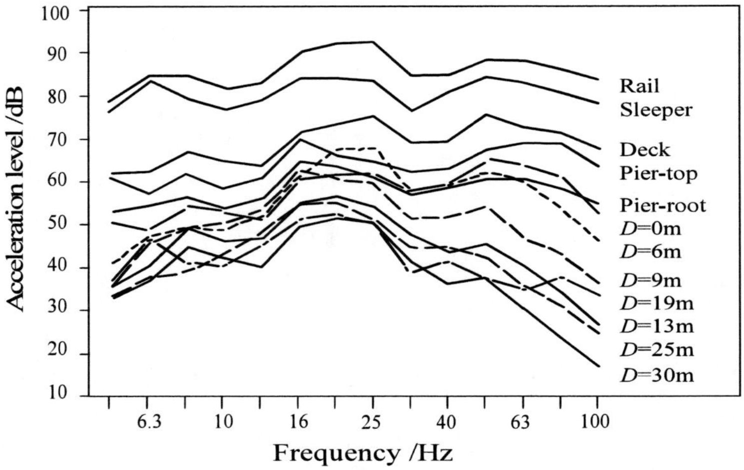

Figure 26. Measured acceleration levels at different parts of a Shinkansen bridge.

4.2. Underground Trains

For ground vibration induced by metro trains, Pan proposed a mathematical expression to simulate the moving train load based on the measured track accelerations, and analyzed the vibration properties of a tunnel and the surrounding soils [28].

Xia et al. [30] established a train-track-tunnel-ground system model to study the metro train induced ground vibrations and their propagation laws. Figure 27 illustrates the contour distribution of the maximum vertical ground accelerations in the soil around the tunnel. It shows on the ground surface, there appear two local vibration magnification zones: one is right over the tunnel, the other is about 35 m away from the tunnel central.

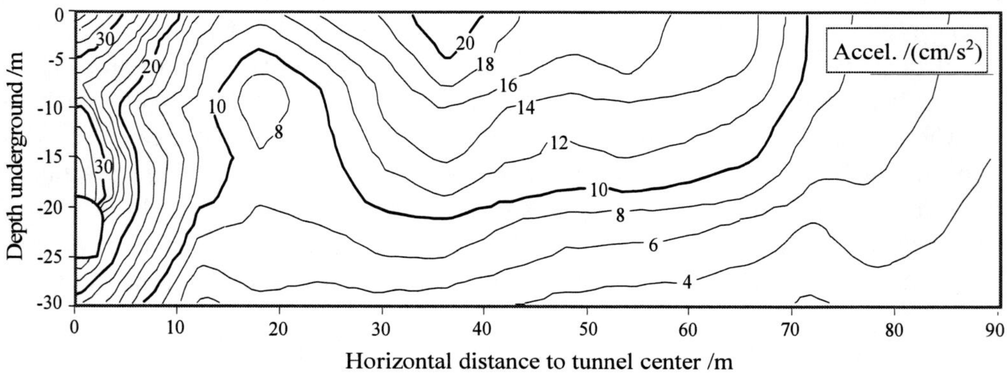

Figure 27. Distribution of metro train induced accelerations in surrounding soils.

Figure 28 shows the distributions of acceleration in different frequency components. For vibrations with low frequencies of 1–3 Hz, there occur three magnifying zones at 0 m, 36 m and 60 m to the tunnel. For vibrations with middle frequencies 5–6 Hz, only two magnifying zones occurred at 0 m and 30m, and they attenuate rapidly with distance. For the frequencies higher than 8 Hz, vibrations show monotone decrease trend versus the distance. The spectra at several distances also explain the attenuation of high-frequency acceleration with distance.

It is investigated that when a metro train runs in the tunnel, the average vibration levels at ground surface are 81 dB at the place of 1.5 m to the track center, while 71.6 dB at 24 m away, which shows the vibration attenuates with the distance to the track center. Train speed and tunnel depth also influence the ground vibration. Normally the higher the train speed, the higher the vibration level and the larger the influencing area; while the deeper the embedded depth of the tunnel, the smaller the influencing area.

According to the regression analysis, the ground vibration levels induced by metro trains can be predicted by the following expression [1, 2]:

$$\mathrm{VL}_1 = K_1 - 12\log L_1 + 25\log\frac{V}{50} - 24\log\frac{W_1}{30} + X \ \text{(dB)} \tag{7}$$

$$\mathrm{VL}_2 = K_2 - 16\log L_2 + 25\log\frac{V}{50} - 24\log\frac{W_2}{60} + X \ \text{(dB)} \tag{8}$$

where: the subscripts 1 and 2 represent, respectively, tunnels with round section and box section. L is minimum distance from the lining side to the ground surface of the prediction point (m). V is train speed (km/h). W is unit weight of tunnel structure (t/m). X is amendment to consider vehicle weight and other factors. K_1 and K_2 are basic vibration levels for different track structures, which are 63 dB and 74 dB for common slab track, 53 dB and 56 dB for vibration mitigation track, and 64 dB and 67 dB for ballast track, respectively.

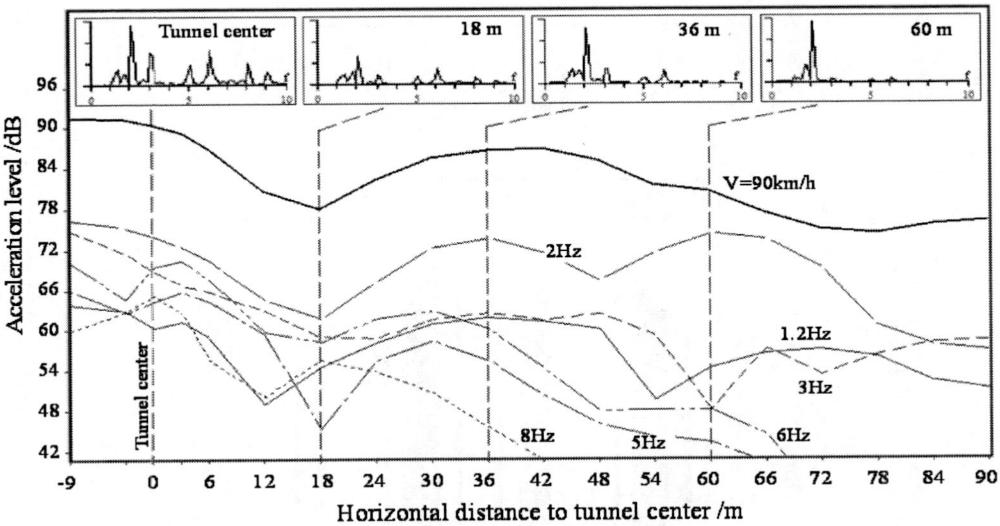

Figure 28. Distribution of ground acceleration in different frequency components versus distance.

4.3. Buildings Near Traffic Lines

The passage of vehicles may induce vibrations of the building close to traffic lines. A measurement near a metro rolling stock depot in Beijing shows that the vibration level of the residence building over the metro lines was as high as 85 dB when the trains ran through beneath the building at 15−20 km/h. It is estimated to have higher levels when the trains run at the normal operation speed of 80−100 km/h.

The phenomenon of vibration attenuation with distance is also applicable to buildings near traffic lines. Because the train-induced ground vibration attenuates faster in horizontal direction, the horizontal vibration of building near railway is smaller than that in vertical. Some measurements on building floors showed tens of decibel vibration smaller in horizontal direction than in vertical. Therefore, the vertical component is mainly considered in evaluating vibrations of buildings near railways.

For low-rise and mediate-rise multi-story buildings, there is a trend of floor vibration increasing at higher floors. The measurements on several 3−5 storey buildings in Shenyang, Beijing and Xi'an showed that at different distances, vibrations at the 3−5th floors are about 3−5 dB higher than that at the first floor (Figure 29) [23, 24].

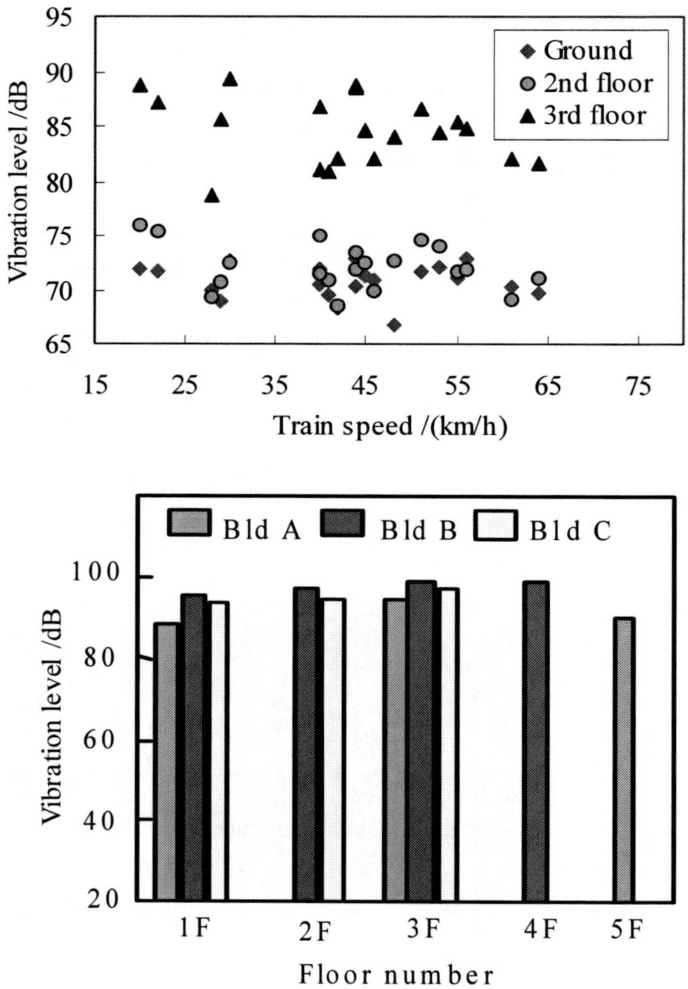

Figure 29. Measured vertical vibration levels at different floors.

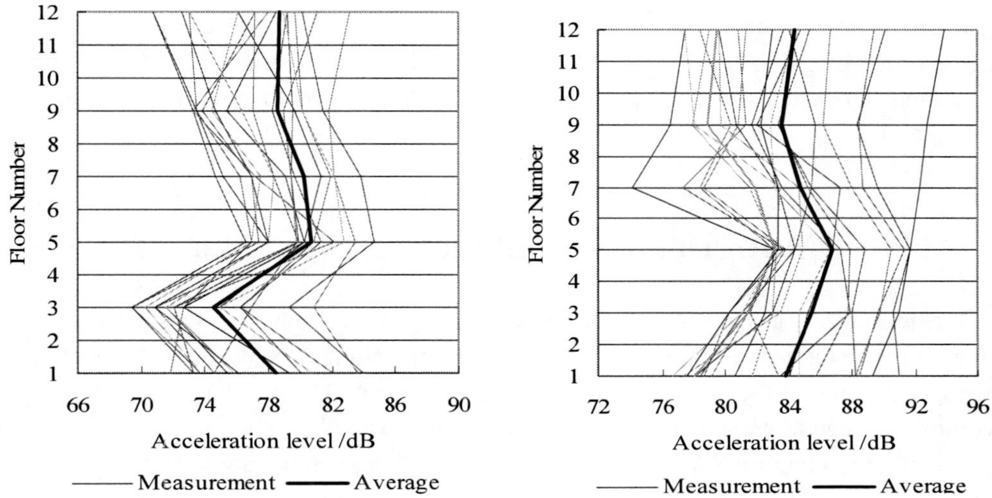

Figure 30. Measured lateral (left) and vertical (right) floor accelerations of a 12-storey building.

For high-rise buildings, the distribution of floor vibration versus the heights is different. A field measurement on a 12-storey building showed the vibrations at the 3rd to 7th floors are relatively bigger than on the ground and the higher floors, as shown in Figure 30 [31].

The types of structure and foundation of a building affect the floor vibrations. For a high-rise RC building with good foundation and huge mass and thus having low frequencies, the vibration level can be 10−20 dB lower than that of the base soil. For a residential building of brick-concrete composite structure and common foundation, the vibration level can be 5−10 dB lower. While for a building with light structure and poor foundation, the amount of attenuation is quite small, thus the indoor is of the same level vibration as the base soil, or even more, the floor vibration is greater than that of outdoor ground.

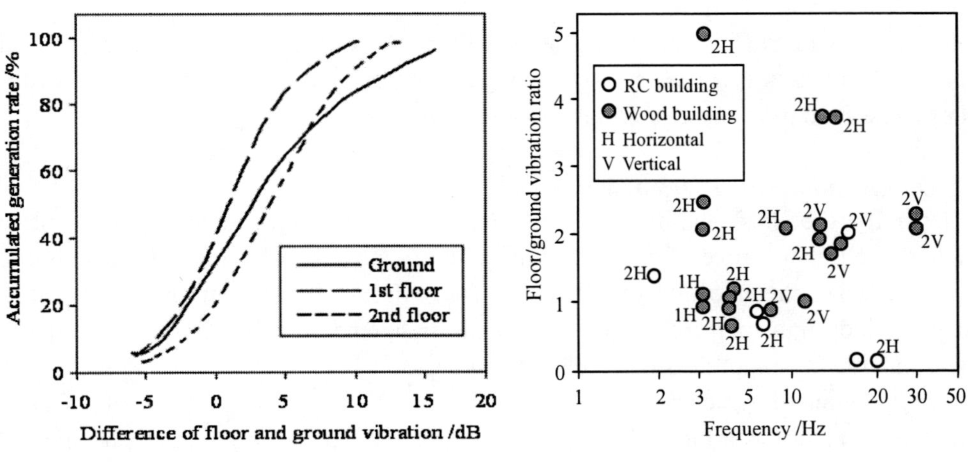

Figure 31. Difference of floor and ground vibrations (left) and difference of wood and RC buildings (right).

According to the investigation near railways [2], the vibrations at floors are often bigger than on the ground, with 50% of them being 5 dB and the maximum 15 dB higher. As shown in Figure 31 (right), the responses of wood buildings are much greater than the RC ones, in which the numbers 1 and 2 represent floor number. The vibration of the 2nd floor of wood buildings may magnify the ground vibration by nearly 5 times (about 15 dB), while that of RC buildings are often smaller than the ground, with the maximum increase less than 2 times (about 5 dB).

5. Control Policies on Environmental Vibrations

Traffic-induced ground vibrations spread through the strata to the surrounding area, which influence the normal live of residents and the operation of precision instruments, and in serious case, can hurt the human health and even damage buildings. Therefore, it is important to control the generation and propagation of traffic-induced vibrations.

The vibration intensity L_r in a building near traffic lines can be estimated as

$$L_r = L_s - L_g - L_{gb} - L_b \text{ (dB)} \quad (9)$$

where L_s is the vibration intensity at the source; L_g is the attenuated vibration in the ground; L_{gb} is the attenuated vibration generated from interaction between soil and structures; L_b is the vibration attenuation in the building.

Therefore, to reduce or to control the negative effects of vibration, measures can be taken via reducing vibration source intensity, cutting off vibration propagation path or mitigating vibration on the path, and through rational plan and design to arrange the buildings outside the vibration impact area.

5.1. Control of Vibration Source Intensity

The sources of traffic-induced vibration include the moving gravity load when vehicles run on a railway track or a road surface, and the vibration load from the supporting structures such as track, road pavement, subgrade, bridge, tunnel lining, etc.

5.1.1. Optimization Measures of Vehicles

Generally, the closer the damping measures to the vibration source, the more effective. Vehicle is the initial source of traffic vibration, so optimization of vehicle plays an important role for reducing the intensity of vibration. The current vibration mitigation measures of vehicles include lightening of vehicles, rational arrangement of axles, reasonable design of suspension system, smoothing of wheels, and so on.

(1) Lightening of vehicles

Vehicle weight has a big influence on vibration source intensity: the greater the vehicle weighs, the bigger the influence will be. This influence can be obviously seen in Figure 32, the distributions of ground vibrations caused by the E300 (axle-load 110 kN) and the original E100 (axle-load 160 kN) Shinkansen trains [32].

From the figure, one can observe that the ground vibration will decrease by about 3 dB if lighter train vehicles are used. Through a statistical analysis, the difference of vibration caused by different train axle-loads can be approximately expressed as:

$$\Delta L_V = 10\lg(W_1 / W_0) \tag{10}$$

where: W_1 and W_0 denote two vehicle axle-loads for comparison.

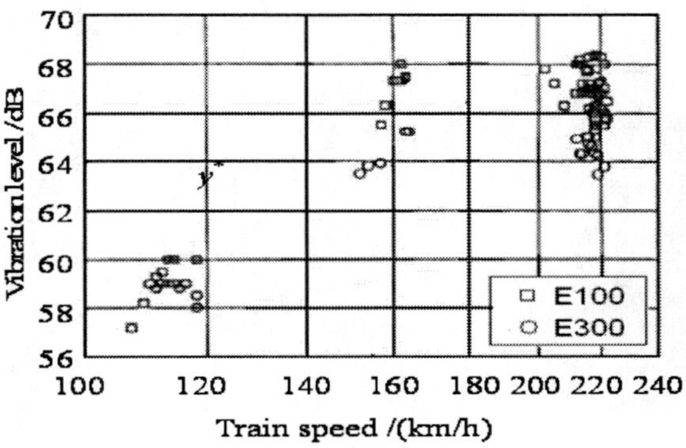

Figure 32. Comparison of ground vibration induced by trains with different axle-loads.

(2) Rational arrangement of vehicle axles

The dynamic impact of a train on the structure has a big relationship with the axle-load arrangement of vehicles. When the train runs at a speed V, due to the regular arrangement of its axle-loads, it generates a periodic dynamic load on the structure, the loading period is d_v/V (d_v is the loading interval). When this period is equal to the vibration period of the structure, a resonance will occur in the structure.

A comparison was performed on the Shinkansen trains with different axle arrangements, with full length L_v of vehicles 20−25 m, distance L_c between bogies 11−19 m, wheelbase L_v 1.3−3.5 m. The results show that the best axle arrangement is L_v=22.1 m, L_c=18.1 m, and L_w=1.1 m, which can reduce the ground vibration at 12.5m to the track about 3−3.5 dB, as shown in Figure 33 [1, 32]. This comparison gives an illustration on the importance of rational design of vehicle axle arrangement in reducing vibration resource intensity.

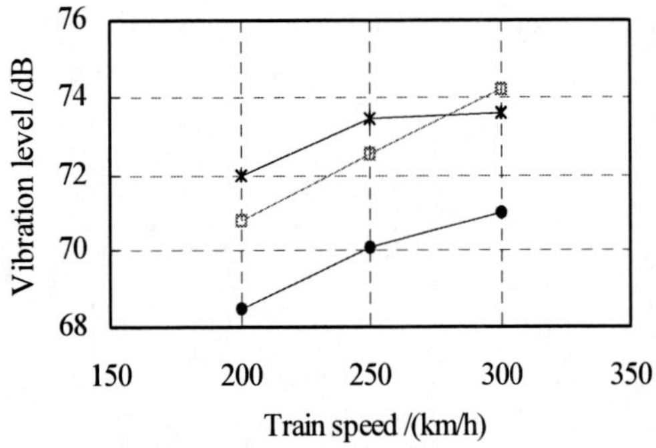

Train type	Vehicle dimension /m		
	Wheelbase	Bogie interval	Full length
* Current	2.50	17.50	25.0
□ E400	2.25	14.15	20.5
• Optimal	2.10	18.10	22.1

Figure 33. Comparison of ground vibrations caused by trains with different axle arrangements.

(3) Rational design of vehicle suspension system

Experiments show that rational design of vehicle suspending stiffness and damping can avoid the resonance between vehicle and track and reduce the vibration intensity by 10−15 dB.

(4) Adoption of resilient and damped wheels

Resilient wheels acquire the vibration mitigation ability by inserting the elastic materiel between the wheel hub and the tire, as shown in Figure 34, which can reduce the track vibration by 6–20 dB in the frequency 40–250 Hz. This type of wheel has been used for more than 50 years in Europe and North American countries and more than 30 years in Japan.

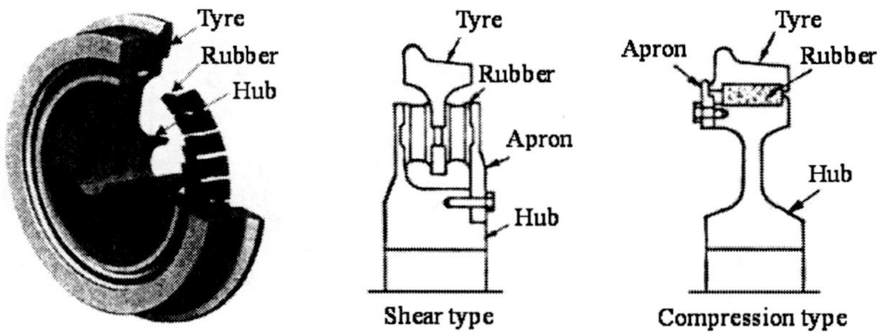

Figure 34. Resilient wheel and its structure.

Damped wheel utilizes the damping material on the tire to mitigate the vibration, through converting the vibration energy of the wheel into heat energy. According to the structure and the function, the damped wheels in use or in experiment include three types: the annular damped wheels, the resonance damped wheels and the constrained damped wheel.

(5) Maintenance status of vehicles

When a vehicle runs on track, the wheel unevenness such as its tire flat defects causes a periodic impact on the rail, which increases the track vibration by 10–15 dB. To reduce the impact on the track, it is necessary to trim the wheels regularly, to keep them in good status.

5.1.2. Optimization of Track Structure

Optimization of track structure is the key to reduce the vibration source intensity. The track consists of under-rail structures such as rails, fasteners, sleepers and ballast bed, and different components of the track contribute different mitigation functions, as shown in Figure 35. Thus, proper mitigation measures can be selected for different requirements.

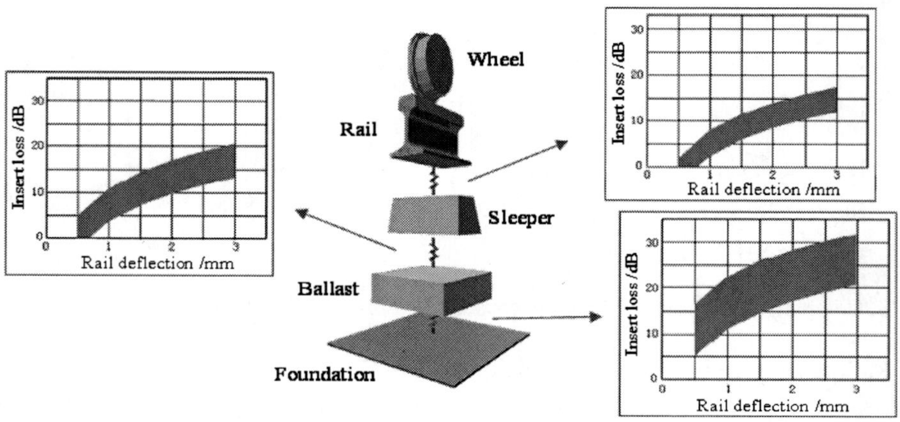

Figure 35. Track structure and mitigation levels for different components.

5.1.2.1. Rails

(1) Heavy rails

Heavy rails are featured of long life, stable performance and good vibration resistance. As shown in Figure 36, heavier rails provide larger flexural stiffness, which can withstand larger load with smaller deformation, and transmit smaller impact to the lower track structures. Compared with the 50 kg/m rail, the 60 kg/m rail can reduce vibration by 2–4 dB.

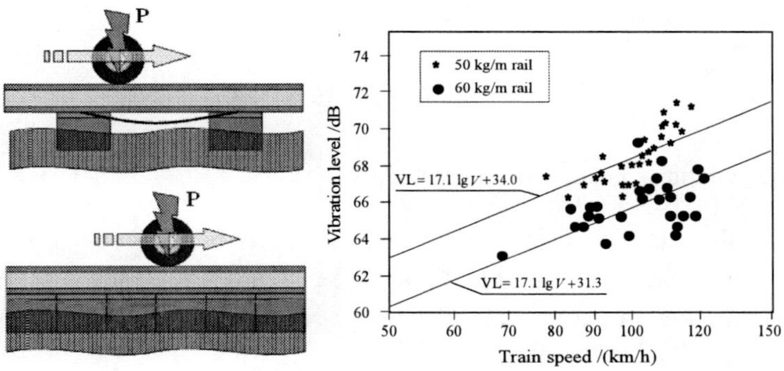

Figure 36. Comparison of mitigation effects of heavy rail and light rail.

(2) Continuously welded rail track (CWR)

Vibration at rail joints is three times that of non-joint parts. CWR track can eliminate the impact of wheels on rail joints, mitigate vibration intensity by about 5 dB, and reduce the track maintenance. Therefore, most main lines of railway adopt CWR track, which greatly reduces the environmental vibration induced by running trains.

(3) Rail Irregularity

Rail irregularity directly excites the wheels to vibrate. Grinding rails regularly to keep it smooth can reduce the track vibration. New rail has initial irregularity on the surface. After a period, due to braking, wheel slipping or other factors, the track surface becomes rugged, the wheels produce flat scars and other defects, these irregularities will intensify the wheel-rail vibration and noise. The investigations show that after rail and wheel grinding, the ground vibration higher than 100 Hz can be reduced by about 10 dB.

5.1.2.2. Track Structure

Adopting suitable type of track bed and structure can increase elasticity of the track and reduce vibration.

For ground and elevated railways, there are mainly two types of tracks: ballast track and ballastless slab track. The ballast track has a good damping effect, which can reduce vibration and noise by 5–8 dB than the ballastless track. However, ballast track has shortcomings such as poor stability, heavy workload for maintenance, large dead weight, large construction height and ballast pollution, so ballastless slab track is often selected, especially for underground lines. Ballastless slab track has insufficient elasticity, which is solved by adopting elastic fasteners with good damping effect or other mitigation measures.

Using new track structure is one of the most effective ways to reduce vibration levels. There are several kinds of vibration reduction tracks. According to the test, adopting vibration reduction track (VR track) can reduce ground vibration levels by 2–10 dB (Figure 37).

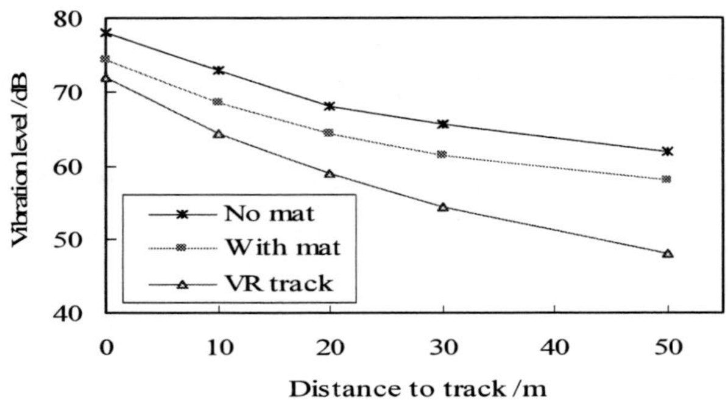

Figure 37. Ground vibrations induced by different types of tracks.

According to the damping measures on different parts of the track shown in Figure 35, vibration reduction tracks can be grouped into three categories: elastic fasteners, elastically-supported block tracks and floating slab tracks.

(1) Low Vibration Track (LVT)

LVT is a ballastless track composed of elastic supporting blocks, track board, concrete base, rubber pads and elastic fasteners (Figure 38). This structure has obvious vibration and noise reduction effects. LVT is an ideal track for vibration and noise-sensitive urban area, especially for the elevated structures.

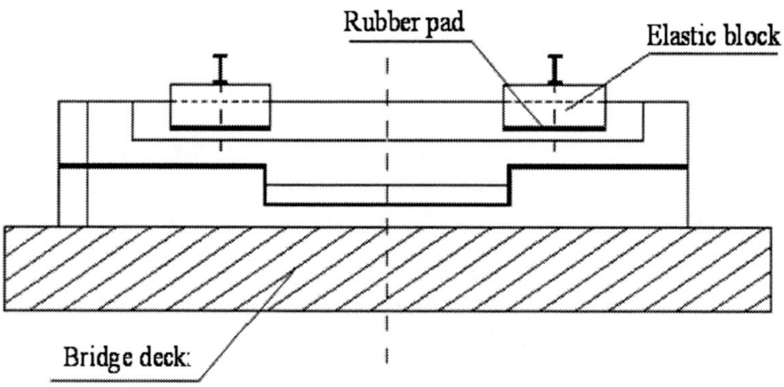

Figure 38. Structure of LVT on elevated bridge.

The China Academy of Railway Sciences conducted a test on the LVT, showing good vibration mitigation effects with 30.8% to 50.7% higher than the rigid slab track. LVT has been applied in the Dongsishitiao Metro Station in Beijing, the Guangzhou Metro Line 1 and the Qinling Railway Tunnel in China.

(2) Embedded Rail Structure

The Embedded Rail Structure (ERS) was developed by the Edilon Company, which adopts continuous longitudinal supports instead of the traditional discrete supports, as shown in Figure 39. This kind of rail structure produces good sound absorption and vibration isolation effects because of a material called Edilon Corkelast around the rail. The Embedded Rail Structures behaved well in metros and ground trams in Brussels, Paris, Athens, Madrid and other cities, and the maintenance workload is considerably less.

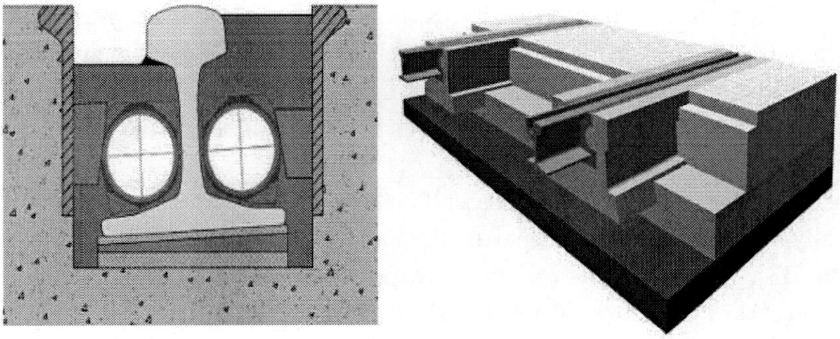

Figure 39. Embedded Rail Structure.

(3) Floating slab track

The floating slab track (FST) is composed of a floating slab, elastic supports under the slab, lateral buffer pads and sleeper mat (Figure 40).

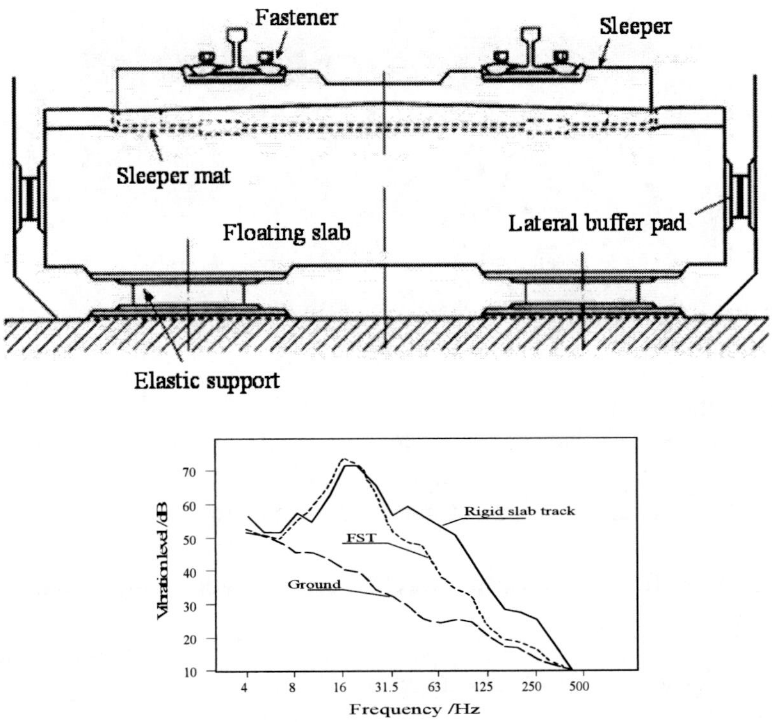

Figure 40. Floating slab track and its vibration reduction effect.

The basic isolation principle of FST is to insert a low-frequency linear resonator between the superstructure and the base of the track. The elastic supports under the slab can be rubber plates or steel springs. Using steel springs to support the floating slab provides better damping effect but costs more, so it is usually adopted as a high-grade isolation measure for some special sensitive places.

In all types of damping measures, floating slab track has the best vibration isolation effect. According to the measurement, a 30 dB vibration can be reduced by the floating slab track with ballast mats, and this effect keeps almost the same for the vertical load varying from 20% to 100%. However, floating slab track may produce large dynamic deflection. For example, under the axle-load 140−160 kN of a light rail train, its dynamic deflection comes up to 4 mm, which seriously worsen influences the smoothness of the track, thus limits its application in high-speed railway.

(4) Elastically-supported Ladder Sleeper Track

Elastically-supported ladder sleeper track (Ladder track) is a new type of low vibration track system [33], which consists of ladder sleepers, resilient isolators and concrete base, as shown in Figure 41. The ladder sleeper is composed of PC longitudinal beams and steel pipe connectors, looking like a ladder. In ladder track, the rails and the PC beams form a "composite track" to carry the train load, which improves the distribution performance of the track to dynamic load, thus possesses good noise and vibration mitigation property. As a high-grade vibration isolating measure, the ladder tracks was originated in Japan, and has been applied in some cities in the United States and China.

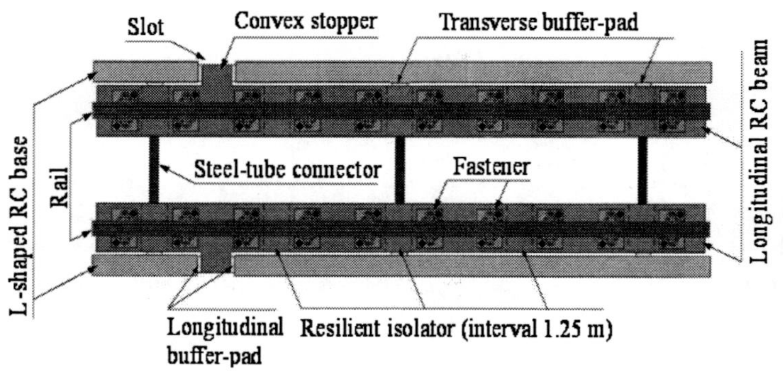

Figure 41. Elastically-supported ladder sleeper track.

5.1.2.3. Damping Fasteners

There are various track damping fasteners with good vibration mitigation effect, such as damping fastener series (Types I−IV), ALT.1 fasteners, Vanguard fasteners, Lord fasteners, and so on, which are commonly used in China, as shown in Table 3.

Table 3. Characteristic parameters of several damping fasteners

Fastener type	Nominal reduction amount /dB	Working frequency /Hz
Type I damping fastener	6−10, max. 25 for high frequency	≥15
Type III damping fastener	10−15, max. 25 for high frequency	≥10
Type IV damping fastener	12−15	≥25
ALT.1 Fastener	5−10	≥42
Vanguard Fastener	8−15	≥10
Lord Fastener	6−7	≥19

5.1.2.4. Elastic Mats

Elastic mats are an important part for increasing the elasticity of track. Laying elastic mats under rails and sleepers can effectively buffer the dynamic action of the train.

Vibration produced by rolling wheels on the rails spreads though the track members under the rails. The farther the member from the rails, the lower the vibration frequency is. To achieve best isolation effect, the elastic coefficients of different track members as the elastic isolating layers should be coordinated. Therefore, the elastic coefficient of elastic mats under the rails, the sleepers, the ballast and the concrete slab of non-ballasted track should decrease in a descending order.

The test results by the Japanese Railway showed that elastic fastener reduced the ground vibration near the track by 3.5−6 dB. In the region 12.5−25 m from the track, elastic sleepers and ballast mats reduced the ground vibration by 2 dB, while the damping effect of elastic fasteners was unobvious. In addition, ballast mats had significant damping effect on the vibration higher than 50 Hz, and reduced the vibration at the subway tunnel wall by 6−7 dB.

5.1.3. Other Control Measures on Vibration Sources

There are other control measures to reduce the intensity of vibration source induced by traffics, including:

(1) Maintain the flatness of road surface. For every 1 mm decrease of standard deviation of pavement irregularity, the vibration can be reduced by 3−4 dB.
(2) Adopt the elevated system. The elevated rail transit system not only induces smaller vibration, but also occupies smaller land area, which is especially suitable for urban areas.
(3) Use concrete beams and structures with good integrity for the bridge in elevated rail transit system. Rational design of spans, structural forms and vibration properties is helpful for the bridge to avoid resonance induced by running trains.
(4) For bridge foundations, adopting pile foundations may achieve better vibration reduction effect than using shallow flat foundations.
(5) Use damping bearings to isolate the vibration, or use vibration control devices to suppress the vibration of the structure.
(6) For underground lines, using deeper tunnel embedment and heavier tunnel structure may achieve better attenuation of ground vibration versus the depth and distance.
(7) For ground lines, various foundation treatment measures can be applied to reinforce the subgrade, such as dual-gray piles under the subgrade, or rigid board or concrete box grilles in the ballast bed.
(8) Adopt integrated floating slab track. In some vibration-sensitive areas, a whole section of track can be floated with steel springs. Applications of integrated floating slab track have achieved ideal vibration mitigation effect.

5.2. Propagation Path Control

The propagation path control of vibration is mainly by means of foundation isolation, which is realized by various foundation barriers such as ditches, rubber mats, sheet pile walls and row-piles to cut, scatter and diffract the stress waves.

According to the isolating principles, there are three types of vibration isolation methods for propagation path control, as shown in Figure 42.

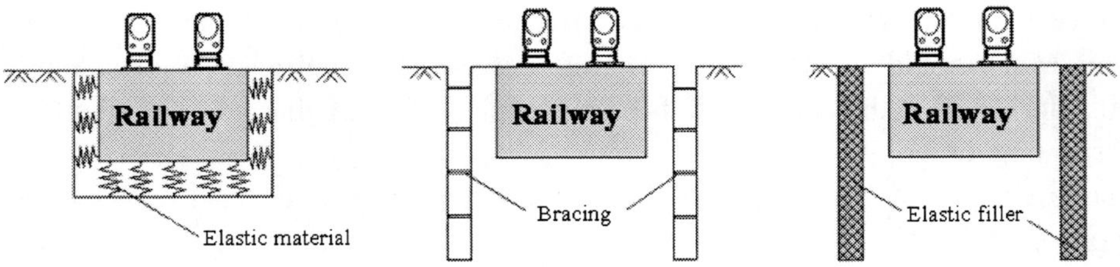

Figure 42. Vibration isolations: elastic foundation (left), open trench (middle) and filling ditch (right).

5.2.1. Foudamental Principle of Elastic Foundation Isolation

The basic mechanical principle of vibration isolation using elastic foundation is illustrated in Figure 43 (a). The governing equation of the system can be written as

$$\ddot{v}(t)+2\xi\omega\dot{v}(t)+\omega^2 v(t)=p_0\sin\overline{\omega}t \tag{11}$$

According to structure dynamics, the steady displacement response of the mass m is given by the following equation

$$v(t)=\frac{p_0}{k}D\sin(\overline{\omega}t-\theta) \tag{12}$$

where D is the dynamic amplification factor

$$D=\frac{1}{\sqrt{(1-\beta^2)^2+(2\xi\beta)^2}} \tag{13}$$

Thus, the forces transmitted from the spring and damper to the foundation are:

$$f_S=kv(t)=p_0D\sin(\overline{\omega}t-\theta) \tag{14a}$$
$$f_D=c\dot{v}(t)=2\xi\beta p_0D\cos(\overline{\omega}t-\theta) \tag{14b}$$

Since the phase difference by between f_s and f_d is 90°, the amplitude of the dynamic reaction on the base is:

$$f_{\max}=\sqrt{f_{S\max}+f_{D\max}}=p_0D\sqrt{1+(2\xi\beta)^2} \tag{15}$$

The ratio between the maximum reaction and the applied force is defined as the transmissibitily of the isolation system

$$\mathrm{TR}=\frac{f_{\max}}{p_0}=D\sqrt{1+(2\xi\beta)^2} \tag{16}$$

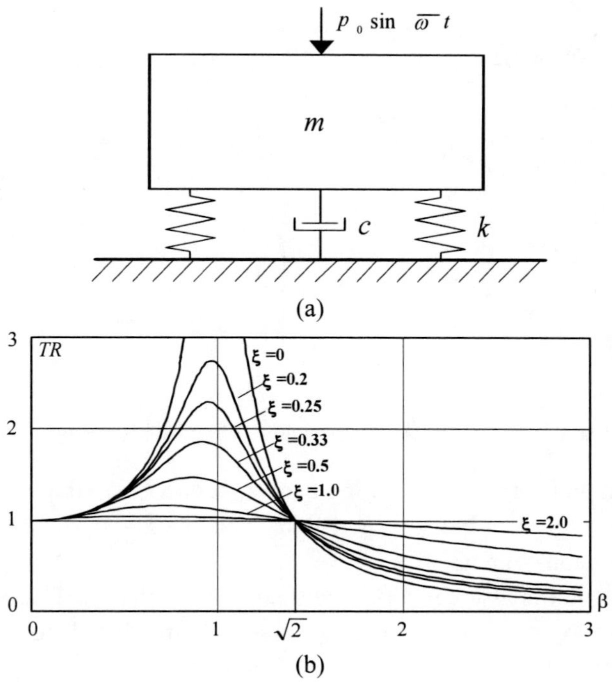

Figure 43. Basic model (a) and transmissibitily of vibration isolation with elastic foundation (b).

Figure 43 (b) shows the trend of transmissibitily TR versus frequency ratio $\beta = \bar{\omega} / \omega$ and damping ratio ξ . Obviously, the transmissibitily is smaller than 1 when $\beta > \sqrt{2}$, namely, the elastic foundation is helpful for isolating vibration.

This vibration-isolation system means setting elastic foundation near the vibration source, so it is also called near-field isolation. Setting rubber pad between rail and sleeper, and mats between track bed and foundation, are examples of using this principle. Therefore, elastic foundation is mainly used for isolation of vibration source.

It is observed from Equation (16) and Figure 43 that for a load with particular frequency, the larger the frequency ratio is, the better the isolation effect will be. While for railway, the problem is not so simple: too small elastic coefficient may induce overlarge deflection of the track, which endangers the running safety of the train, and should be considered in the vibration-isolation design of tracks.

5.2.2. Isolation Barriers

The vibration produced by moving train loads propagates along the ground and in subsoil, so it can be isolated by cutting the propagating path, namely, setting vibration-isolating barriers.

(1) Active isolation and passive isolation

Isolation barriers can be divided into two categories, active isolation and passive isolation, corresponding to their location with respect to the vibration source and the isolated building, as shown in Figure 44.

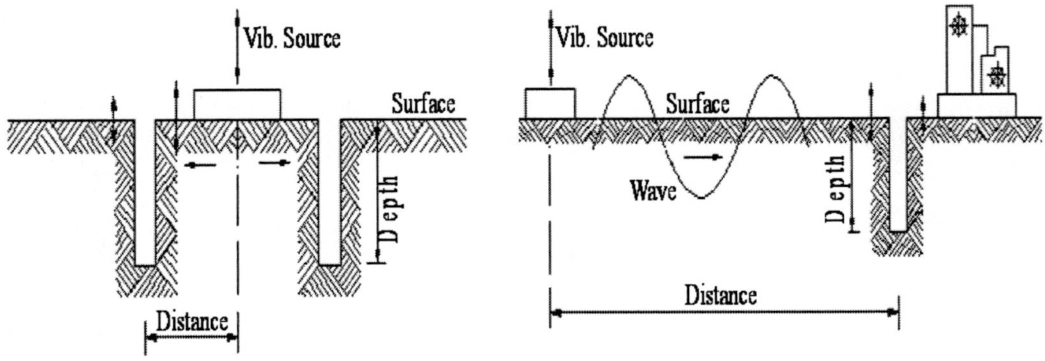

Figure 44. Near-field active isolation (left) and far-field passive isolation (right).

The active isolation is realized by setting barriers around the vibration source, to separate or damp the wave energy from the source. The barriers mainly separate body waves because they are close to vibration source.

The passive isolation is also called far-field negative isolation. The barriers are set away from the vibration source but close to the building to be protected, to separate or damp the wave energy transmitted to the building. The barriers mainly separate Rayleigh waves because they are far from vibration source.

(2) Continuous and discontinuous isolation barriers

There are continuous and discontinuous barriers corresponding to the barrier's structure. The continuous barriers include open trench, filling ditch (filled with bentonite mud, saw scrap, sand and coal ash), reinforced concrete wall, etc. The discontinuous barriers include hole rows, row-piles and sheet piles.

For continuous barriers, the isolation effect relies on the barrier's depth, which should be larger than the wavelength of the Rayleigh wave. Due to the soil stability, groundwater level and construction cost, the continuous barriers cannot be very deep, and so they are only suitable for isolating vibrations with medium-high frequencies. For low frequency vibrations, the wavelength of Rayleigh wave produced by traffic-induced vibration may be 50−100 m as long. In this case, the continuous barriers are inapplicable because of the depth limit, while the discontinuous barriers like row-plies are free from the restriction.

When the vibration wave propagates to the barrier, it induces transmission and reflection effects, and diffraction at both ends and bottom of the barrier. Thus, the remained vibration energy behind the barrier consists mainly of transmission and diffraction waves, and is usually smaller than that of the incident wave, which leads to decline of ground movement. However, there may appear some local areas before the barrier, where the ground movement is amplified because of the wave reflection. Woods [34] studied the vibration isolation method using open trenches. Based on test data, he proposed the basic principle for isolation design and suggested an index for evaluating the isolation effect of barrier, i.e. the amplitude reduction index A_r of ground movement:

$$A_r = \frac{A_2}{A_1} \tag{17}$$

where A_1 and A_2 indicate the displacements before and behind the barrier, respectively.

The index has been applied to assess the vibration-isolating barriers. Obviously, the isolation effect can be expected only when $A_r<1$, and the smaller the A_r, the better the effect.

(3) Wave impeding barriers

Wave impeding barrier is another kind of isolation barrier by embedding elastic or rigid plates at certain depth under the vibration source or the structure to be protected.

The elastic wave demonstrates different features when propagating in different soils. When a single soil layer on the bedrock is subjected to a harmonic line load, there exists a cut-off frequency for the forced vibration of the soil. If the frequency of the harmonic load is lower than the cut-off frequency, no wave spreads in the soil. Based on this principle, Chouw et al. [35] proposed to imbed a hard interlayer in the soil, which acts as a manual bedrock with finite dimension, to isolate the vibration. They named the hard interlayer as Wave Impedance Block (WIB), as shown in Figure 45.

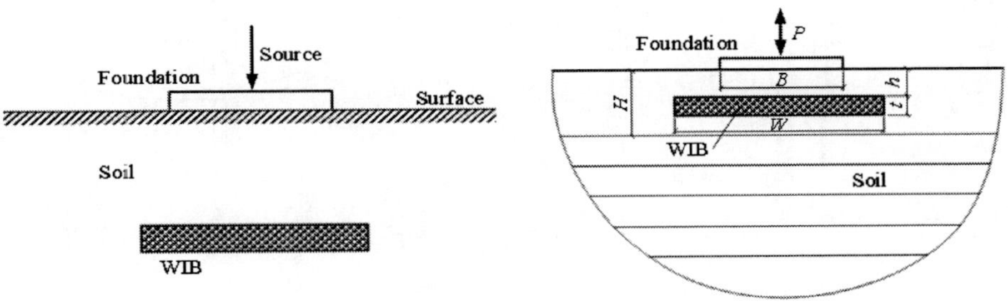

Figure 45. Isolation system with elastic plate.

Compared with open trench, WIB isolation is more effective for low-frequency vibration, while it may lead to amplification of high-frequency vibration in the far-field, and its isolation effect is very sensitive to the field property. Increasing WIB's stiffness can improve the effect of active isolation.

Many scholars studied WIB isolation for ground vibration induced by high-speed railways, and described in detail its allocation form and mechanic properties. Peplow and Kaynia [18] investigated the isolation effect of rigid slab (can be regarded as a WIB) embedded in or under the ballast bed, by FEM with no reflection boundary conditions, showing the rigid slab under ballast can effectively limit those vibrations along the railway. There are also some new WIB isolation systems, such as the cylindrical wave resistance barrier arranged with honeycomb form, as shown in Figure 46, which is defined as Honeycomb WIB [17]. These techniques have been validated by vibration tests at railway sites, and they will be included in the Code for Design of Vibration Isolation of China.

For a point vibration source, isolation materials can be set around it to form an annular WIB, which can absorb the energy of the vibration and effectively block the wave inside the barrier. The isolation effect of WIB is associated with the property of the material, the structure and the buried depth of the barrier. Shown in Figure 47 is a viaduct system with an annular WIB embedded around the pier foundation, by which the vibration outside the barrier dropped by 5−15 dB [36].

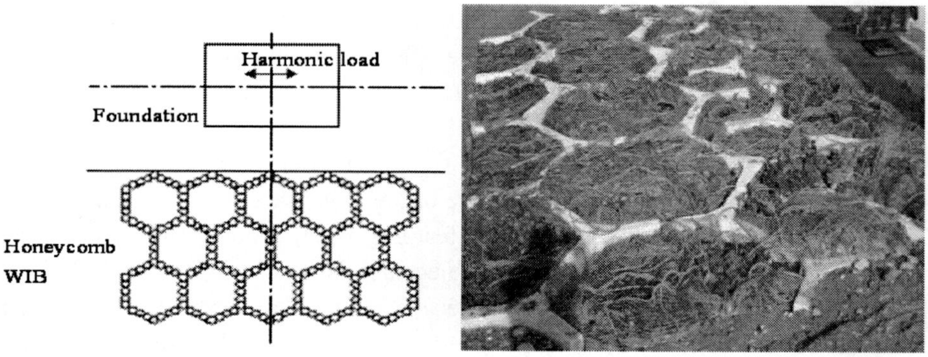

Figure 46. Honeycomb WIB isolation system.

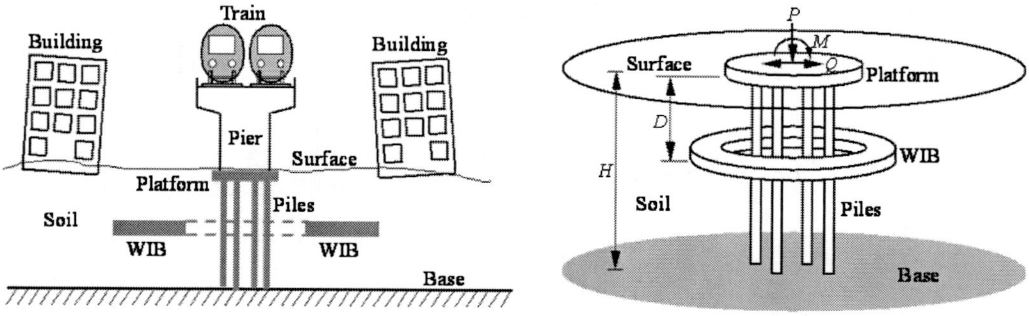

Figure 47. A bridge with WIB and the isolation principle.

(4) Isolation for metro induced vibration

Unlike the isolation on ground vibration source where mainly Rayleigh wave considered, metro belongs to inter-soil source where the vibration propagates to the ground receiver in form of body wave. Thus, a particular investigation is necessary on isolation measures for ground vibration induced by metro trains.

The effect of vibration isolation barriers is related to the depth, thickness and material properties of the barriers. Isolation analyses of concrete continuous walls, open trenches and filling trenches show that the vibration isolation effect is unsatisfactory when a rigid barrier or a flexible barrier used alone. Thus, it is necessary to find a better underground vibration barrier that not only has the advantages of rigid material, flexible material and open trench, but also is feasible in engineering. Such underground continuous barrier is called composite isolation barrier.

According to the characteristics of underground vibration source induced by metro trains, Zhang et al. [4] proposed three kinds of composite isolation barriers.

1) Underground concrete hollow continuous wall (see Figure 48). This composite isolation barrier, due to the cavity and the concrete material, combines the advantages of both concrete continuous wall and open trench.

Herein, the hollow concrete wall is completely vacant between two walls, instead of the honeycomb structure used in common engineering, which can fully block the passage of the vibration wave.

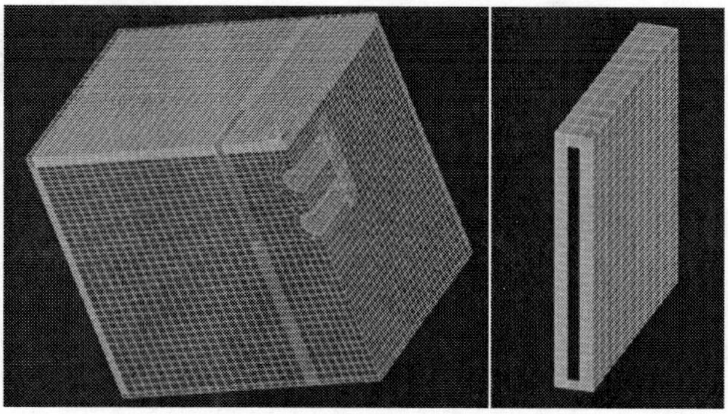

Figure 48. Underground concrete hollow continuous wall isolation barrier.

2) Underground concrete continuous wall with cushions. In this composite isolation barrier, cushions made of flexible material with a certain thickness are added to both sides of the solid wall. Therefore, it has the isolation behavior of both barriers with rigid and flexible materials.

3) Concrete hollow wall with cushions. In this composite isolation barrier, cushions made of flexible material are added to both sides of the hollow wall. Therefore, it combines the advantages of both barriers of rigid material, flexible material and open trench.

The concrete continuous wall magnifies the ground vibration behind it considerably because of its own vibration. Hence, adding flexible cushion on the wall at the source side can help reduce the wall vibration. Since the flexible material is effective for the high-frequency vibration, the high-frequency component through the flexible cushion to the wall will decline greatly.

Adding flexible cushion on the wall at the other side is to take advantage of wide isolation range of flexible material. Thus, when the vibration passes the barrier, the low-frequency components will be damped by the rigid wall while the high-frequency components by the flexible material, which extends the frequency range of isolation.

Therefore, adding flexible cushions to both sides of continuous wall not only reduces the vibration of the wall itself but also extends the low-frequency isolation effect to larger area behind the wall, and as well strengthens the isolation effect of high-frequency components.

5.3. Foundation Isolation for Buildings

5.3.1. Fundamental Principle of Foundation Isolation for Buildings

The fundamental principle of foundation isolation for building is illustrated with a SDOF vibration isolation system, as shown in Figure 49, where the isolated mass lays on the base with simple harmonic movement $v_g(t) = v_{g0} \sin\overline{\omega}t$. The motion equation of the system is:

$$\ddot{v}(t) + 2\xi\omega\dot{v}(t) + \omega^2 v(t) = -\frac{1}{m}\ddot{v}_{g0} \sin\overline{\omega}t \tag{18}$$

According to structural dynamics, the relative displacement of the mass with respect to the base bottom is defined as:

$$v(t) = v_{g0}\beta^2 D\sin(\overline{\omega}t - \theta) \tag{19}$$

where $D = 1/\sqrt{(1-\beta^2)^2 + (2\xi\beta)^2}$ is the dynamic amplification factor.

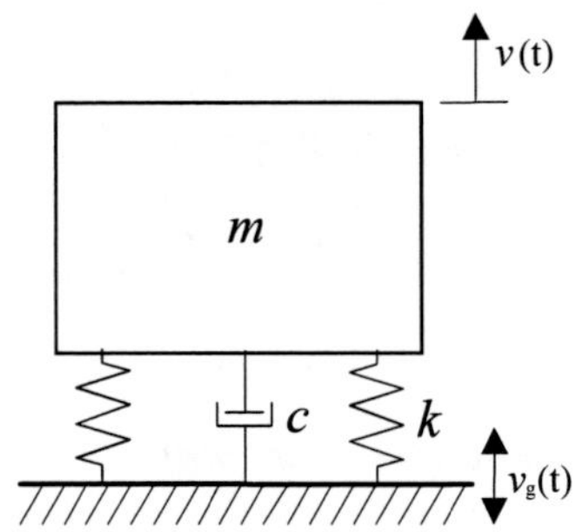

Figure 49. Fundamental principle of foundation isolation for building.

The overall mass movement can be expressed by adding the vector of base bottom movement to Equation (19) as:

$$v^{t}(t) = v_{g0}\sqrt{1+(2\xi\beta)^2}\,D\sin(\overline{\omega}t - \theta) \tag{20}$$

In this case, the transmissibitily of the isolation system is defined as the amplification ratio of the mass movement to the base bottom movement:

$$\mathrm{TR} = \frac{v^{t}_{max}}{v_{g0}} = D\sqrt{1+(2\xi\beta)^2} \tag{21}$$

It is noticed the expression of the transmissibitily TR herein is the same with that in the near-field isolation. The trend of TR versus the frequency ratio β and the damping ratio ξ is the same as what is illustrated in Figure 43. Hence, the isolation effect can be expected only when $\beta > \sqrt{2}$. However, the train-induced vibration is featured mainly of low frequency when filtered by soil, so it is quite difficult to design the isolation system meeting this condition. Therefore, the far-field isolation of foundation is mainly applied in the isolation for precise devices.

5.3.2. Isolation for Building Foundations

The fundamental principle of vibration isolation for building is to place the building on elastic foundation. In practice, isolation foundation is usually adopted by inserting steel springs or rubber blocks between the building and the foundation to separate vibration from ground.

Elastic floating isolations can reduce the influence of vibration on structures and eliminate the subsequent secondary noise, when applied to residential buildings, office buildings, hotels, concert halls and conference centers close to railways, metro lines and roads with heavy truck traffics. A wide use of the elastic floating isolation technique can be expected for vibration-sensitive real estates, public houses, theaters and concert halls in urban downtown with convenient traffic transit center, huge prosperity but exorbitant land price.

The first floating building with elastic bearings was built by the Gerb Company in Berlin, 1986, which was a residence building mainly for vibration isolation as well as for aseismic function. The whole building floated on spiral springs with the natural frequency of 4 Hz. Afterwards, floating buildings appeared gradually in Bonn, Munich, London, Manchester, Paris, Prague, Sydney, Hong Kong and many other cities. Shown in Figure 50 (top) is a floating concert hall with 2400 seats, which is built above the metro station in Manchester, where the GP spring isolators with the system frequency 3.5 Hz are adopted and the total supporting load reaches 266,000 kN.

In France, the floating isolation technique was applied to the Sheraton Hotel to prevent it from the vibration disturbances of metro trains, road traffics and airplanes' taxing and landing on runways in the Roissy Airport in Paris (see Figure 50 bottom). In Korea, the Cheonan Station of the Seoul-Busan high-speed railway is isolated from vibration using the floating technique, with the floating area 4800 m^2 and the floating load 1,000,000 kN, respectively.

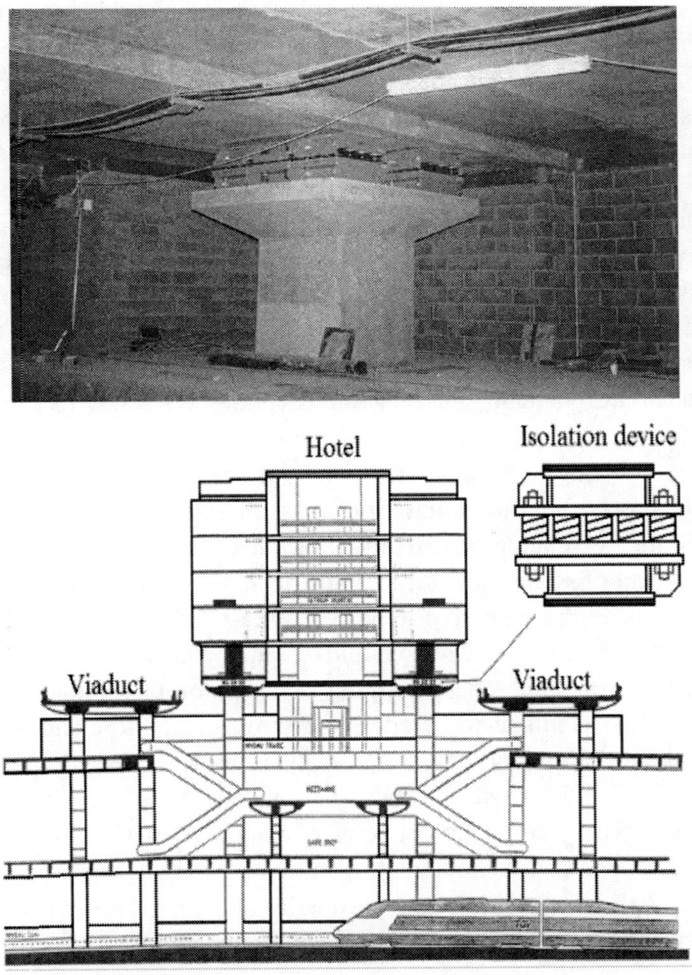

Figure 50. Buildings with floating isolation: (top) the concert hall in Manchester; and (bottom) the Sheraton hotel in Roissy airport in Paris.

The noted buildings adopting floating isolation technique in China include the Shanghai Concert Hall, the Wuhan Grand Theater, the National Grand Theater, and so on. Therein the most worthwhile mentioned is the Shanghai Concert Hall, a heritage building of Shanghai built in 1930's. Since it was too close to the Yan'an Viaduct, the background noise during performance was high, which was extremely discordant with the internal environment, the relevant department decided to move it by 66.4 m to a new place.

However, the new place is closer to Metro Line 1, the metro train-induced vibration and the secondary noise became the new nuisance. As a result, the floating floor isolation technique is adopted to reform the hall: the first floor with the bottom beams and floor slabs are floated on 24 spring-damper isolators installed between the concrete bottom beams and the fixed pillars. The test on the refitted hall shows the vibration produced by the metro train declines by 70%, the background noise reduced by 13 dB(A), and the intermediate frequency reverberation time increases from 1.46 s to 1.83 s, meaning a very good acoustic effect.

5.3.3. Microvibration Control for High-tech Devices

The control of traffic-induced vibration for high-tech devices is quite different from seismic control of building structures. The isolation system should be carefully designed to make the frequency of the system as far as possible from the disturbing frequency.

Passive isolation system is the most popular and effective method for micro vibration control, which is usually installed between the precise equipment and the support structure, or a massive pedestal under the precise equipment, commonly used for isolation of large or ultra-large precise equipment. For small-size precise devices, they are often equipped with their own passive vibration isolation system.

Using passive isolation (spring-damper system) has been commonly applied to protect the high-tech devices from vibrations due to ground movement, but its inherent characteristics limit its vibration-reducing capacity. To reduce the vibration sufficiently, the frequency of spring-damper system should be as low as possible, which often leads to unacceptable displacement under the operating loads. Besides, if the system is too flexible, the static stability of the device will be threatened. Thus, in recent years, the hybrid control has been investigated by combining the active control with the passive control, which is more adaptable than the passive control because it can response any real-time structural changes of the objective building or isolation system through feedback control.

With development of science and technology, operation of high-tech devices and modern machining techniques has become precise or even ultra-precise. Xu et al. [37] investigated a microvibration control platform to isolate high-tech equipments from a building subjected to traffic-induced ground motion. The governing equation of the coupled platform-building system was derived to facilitate the feedback control and performance evaluation of the platform based on the BBN vibration criteria. A hybrid control system composed of passive mounts and active hydraulic actuators with a sub-optimal control algorithm was designed to control the platform. The performance of actively controlled platform was assessed through comparisons with the building without control and with passive control platform. Simulation results indicate the hybrid control platform is superior to the passive control platform because of its high performance and robustness, as shown in Figure 51.

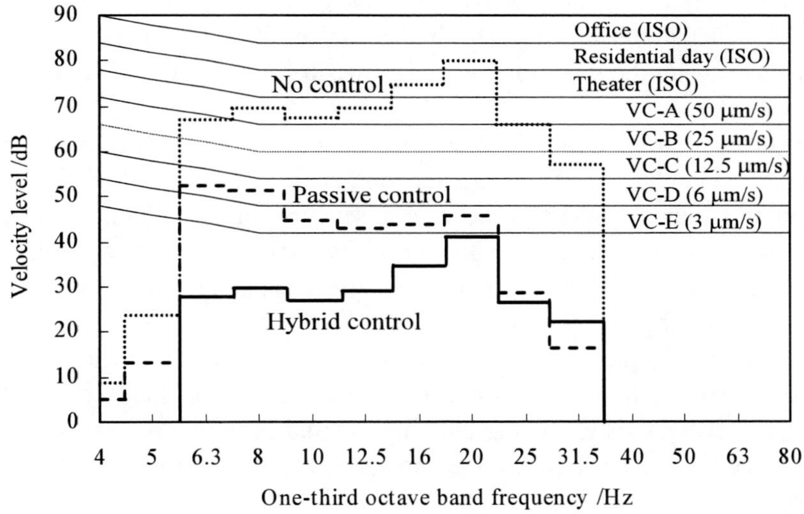

Figure 51. Comparison of control effects for high-tech instruments with the VC-A−VC-E curves in BBN vibration criteria.

Yang and Agawal [38] studied the applicability of various protective systems adopted in traffic-induced micro vibration control for high-tech facilities, by considering the dynamic interaction of isolation system and structure. These protective systems include passive control of base isolation, hybrid base isolation, passive platform isolation, hybrid platform isolation, passive energy dissipation system and active control system. The result shows that the hybrid platform isolation is the most effective and feasible method for micro vibration reduction of high-tech facilities.

5.4. Principles for Planning and Designing Traffic System

Traffic-induced vibration is a significant issue in planning an urban traffic system. The influences of vibration on the environment and the corresponding mitigation measures should be taken into account at the initial stage of plan and design. The main planning and design principles include:

(1) In design of a traffic line, a comprehensive consideration is necessary on the urban development plan, traffic flow, geological and construction conditions, current and long-term economic benefits, as well as environmental vibration. Generally, the alignment of a traffic line is not affected by environmental vibration. However, for some buildings of significant cultural and historic values, the route should be as far away as possible, to protect them from damage. In practice, quite a few cultural relics are separately located in the downtown area, and it is difficult to alter the alignment. In such a case, a particular ancient building may become a control factor for the scheme comparison, thus a special investigation should be done on the vibration environment.

(2) It is necessary to stipulate the shortest distance of ground buildings to the rail transit line. When the subway is to be built near sensitive buildings, try to use deeper tunnels to damp the vibration by energy dissipation during the wave propagation in the soil.
(3) In planning a route, fully consider utilization of natural barriers such as rivers, valleys and tall buildings to isolate the vibration.
(4) For new buildings, try to locate them away from the vibration amplification zone near the existing traffic line. While for a new traffic line, try to arrange the vibration amplification zone away from the existing ancient buildings or other vibration-sensitive buildings.
(5) Buildings along the subway or elevated railway should be designed with firm structure and foundation, instead of light structure or shallow foundation. The vibration characteristics of the buildings should be properly designed to avoid the resonance due to the frequency coincidence of the structure and the train-induced vibration.
(6) Laboratories and factories equipped with precise instruments sensitive to vibration should be located away from the areas with vibration disturbances, such as the vibration active areas, geological fault zones, and road, railway and subway nearby areas. Besides, prediction should be done on other possible environmental vibrations such as current and long-term constructions. For the existing projects with strict vibration allowance, the micro-vibration test is indispensable, including the measurements on wave velocities, dynamic stiffnesses and damping ratios of site soil, engineering geological conditions, and so on. Necessary measures should be investigated beforehand to reduce and isolate the vibration.

Conclusion

The traffic-induced environmental vibrations have become more concerned worldwide in recent years. In this chapter, a general review is made on the research background of this subject in theoretical analyses, numerical simulations and experimental measurements. The characteristics of traffic-induced ground vibrations and their propagations from the aspects of duration, intensity distribution, vehicle load and speed are introduced. The influences of vibrations on the daily-life of the people, the normal operation of high-tech devices and the serviceability of buildings, the vibration properties of ground soil and buildings adjacent to underground, ground and elevated railway systems are summarized. The reduction and isolation measures against traffic-induced vibrations and the corresponding principles for planning and designing traffic systems are introduced.

Acknowledgments

This study is sponsored by the Natural Science Foundation of China (51078029) and the Fundamental Research Funds for the Central Universities (2009JBM078).

REFERENCES

[1] Japan Institute of Noise Control. (2001). *Regional Vibrations of Environment.* Tokyo: Gihodo Press.

[2] Environment Bureau of Japan, MITI. (2002). *Technologies and Laws for environmental Pollution Preventions: Vibration.* Tokyo: Maruzen Publishing Co., Ltd.

[3] Bata M. (1971). Effect on buildings of vibrations caused by traffic, *Building Science*, 6: 221−246.

[4] Xia H. (2010). *Traffic Induced Environmental Vibrations and Control.* Beijing: Science Press.

[5] Yasutoshi K. (2003). Simulation and observation of ground vibration caused by road traffic, *Proc. ISEV'2003*, Hangzhou, p. 202−216.

[6] Takemiya H., Bian X.C. (2007). Shinkansen high-speed train induced ground vibrations in view of viaduct–ground interaction. *Soil Dynamics and Earthquake Engineering*, 27(6): 506−520.

[7] Krylov V.V. (1997). Spectra of low-frequency ground vibrations generated by high-speed trains on layered ground. Journal of Low Frequency Noise, *Vibration and Active Control.* 6(4): 251−262.

[8] Wilson G.P., Saurenman H.J., Nelson J.T. (1983). Control of ground-borne noise and vibration. *Journal of Sound and Vibration*, 87(2): 339−350.

[9] Kurzweil L. (1979). Ground-borne noise and vibration from underground system. *Journal of Sound and Vibration*, 66(3): 173−182.

[10] Degrande G., Lombaert G. (2001). free field vibrations during the passage of a thalys high-speed train at variable speed, Journal of Sound and Vibration, 247(1): 131−144.

[11] Wolf S. (2003). Potential low frequency ground vibration impacts from underground LRT operations, *Journal of Sound and Vibration*, 267: 651−661.

[12] Volberg G. (1984) Propagation of ground vibrations near railway tracks, *Journal of Sound and Vibration*, 87(2): 371−376.

[13] Metrikine A.V., Vrouwenvelder A.C.W.M. (2000). Surface ground vibration due to moving train in a tunnel: two-dimensional model, *Journal of Sound and Vibration*, 234(1): 43−66.

[14] Crispino M., D'apuzzo M. (2001). Measurement and prediction of traffic-induced vibrations in a heritage building, *Journal of Sound and Vibration*, 246(2): 319−335.

[15] Bahrekazemi M. (2004). *Train-Induced Ground Vibration and Its Prediction.* Royal Institute of Technology, Stockholm.

[16] Kattis S.E., Polyzos D., Beskos D.E. (1999). Modelling of pile wave barriers, *Soil Dynamics and Earthquake Engineering*, 18(1): 1−10

[17] Takemiya H., (2004). Field vibration. *Soil Dynamics and Earthquake Engineering*, 24(1): 69−87.

[18] Peplow A.T., Kaynia A.M. (2007). Prediction and validation. *Soil Dynamics and Earthquake Engineering*, 27(8): 793−802.

[19] Yang Y.B., Lin C.W. (2005). Vehicle–bridge interaction dynamics and potential applications. *Journal of Sound and Vibration*, 284(1–2): 205−226.

[20] Ju S.H., Liao J.R., Ye Y.L. (2010). Behavior of ground vibrations induced by trains moving on embankments. *Soil Dynamics and Earthquake Engineering*, 30(11): 1237−1249

[21] Richart F.E. (1970). *Vibration of Soils and Foundations*, New Jersey: Prentice-Hall Inc..

[22] Ministry of Mechinery Industries of China. (1996). *Design Code for Power Machine Foundations (GB50040-96)*. Beijing: China Plan Press.

[23] Gu X.A., Ren J.F., Liu Y. (2003) Vibration environment of railways in China: an overview. *Railway Working Safety and Labor Health*, 30(5): 206−209.

[24] Xia H., Zhang N., (2005). Experimental study of train-induced vibrations of environments and buildings, *Journal of Sound and Vibration*, 280: 1017−1029.

[25] Fujikake T., (1986). A prediction method for the propagation of ground vibration from railway trains, *Journal of Sound and Vibration*, 111(2): 357−360.

[26] Volberg G. (1983). Propagation of ground vibrations near railway tracks. *Journal of Sound and Vibration*, 87(2): 371−376.

[27] Yoshioka O. (2000). Basic characteristics of Shinkansen-induced ground vibration and its reduction measures, *Proc. Wave'2000*, Bochum, p. 219−239.

[28] Pan C.S., (1990). Measurement and analysis of train induced vibrations in metro tunnel. *Civil Engineering*, 23(2): 21−28.

[29] Xia H., Cao Y.M. (2002). Vibration effects of light-rail-train-viaduct system on surrounding environment, *International Journal of Structural Stability and Dynamics*, 2(2): 227-240.

[30] Xia H., Cao Y.M., Zhang N. (2007). Numerical analysis of vibration effects of metro trains on surrounding environment, *International Journal of Structural Stability and Dynamics*, 7(1): 154−166.

[31] Xia H., Han Y. (2009). Experimental investigation of railway train-induced vibrations of surrounding ground and nearby multi-story buildings, *Journal of Earthquake Engineering and Engineering Vibration*, 8(1): 137−148.

[32] Enoshima S. (1982). *Countermeasures for Foundation Vibrations*. Tokyo: Yoshii Bookstore.

[33] Wakui H., Matsumoto N., Okuda H. and Asanuma K. (2002). Performance and application of ballast ladder track. *New Railway structure*, 52(4): 32−34.

[34] Woods R.D. (2005). Soil dynamics considerations in the automobile industry. *Proc. ISEV'2005*, Okayama, p. 235−241.

[35] Chouw N., Le R., Schmid G. (1998). An approach to reduce foundation vibrations and soil waves using dynamic transmitting behavior of a soil layer. *Bauingenieur*, 66: 215−221.

[36] Jiang J.Q. (1993). *Studies on Dynamic Soil-structure Interaction of Extended High Pier Bridges*. Doctoral Dissertation, Okayama University, Japan,.

[37] Xu Y.L., Yang Z.C., Chen J., Liu H.J. (2003). Microvibration control platform for high technology facilities subject to traffic-induced ground motion. *Engineering Structures*, 25: 1069–1082

[38] Yang J.N., Agrawal A.K. (2000). Protective systems for high-technology facilities against microvibration and earthquake, *Journal of Structural Engineering and Mechanics*, 10(6): 561–567.

In: Traffic Induced Environmental Vibrations and Controls ISBN: 978-1-62618-240-0
Editors: H. Xia and R. Calçada

Chapter 2

VIBRATIONS INDUCED BY RAILWAY TRAFFIC: PREDICTION, MEASUREMENT AND MITIGATION

Pedro Alves Costa**, Rui Calçada and António Silva Cardoso
Faculty of Engineering, University of Porto, Porto, Portugal

ABSTRACT

This chapter is about the vibrations induced by railway traffic, comprising the prediction models, its experimental validation and its application on the analysis of mitigation measures. Due to the specific aspects of this kind of problems, where very large domains need to be considered, a 2.5D approach is adopted in the developing of the numerical model. Moreover, the ability of the model for dealing efficiently with the train-track-ground dynamic interaction is highlighted. After the description of the numerical model main characteristics, its ability for dealing with real situations is evaluated through the comparison between prediction and experimental results for a case study developed by the authors. The experimental activities performed in that case study are presented, being the measurements of the vibrations induced by the train passage compared with the predicted results. Finally, after the experimental validation of the numerical model, its potentialities are highlighted namely in the analysis of the efficiency of ballast mat solutions on the abatement of vibrations induced by traffic.

1. INTRODUCTION

The new challenges imposed to modern societies demand for the development of the railway transportation systems lines, namely for high-speed, in order to minimize the regional asymmetries and to improve the effectiveness and the economy of the passenger transportation systems. This demand of the society has also stimulated the scientific community to perform research focused on the analysis and mitigation of some adverse problems arising from railway traffic. One particular and important concern is related with the environmental impact that railway traffic can present on the facilities near to the track, mainly

* E-mail address: pacosta@fe.up.pt.

in urban environment, comprising noise and vibrations induced by the passage of the train. Apart from other consequences, the railway traffic can cause discomfort to inhabitants or affect the regular use of sensitive equipments in surrounding buildings.

Currently, it is recognized that the geometrical imperfections of wheels and tracks are the major cause of dynamic loads developed during the rolling of trains. These dynamic loads act and excite waves propagating from the track to the free-field and to nearby buildings, giving rise to the commonly called train-induced ground vibrations. Therefore, contrarily to what happens in the road traffic, the train-track dynamic interaction is an important issue that must be addressed on the analysis and prediction of the problem.

Although the great progress performed in the last years in order to obtain a better understanding of the problem, several questions and doubts remain. The progress on the developing of new countermeasure techniques for the mitigation of ground-borne vibrations induced by traffic was not always followed by a scientific and technical decision about its efficiency. Therefore, due to the aspects mentioned above, the need of efficient prediction tools and procedures, which allow the performance of a mechanical design of vibration mitigation countermeasures, are demanded by the engineering practitioners who deal with this kind of problems. Nevertheless, it cannot be neglected that a great effort has been developed by the technical community, mainly during the last decade, for reaching the expressed goal. The visible face of those efforts has been proved by the significant proposals of distinct classes of prediction models.

Regardless of the indisputable relevance of empirical and scoping models, advanced design stages of new railway lines or mitigation countermeasures for existing railway lines, demand the usage of advanced numerical models in order to predict, quantify and mitigate the potential problems, without forgetting the life-cycle cost of the infrastructure. It is well known that the dynamics of the complete system is complex, dominated by the dynamic interaction between the train, the track and the ground [1].

The scientific community has allocated a considerable effort to this subject during recent years. Consequently, several numerical and semi-analytical models have been proposed. The semi-analytical approaches present a clear supremacy in terms of computational efficiency and capability for the understanding of the essence of the phenomena. Many research studies about this subject have been carried out since the introduction of the pioneer concept of equivalent stiffness of the ground by Dieterman and Metrikine [2, 3]. An efficient and comprehensive three dimensional model that fully takes into account the train-track-ground interaction was presented by Sheng et al. [4]. In spite of the efforts of many researchers to improve the semi-analytical models in order to incorporate more complex geometries, namely regarding the embankment geometry or the presence of ground inclusions such as trenches [5], the strictness inherent to these solutions does not allow obtaining general formulations capable of dealing with the complex geometries usually encountered in practical applications [6].

The above-mentioned limitations, intrinsic to the semi-analytical approaches, combined with the development of computational capabilities, have led to the development of numerical approaches especially designed for the simulation of large domains subjected to moving loads. Almost all these models are based on the well-known finite elements (FEM) and boundary elements (BEM) methods [7]. In spite of the amazing evolution of the computational capabilities, the intensive usage of fully 3D analysis, including the train-track-ground interaction, still remains a difficult challenge, at least if one must consider a reasonable computation time [6]. Nevertheless, some progresses have been reached recently, namely in the BEM context [8−10] or through the usage of special techniques such as the moving finite elements method [11, 12].

To overturn this drawback, and without losing the 3D character of the problem, a number of assumptions is often made in an attempt to reduce the computational efforts. Actually, transport infrastructures, as for instance, roads and railways, fit particularly well to these assumptions, since they can be assumed as infinite structures with periodic or invariant geometries.

Assuming a periodic domain, advantage can be taken of the properties of the Floquet transforms, being only necessary the simulation of a unique cell [13−16]. On the other hand, if the domain is invariant along the longitudinal direction, a computationally efficient 2.5D numerical scheme can be adopted, where the Fourier transform of the longitudinal coordinate allows representing the 3D response and the radiated wave field using only numerical discretization along the cross-section of the domain [17]. This is the approach followed in the present paper. Actually, the 2.5D concept can be applied to distinct numerical approaches, as for instance FEM or BEM [18−27]. The option by one of those methods in detriment of the other depends of the specificity of the problem. An efficient computational scheme can also be achieved by coupling both methods. In that case, the irregular domain is simulated by 2.5D FEM and the ground is modelled by 2.5D BEM [16, 27−30]. Very recently, François et al. [27] presented an efficient and comprehensive approach for coupling both methods, which was also applied in a study performed by Galvín et al. [30]. Following a similar numerical procedure, Alves Costa et al. [31] presented the experimental validation of 2.5D FEM-BEM model.

Apart from the issues related with the modeling of the track-ground response, a general theoretical model must also attend to the train dynamics. As already pointed out, the excitation induced by the train passage can be decoupled into two main components: i) the quasi-static-mechanism; ii) the dynamic mechanism [1, 32]. The quasi-static mechanism, due to the moving of the dead loads of the train, is very easily considered on a numerical model. However, the same statement cannot be extended to the dynamic excitation mechanism. This mechanism is generated by the train-track interaction, which requires the consideration of the dynamics of the train.

Several models for the simulation of the train have been proposed during the last decade. However, it is well known that, at least for the purposes under consideration, multi-body models suffice for the assessment of the dynamic loads [8, 33, 34].

Despite the great developments occurred in the numerical simulation field, the experimental validation of the numerical progresses highlighted above has not been accomplished. The focus of the present work is exactly this concern, i.e., the presentation of robust and efficient numerical models, developed by the authors, its experimental validation and, finally, its application for the study of mitigation countermeasures efficiency.

The selected case study covers an important aspect of the dynamic of transportation infrastructures: the prediction of vibrations induced by traffic and radiated by the ground. The experimental assessment of the train-track-ground dynamics is highlighted, since an extensive and comprehensive experimental campaign was designed, implemented and developed by the authors of the present chapter.

Since the numerical model is experimentally validated, being its robustness and accuracy demonstrated, its potentiality for the study of specific mitigation measures is highlighted.

Regarding this aspect, it should be mentioned that the countermeasures against vibrations induced by traffic can assume distinct configurations. Actually, it is common to group them into three distinct groups, depending on the location: i) at the source; ii) along the wave propagation path; iii) at the receiver. The latter, comprising the isolation of the buildings, does not change the pattern or the amplitude of the waves that propagate from the track. Alternatively, it is also possible to interrupt the wave path, for instance by the construction of filled trenches and barriers in order to prevent the waves from reaching the buildings [35].

In the present chapter, only countermeasures adopted at the source level, i.e., at the track, are addressed. Generally, in that situations, the mitigation effect is achieved by changing the dynamic behavior of the track through the insertion of resilient elements or additional masses to the track structure [36].

The former, more usual than the latter, can be placed at different location levels of the railway track cross-section. The main idea is to induce a lower natural frequency to the railway track, which is conditioned by the mass above the resilient element and by its stiffness (lower than that of the remaining elements). As is evident, the effectiveness of this isolation technique is related to the value of the resonance frequency. Insertion of resilient elements immediately beneath the rails or the sleepers, such as railpads or under sleeper pads, gives rise to high values for the resonant frequency due to the lower mass above the isolation.

On the other hand, lower resonant frequencies can be achieved in heavy systems, such as in floating slab tracks or ballasted tracks with ballast mats, which allow increasing the effectiveness of the isolation for frequencies above few tens of Hertz.

In the middle frequency range, ballast mats can be a suitable solution for mitigation of the vibrations that propagate from the track through the ground [37]. Indeed, the introduction of the resilient elements beneath the ballast or the sub-ballast, allows achieving a low resonant frequency, i.e., from 20 Hz to 50 Hz, due to the heavier mass above the resilient element [38]. As expected, the efficiency of the ballast mat on the abatement of the vibrations is also dependent of several aspects, as for instance the support stiffness provided by the ground beneath the track [39].

Nevertheless, despite the studies previously performed by other authors and reported at the bibliography, this is one of the cases where the construction industry is one step ahead of the research, i.e., there are several products and procedures in contrast to the lack of comprehensive studies about its particular performance. In the authors' knowledge, the existing numerical studies about this topic are reduced to simple one-dimensional formulations [37, 40], with some exceptions concerning the studies performed on floating slab tracks [8, 41], which can be regarded as a similar system, and to a research paper dedicated to the topic presented by Auersch [38].

The present chapter aims to contribute for fulfilling the lack of comprehensive knowledge about the particular performance of mitigation measures installed at the source level. That objective is achieved by a comprehensive approach, where a numerical model is presented, validated and used for a deeper understanding of the main aspects that persuade the performance of ballast mats for the attenuation of vibrations induced by railway traffic.

2. NUMERICAL MODELLING

2.1. Generalities

The modeling of vibrations induced by traffic is a complex task, where the train-track-ground dynamic interactions must be taken into account. Following that goal, the global model proposed by the authors is divided into two main modules corresponding each one to the "stand-still" part, i.e., the track ground-system and to the "moving" part, which comprises the train. The latter is simulated by a multi-body formulation where the main masses and suspensions of the vehicle are taken into account. The interaction between both models is established by a compliance formulation, which imposes displacement compatibility between the contact points of the vehicles with the remaining system. The basic formulation of the interaction scheme is summarized in Figure 1.

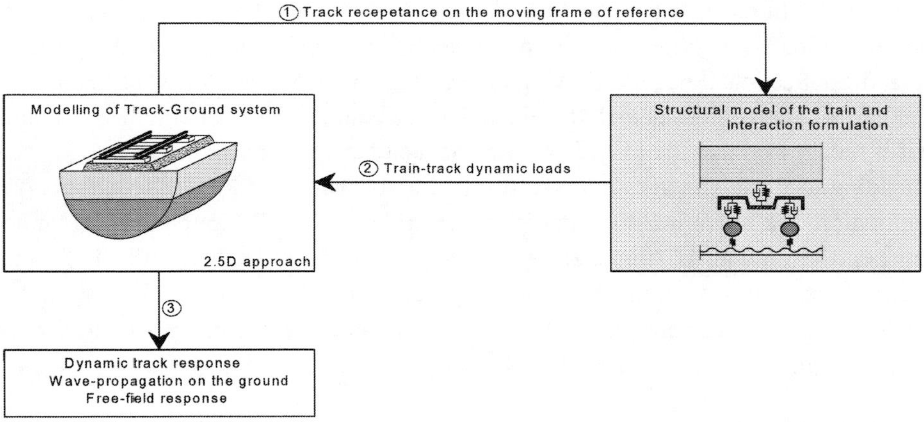

Figure 1. Computational scheme adopted for the train-track interaction.

In the following sections, the main aspects and assumptions inherent to the numerical formulation of both parts of the model are presented and discussed.

2.2. Numerical Modeling of the Track-ground-system

2.2.1. 2.5D Approach - Basic Concept

In order to avoid extremely high costs of computation intrinsic to a full 3 D simulation, several attempts have been developed in order to respect the tri-dimensionally of the domain by one hand, but imposing some restrictions by the other hand in order to minimize computational effort. Tri-dimensional structures with infinite development and invariable properties (geometrical and mechanical) can be dealt by a 2.5D formulation. In these cases, the structure main properties can be represented by its cross-section, being, by that way, faced as a two-dimensional medium. However, despite of that condition, it should not be forget that the loading is not two-dimensional, which demands the solution of a 3D problem. The main concept behind the proposed solution to the problem is the use of a method which is between the two and the three dimensional domain. This method was firstly proposed by Hwang and Lysmer [42], in the FEM context, for the study of underground structures under travelling seismic waves. Subsequently, the method has been applied by a few researchers to the study of vibrations induced by traffic. In this field, special attention should be dedicated to the following works: Yang and Hung [17], Alves Costa et al. [23], Alves Costa et al. [31], Sheng et al. [43], Muller [44], and Alves Costa [45].

Assuming that the response of the structure is linear, the analysis can be carried out on the wavenumber/frequency domain. All the variables, i.e., loads (action) and displacements (response), must be transformed to the wavenumber/frequency domain by means of a double Fourier transform, related with the direction along the track (x-direction) and with time. Transformed quantities are denoted as functions of the Fourier images of x and t, defined as wavenumber and frequency and represented by k_1 and ω, respectively.

From the numerical point of view, the 2.5D approach can be extended to the most popular numerical methods, as for instance the FEM or the BEM [27, 28, 43]. Due to its powerful capabilities, when dealing with complex geometries, the FEM has considered the method of choice by a large number of geotechnical engineers. However, despite of its widespread application field, the FEM also presents some drawbacks when compared with other numerical approaches, namely for the numerical simulation of unbounded domains [46], where it is recognized that the FEM is not the most suitable approach.

Attending to the advantages but also to the drawbacks of the FEM, the authors opted by the formulation of a numerical tool that is based on the 2.5D FEM, but where the unbounded domain, i.e., the ground, is simulated by the 2.5D BEM in order to fulfill the Sommerfeld condition. General information about the application of BEM to the simulation of more complex domains than the ones treated in this paper can be found in Galvin and Dominguez [10], Galvin et al. [9] or in Andersen and Nielsen [47].

2.2.2. 2.5D FEM-BEM Formulation for the Simulation of the Track-Ground System

Following the assumptions mentioned in the latter section, domain decomposition is used to solve the dynamic problem. As shown in Figure 2, the irregular geometry, i.e., the domain above the ground surface is attended by a 2.5D FEM modelling, being the regular geometry, i.e., the layered ground, simulated by the 2.5D BEM approach.

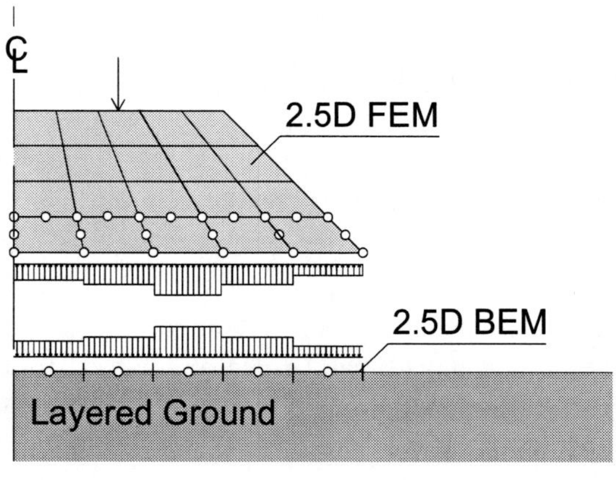

Figure 2. The 2.5D FEM-BEM coupling.

Subsequent to the finite elements formulation, the flexibility matrix that describes the dynamic behavior of the BEM domain is inverted in order to give rise to a dynamic stiffness matrix, which, in turn, is assembled to the remaining matrices that describe the equation system of the finite elements domain.

Adopting the proper formalism of the 2.5D FEM, the dynamic equilibrium equations of the medium can be determined by the formulation in the variation form. So, for a conservative system, the sum of the virtual work of the internal forces and of the inertial forces is equal to the virtual work of the external forces. After introducing the discretization of the cross-section into 2.5D finite elements and applying a Fourier transform regarding the x coordinate, the dynamic equilibrium is described by the following system of equations:

$$[K_1^{global} + ik_1K_2^{global} + k_1^2K_3^{global} + k_1^4K_4^{global} - \omega^2M^{global} + K_5^{global}(k_1,\omega)]u_n(k_1,\omega) = p_n(k_1,\omega) \quad (1)$$

where K_1^{global} to K_4^{global} are the stiffness matrices of the domain described by finite elements, M^{global} is the corresponding mass matrix, k_1 is the Fourier image of the coordinate x, ω is the frequency, un is the vector of the nodal displacements, p_n is the vector of the external forces, and K_5^{global} is the matrix that collects the impedance terms of the layered ground. Solving this system of equations, the displacements on the transformed domain are obtained, which are changed to the space/time domain through a double inverse Fourier transform.

The continuous medium, i.e., the embankment and the granular layers of the track are simulated by 8-node finite elements, with the exception of the elements that establish the connection between the FEM domain and the BEM domain. In the latter, 7-node elements are adopted in order to have two nodes along the edge of the FEM-BEM connection.

Regarding the particular elements of the track, as for instance the sleepers and the rails, a brief note should be made. Since one of the intrinsic assumptions of the 2.5D technique is the invariability of the domain along the track direction, an anisotropic constitutive model was adopted for the description of the dynamic behavior of the sleepers by an equivalent continuous formulation, which is acceptable, in general, for excitations with frequencies up to 500 Hz [48]. A detailed description of that approach can be found in Alves Costa et al. [23].

Concerning the rails, Euler-Bernoulli beams are adopted for the simulation of these elements. This approach suffices for the description of the dynamic behavior of the rail for frequencies up to 500 Hz, thus the adoption of more complex approaches, as for instance, Timoshenko beams is unjustified [49]. Finally, the connection between the rails and the sleepers is established by spring-dashpot elements, which are introduced in order to take into account the presence of the railpad.

Since the 2.5D FEM is well explained in several technical documents, including in one paper previously presented by the authors, the deduction of the above mentioned matrices is not presented here again; the reader is advised to consult Alves Costa et al. [23, 50]. In spite of that a special attention is given to the coupling between the 2.5D FEM-BEM, which is described by the matrix K_5.

The matrix K_5 present in Equation (1) corresponds to the impedance of the ground. This matrix, which is k_1 and ω dependent, is computed from the flexibility matrix that governs the dynamic behavior of the domain described by 2.5D BEM.

Several proposals for the 2.5D boundary integral equations can be found in the specialized bibliography. Sheng et al. [16] have proposed a 2.5D boundary integral equation based on the reciprocity theorem. More recently, François et al. [27] proposed a new approach based on a regularized version of the boundary integral equation in order to avoid the analytical integration of singular terms. However, the 2.5D boundary integral equations can be considerably simplified if assumed that the coupling between the FEM and BEM domains only occurs along the ground surface, not allowing the embedment of the finite element mesh in the BEM domain. In that case, taking the reciprocity theorem and the Somigliana identity into account, the boundary integral equation assumes the following configuration:

$$u_j(x,\omega) = \int_{\Sigma} u_{ji}^G(x,y,\omega) p_i(y,\omega) d\Sigma \qquad (2)$$

where u_j corresponds to the displacement of the point with coordinates x when a pressure, pi, is applied along the surface Σ (belonging to the ground surface). On the other hand, u_{ji}^G is the tensor of the Green's functions of the displacements.

For the computation of the Green's functions of the displacements, several methodologies can be followed. Since the objective is to find the BEM matrices in the 2.5D domain, it is only natural and, apparently more efficient, to opt for the 2.5D Green's function for a halfspace proposed by Tadeu and Kausel [51], instead of other methodologies. However, despite the analytical elegance of that approach, the ground is almost always formed by layers with distinct properties, in which case, all interfaces between different layers must be discretized into boundary elements. In order to avoid such scenario, in the present study the 3D Green's functions are adopted, which are computed in the transformed domain using a hybrid formulation of the thin-layer method, as proposed by Kausel [52, 53]. This method presents several advantages, as previously discussed by François et al. [27], namely the fact that boundary elements discretization is only necessary along the surface that establishes the connection between the layered ground and finite elements domain (see Figure 2).

In the approach followed in the present work, only linear boundary elements are used, with one collocation point (in the middle of the element) and with its dimension defined by the length of the edge of the neighboring finite element. A detailed description of the procedure followed in the determination of the terms of matrix K_5 is described in Alves Costa et al. [31].

Another advantage of the adopted procedure regards to the introduction of moving loads. Indeed, since the motion and equilibrium of the domain are formulated by a 2.5D approach, where Fourier transforms are applied on the x coordinate and time, the simulation of moving loads is easily attended by the shift property of the Fourier transform [43, 54–56]. This property leads to a direct relationship between wavenumber, k_1, and the frequency, ω, through the knowledge of the speed of the moving load [23, 31] .

Solving the system of equations presented in Equation (1), the nodal displacements in the transformed domain are obtained, as well as the tractions along the coupling boundary. Once the tractions and the Green's functions of the displacements are known, the computation of the free field response is a trivial step.

2.3. Train-track Interaction

The load applied by the train on the track can be divided into two components: i) the static load, resulting from the weight of the train; ii) the dynamic load, due to the dynamic interaction between the train and the track. The latter component requires the solution of the dynamic train-track interaction problem. This problem is solved by a compliance procedure formulated in a frame of reference that moves with the train, as suggested by several authors [4, 32, 33, 57] and schematically depicted in Figure 1.

A 2D vehicle model is adopted in the present study. So, only the vertical movement of the train is taken into account, i.e., the dynamic loads induced by the movement of the train in the directions different from the vertical are neglected. Consequently, assuming perfect contact between both structures, i.e., the train and the track, the following relationship must be adhered to at any temporal instant for all connection points between the train and the track:

$$u_{c,i} = u_r\left(x = ct + a_i\right) + \Delta u\left(t + \frac{a_i}{c}\right) + \frac{P_i(t)}{k_H} \tag{3}$$

where $u_{c,i}$ represents the vertical displacements of the contact point i of the vehicle; u_r is the vertical displacements of the track at the same location; Δu is the rail unevenness; t is the time; a_i is the location of contact point i at $t=0$ and c is the vehicle speed, k_H is the Hertzian stiffness and P_i is the dynamic interaction load developed at the connection point i.

Equation (3) can be written in matrix form in the frequency domain using the transformation of the unevenness for that domain. So, the train-track interaction force at the frequency domain is given by:

$$\mathbf{p}(\Omega) = -\left(\mathbf{F} + \mathbf{F}^{H} + \mathbf{A}\right)^{-1} \mathbf{\Delta u}(\Omega) \tag{4}$$

where **F** is the train compliance at the contact points with the track, $\mathbf{F}^H$ is a diagonal matrix where the terms are equal to $1/k_H$, **A** is the compliance matrix of the track and Ω is the driven frequency, i.e., the frequency of oscillation of wheelset due to the unevenness with a wavelength λ ($\Omega=2\pi c/\lambda$).

All matrices of Equation (7) are square with a dimension equivalent to the number of wheelsets. Matrix F is computed from the vehicle model and the terms of matrix A are evaluated from the track ground model. A detailed description of the mathematical formalism inherent to the evaluation of these matrices can be found in Alves Costa et al. [31, 34].

Analyzing Equation (5), one can conclude that the influence of the vehicle properties and of the modeling strategy used only affects matrix F. A simple but general vehicle model, which takes into account the main structural aspects of the train dynamics, was proposed by Zhai and Cai [58]. Despite the conceptual simplicity of that model, the difficulty generally encountered for the determination of the mechanical properties of the vehicle, has given rise to proposals of more simplified versions where some masses are discarded [30, 32, 33].

3. THE CASE STUDY

3.1. General Description

In order to validate the numerical models developed by the authors, an experimental test site was selected and implemented in the Portuguese railway network.

Regarding the main aspects of the site, it must be referred that the track presents a straight alignment, corresponding to a renewed part of the railway connection between Porto and Lisbon. The line corresponds to a double track on ballast, as illustrated in Figure 3, in which distinct types of traffic are allowed, ranging from freight trains to passenger trains that circulate at speeds up to 220 km/h (Alfa-Pendular).

Figure 3. General view of the track at the test site.

In that location, two sets of experiments were performed: i) the first, to evaluate the dynamic properties of the track and of the ground; ii) the second, to measure the vibrations induced by the passage of the Alfa-Pendular train. Additionally, the track unevenness, key parameters for performing a dynamic train-track interaction analysis, were also measured.

The first set of experiments was used for the calibration of the numerical model, whilst providing a deeper understanding of the dynamic behavior of the track-ground system. On the other hand, the second set of experiments, comprising the response of the track-ground system during the passage of Alfa-Pendular train, was used for the experimental validation of the numerical model presented above. Moreover, some identification tests were also performed regarding the assessment of the dynamic properties of the train Alfa-Pendular.

3.2. Geodynamic Characterization of the Ground

For an accurate prediction of vibrations induced by traffic, a detailed attention must be given to the geotechnical investigation. Among typical geotechnical tests, as for instance CPT and SPT, two cross-hole tests were performed in the test site.

Taking into account the strain level induced by elastic wave propagation, the results of the cross-hole tests, i.e., the shear and the compression wave velocities were used as quantitative information, being complemented by the information provided by the remaining tests. Moreover, intact soil samples were collected, at different depths along the borehole, in order to perform some laboratory tests for the classification and identification of the soil.

Figure 4 shows the measured wave velocities profile as well as the values adopted in the numerical model (average value lines). Unfortunately, it was not possible to evaluate the seismic wave velocities for depths below 9.0 m. Due to this limitation, empirical correlations between CPT results and shear wave velocities, provided in the literature, were employed for the estimate of the shear wave velocity for depths greater than 9.0 m [59].

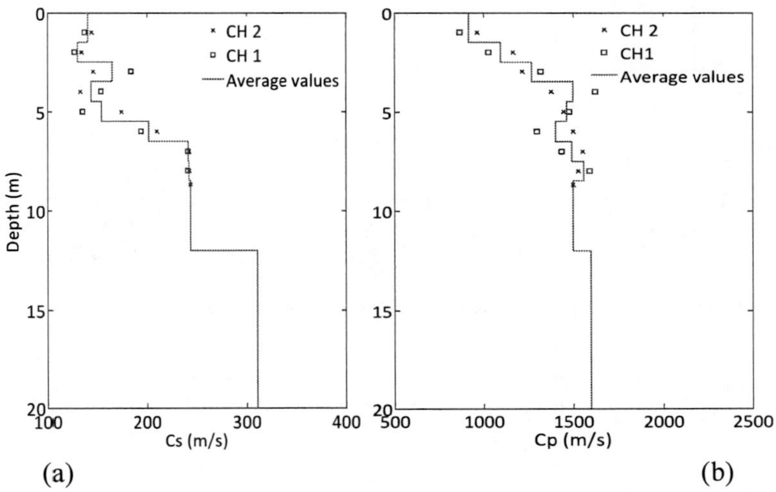

Figure 4. Seismic wave velocities: (a) S waves; (b) P waves.

A complete geodynamic characterization, for the purpose of the present study, should attend not only to the elastic properties of the ground but also to an estimate of the damping profile in depth. In the present study, the damping profile was estimated using an inversion procedure based on the calibration of the numerical model, in order to obtain a reasonable fit between measured and computed mobilities of the ground due to an excitation induced by a controlled source. The experimental setup is shown on Figure 5. The source of excitation was assured by an electromagnetic oscillator capable of applying a harmonic load on the ground in the frequency range between 5 Hz and 600 Hz. On the other hand, accelerometers were used to measure the vertical response of the ground surface at distinct distances from the source.

From the acquired data, the experimental mobility was computed for the set of measuring points. On the other hand, a theoretical model was developed, taking into account the ground layering and stiffness properties described in Figure 4, where the damping profile was adjusted in order to obtain a reasonable fit between the measured and computed mobilities. Thus, the only unknown variable in the numerical model was the damping of the ground, resulting in a simple inversion procedure. Figure 6 shows the modulus of the mobility measured and computed, for the 4 points under analysis. It is clear in the figures that a reasonable fit between measured and computed results is achieved when the damping profile illustrated in Figure 7b is adopted. In Figure 6, the computed mobility obtained when a uniform damping profile is assumed with an average value of 0.03 (see Figure 7a) is also represented. In that case, there is a great discrepancy between measured and computed results, mainly for lower distances between the source and the receiver.

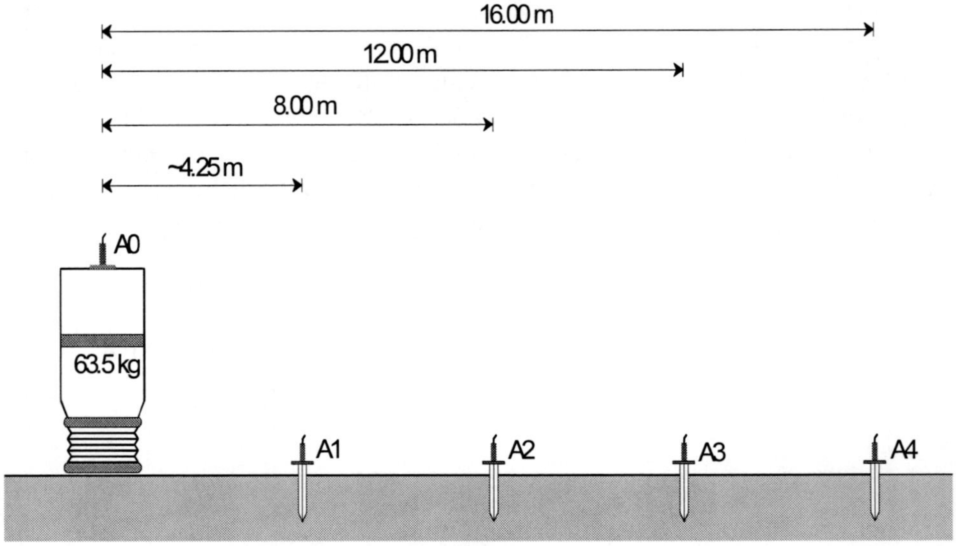

Figure 5. Experimental setup adopted in the estimate of the damping profile.

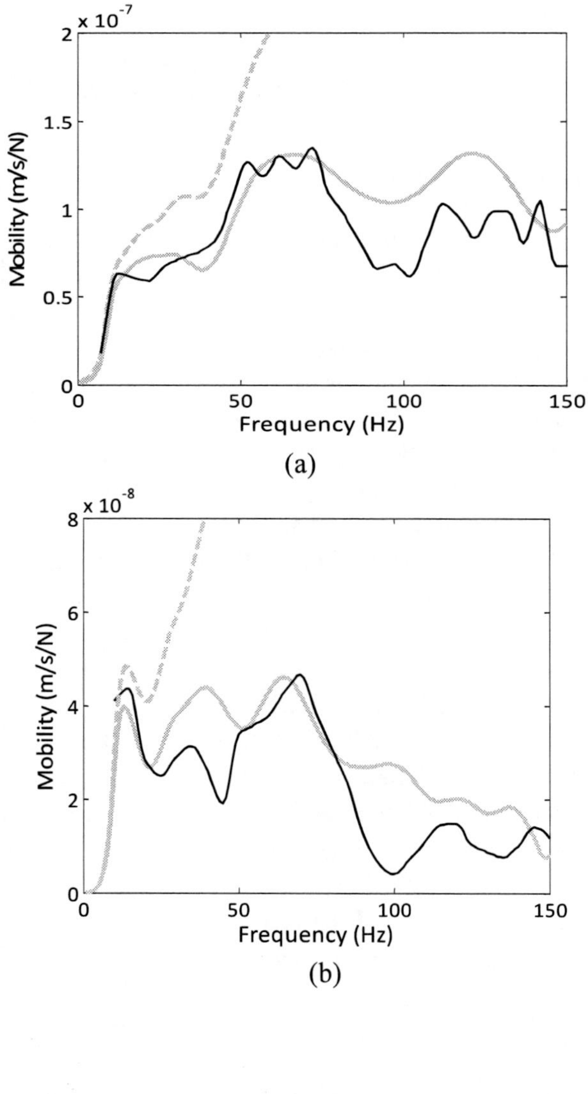

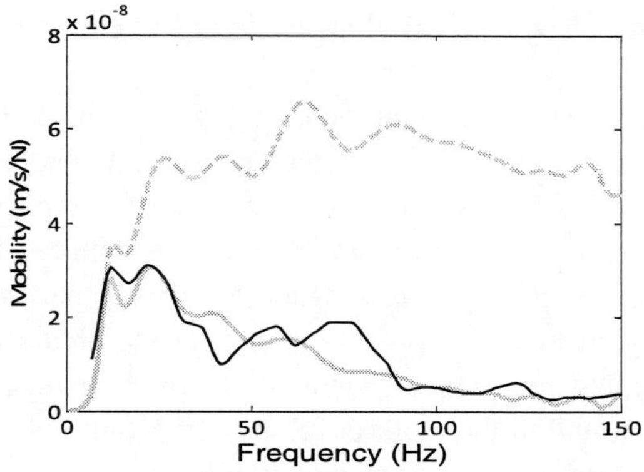

(c)

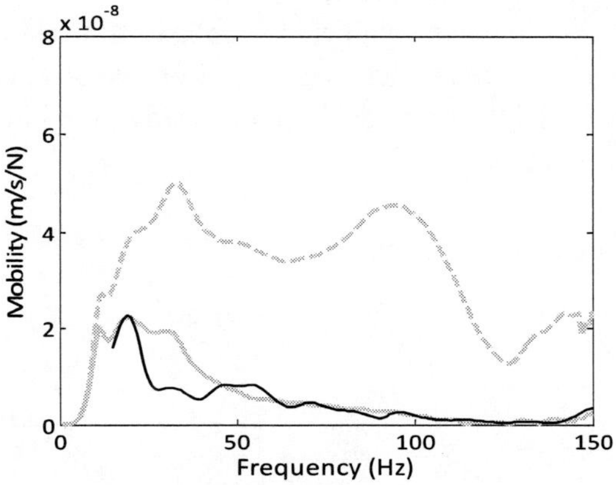

(d)

Figure 6. Computed and measured mobilities for distinct distances receiver-source: (a) 4.25 m; (b) 8.0 m; (c) 12.0 m; (d) 16.0 m (black solid line – measured; gray solid line – damping profile showed in Figure 7b; gray dashed line – damping profile shown in Figure 7a).

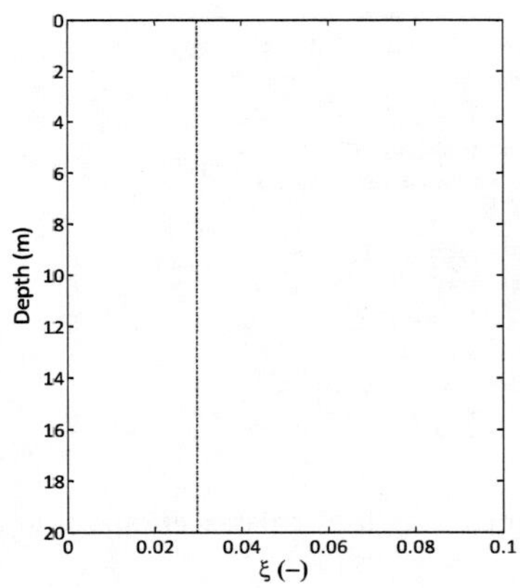

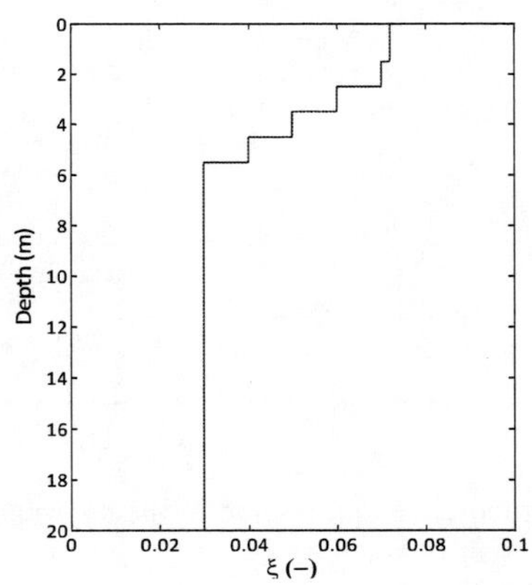

a b

Figure 7. Damping profiles: (a) constant value (before fitting); (b) after fitting. In the absence of more detailed information, the damping profile shown in Figure 7b was assumed for both deviatory and volumetric deformation.

3.3. Assessment of the Mechanical Properties of the Track

Another challenging issue for the numerical analysis of railway tracks is the difficulty in the assessment of reliable properties of the structural system. The properties of the rail and sleepers are well defined, however the same statement cannot be extended to other components such as the ballast or the railpads. In order to minimize the uncertainty associated with the properties of those elements, receptance tests were performed and the numerical model was calibrated, by an inversion procedure, in order to obtain a reasonable fit between the measured and computed receptances. General information regarding the application of this kind of tests can be found in the work developed by Mann [49] or, more recently, in the studies developed by Kaewunruen and Remennikov [60, 61].

The adopted setup for the receptance tests is shown in Figure 8. As can be seen, the impact load (provided by an impulse hammer) is applied in the mid span of the sleeper and the response is measured by accelerometers installed on the sleeper ends and on the rails. Therefore, since the source and the receiver are not located in the same vertical alignment, the receptance is called cross-receptance.

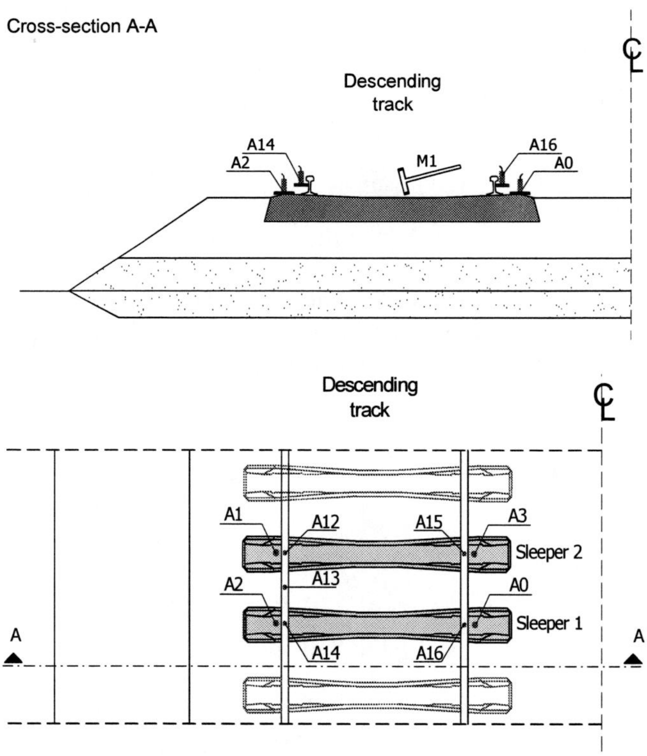

Figure 8. Experimental setup adopted in the receptance tests: (top) cross-section; (bottom) plan overview.

Figures 9 (a) and (b) show the cross-receptance of the rails and sleeper, respectively. In the same figures is also indicated the range of frequencies of interest, i.e., the range where coherence values greater than 0.95 were obtained.

A preliminary inspection of the results enables to confirm that very stiff railpads are used in this railway track. This fact is denoted by the peak of the receptance function of the rails for frequencies around 650 Hz. A simplified analysis enables to conclude that this resonance frequency is related to a railpad stiffness of 600 kN/mm, which is in correspondence with the value previously pointed out by the railpad supplier

As previously mentioned, several variables (properties of the track elements) are very difficult to characterize, namely the stiffness, the damping and the mass of the ballast and subballast. These properties were evaluated by solving a nonlinear least squares optimization problem with a residual value that is a function of the track characteristics. For the determination of those unknown variables, the discrepancy between measured and computed sleeper receptance was minimized in the range of frequencies between 30 Hz and 250 Hz. As evidenced in Figure 9a, only experimental data with coherence values greater than 0.95 were considered.

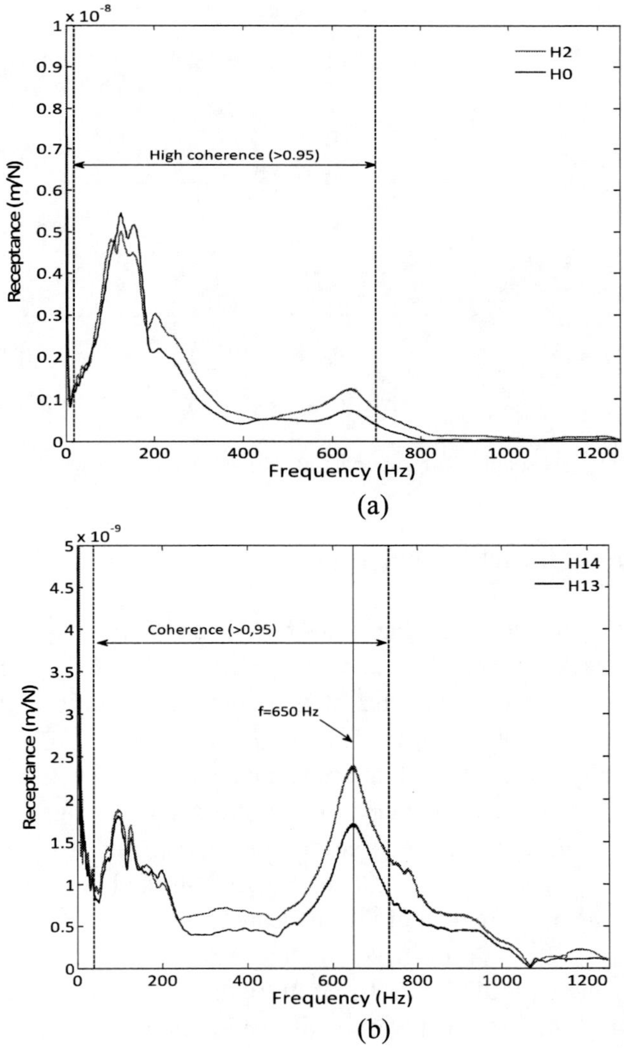

Figure 9. Cross-receptances of: (a) the sleeper; (b) the rails.

Figure 10 shows the geometry of the numerical model, as well as the properties adopted after optimization [31]. It should be noticed that the rail and sleepers were simulated according to the physical properties of the materials and the layered ground was simulated by the 2.5D BEM procedure taking into account the stiffness, mass and damping assessed by the set of tests described in the previous section. Concerning the computation of the Green's function needed for the construction of the 2.5D BEM approach, those were computed by the hybrid thin-layer method referred above, assuming the maximum dimension of each thin-layer smaller than 0.10 m in order to obtain an accurate prediction of the ground dynamics for frequencies up to 250 Hz.

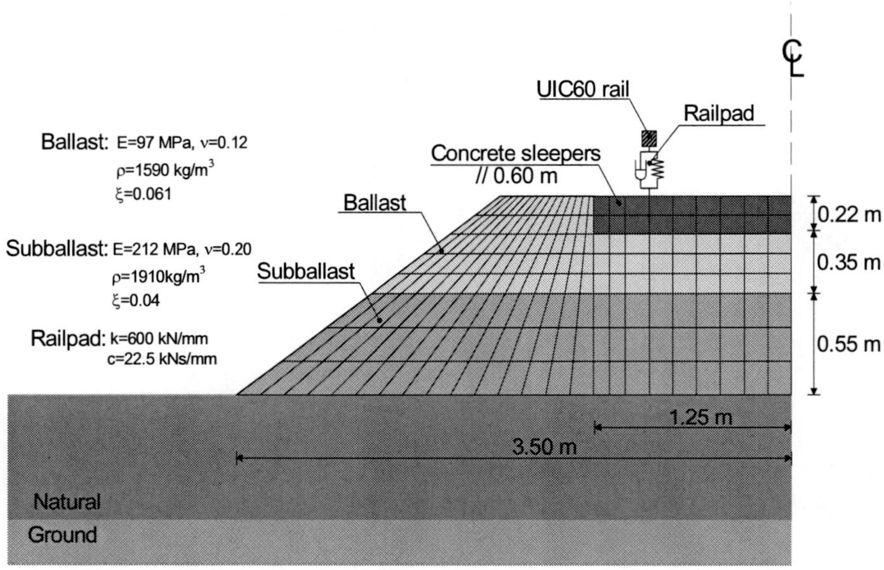

Figure 10. Numerical models adopted for the updating of the track properties.

3.4. Measurement of Track Unevenness

A key aspect to perform a reliable train-track interaction analysis resides on a good assessment of the track unevenness, as was indeed emphasized by Equation (5). This important aspect was addressed in the present study, where a specific and detailed measuring campaign of the track unevenness was carried out.

Since it is usually assumed that the vibrations induced by traffic are relevant for frequencies up to 150 Hz, it is necessary to measure a broad range of wavelengths. This aspect was taken into consideration, and a recording car was used to measure the unevenness of the track for wavelengths between 1.0 m and 25.0 m, and a corrugation measurement device was used for the measurement of the unevenness in the wavelength range between 0.4 m and 1.0 m.

The track unevenness profile is illustrated in Figure 11(a), for a length of 100 m before and after the section of the track that was monitored during the passage of trains. Moreover, Figure 11(b) shows the power spectral density of the rail on the right side in the range of wavelengths between 0.4 m and 25 m.

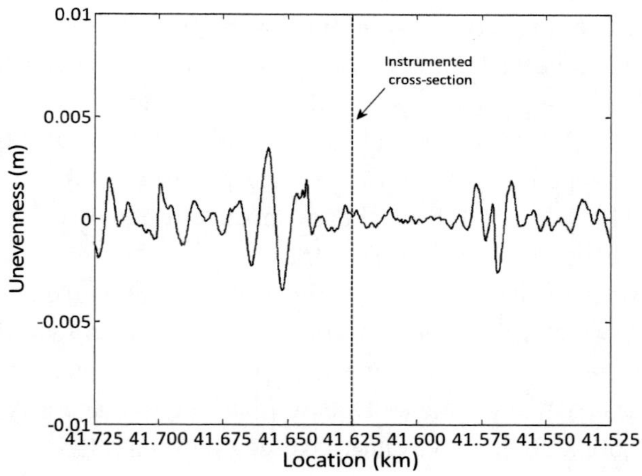

(a)

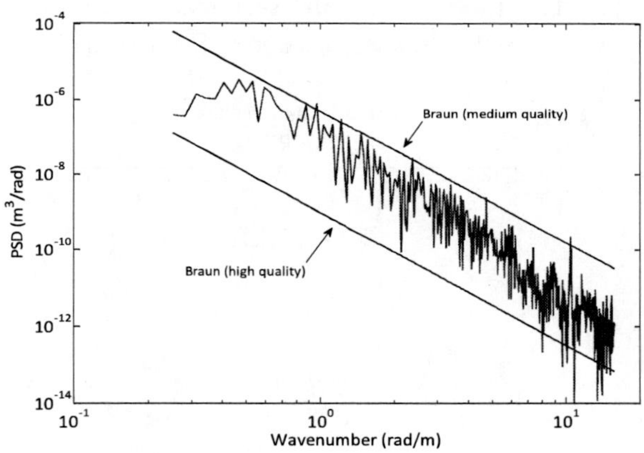

(b)

Figure 11. Rail unevenness profile: (a) spatial representation; (b) power spectrum density.

3.5. Assessment of the Dynamic Properties of Alfa-Pendular

The Alfa-Pendular is the fastest train operating in Portugal. It is a conventional train composed of six vehicles, as indicated in Figure 12.

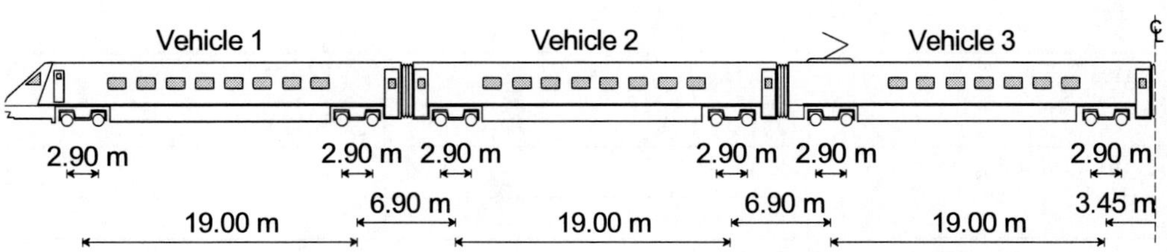

Figure 12. Geometry of the Alfa-Pendular train.

The main mechanical properties of the train were provided by the operator. In spite of this information, some identification modal tests for the structural characterization of the train were also developed and the properties adopted in the numerical model were adjusted in order to obtain a good fit between numerical and experimental natural frequencies. Table 1 summarizes the main properties of the train, in correspondence with the values adopted in the numerical analyses. The vehicles of the train are not exactly equal. Table 1 indicates the range of values found for the distinct vehicles of the train.

The first two modes of the vehicle are only related to the movement of the car box and correspond to quite low natural frequencies, i.e., 1.34 Hz and 1.59 Hz [34]. Therefore, these modes are more important for the requirements of passengers' comfort than for the generation of relevant train-track dynamic loads. Due to that, in the analysis presented below, a simplified model is adopted where only the unsprung and semi-sprung masses are taken into account, as depicted in Figure 13. A recent study presented by Alves Costa et al. [34] shows that this model suffices for the study of vibrations induced by railway traffic. This conclusion is valid for current passenger trains, where the secondary suspension is responsible for the dynamic isolation of the car box for frequencies above few hertz.

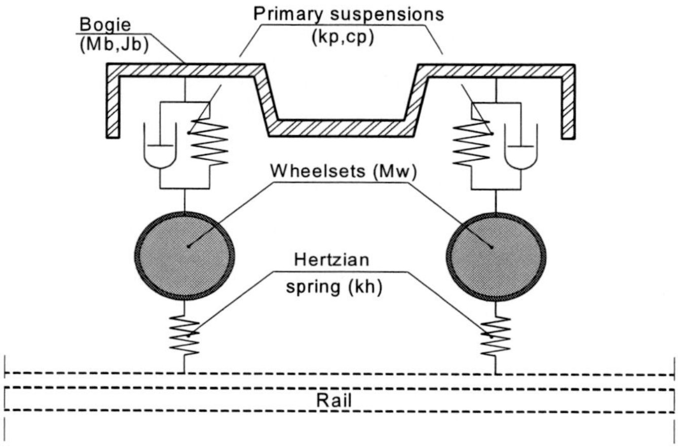

Figure 13. Elementary structural unit of the train adopted in the present study.

Table 1. Train properties

Axles	M_w (kg)	1538–1884
Primary suspension	k_p (kN/m)	3420
	c_p (kNs/m)	36
Bogies	M_b (kg)	4712-4932
	J_b (kg/m^2)	5000-5150
Secondary suspension	k_s (kN)	1320
	c_s (kNs/m)	36
Car body	M_c (kg)	32900-35710
	J_c (kg/m^2)	2083600

3.6. Measurement of Vibrations Induced by the Train Passage

After the initial characterization experiments, focus was given to the measurement of track and free-field vibrations induced by traffic. Bearing this in mind, the setup illustrated in Figure 14 was implemented in order to experimentally evaluate the track-ground response due to traffic. As can be seen, the experimental setup comprised the measurement of vertical accelerations on the sleepers (see Figure 15a) and on the ground surface at several distances from the track (Figure 15b). Complementarily, the dynamic displacement of the rail due to traffic was also measured by a laser device, installed 9.0 m away from the track (Figure 15c).

A vast amount of data was acquired comprising the passage of several distinct trains running at different speeds. Nevertheless, in the following analysis only a fraction of the collected data is used, always regarding the passage of the Alfa-Pendular train at the running speed of 212 km/h.

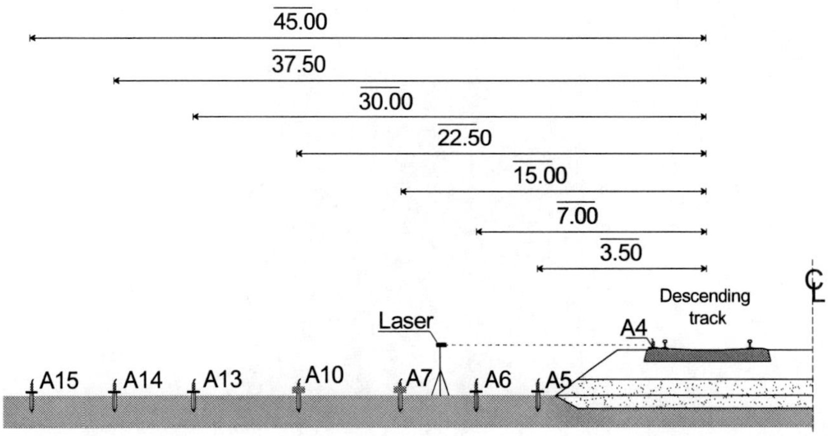

Figure 14. Experimental setup adopted for the measurement of vibrations induced by traffic.

(a)

Figure 15. (Continued)

(b)

(c)

Figure 15. Photographs of the devices used in the test: (a) accelerometer installed on the track; (b) accelerometer used to measure the free-field dynamic response; (c) laser device.

3.7. Validation of the 2.5D FEM-BEM Model

3.7.1. Objectives and Model Description

For the experimental validation, a numerical model comprising the train-track-ground dynamic interaction was constructed, in accordance to the previously described formulation.

The numerical model for the track-ground system is consistent with the information illustrated in Figure 10, which is why no further comments or justifications are required

Concerning the rolling stock properties and the interaction mechanisms, in the following sections is considered the passage of Alfa-Pendular train at the speed of 212 km/h.

3.7.2. Dynamic Response of the Track

Figure 16(a) compares the computed rail displacement with the measurement supplied by the laser device. In order to minimize the noise present in the measured signal, a low-pass filter with a cut-off frequency of 60 Hz was applied to both results. The comparison in terms of frequency content is available in Figure 16(b).

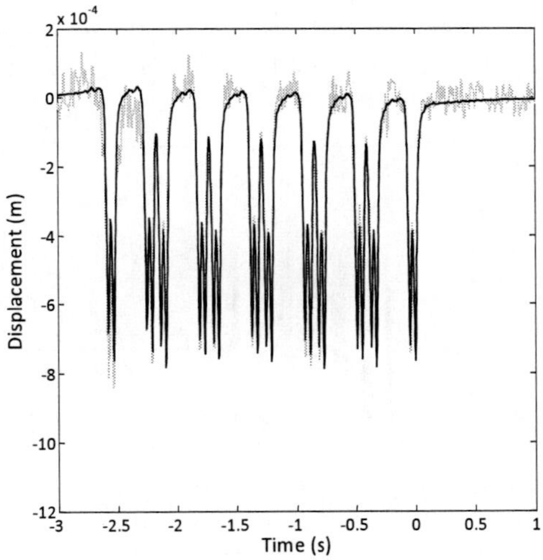

(a)

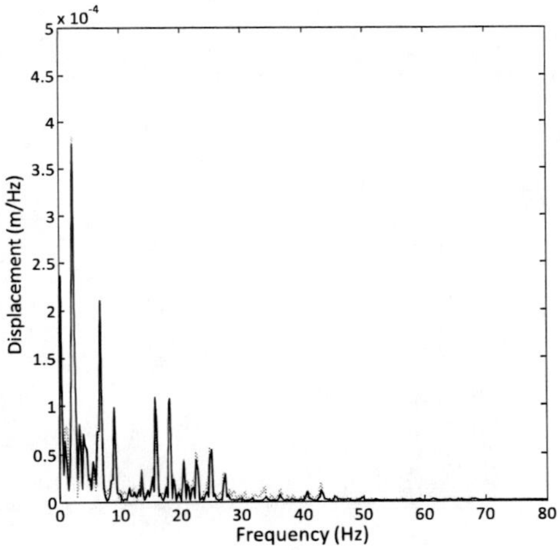

(b)

Figure 16. Vertical rail displacement due to the train passage: (a) time record; (b) frequency content (gray line – measured; black line – computed).

A very good fit between rail displacement prediction and measurement was achieved, as demonstrated in the figure. In fact, excluding a minor discrepancy during the passage of the first vehicle, the ability showed by the numerical model to reproduce the rail displacement is remarkable and is evident in both time and frequency domains.

Concerning the sleeper response, Figure 17 compares experimental and numerical time histories of the vertical velocity of the sleeper induced by the passage of the Alfa-Pendular train at 212 km/h.

Once again, a very good match between both results is evidenced in Figure 17. A clear distinction of the passage of each bogie is identified in the temporal record, and a very good agreement between numerical and experimental results is found. Despite the high quality of the results, the slight differences between measurement and prediction for frequencies above 75 Hz must be noted (Figure 17b). These differences are due to the dynamic excitation mechanism, and can be justified by the fact that only the excitation induced by the track unevenness is taken into account in the numerical modeling. Actually, several other causes, discarded in the numerical model, can provoke excitation of the rolling stock, such as the irregularities of the wheels or the inhomogeneities of the track support.

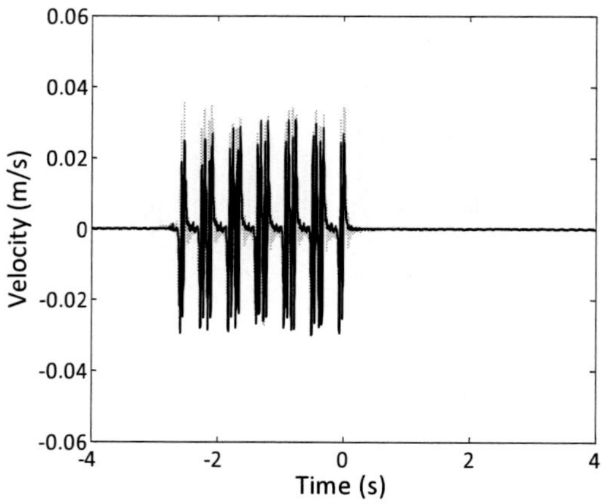

(a)

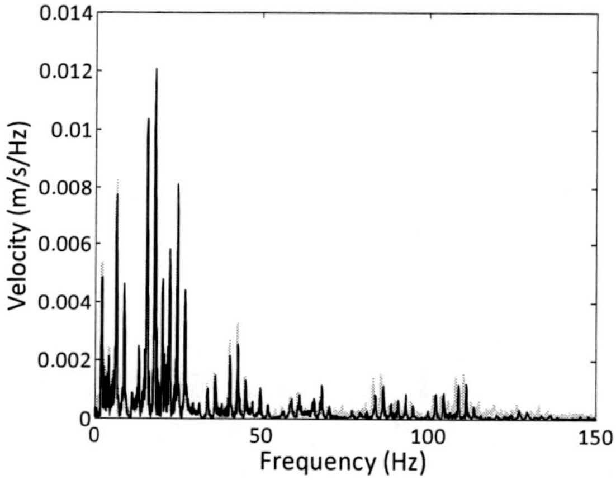

(b)

Figure 17. Vertical sleeper velocity induced by the train passage: (a) time record; (b) frequency content (gray line – measured; black line – computed).

3.7.3. Dynamic Response of the Free-Field

Figure 18 shows the time records of the vertical velocity of the points monitored during the passage of the Alfa-Pendular train at 212 km/h. The numerical prediction is overlapped with the experimental record. Several remarks can be pointed out:

i) The passage of the axles and bogies in the instrumented section becomes more and more imperceptible as the distance to the track increases;
ii) The amplitude of the vertical velocity decreases with the increase of the receiver-track distance while the duration of the event increases;
iii) A very good fit between computed and measured results is achieved in the receiver points closest to the track;

iv) The quality of the fit between prediction and measurement decreases with the increase of the distance between the receiver location and the track, and it is possible to identify a clear loss of accuracy when that distance is greater than 22.5 m.

However, in spite of the differences found, the prediction model leads to a very reasonable assessment of the main aspects of the response. Moreover, the main trends of the problem are well estimated by the proposed model, with more emphasis to the evolution of the amplitude with the increasing distance to the track.

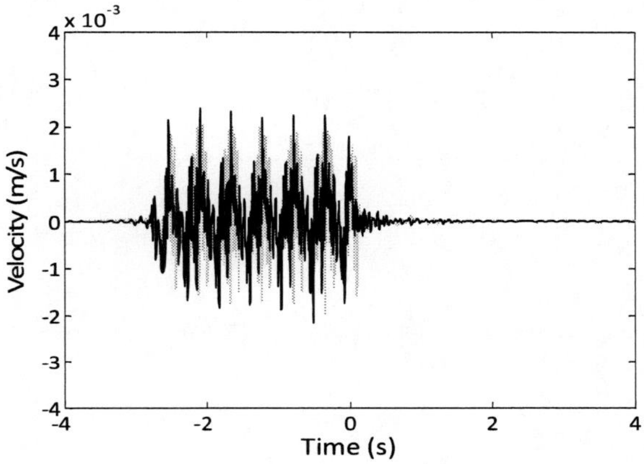

(a)

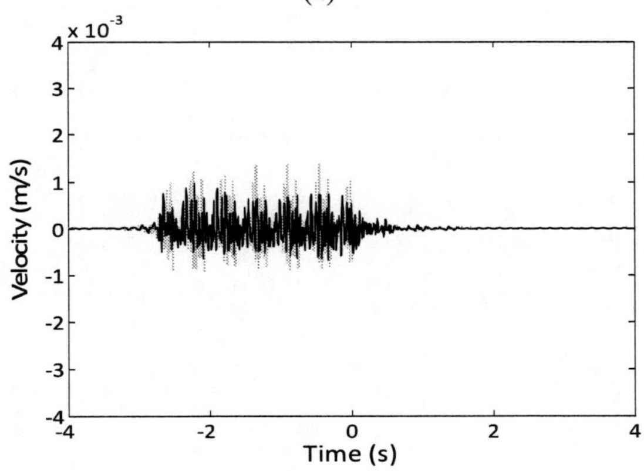

(b)

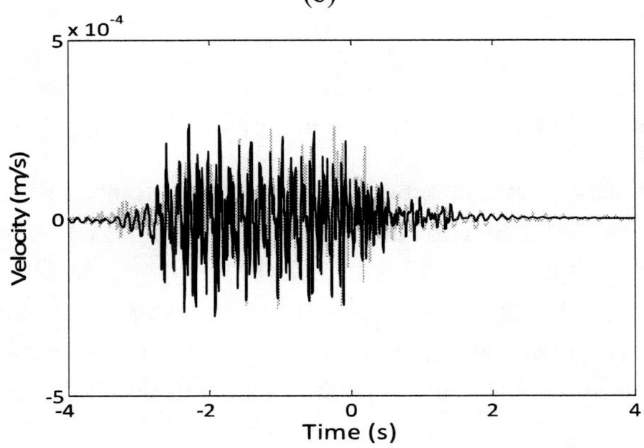

(c)

Figure 18. (Continued)

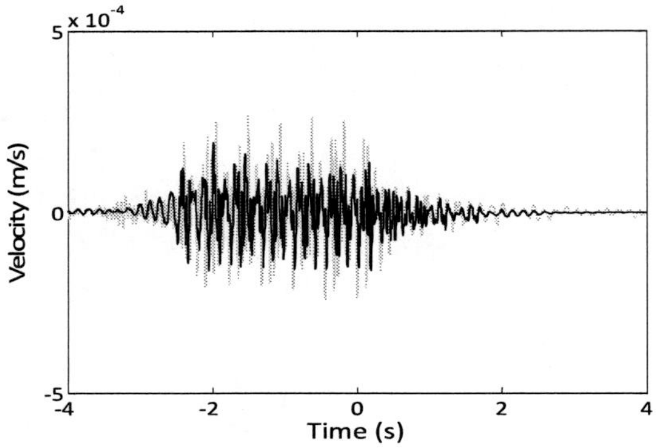

(d)

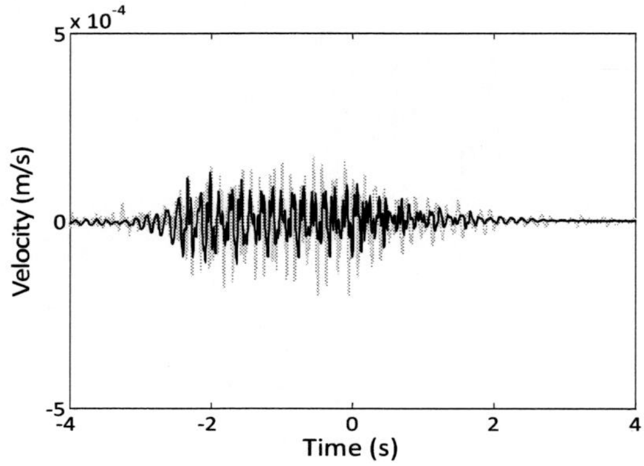

(e)

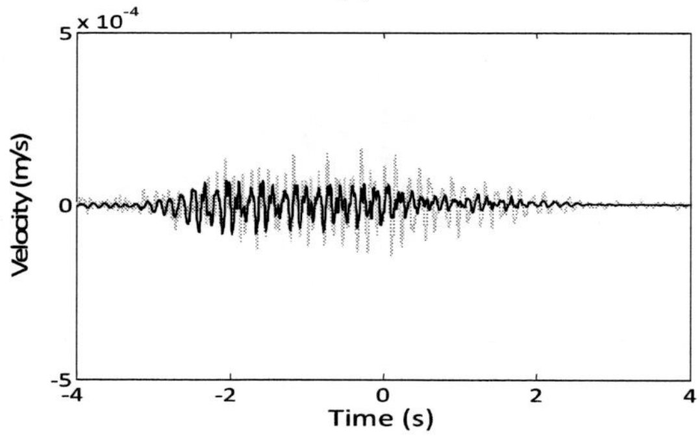

(f)

Figure 18. Time record of the vertical velocity at distinct distances at the free-field: (a) 3.5 m; (b) 7.0 m; (c) 15.0 m; (d) 22.5 m; (e) 30 m; (f) 37.5 m (gray line – measured; black line – computed).

Despite the satisfactory agreement achieved, mainly considering the complexity and uncertainty intrinsic to the problem under analysis, the underestimate of the predicted response for distances higher than 22.5 m cannot be ignored. This remark leads to the conclusion that the soil damping was possibly overestimated, which gives rise to an underestimate of the response with the increase of the receiver-track distance. A detailed discussion about the causes of these discrepancies can be found in Alves Costa et al. [31].

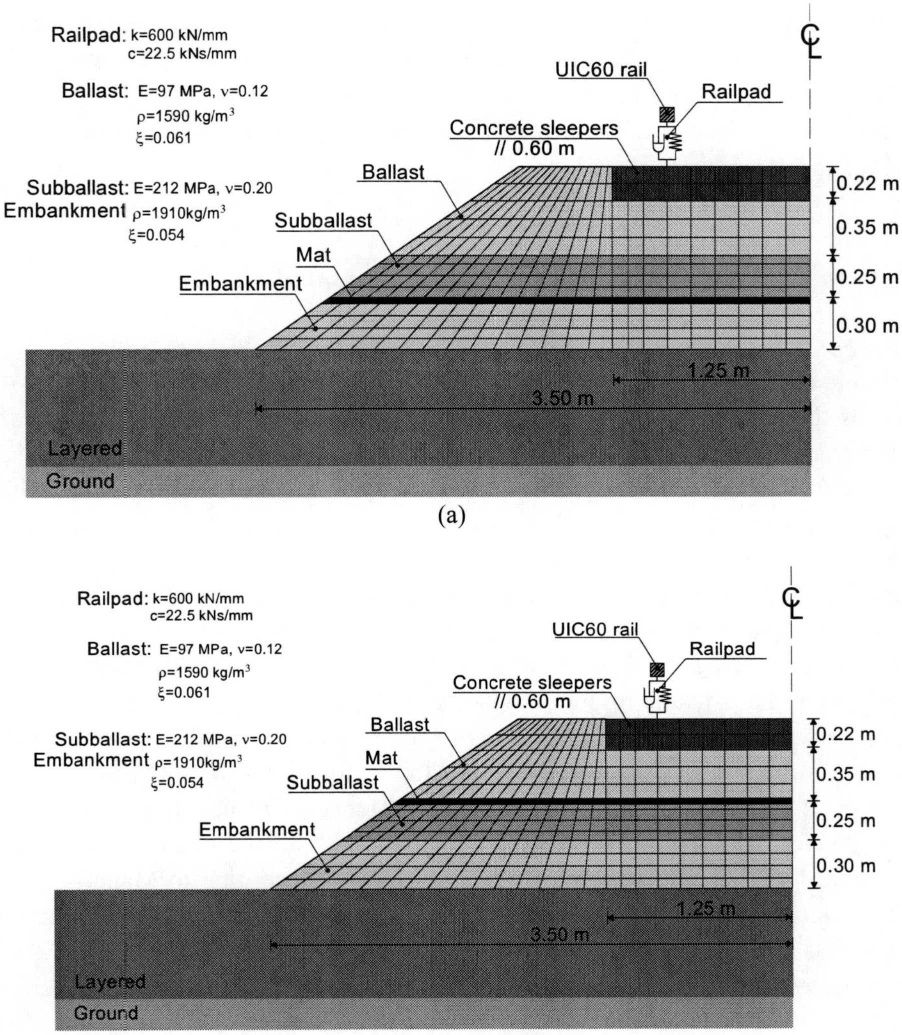

Figure 19. Geometry and properties of the isolated scenarios: (a) scenario 1 – mat beneath subballast layer; (b) scenario 2 – mat beneath ballast layer.

4. Ballast Mats Efficiency

4.1. Objectives and Model Description

In present section, a discussion is preformed about the efficiency of ballast mats on the mitigation of vibrations induced by railway traffic. The study here present, of numerical character, is based on the application of the proposed numerical models, which was previously validated, as shown above, on the evaluation of the dynamic consequences of mitigation of vibrations by introduction of ballast mats. The case study presented above is used as reference scenario, being now considered two new scenarios as depicted in Figure 19.

For isolated systems, the mat is introduced immediately below the subballast layer (scenario 1), as depicted in Figure 19(a), or beneath the ballast layer (scenario 2), as shown in

Figure 19(b). Since the embankment has a thickness of 0.30 m and the same mechanical properties as the subballast layer, a proper foundation for the mat is guaranteed when it is placed beneath the subballast layer.

Different values for the ballast mat stiffness were considered, always assuming a constant value of 0.1 for its hysteretic damping.

4.2. Dynamic Behavior of Non-isolated and Isolated Tracks

4.2.1. Receptance Analysis

A simple but comprehensive approach for understanding how the mat affects the dynamic behavior of the train-ground system is based on the analysis of its receptance. Figure 20 shows the receptance of the rail for both scenarios under consideration and for two values of the ballast mat stiffness. The receptance curves are computed assuming a half-unit stand-still load applied in each rail, and the displacements are evaluated at the rail under the loading position.

Considering the dynamic response of the non-isolated scenario as reference solution, which is also depicted in both Figure 20(a) and 20(b), it is possible to identify two main resonances of the track-ground system as follows: i) the first peak of the receptance, around 9 Hz, is related to the dynamic response induced by the layering of the ground; ii) the highest frequency peak, around 82 Hz, corresponds to the resonance of the track system above its foundation. As previously discussed by Knoth and Wu [48], the full-track resonance can be highly damped when its support is soft and homogeneous, being more pronounced if the ground is layered, as in the present case. The value found for the frequency of the full-track resonance, 82 Hz, is within the typical range of values. According to Mann [49], ballasted tracks have full-track resonant frequencies in the range between 40 Hz and 140 Hz. The same conclusion was also achieved by Dahlberg [62].

As expected, the introduction of the mat gives rise to a new natural frequency of the system, here called as cut-on frequency. Actually, the resonance induced by the presence of the mat is very damped in general situations, but not fully perceptible when the stiffer mat is located beneath the subballast layer (Figure 20a). However, for frequencies sufficiently higher than the resonance induced by the mat and for all the scenarios under analysis, a reduction of the receptance of the rail is clear, denoting the efficiency of the isolation induced by the ballast mat, which attenuates the resonance of the track on the ground.

A more detailed analysis allows to point out the following main remarks: i) independently of the mat position, the increase of its stiffness leads to an increase of the frequency for which the attenuation effects starts to be effective; ii) for the same mat stiffness, the cut-on frequency induced by the presence of the mat is as low as deep its position is; iii) for the same mat stiffness and in the range of lower frequencies, the response of the rail is emphasized when the mat is located beneath the ballast. This last finding is particularly important since the usage of softer mats is usually not recommended due to the required limitation of the deflection of the rail. However, it seems that this limitation can be overcome by a meticulous selection of the mat position on the cross-section of the track.

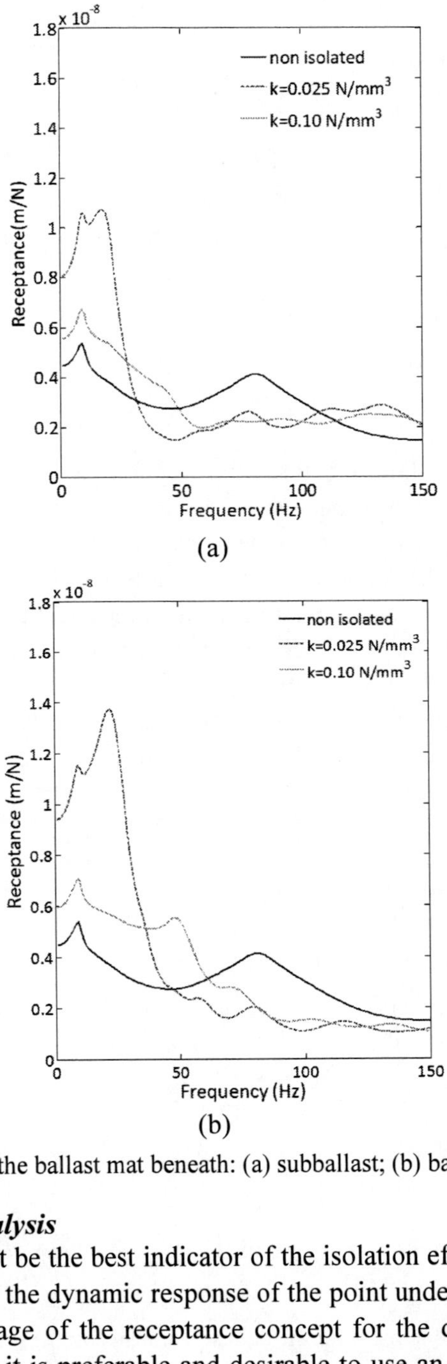

Figure 20. Rail receptance for the ballast mat beneath: (a) subballast; (b) ballast.

4.2.2. Force Transfer Analysis

The receptance may not be the best indicator of the isolation efficiency, since it is a local measurement depending on the dynamic response of the point under consideration. Indeed, in spite of the widespread usage of the receptance concept for the discussion of the dynamic behavior of railway tracks, it is preferable and desirable to use an indicator that can give an overview of the dynamic behavior of the system based on its global performance. Therefore, Auersch [63] recently proposed the concept of dynamic force transfer as a global indicator of the efficiency of the isolation solutions. According to this author, the force transfer corresponds to the ratio between the total force transmitted by the track to the ground, F_s, and the force applied to the rails, F_t. Being t_z the vertical pressures transmitted by the embankment to the ground, Fs can be computed as follows:

$$F_s(\omega) = \int_{-\infty}^{+\infty}\int_{-b}^{b} t_z(x,y,\omega)dydx \quad (5)$$

where b corresponds to the half-width of the contact section of the embankment with the ground. Taking advantage of the Fourier transform operations and of the transformed domain analysis, the above integral operation can be computed in an easy and accurate way as follows:

$$F_s(\omega) = \int_{-b}^{+b} t_z(k_1 = 0, y, \omega)dy\,. \quad (6)$$

Figures 21(a) and (b) show the force transfer for scenarios 1 and 2, respectively. The force transfer curve of the non-isolated scenario is also shown in both figures as a reference scenario and the following remarks are noteworthy: i) for 0 Hz, the force transfer is equal to 1, meaning that the load transmitted to the ground is equal to the load applied on the rails (which is also valid for all scenarios under consideration); ii) the force transfer exhibits a first peak, around 9 Hz, which corresponds to the ground resonance, presenting a decrease to values slightly lower than 1 in the frequency range between 12 Hz and 45 Hz; iii) a strong peak of the curve is noticed for frequencies around 82 Hz, denoting resonance of the track and leading to a decrease of the force transfer for frequencies higher than such value.

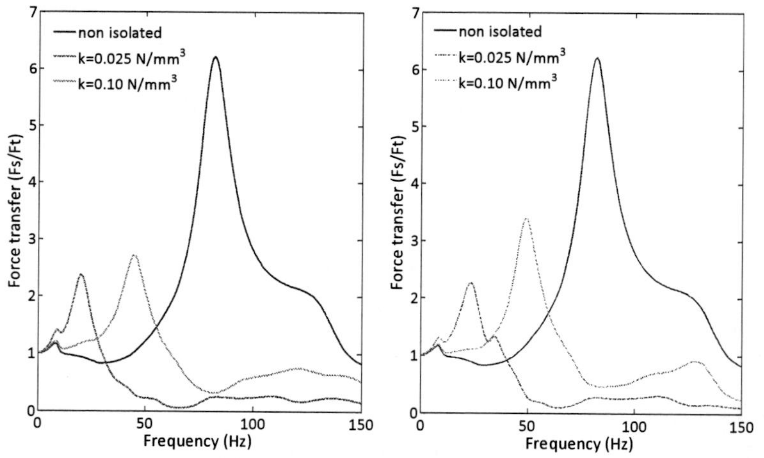

(a) (b)

Figure 21. Track-ground force transfer for the ballast mat beneath: (a) subballast; (b) ballast.

Great differences in the force transfer evolution are observed when the non-isolated scenario is compared with the solutions with mats. In fact, the strong peak associated with the track resonance present in the non-isolated scenario is no longer perceptible when the mats are installed, independently of their position or stiffness. However, the softest mat leads to a significantly more effective reduction of the force transfer, lower than 0.25 for frequencies higher than 60 Hz, for both scenarios under consideration. On the other hand, the installation of the mat beneath the subballast (Figure 21a) seems to be more effective for two main reasons: i) the reduction effect of the force transfer is enhanced for lower frequencies; ii) the amplification effect for frequencies around the resonance of the mass above the mat is less pronounced. This parametric study allowed concluding that the efficiency of the vibration mitigation measure is much greater with lower stiffness mats. Nevertheless, there is a lower bound for the stiffness of the mat due to the need of limiting rail deflection. On the other hand, taking into account the results depicted in Figure 19, it seems that this issue can be partially overcome by the installation of the mat beneath the subballast instead of its location beneath the ballast. In the following section, this issue is evaluated by the analysis of the train-track-ground response due to traffic in both scenarios. Taking into account the conclusions here found as well as the length limitations of the paper, only the softest mat ($k=0.025$ N/mm^3) is considered in the following studies.

4.3. Track Dynamic Response

Figure 22 compares the computed vertical displacement at the rail for both considered scenarios during the passage of the Alfa-Pendular train at 212 km/h. As expected, the observed displacements of the rail containing a mat are much higher than the homologous for the non-isolated case, especially if the mat is beneath the ballast layer; in this case the rail displacements can produce values as high as double those observed for non-isolated conditions. From Figure 23, where the frequency content of the rail displacement is depicted, it is possible to notice that the main contribution for the response is concentrated in the lower frequency range. This effect was already expected, since the rail displacements are mainly due to the quasi-static excitation mechanism (movement of the dead load of the train) [34, 64]. As previously presented (see Figure 19), in the lower frequency range, the rail receptance is increased by the presence of the mats, and, consequently, higher displacements are also induced by the train passage. Indeed, for the lower frequency range, the displacements induced by the train passage, when the mat is introduced beneath the ballast layer, are almost twice those observed for the non-isolated scenario.

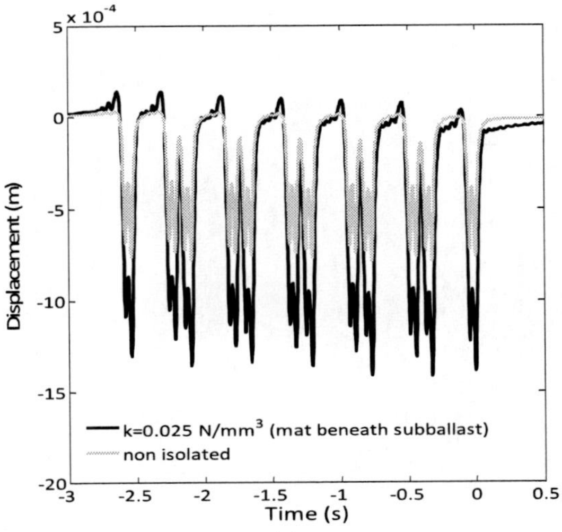

(a)

Figure 22. (Continued)

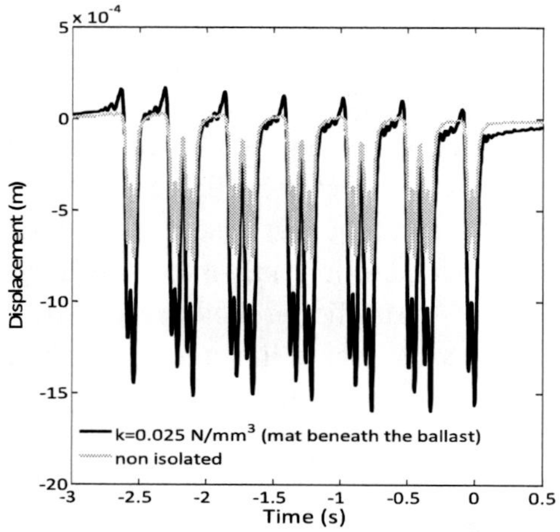

(b)

Figure 22. Rail displacements due to the passage of Alfa-Pendular at 212 km/h for ballast mat beneath: (a) subballast; (b) ballast.

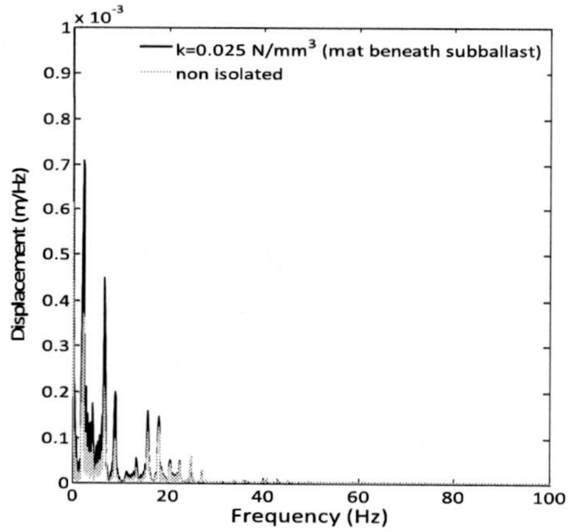

(a)

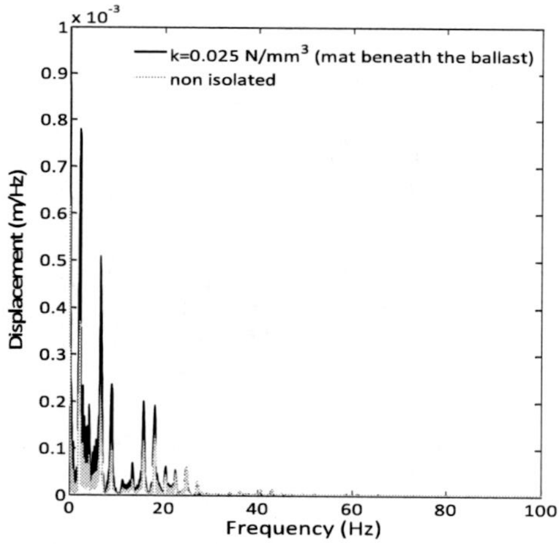

(b)

Figure 23. Frequency content of the rail displacements due to the passage of Alfa-Pendular at 212 km/h for ballast mat beneath: (a) subballast; (b) ballast.

Similar conclusions are also valid for the response of the sleeper, as can be observed by Figure 24, where the one-third octave spectrum of the insertion loss (IL) achieved by each mitigation solution is shown. As the figure shows, the introduction of the mat beneath the ballast layer leads to an average increase of the vertical velocity of more or less 7–9 dB for the frequency range up to 25 Hz. Moreover, independently of the mat position in depth, the mitigation effect is only pronounced for frequencies above 40 Hz, never reaching values higher than 10 dB for the frequency range under consideration.

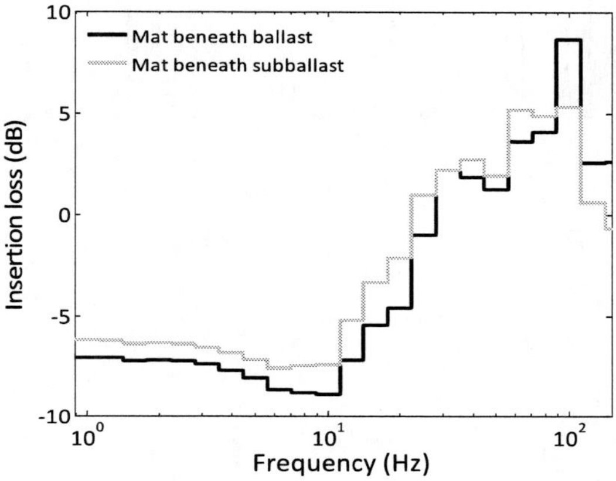

Figure 24. Insertion loss assessed at the sleeper for the passage of Alfa-Pendular at 212 km/h.

4.4. Free-field Dynamic Response

In opposition to what was observed in the dynamic response of the track, the effects induced by the dynamic excitation mechanism are predominant in the response of the free-field [32]. So, in this case, it is expected that the presence of the ballast mat could achieve a non-negligible relevance in terms of attenuation of vibrations.

Figure 25 shows the frequency spectra of the vertical velocity at 3.50 m, 7.0 m and 15.0 m away from the track. An overview of the results enables to conclude that the introduction of the mat, with the properties indicated in the figures, is an efficient mitigation measure for the attenuation of vibrations with frequencies above to, more or less, 40 Hz. Comparing both mitigation solutions, it was found that the attenuation efficiency is, apparently, not much different in either case. Indeed, the changes of the dynamic behavior of the track induced by the presence of the mat, and the subsequent change of the frequency content of the dynamic loads on the train, are responsible for a dual and sometimes contradictory effect. On the one hand, the mat attenuates the energy transmitted to the ground in the higher frequency range, while on the other hand, it is responsible for the increase of the dynamic loads at the same frequency range. In spite of the results depicted in Figure 25, the mitigation efficiency is better assessed from the analysis of the one-third octave spectrum of the insertion loss, as depicted in Figure 26. As can be seen, the attenuation effect is only effective for frequencies above 40 Hz. Despite of the differences denoted for the distinct points, the mat beneath the ballast layer seems to be more efficient, since gains up to 20 dB can be achieved for frequencies above 50 Hz.

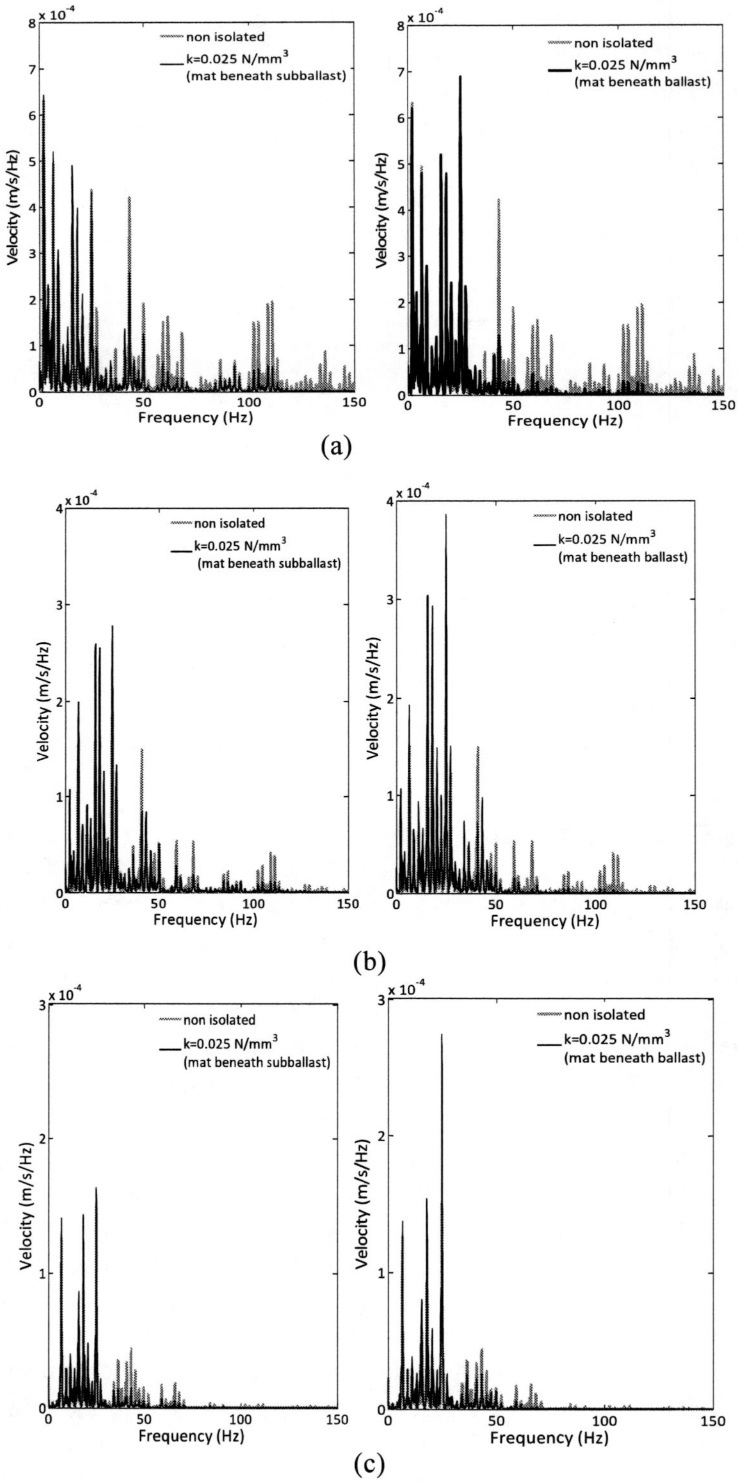

Figure 25. Frequency content of vertical velocity at the free-field: (a) 3.5 m; (b) 7.0 m; (c) 15.0 m (left side – mat beneath the subballast layer; right side – mat beneath ballast layer).

Apart from the attenuation effect, for the frequency range between 10 Hz and 40 Hz, the presence of the mat leads to an increase of the vibration velocity levels, mainly if this element is introduced beneath the ballast layer. Indeed, for this frequency range, an insertion loss of −5 dB to −8 dB can be observed when the mat in placed beneath the ballast layer. This corresponds to vibration levels around 2 to 2.5 times the vibration levels observed in the absence of mitigation measures. However, this effect is minimized when the mat is placed beneath the subballast layer.

To better illustrate the correlation between force transfer and vertical velocity at the free field, Figure 27 presents, in the same plot, the IL curves for vertical velocity at the free field and for the force transfer when the mat is placed beneath the ballast layer. Despite the differences, a global trend between IL of the force transfer and of the homologous for the vertical velocity at the free field can be observed.

From the results and remarks pointed out above, it can be stated that ballast and subballast mats can be faced as an effective vibration countermeasure if the effects to be minimized are related to higher frequency contents, as for instance reradiated noise in buildings, but this solution can also be responsible for vibration nuisance (depending of the modal behavior of the building), which is more sensitive to the lower frequency range.

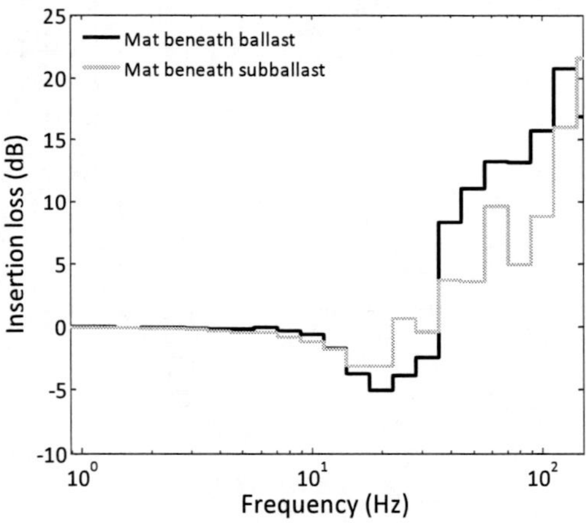

(a)

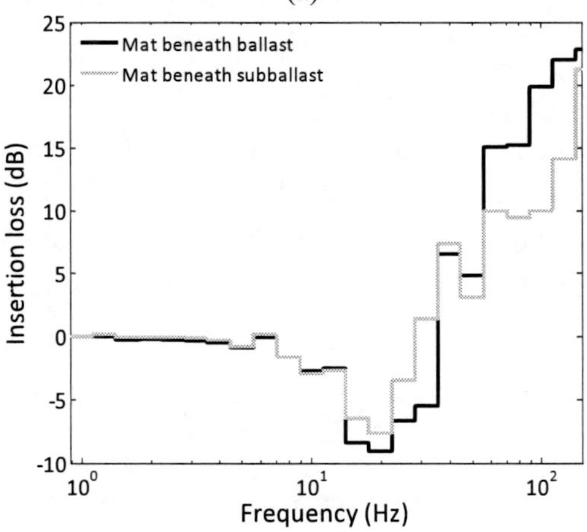

(b)

Figure 26. (Continued)

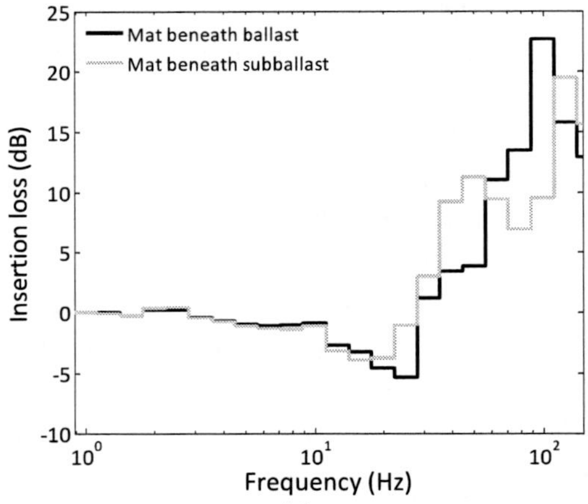

(c)

Figure 26. One-third octave spectrum of the insertion loss of vertical velocity for the passage of Alfa-Pendular at 212 km/h at the distances of: (a) 3.5 m; (b) 7.0 m; (c) 15.0 m.

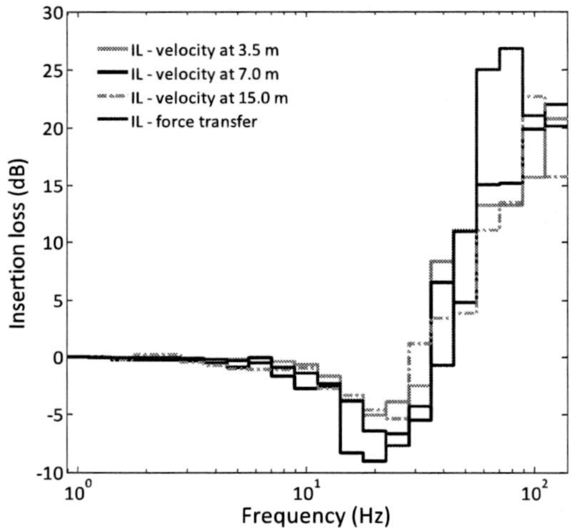

Figure 27. One-third octave spectrum of the insertion loss for the vertical free field velocity and for force transfer when the mat is placed beneath the ballast layer.

In sum, independently of the remarks expressed above, the results here presented seem to indicate that the solution of mats beneath the subballast leads to better global results. The reduction effect for higher frequencies, although not as pronounced as found when the mat is placed beneath the ballast layer, can be achieved without such expressive increase of the response in the lower frequency range.

CONCLUSION

In the present chapter, the main trends for the efficient numerical modeling and measuring of the dynamic behavior of transportation infrastructures were presented, with emphasis on the simulation of vibrations induced by railway traffic.

Taking into account the great longitudinal extension of the railway track and its invariant characteristics, a 2.5D approach is applied, which allows for a significant reduction of the computational effort, without sacrificing the 3D behavior of the domain. Following this approach, a numerical model was developed based, for the simulation of the railway track, on the finite elements method and, for the modeling of the ground a 2.5D boundary elements formulation. Moreover, due to the complexity of the problem of vibrations induced by traffic, the simulation of rolling stock itself is also required in several situations. To overcome this issue, the train-track interaction problem is solved by a compliance formulation where the structural behavior of the train is simulated by a multi-body approach.

After the presentation of the numerical model a case study is analyzed, being the predictions compared with the measured data.

The proposed case study comprises the analysis of vibrations induced by traffic in the track and in free-field. To achieve that goal, the authors developed a comprehensive experimental campaign in a stretch of the Portuguese railway network, near to Carregado town.

As exposed along the chapter, experimental and numerical approaches should be considered as complementary tools for a deeper understanding of this complex and difficult problem. The experimental measurements, especially the identification of the dynamic properties of both ground and track, enabled to build up a reliable and comprehensive calibration of the numerical model. After that, the vibrations measured during the passage of the Alfa-Pendular train were compared with the prediction. The comparison revealed a very good agreement in terms of the track response as well as of the ground vibrations.

Finally, after the experimental validation of the numerical model, a study about the efficiency of ballat mats on the abatement of vibrations was presented. From this study it was possible to conclude that the efficiency of the use of mats for the mitigation of vibrations induced by railway traffic is clearly dependent on its stiffness. In order to obtain a considerable reduction of vibrations, softer mats are required. However, softer mats imply higher track displacements, limiting the field of application of this mitigation solution. Alternatively, the mats can be inserted beneath the subballast layer, which leads to a double beneficial effect: i) a reduction of the cut-on frequency and consequent increase of effectiveness of the mitigation countermeasure; ii) a reduction of the track displacements related with the lower frequency range. Moreover, a correlation between the force transfer insertion loss and the free-field velocity insertion loss. From the results here presented it was found that the insertion loss computed from the force transfer curves can be used for a preliminary estimation of the vibration reduction effect induced by the introduction of mats.

ACKNOWLEDGMENTS

This paper reports research developed under the financial support of "FCT - Fundação para a Ciência e Tecnologia", Portugal, namely from the research project - PTDC/ECM/ 114505/2009.

References

[1] Standard I. (2005). *Mechanical vibration – Ground-borne noise and vibration arising from rail systems*, in *ISO 14837-1*.

[2] Dieterman H.A., Metrikine A. (1996). The equivalent stiffness of a half-space interacting with a beam. Critical velocities of a moving load along the beam. *European Journal of Mechanics A/Solids*, 15(1): 67–90.

[3] Dieterman H.A., Metrikine A. (1997). Steady-state displacements of a beam on an elastic half-space due to a uniformly moving constant load. *European Journal of Mechanics A/Solids*, 16(2): 295–306.

[4] Sheng X., Jones C., Thompson D. (2003). A comparison of a theoretical model for quasi-statically and dynamically induced environmental vibration from trains with measurements. *Journal of Sound and Vibration*, 267(3): 621–635.

[5] Karlstrom A. (2006). An analytical model for ground vibrations from accelerating trains. *Journal of Sound and Vibration*, 293: 587–598.

[6] Kausel E. (2008). *Commentaries on methods to estimate ground vibrations elicited by fast moving loads*, in *Noise and Vibration on high-speed railways*, R.D. Calçada, R.; Carvalho, A., Degrande, G., Editor: Porto.

[7] Andersen L., Nielsen S., Krenk S. (2007). Numerical methods for analysis of structure and ground vibration from moving loads. *Computers and Structures*, 85: 43–58.

[8] Galvín P., Romero A., Domínguez J. (2010). Vibrations induced by HST passage on ballast and non-ballast tracks. *Soil Dynamics and Earthquake Engineering*, 30(9): 862–873.

[9] Galvín P., Romero A., Domínguez J. (2010). Fully three-dimensional analysis of high-speed train-track-soil-structure dynamic interaction. *Journal of Sound and Vibration*, 329: 5147–5163.

[10] Galvín P., Domínguez J. (2009). Experimental and numerical analyses of vibrations induced by high-speed trains on Córdoba-Málaga line. *Soil Dynamics and Earthquake Engineering*, 29(4): 641–657.

[11] Koh C.G., Chiew G.H., Lim C.C. (2007). A numerical method for moving load on continuum. *Journal of Sound and Vibration*, 300: 126–138.

[12] Lane H., Kettil P., Wiberg N. (2008). Moving finite elements and dynamic vehicle interaction. *European Journal of Mechanics A/Solids*, 27: 515–531.

[13] Clouteau D., Aubry D., Elhbre M. (2000). Periodic BEM and BEM-FEM coupling: applications to seismic behaviou of very long structures. *Computation Mechanics*, 25(6): 567–577.

[14] Gupta S., Hussein M., Degrande G., Hunt H., Clouteau D. (2007). A comparison of two numerical models for the prediction of vibrations from underground railway traffic. *Soil Dynamics and Earthquake Engineering*, 27(7): 608–624.

[15] Chebli H., Othman R., Clouteau D., Arnst M., Degrande G. (2007). 3D periodic BE-FE model for various transportation structures interacting with soil. *Computes and Geotechnics*, 35: 22–32.

[16] Sheng X., Jones C., Thompson D. (2005). Responses of infinite periodic structures to moving or stationary harmonic loads. *Journal of Sound and Vibration*, 282: 125–149.

[17] Yang Y.B., Hung H.H. (2001). A 2.5D finite/infinite element approach for modelling visco-elastic body subjected to moving loads. *International Journal for Numerical Methods in Engineering*, 51: 1317−1336.

[18] Bouchon M., Aki K. (1977). Discrete wave-number representation of seismic-source wave fields. *Bulletin of the Seismological Society of America*, 67(2): 259−277.

[19] Gavric,L. (1995). Computation of propagative waves in free rail using a finite element technique. *Journal of Sound and Vibration*, 185(3): 531−543.

[20] Tadeu A., António J., (2001). 2.5D Green's functions for elastodynamic problems in layered acoustic and elastic formations. *CMES - Computer Modeling in Engineering and Sciences*, 2(4): 477−495.

[21] António J., Tadeu A. (2002). 3D seismic response of a limited valley via BEM using 2.5D analytical Green's functions for an infinite free-rigid layer. *Soil Dynamics and Earthquake Engineering*, 22(8): 659−673.

[22] François S., Schevenels M., Degrande G., Borgions J., Thyssen B. (2008). A 2.5D finite element-boundary element model for vibration isolating screens, in *ISMA 2008 - International Conference on Noise and Vibration Engineering*, P. Sas and B. Bergen, Editors: Leuven. p. 2765−2776.

[23] Alves Costa P., Calçada R., Silva Cardoso A., Bodare A. (2010). Influence of soil non-linearity on the dynamic response of high-speed railway tracks. *Soil Dynamics and Earthquake Engineering*, 30(4): 221−235.

[24] Yang Y., Hung H., Chang D. (2003). Train-induced wave propagation in layered soils using finite/infinite element simulation. *Soil Dynamics and Earthquake Engineering*, 23: 263−278.

[25] Yang Y., Hung H. (2008). Soil Vibrations Caused by Underground Moving Trains. *Journal of Geotechnical and Geoenvironmental Engineering*, 134(11): 1633−1644.

[26] Alves Costa P., Calçada R., Couto Marques J., Cardoso A. (2010). A 2.5D finite element model for simulation of unbounded domains under dynamic loading, in *Numerical Methods in Geotechnical Engineering - Proceedings of the 7th European Conference on Numerical Methods in Geotechnical Engineering* T. Benz and S. Nordal, Editors: Trondheim. p. 397−404.

[27] François S., Schevenels M., Galvín P., Lombaert G., Degrande G. (2010). A 2.5D coupled FE–BE methodology for the dynamic interaction between longitudinally invariant structures and a layered halfspace. *Computer Methods in Applied Mechanics and Engineering*, 199(23−24): 1536−1548.

[28] Calçada R., Delgado R., Cardoso A., Alves Costa P., Santos N., Ribeiro C., Coelho B. (2008). Train-track-ground interaction on high speed lines. in *Noise and Vibration on High-Speed Railways*. Porto: FEUP.

[29] Muller K., Grundmann H., Lenz S. (2008). Nonlinear interaction between a moving vehicle and a plate elastically mounted on a tunnel. *Journal of Sound and Vibration*, 310: 558−586.

[30] Galvín P., François S., Schevenels M., Bongini, E. Degrande G., Lombaert G. (2010). A 2.5D coupled FE-BE model for the prediction of railway induced vibrations. *Soil Dynamics and Earthquake Engineering*, 30(12): 1500−1512.

[31] Alves Costa P., Calçada R., Cardoso A. (2012). Track–ground vibrations induced by railway traffic: In-situ measurements and validation of a 2.5D FEM-BEM model. *Soil Dynamics and Earthquake Engineering*, 32: 111−128.

[32] Lombaert G., Degrande G. (2009). Ground-borne vibration due to static and dynamic axle loads of InterCity and high-speed trains. *Journal of Sound and Vibration*, 319(3−5): 1036−1066.

[33] Lombaert G., Degrande G., Kogut J., François S. (2006). The experimental validation of a numerical model for the prediction of railway induced vibrations. *Journal of Sound and Vibration*, 297: 512−535.

[34] Alves Costa P., Calçada R., Cardoso A. (2012). Influence of train dynamic modelling strategy on the prediction of track-ground vibrations induced by railway traffic. *Journal of Rail and Rapid Transit*, 226: 434−450.

[35] Roma V., Abate G., Pantaleo M., Pezzoli P. (2010). High Speed Train Vibrations along the Turin-Milan Railway and mitigation systems. *Scienza e Tecnica*, 5/2120: 427−445.

[36] Lakusic S., Ahac M., Haladin I. (2010). Experimental investigation of railway track with under sleeper pads, in *Slovenski Kongres o Cestah in Prometu* 2010: Portoroz. p. 386−393.

[37] Hanson C., Singleton J. (2006). Performance of ballast mats on passenger railroads: measurement vs. projections. *Journal of Sound and Vibration*, 293: 873−877.

[38] Auersch L. (2006). Dynamic axle loads on tracks with and without ballast mats: numerical results of three-dimensional vehicle-track-soil models. *Journal of Rail and Rapid Transit*, 220: 169−183.

[39] Muller R. (2007). Mitigation measures for open lines against vibration and ground-borne noise: a swiss overview, in *9th International Workshop on Railway Noise*, B. Schulte-Werning, D. Thompson, P. Gautier, C. Hanson, J. Nelson, T. Maeda, and P. Vos, Editors: Munich, Germany. p. 264−270.

[40] Madshus C., Rothschild K., Cleave R., Brekke A. (2008). Full scale tests on mitigation of ground borne noise from railway tunnels in rock, in *Noise and vibration on high-speed railways*, R. Calçada, R. Delgado, A. Carvalho, and G. Degrande, Editors. Porto. p. 217−240.

[41] Lombaert G., Degrande G., Vanhauwere B., Vandeborght B., François S. (2006). The control of ground-borne vibrations from railway traffic by means of continuous floating slabs. *Journal of Sound and Vibration*, 297: 946−961.

[42] Hwang R., Lysmer J. (1981). Response of buried structures to travelling waves. *Journal of Geotechnical Engineering Division*, 107(2): 183−200.

[43] Sheng X., Jones C., Thompson D. (2006). Prediction of ground vibration from trains using wavenumber finite and boundary element method. *Journal of Sound and Vibration*, 293: 575−586.

[44] Muller K. (2007). Dreidimensionale dynamische Tunnel-Halbraum-Interaktion, in *Lehrstuhl fur Baumechanik* 2007, Technische Universitat Munchen: Munchen.

[45] Alves Costa P. (2008). Moving loads on the ground: a 2.5D transformed finite element code for train-track-soil interaction, in *(Research Report - FEUP)*2008, FEUP: Porto.

[46] Wolf, J.P., Song C. (1996). *Finite-element modeling of unbounded media*. New York: Wiley.

[47] Andersen L., Nielsen S. (2003). Boundary element analysis of the steady-state response of an elastic half-space to a moving force on its surface. *Engineering Analysis with Boundary Elements*, 27(1): 23−38.

[48] Knothe K., Wu Y. (1998). Receptance behaviour of railway track and sub-grade. *Archive of Applied Mechanics*, 68: 457−470.

[49] Mann A. (2002). *DYNATRACK: A survey of dynamic railway track properties and their quality*, Technische Universiteit Delft: Delft.

[50] Alves Costa P., Calçada R., Silva Cardoso A. (2012). Track-ground vibrations induced by railway traffic, in *Applications of Computational Mechanics in Geotechnical Engineering*, L. Sousa, M. Fernandes, E. Vargas, and R. Azevedo, Editors. Balkema. p. 125−159.

[51] Tadeu A., Kausel E. (2000). Green's functions for two-and-a-half-dimensional elasto-dynamic problems. *Journal of Engineering Mechanics*, 126(10): 1093–1096.

[52] Kausel E. (1981). *An explicit solution for the green functions for dynamic loads in layered media*, R.R. R81-13, Editor, MIT: Boston.

[53] Kausel E. (1974). Forced Vibrations of Circular Foundations on Layered Media, in *Soils Publication No. 336; Structures Publication No. 384*, R.R. R74-11, Editor, MIT: Boston.

[54] Grundmann H., Lieb M., Trommer E. (1999). The response of a layered half-space to traffic loads moving along its surface. *Archive of Applied Mechanics*, 69: 55−67.

[55] Auersch L. (2005). The excitation of ground vibration by rail traffic: theory of vehicle–track–soil interaction and measurements on high-speed lines. *Journal of Sound and Vibration*, 284: 103−132.

[56] Lombaert G., DeGrande G., Clouteau D. (2000). Numerical modelling of free field traffic-induced vibrations. *Soil Dynamics and Earthquake Engineering*, 19: 473−488.

[57] Lopes P., Alves Costa P., Calçada R., Silva Cardoso A. (2010). Análise numérica de vibrações induzidas por tráfego ferroviário em túneis baseada em modelos 2.5D, in *12º Congresso Nacional de Geotecnia*, G. Correia, Editor: Guimarães. p. CD-ROM publication.

[58] Zhai W., Cai Z. (1997). Dynamic Interaction between a lumped mass vehicle and a discretely supported continuous rail track. *Computers and Structures*, 63(5): 987−997.

[59] Mayne P., Rix G. (1995). Correlations between shear wave velocity and cone tip resistance in natural clays. *Soils and Foundations*, 35(2): 107−110.

[60] Kaewunruen S., Remennikov A. (2010). Dynamic properties of railway track and its components: recent findings and future research direction. *Insight-Non-Destructive Testing and Condition Monitoring*, 52(1): 20−22.

[61] Kaewunruen S., Remennikov A. (2007). Field trials for dynamic characteristics of railway track and its components using impact excitation technique. *NDT&E international - Independent Nondestructive Testing and Evaluation*, 40: 510−519.

[62] Dahlberg T. (2006). *Track issues*, in *Handbook of Railway Vehicle Dynamics*, S. Iwnicki, Editor. Taylor & Francis.

[63] Auersch L. (2011). The dynamic force transfer of slab tracks and floating slab tracks and the corresponding ground vibration, in *EURODYN 2011*, G. De Roeck, G. Degrande, G. Lombaert, and G. Muller, Editors: Leuven. p. 820−827.

[64] Alves Costa P., Calçada R., Silva Cardoso A. (2011). Vibrations induced by railway traffic: experimental validation of a 2.5D numerical model. in *COMPDYN 2011 - 3rd International Conference on Computational Methods in Structural Dynamics and Earthquake Engineering*, M. Papadrakakis, M. Fragiadakis, and V. Plevris, Editors. Corfu, Greece. CD-ROM.

In: Traffic Induced Environmental Vibrations and Controls ISBN: 978-1-62618-240-0
Editors: H. Xia and R. Calçada

Chapter 3

BASIC THEORY AND ANALYSIS APPROACHES IN TRAFFIC-INDUCED ENVIRONMENTAL VIBRATIONS

Yanmei Cao*, Zhu Yu and Kunpeng Cui
School of Civil Engineering, Beijing Jiaotong University
Beijing, China

ABSTRACT

This chapter presents the basic theory and analysis approaches in traffic-induced environmental vibrations, mainly including vibration source, wave propagation, dynamic characteristics of soil, analytical model of train-track-ground, Green's functions of ground displacements, and so on. By application of the Betti-Rayleigh Dynamic Reciprocal Theorem, the reciprocal property of Green's function and the solution of soil/ground vibration subjected to moving loads are derived. Case studies show for receivers near the source, the R-wave contributes to ground vibration much more than the P-wave and the S-wave; while for receivers far from the source, the contribution of the P-wave is more obvious. For the source of moving constant-load, the ground vibration is featured of typical low frequency; while for moving harmonic-load, its frequency range is controlled by the Rayleigh's wave velocity of the top layer soil.

1. INTRODUCTION

The primary step of studying environmental vibrations is to understand and master the source of vibration, wave propagation characteristics in the soil, nature of layered subsoil, and soil vibration induced by moving load.

The soil vibration problem was first studied by Lamb [1], whose pioneering research concentrated upon the response of elastic halfspace and elastic-body with infinite boundary generated by stationary loads or approximately stationary loads. In the late 20th century, the dynamic behavior of soil induced by moving loads has been paid more attention, especially in

* E-mail address: ymcao@bjtu.edu.cn.

railway and urban transit traffic systems. Grundmann et al. [2] studied the response of a layered half-space subjected to a single moving periodic load and furthermore a simplified train load. Hung and Yang [3] studied the steady response of viscoelastic halfspace to various vehicle loads. Recently, the theoretical derivation and the numerical calculation were combined and used in many research. Bierer and Bode [4] proposed a semi-analytical model, directly defined in space and time domain, to study the road traffic induced ground motions. Martínez-Castro et al. [5] analyzed the response of Bernoulli-Euler beams subjected to flexural vibrations under concentrated moving forces. Zhou and Jiang [6] derived the analytical solution of the dynamic responses of viscoelastic half-space under high-speed moving loads and gave the calculated results by numerical method. Galvin et al. [7] proposed a 2.5D coupled FE-BE model based on a regularized boundary integral equation to predict the environmental vibrations induced by ground and underground trains. Hamid et al. [8] numerically simulated the train movement and calculated the ground vibration induced by underground trains. Motazedian et al. [9] studied train-induced ground vibrations in a soil layer with low shear wave velocity overlaying bedrock with high shear wave velocity in eastern Canada.

2. Vibration Sources Induced by Traffics

For traffic induced environmental vibrations, the sources mainly include trains running on the track (ordinary railways, high-speed railways, underground railways and urban elevated rail transit lines) and automobiles or other vehicles running on the road (highways, urban roads, etc.).

2.1. Vibration Source Model of Rail Traffic System

2.1.1. Development of Vehicle Models

Vehicles vibrate on the track due to the moving gravity loading and track irregularity, which are transmitted to the track structure via wheel-sets, forming the process of dynamic interaction of the entire vehicle-track system. The vehicle can be considered as a system composed of a car-body, bogies and wheel-sets, connected by suspension systems consisting of springs, dampers, and linkages. The wheel-sets are connected to the bogies by the primary suspension system, while the car-body and the bogies are connected by the secondary suspension system. Various vehicle models have been developed for the dynamic analysis of vehicle-track dynamic coupling system, as shown in Figure 1.

In Figure 1, (a) is the earliest model, which simplifies the excitation of vehicle as a moving constant load or a load series. (b) considers the variance of the moving force, such as a harmonic force or an impulse force changing with time; (c) regards the vehicle load as a moving mass, considering the action of inertia force due to the mass during vibration. These three models did not take into account the dynamic interaction of car-body, bogies, wheel-sets and various components of the track, which has some limitation in simulating the actual conditions of trainloads.

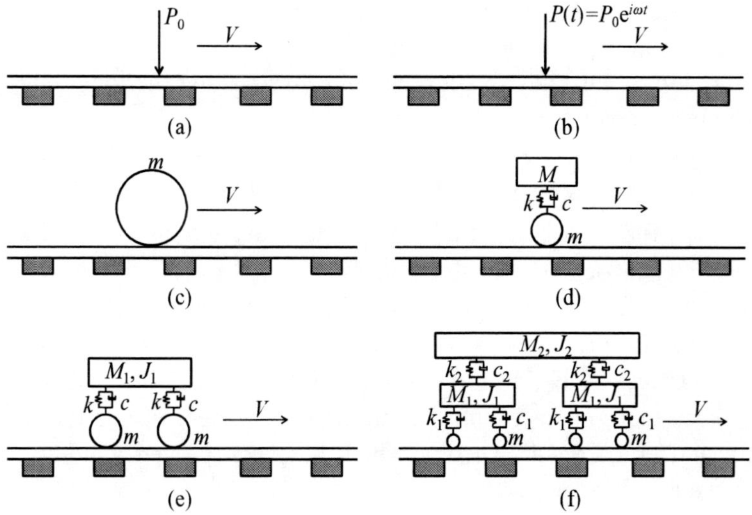

Figure 1. Development of moving vehicle model in train-track interaction analysis.

The model in Figure 1 (d) uses springs and dampers to connect unsprung mass with sprung mass, which can consider the inertia forces and the interaction of two masses. In Figure 1 (e), two inertia forces of translation and rotation are considered for the sprung mass, which can simulate the vertical vibration of a two-axle vehicle or a bogie. In the 1970's, the research of train-track dynamic interaction broke the traditional framework into the stage of dynamics. The model shown in Figure 1 (f) was established, which is a multiple-DOF vibration system of train-track structure, reflecting not only the vibration of car-body, bogies and wheel-sets, but also the dynamic interaction of train-track by wheel-rail contact spring. This model is adopted as the source model of environmental vibration in the flowing analysis.

2.1.2. Train Model

The train model is composed of locomotives and passenger or freight vehicles (Figure 2), and each vehicle is a multiple-DOF vibration system consisting of a car-body, bogies, wheel-sets, and spring-damper suspension system.

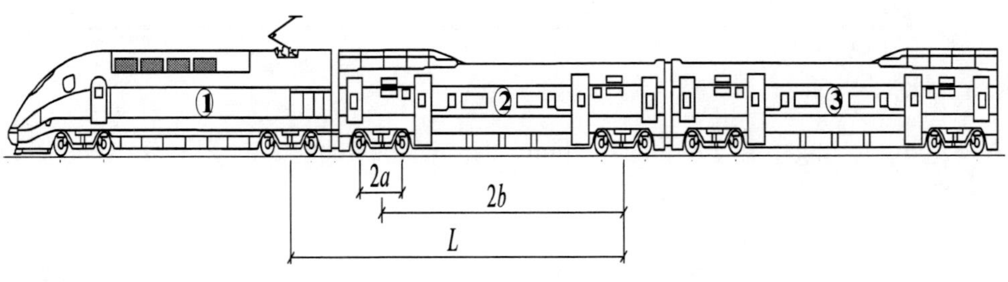

Figure 2. Schemetic composition of a train.

In studying the environmental vibrations induced by trains, the moving excitation of vehicles on the substructure (track, roadbed, tunnel lining, elevated bridges, etc.) is mainly in the vertical plane. Therefore, in the following analysis, only the vibrations in the vertical-longitudinal plane are considered. In the vehicle model, the following assumptions are considered:

(1) The car-body, bogies and wheel-sets are regarded as rigid-bodies, namely, neglecting their elastic deformation during vibration.
(2) The vibration of car-body, bogies and wheel-sets along the longitudinal direction of the vehicle are not considered.
(3) Five vibration DOFs are considered for each vehicle model, including two DOFs (floating Z_{ci} and pitching φ_{ci}) for the *i*-th car-body, two DOFs (floating Z_{tij} and pitching φ_{tij}) for the *j*-th bogie of the *i*-th car-body, and only one DOF (floating Z_{wijl}) for the *l*-th wheel-sets of the *j*-th bogie.
(4) For the *i*-th vehicle, the masses and the pitching inertias are represented, respectively, by M_{ci} and J_{ci} for the car-body, M_{ti} and J_{ti} for the bogie, and M_{wi} for the wheel-sets.
(5) Only the vertical excitation force of the train on the rail is considered, assuming the wheels keep close contact with the rail, and the wheel-rail relationship accords with the Linear Hertz Contact Theory, which has been proved valid for environmental vibration analysis, expressed as:

$$f_H = k_H[Z_w(t) - v(x,t) - Z_s(t)]^{3/2} \tag{1}$$

where k_H is equivalent stiffness coefficient of Hertz contact; $Z_w(t)$ is the vertical displacement of wheel-sets; $v(x,t)$ is the vertical displacement of track at the position of wheel-set; $Z_s(t)$ is the wheel-rail relative displacement caused by track or wheel irregularity.

Based on the above assumptions, the vehicle model for analyzing the motion of rail vehicle is established, shown in Figure 3.

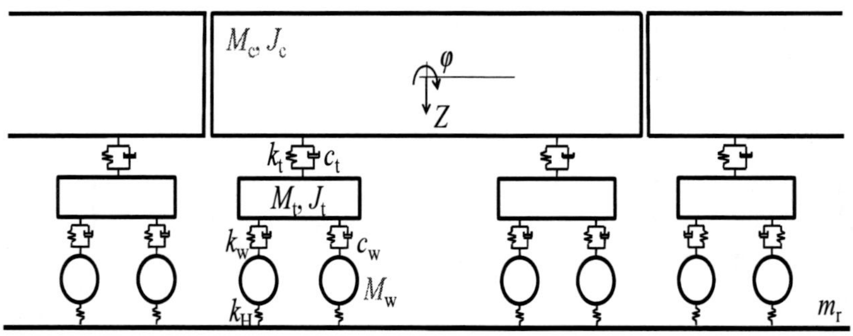

Figure 3. Sketch of vehicle model.

According to Lagrange motion equations, the dynamic equation of the *i*th vehicle can be written as:

$$\mathbf{M}_i\ddot{\mathbf{Z}}_i + \mathbf{C}_i\dot{\mathbf{Z}}_i + \mathbf{K}_i\mathbf{Z}_i = \mathbf{P}_i(t) \ \ (i = 1, 2, \cdots, N_v) \tag{2}$$

where $\mathbf{Z}_i(t)$, $\dot{\mathbf{Z}}_i(t)$ and $\ddot{\mathbf{Z}}_i(t)$ represent the displacement, velocity and acceleration vectors, respectively. The displacement vector consists of the movements of car-body, bogies and wheel-sets as follows:

$$\mathbf{Z}_i(t)=[Z_{ci} \quad \varphi_{ci} \quad Z_{ti1} \quad \varphi_{ti1} \quad Z_{ti2} \quad \varphi_{ti2} \quad Z_{wi1} \quad Z_{wi2} \quad \cdots \quad Z_{wiN_w}]^T \tag{3}$$

in which N_w is the number of wheel-sets of a vehicle, and N_v is the total number of vehicles in a train.

For a four-axle vehicle, the mass matrix, stiffness matrix and damping matrix of the i-th vehicle can be expressed as follows:

$$\mathbf{M}_i = \mathrm{diag}[M_{ci} \quad J_{ci} \quad M_{ti} \quad J_{ti} \quad M_{ti} \quad J_{ti} \quad M_{wi} \quad M_{wi} \quad M_{wi} \quad M_{wi}] \tag{4}$$

$$\mathbf{K}_i=\begin{bmatrix} 2k_{ti} & & & & & & & & & \\ 0 & 2k_{ti}b^2 & & & & & \text{sym} & & & \\ -k_{ti} & -k_{ti}b & k_{ti}+2k_{wi} & & & & & & & \\ 0 & 0 & 0 & 2k_{wi}a^2 & & & & & & \\ -k_{ti} & k_{ti}b & 0 & 0 & k_{ti}+2k_{wi} & & & & & \\ 0 & 0 & 0 & 0 & 0 & 2k_{wi}a^2 & & & & \\ 0 & 0 & -k_{wi} & -k_{wi}a & 0 & 0 & k_{wi}+k_H & & & \\ 0 & 0 & -k_{wi} & k_{wi}a & 0 & 0 & 0 & k_{wi}+k_H & & \\ 0 & 0 & 0 & 0 & -k_{wi} & -k_{wi}a & 0 & 0 & k_{wi}+k_H & \\ 0 & 0 & 0 & 0 & -k_{wi} & k_{wi}a & 0 & 0 & 0 & k_{wi}+k_H \end{bmatrix} \tag{5}$$

$$\mathbf{C}_i=\begin{bmatrix} 2c_{ti} & & & & & & & & & \\ 0 & 2c_{ti}b^2 & & & & & \text{sym} & & & \\ -c_{ti} & -c_{ti}b & c_{ti}+2c_{wi} & & & & & & & \\ 0 & 0 & 0 & 2c_{wi}a^2 & & & & & & \\ -c_{ti} & c_{ti}b & 0 & 0 & c_{ti}+2c_{wi} & & & & & \\ 0 & 0 & 0 & 0 & 0 & 2c_{wi}a^2 & & & & \\ 0 & 0 & -c_{wi} & -c_{wi}a & 0 & 0 & c_{wi} & & & \\ 0 & 0 & -c_{wi} & c_{wi}a & 0 & 0 & 0 & c_{wi} & & \\ 0 & 0 & 0 & 0 & -c_{wi} & -c_{wi}a & 0 & 0 & c_{wi} & \\ 0 & 0 & 0 & 0 & -c_{wi} & c_{wi}a & 0 & 0 & 0 & c_{wi} \end{bmatrix} \tag{6}$$

in which k_t and k_w are the stiffness of springs on bogie and on axle box, respectively; c_t and c_w are the damping coefficients of bogies and axle box; b is half of the distance between adjacent vehicles and a is that between adjacent bogies.

$\mathbf{P}_i(t)$ is the excitation vector acting to the ith vehicle, expressed as:

$$\mathbf{P}_i(t)=\{0 \quad 0 \quad 0 \quad 0 \quad 0 \quad 0 \quad -P_{i1}^{w/r}(t) \quad -P_{i2}^{w/r}(t) \quad \cdots \quad -P_{iN_{wi}}^{w/r}(t)\}^T \tag{7}$$

in which $P_{ij}^{w/r}(t)$ represents the wheel-rail interaction force due to track irregularity of the ij-th wheel-set (short for the j-th wheel-set of i-th vehicle):

$$P_{ij}^{\mathrm{w/r}}(t) = k_{\mathrm{H}} \sum_{l=-1}^{+1} \phi_{n+l} v_{n+l}^{ij} + k_{\mathrm{H}} Z_{\mathrm{s}}(x_{ij}) \tag{8}$$

where ϕ_n represents the interpolation function corresponding to the nth node of track; v_{n+l}^{ij} represents nodal displacements of rail near the ij-th wheel-set, which is used to construct the interpolation function; $Z_s(x_{ij})$ means the track irregularity at the position of the ij-th wheel-set.

2.1.3. Track Irregularities

The track irregularities, which are regarded as one of the main self-excitations, play an important role on the interaction of train-track system. Track irregularities refer to the deviations of the rails that support and guide the wheels from the ideal rail of perfect geometry, which includes static irregularity at unloaded status and dynamic irregularity under loading. Track irregularities may occur because of initial installation errors, degradation of support materials, and dislocation of track joints, etc. The four commonly used geometrical parameters to define tangent track irregularities are illustrated in Figure 4, which are vertical profile, cross-level, alignment and gauge.

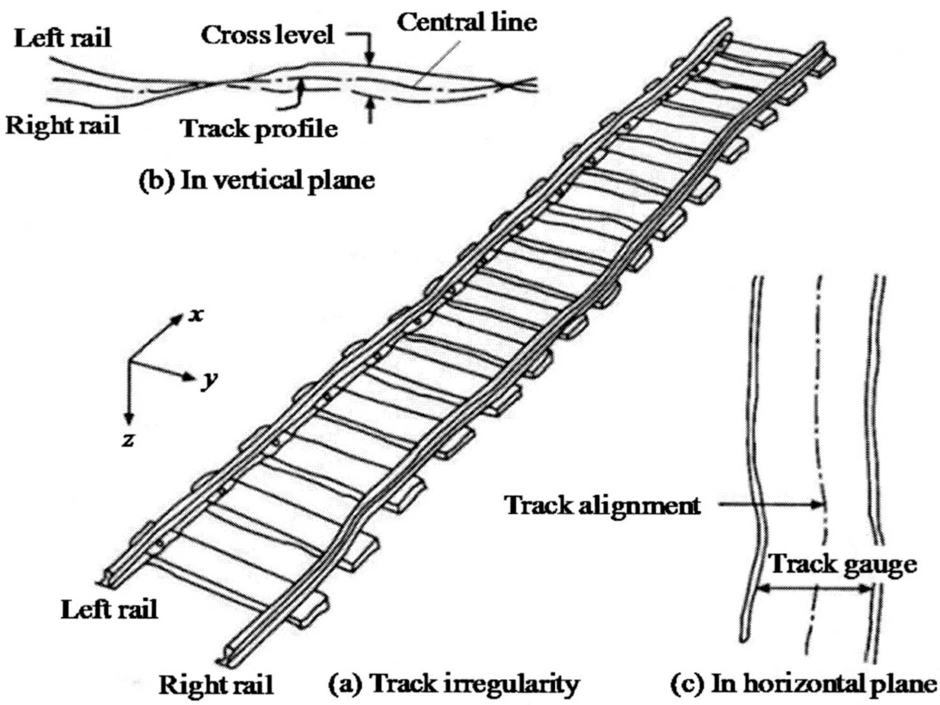

Figure 4. Sketch of track irregularities.

For dynamic analysis of coupled train-track system, the track irregularity input can be the measured irregularities from the actual tracks [10], or the simulated ones based on the power spectral density functions of the track irregularity [10−12].

Recently, the China Academy of Railway Sciences measured the track irregularities on several high-speed railways, using the special track inspection train. Shown in Figure 5 are the power spectrum densities (PSD) of ballast-less slab track measured on the Beijing-Tianjin intercity high-speed railway.

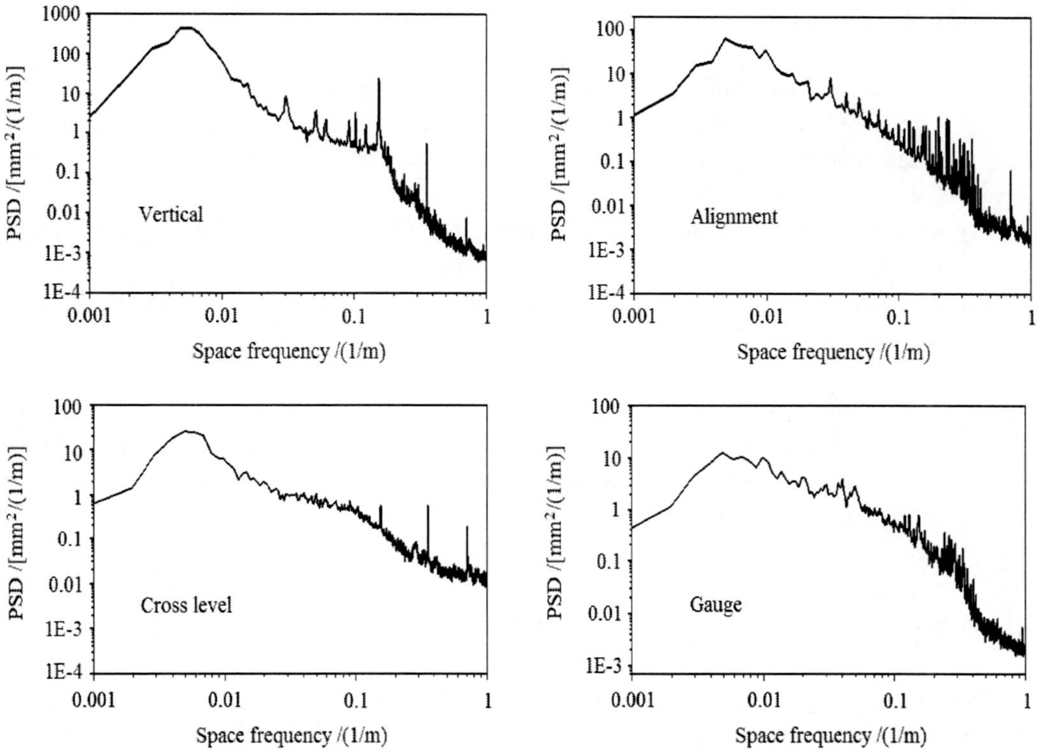

Figure 5. Track spectrum densities measured on the Beijing-Tianjin intercity high-speed railway.

Using the PSD functions of track irregularity for different grade of railways, the sample of track irregularities can be obtained by numerical simulation method. For example, when the trigonometric series superposition method is adopted, the track irregularity series can be generated as follows:

$$w(x) = \sqrt{2}\sum_{k=1}^{N}\sqrt{S(\omega_k)\Delta\omega}\cos(\omega_k x + \phi_k) \tag{9}$$

where $w(x)$ is the sequence of track irregularity; $S(\omega_k)$ is the PSD function for a given track irregularity; ω_1 and ω_N correspond to, respectively, the lower and upper limits of the frequency components considered; $\Delta\omega$ is the width of frequency interval; ϕ_k is the phase corresponding to the kth frequency component, with a uniform distribution between 0 and 2π.

In addition, for the simulated sample, it must be checked whether it has the same characteristics with the given PSD function $S(\omega_k)$.

3. Propagation of Vibration Wave in Soils

The excitation of railway traffic-induced environmental vibrations mainly include the following four categories:

(1) Moving axle loads of vehicles, also called quasi-static loads;
(2) Transient loads at fixed acting points (or called impulsive loads), which are dynamic loads generated by the impact of moving wheels on rail joints or welding points of CWR (continuously welded rail track);
(3) Moving dynamic loads induced by track irregularities, which are a random load as self-excitation of the train-track vibration system;
(4) Parametric excitations, which are caused by the variation of dynamic stiffness of the track due to periodic discrete supporting of sleepers.

In recent years, with the increase of train speed, the propagation of railway-induced vibration waves in soil layers have been more concerned. The vibration of foundation soil under moving loads is studied in this section, considering two kinds of vibration sources: moving axle loads and moving harmonic loads.

3.1. Wave Propagation and Attenuation in Soil

During the propagation of vibration waves, there exist two kinds of energy loss or damping. One is the geometric damping or radiation damping due to the continuous outward diffusing of the wave itself, whose feature is that the energy density (or displacement amplitude) of the wave decreases with the distance to the source. The other kind of energy loss comes from material damping or viscous damping, which is due to the energy absorbing of the material in soil medium during the propagation (such as the friction and viscous effects when soil article moves). Therefore, the attenuation of vibration is not only associated with the characteristics of the soil medium, but also with the frequency and characteristics of vibration source (high frequency or low frequency; transient or steady), vibration energy, body wave effect (particularly in the near-field) and other factors. When the excitation from vibration source is applied to the surface or interior of subsoil, the vibration spreads outward in the form of longitudinal wave (P-wave), transverse waves (S-wave) and surface wave (also called Rayleigh wave or R-wave). In the near-field, both the S-waves, P-waves and surface waves contribute to the vibration. In the far-field region, with the wave propagation in soil, the surface wave plays a dominant role and accounts for about 67% of the total energy, while about 26% comes from the S-waves, only 7% from P-waves. Figure 6 shows, respectively, the attenuations of P-wave, S-wave and R-wave with the distance r between source and receiver point when a vertical point (area) or a line load is applied to the surface of the elastic half-space soil, in which P. M. means particle motion.

3.2. Wave Equations in Infinite Medium

In studying wave propagation in isotropic elastic medium, it is more convenient to use the differential equation to describe a wave problem. If the influence of body-force is neglected, the differential equation of wave can be expressed as:

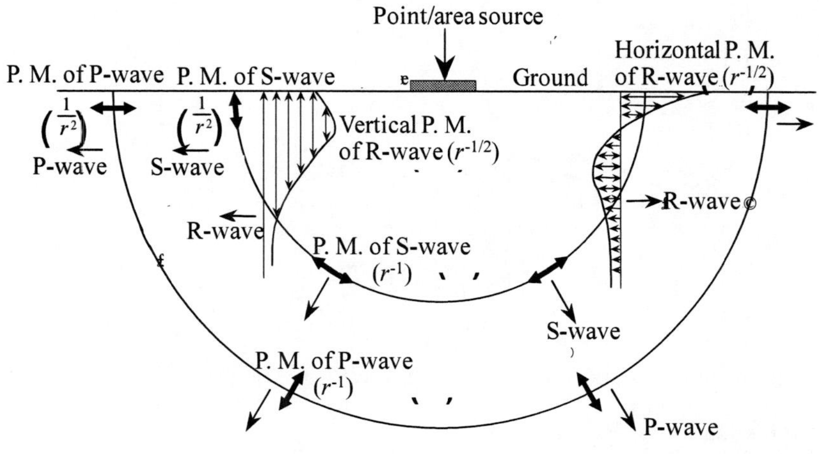

(a) Point source or area source

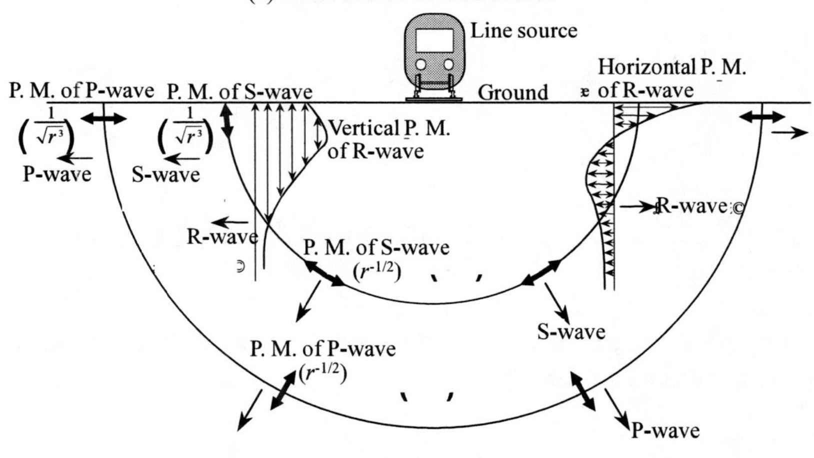

(b) Line source

Figure 6. Wave transmissions on ground surface and in elastic half-space soil.

$$(\lambda + G)\nabla\theta + G\nabla^2\mathbf{U} = \rho\frac{\partial^2\mathbf{U}}{\partial t^2} \tag{10}$$

in which $\theta = \frac{\partial u}{\partial x} + \frac{\partial v}{\partial y} + \frac{\partial w}{\partial z}$; ∇ is the Hamilton operator; ∇^2 is the Laplace operator.

Obviously, the displacement $\mathbf{U} = (u, v, w)$ at any point in elastic medium should satisfy Equation (10). The displacement forms of P-wave and S-wave in soil are shown in Figure 7.

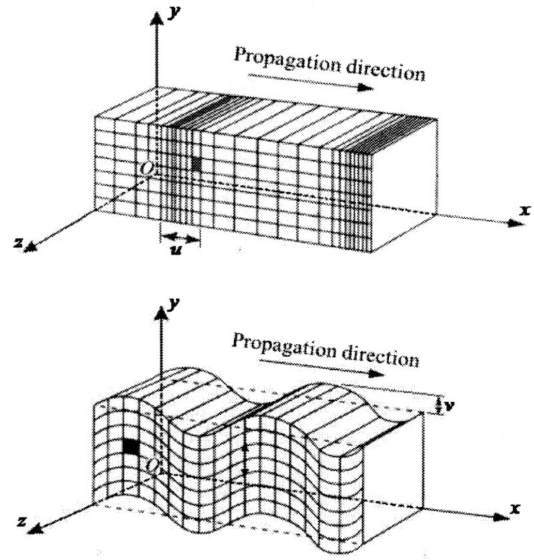

Figure 7. Displacement forms of P-wave (left) and S-wave (right) in medium.

(1) Motion equation of P-wave

It is seen from Figure 7 (left) that the P-wave has the following form of displacement:

$$u=(x,t)\ ;\ v=0;\ w=0 \tag{11}$$

And thus

$$\theta=\nabla\cdot\mathbf{U}=\frac{\partial u}{\partial x}\ ;\ \frac{\partial\theta}{\partial x}=\frac{\partial^2 u}{\partial x^2}\ ;\ \frac{\partial\theta}{\partial y}=\frac{\partial\theta}{\partial z}=0\ ;\ \nabla^2 u=\frac{\partial^2 u}{\partial x^2}\ ;\ \nabla^2 v=\nabla^2 w=0 \tag{12}$$

Substituting the above expressions into Equation (11), the displacement equation along x-direction of P-wave can be written as:

$$(\lambda+2G)\frac{\partial^2 u}{\partial x^2}=\rho\frac{\partial^2 u}{\partial t^2}\ \text{ or }\ \frac{\partial^2 u}{\partial x^2}=c_{\mathrm{p}}^2\frac{\partial^2 u}{\partial t^2} \tag{13}$$

in which $c_{\mathrm{p}}=\sqrt{(\lambda+2G)/\rho}=\sqrt{E(1-\nu)/[(1+\nu)(1-2\nu)\rho]}$ is the P-wave velocity, which is only related to the elastic parameters and density of soil medium.

(2) Motion equation of S-wave

Seen from Figure 7 (right), the displacement form of S-wave is as follows:

$$u=0;\ v=v(x,t)\ ;\ w=0 \tag{14}$$

By substituting Equations (14) and (12) into (10), the displacement equation along y-direction becomes:

$$G\frac{\partial^2 u}{\partial x^2} = \rho\frac{\partial^2 u}{\partial t^2} \text{ or } \frac{\partial^2 u}{\partial x^2} = c_s^2\frac{\partial^2 u}{\partial t^2} \tag{15}$$

in which $c_s = \sqrt{G/\rho} = \sqrt{E/[2(1+\nu)\rho]}$ is the S-wave velocity, which is also related to elastic parameters and density of soil medium. It is obvious that the velocity c_s of S-wave is always less than the propagation velocity c_p of P-wave.

(3) Motion equation of R-wave in half-infinite medium

If the medium is an elastic half-infinite body with free surface, other types of waves may also produce in the thin layer near the surface, just like those generated by throwing a stone into the water, the so-called surface waves, in which R-wave is mostly used. The R-wave attenuates rapidly with the depth in soil, while slowly with distance, so it behaves more dominantly in the far-field region.

Since the deformation caused by R-wave can be regarded as a plane deformation in the far-field, for simplicity the surface of the half-infinite body is taken as *xoz* plane, assuming the surface wave propagates along the positive *ox*-axis. The displacement of the plane is taken as a superposition of P-wave displacement and S-wave displacement on the surface of the half-infinite medium, and can be respectively expressed as:

$$u_1 = Ase^{-ay}\sin(pt - sx)\,;\; v_1 = -Aae^{-ay}\cos(pt - sx)\,;\; w_1 = 0 \tag{16}$$

$$u_2 = Bbe^{-by}\sin(pt - sx)\,;\; v_2 = -Bse^{-by}\cos(pt - sx)\,;\; w_2 = 0 \tag{17}$$

in which A, a, s, p, B and b are constants. In order to satisfy the feature of R-wave that the displacement decreases rapidly with the depth in soil, a and b are required to be positive real numbers.

Rewriting the $(pt-sx)$ in Equations (16) and (17) as $-s(x-c_Rt)$, where $c_R=p/s$, a and b can be expressed by substituting two equations into Equation (10) as:

$$a^2 = s^2 - \frac{\rho p^2}{\lambda+2G} = s^2\left(1 - \frac{G}{\lambda+2G}\frac{c_R^2}{c_s^2}\right);\; b^2 = s^2 - \frac{\rho p^2}{G} = s^2\left(1 - \frac{c_R^2}{c_s^2}\right) \tag{18}$$

The boundary conditions of wave propagating on subsoil surface are as follows:

$$(\sigma_y)_{y=0} = \left(\lambda\theta + 2G\frac{\partial v}{\partial y}\right)_{y=0} = 0\,;\; (\tau_{xy})_{y=0} = G\left(\frac{\partial v}{\partial x} + \frac{\partial u}{\partial y}\right)_{y=0} = 0 \tag{19}$$

By putting Equations (17) to (19) together, the following expression can be obtained:

$$\begin{cases}[(\lambda+2G)a^2 - \lambda s^2]A + 2GsB = 0\\ 2asA + (b^2+s^2)B = 0\end{cases} \tag{20}$$

Equation (20) is a linear homogeneous algebraic equations about A and B. In order to make existence of the R-wave, it is necessary that:

$$\begin{vmatrix} (\lambda+2G)a^2-\lambda s^2 & 2Gsb \\ 2as & b^2+s^2 \end{vmatrix}=0 \tag{21}$$

By substituting $\lambda=\dfrac{E\nu}{(1+\nu)(1-2\nu)}$, $G=\dfrac{E}{2(1+\nu)}$ and the expressions of a and b into Equation (21), the following equation is acquired:

$$\alpha^6-8\alpha^4+8(\frac{2-\nu}{1-\nu})\alpha^2-\frac{8}{1-\nu}=0 \tag{22}$$

in which $\alpha=c_R/c_s$.

3.3. Analytical Model of Harmonic Line Source on Elastic Half-space

Gutowski [13] indicated that when the distance between source and receiving point A is less than $1/\pi$ times of the train length, or is big enough to be regarded as a far-field, the train source can be approximately simulated as a harmonic line load $Q_0e^{i\omega t}$, in which Q_0 is the magnitude of line load (N/m), and ω is the frequency of loading (rad/s), as shown in Figure 8.

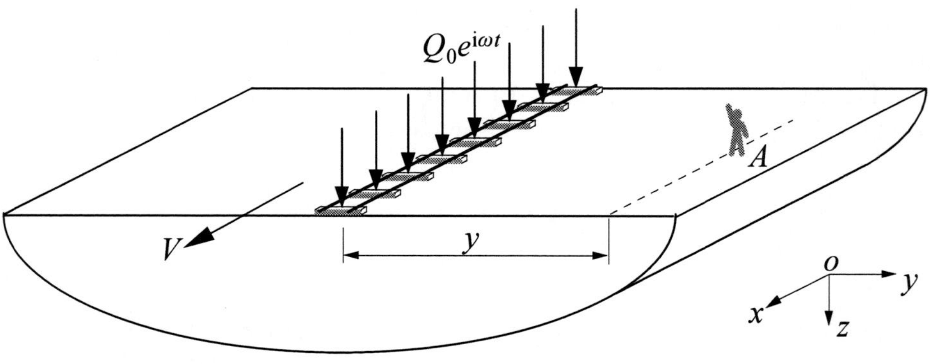

Figure 8. Analytical model of harmonic line source on elastic half-space.

In the analytical model, the foundation soil is simulated as homogeneous and isotropic elastic half-space, and the soil vibration is solved on the basis of plane strain. As for the vertical yoz plane, the horizontal displacement and vertical displacement on the surface of elastic half-space are respectively expressed by u_y and u_z. According to Equation (10), the wave equation of the vibration system can be written as:

$$\rho\frac{\partial^2 u_y}{\partial t^2}=(\lambda+G)\frac{\partial\Delta}{\partial y}+G\nabla^2 u_y; \qquad \rho\frac{\partial^2 u_z}{\partial t^2}=(\lambda+G)\frac{\partial\Delta}{\partial z}+G\nabla^2 u_z \tag{23}$$

where $\Delta = \frac{\partial u_y}{\partial y} + \frac{\partial u_z}{\partial z}$ is the strain of soil; $\nabla^2 = \frac{\partial^2}{\partial y^2} + \frac{\partial^2}{\partial z^2}$ is the Laplace operator; $\lambda = \frac{\nu E}{(1-\nu)(1-2\nu)}$ is the Lamé constant; E, G, ν and ρ are respectively elastic modulus, shear modulus, Poisson ratio and material density.

$$u_y = \frac{iQ_0 e^{i\omega t}}{2\pi G} \int_{-\infty}^{+\infty} \frac{\xi(2\xi^2 - k_s^2 - 2\alpha\beta)e^{i\xi y}}{F(\xi)} d\xi; \qquad u_z = \frac{Q_0 e^{i\omega t}}{2\pi G} \int_{-\infty}^{+\infty} \frac{k_s^2 \alpha e^{i\xi y}}{F(\xi)} d\xi \tag{24}$$

where ξ indicates wavenumber along y-direction, $\alpha^2 = \xi^2 - k_p^2$ is wavenumber of P-wave, with $k_p = \omega / c_p$, and c_p is P-wave velocity; $\beta^2 = \xi^2 - k_s^2$ is wavenumber of S-wave, with $k_s = \omega / c_s$, and c_s is S-wave velocity; $F(\xi) = (2\xi^2 - k_s^2)^2 - 4\alpha\beta\xi^2$ is called Rayleigh function, and $F(\xi) = 0$ is well known as the Rayleigh equation.

Equation (24) has the indefinite integral solution. The horizontal displacement and the vertical displacement on the surface of elastic half-space can be derived as follows:

$$u_y = \frac{Q_0}{G} He^{i(\omega t - k_R y)} e^{-\beta_c k_R y} + i\frac{Q_0}{G}\sqrt{\frac{2}{\pi}} \cdot k_p^3 (1 - \frac{k_p^2}{k_s^2})^{1/2} (1 - 2\frac{k_p^2}{k_s^2})^{-3} (k_p y)^{-3/2} e^{i(\omega t - k_s y - \pi/4)} e^{-\beta_c k_p y}$$

$$-\frac{Q_0}{G}\sqrt{\frac{2}{\pi}} \cdot (1 - \frac{k_p^2}{k_s^2})^{1/2} (k_s y)^{-3/2} e^{i(\omega t - k_s y - \pi/4)} e^{-\beta_c k_s y} + O(y^{-5/2}) \tag{25}$$

$$u_z = i\frac{Q_0}{G} Ne^{i(\omega t - k_R y)} e^{-\beta_c k_R y} - i\frac{Q_0}{G}\sqrt{\frac{2}{\pi}} \cdot \frac{k_p^2}{k_s^2} (1 - 2\frac{k_p^2}{k_s^2})^{-2} (k_p y)^{-3/2} e^{i(\omega t - k_s y - \pi/4)} e^{-\beta_c k_p y}$$

$$-i\frac{2Q_0}{G}\sqrt{\frac{2}{\pi}} \cdot (1 - \frac{k_p^2}{k_s^2})(k_s y)^{-3/2} e^{i(\omega t - k_s y - \pi/4)} e^{-\beta_c k_s y} + O(y^{-5/2}) \tag{26}$$

in which the three parts on the right side correspond, respectively, to the vibrations of R-wave, P-wave and S-wave, and k_R, k_p and k_s are the corresponding wavenumbers; β_c is the viscous damping ratio; $H = \frac{k_R(2k_R^2 - k_s^2 - 2\alpha\beta)}{F'(k_R)}$; and $N = \frac{k_s^2 \alpha}{F'(k_R)}$.

If the influence of Q_0 is neglected, the case corresponds to a unit harmonic line load $e^{i\omega t}$ applied to the surface of elastic half-space. In this case, the displacement attenuation curves of P-wave, S-wave and R-wave with the distance y calculated with Equations (25) and (26) are plotted in Figure 9, in which the Poisson ratio is given by 0.33, and the viscous damping ratio β_c equals to 0.05 [14].

The wave propagating from source contains R-wave, P-wave and S-wave, and the ground displacement is the superposition of the displacements of three kinds of waves. From Equations (25) and (26) and Figure 9, it can be seen that the intensities of three kinds of waves decay exponentially. If the influence of phase difference and damping are not considered, the amplitudes of P-wave and S-wave attenuate with distance y by a law of $y^{-3/2}$, while the attenuation of R-wave with distance does not obey the same law, due to the superposition effect of R-wave produced by the infinite line source during propagation on ground surface. Near the source, the influence of P-wave and S-wave is bigger, while with the increase of distance, the R-wave becomes dominant.

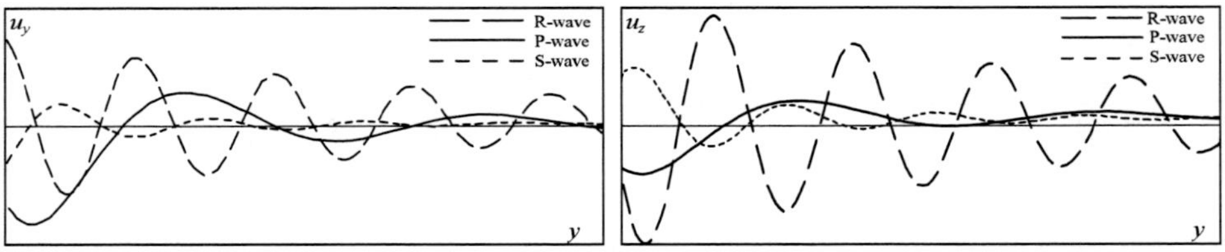

Figure 9. Wave attenuation on surface of elastic half-space subjected to unit harmonic line load.

3.4. Analytical Model of Train-track-ground System (Point Array Source)

When a train runs on the track with discrete supports (sleepers), there appears another type of excitation, called parametric excitation, whose fundamental frequency is $f_s = V/d$ (V is train speed and d is sleeper spacing). Based on the flexibility function of rail between sleepers, Wu and Thompson [15] established an equivalent track model in the time domain to study the influence of parametric excitation on the wheel-track interaction. The result shows that the wheel-track interaction force induced by parametric excitation increases with train speed, and is more obvious than that induced by track irregularity, especially in case of low frequency.

The measured data of Japanese Shinkansen showed that the ground vibration was mainly caused by the moving axle-loads of train [16]. Krylov and Ferguson [17] established an analytical model by using quasi-static axle-loads to solve the ground vibration. Jones and Block [18] investigated ground vibrations caused by freight trains and found that in low frequency range, the quasi-static excitation contributed more, while in higher frequency, the wheel-rail dynamic excitation played a major role. Later, Lombaert [19] proposed that the quasi-static forces are main loads for track deformation, while the dynamic forces are main excitations for free-field vibration. Therefore, in the analytical model of this section, rows of vertical loads (point array) are adopted, as shown in Figure 10.

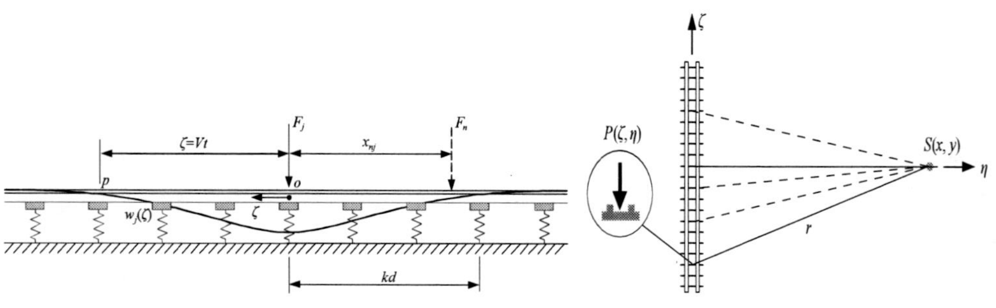

Figure 10. Analytical model of moving axle-loads on track (left) and ground vibration subjected to load array of a train (right).

In this model, the track is simulated as an Euler beam with bending stiffness EI on the Winkler elastic foundation, and the mass per unit length is $\bar{m}$. The array of vertical loads are expressed by F_1, F_2, ···, F_N (N is the number of wheel-sets in a train) with the speed of V. The position on the rail of the j-th concentrated load F_j is defined as the shifting coordinate origin o. The vertical displacement $w_j(\zeta)$ of the rail subjected to F_j can be expressed as:

$$w_j(\zeta) = \frac{F_j}{8EI\beta^3\mu} e^{-\beta\mu|\zeta|}[\cos(\beta\gamma\zeta) + \frac{\mu}{\gamma}\sin(\beta\gamma|\zeta|)] \tag{27}$$

where $\beta = \sqrt[4]{k_s / 4EI}$, k_s is elastic coefficient of roadbed; $\zeta = Vt$ is the relative distance from the shifting coordinate origin o to the point p; $\mu = \sqrt{1-(V/c_{min})^2}$ and $\gamma = \sqrt{1+(V/c_{min})^2}$ are two parameters related to load speed, and when the train speed is much lower than the minimum velocity of free bending vibration of track $c_{min} = \sqrt[4]{4k_s EI/\bar{m}^2}$, μ and γ are approximately equal to 1.

The force between sleeper and roadbed at p point can be derived from Equation (27):

$$P_j(t) = F_j \left[\frac{2w(Vt)}{w_{max}^{st}}\right]\left(\frac{d}{\Omega_0}\right) \tag{28}$$

in which $w_{max}^{st} = F_j/(8EI\beta^3\mu)$ is the maximum static deflection of w; t is time; $\Omega_0 = \pi/\beta$ is the wavelength of static deflection of the track.

The Fourier transform about t is carried to Equation (28), we obtain:

$$P_j(\omega) = \frac{F_j d}{\pi\Omega_0}\left[\frac{(\mu+\gamma)\beta V + \omega}{(\mu\beta V)^2 + (\gamma\beta V + \omega)^2} + \frac{(\mu+\gamma)\beta V - \omega}{(\mu\beta V)^2 + (\gamma\beta V - \omega)^2}\right] \tag{29}$$

Setting the distance from the n-th wheel to the j-th wheel as x_{nj}, the distance from the n-th wheel to the point p can be expressed by $\zeta+x_{nj}$. The impact of the whole load array on the point p of the track is the superposition of the actions of all wheels, that is

$$P(\zeta,\omega) = \sum_{n=0}^{N-1} P_j(\omega) e^{i\omega\left(\frac{\zeta+x_{nj}}{V}\right)} \tag{30}$$

According to the superposition principle, the reaction at the point p in frequency domain when the train runs through all the sleepers can be expressed as:

$$\tilde{P}(\zeta,\eta,\omega) = \sum_{k=-\infty}^{\infty}\sum_{n=0}^{N-1} P_j(\omega)\delta(\zeta - Vt - kd)\delta(\eta) e^{i\omega(\zeta+x_{nj})/V} \tag{31}$$

As a result, for any point $S(x, y, 0)$ on ground surface shown in Figure 10 (right), the accelerations subjected to a single force $\tilde{P}(\zeta,\eta,\omega)$ in three directions can be described in the following form:

$$\begin{cases} a_{\mathrm{x}}(x,y,\omega)=\int_{-\infty}^{+\infty}\int_{-\infty}^{+\infty}\tilde{P}(\zeta,\eta,\omega)G_{\mathrm{x}}(r,\omega)\mathrm{d}\zeta\mathrm{d}\eta \\ a_{\mathrm{y}}(x,y,\omega)=\int_{-\infty}^{+\infty}\int_{-\infty}^{+\infty}\tilde{P}(\zeta,\eta,\omega)G_{\mathrm{y}}(r,\omega)\mathrm{d}\zeta\mathrm{d}\eta \\ a_{\mathrm{z}}(x,y,\omega)=\int_{-\infty}^{+\infty}\int_{-\infty}^{+\infty}\tilde{P}(\zeta,\eta,\omega)G_{\mathrm{z}}(r,\omega)\mathrm{d}\zeta\mathrm{d}\eta \end{cases} \tag{32}$$

where $G_{\mathrm{x}}(r,\omega)$, $G_{\mathrm{y}}(r,\omega)$ and $G_{\mathrm{z}}(r,\omega)$ are respectively Green's functions of ground vibration along *x*, *y* and *z* directions, meaning the acceleration of $S(x, y, 0)$ induced by unit impulsive load with frequency ω applied to the shifting coordinate origin *o*; the distance from the point *p* to *S* is $r=\sqrt{(x-\zeta)^2+(y-\eta)^2}$.

Finally, the ground vibration induced by the array of loads generated from a running train can be obtained by superposition of those subjected to all axle-loads.

4. Green's Functions of Ground Surface Displacements

The Green's functions of ground surface displacement, which will be solved in this section, refers to the vibration displacement of any point on the surface generated by a unit impulsive load $F(t)=\delta(t)$ applied at the origin of coordinates.

4.1. Soil Model and Dynamic Parameters

The natural soil can be divided into three categories based on the strain value ε generated by external forces: the deformation of soil remains elastic when $\varepsilon<10^{-4}$, and it is elastic-plastic when $\varepsilon=10^{-4}-10^{-2}$, and plastic when $\varepsilon>10^{-2}$. For traffic-induced vibrations, the soil strain is generally 10^{-5} or even much smaller, so the elastic soil model is applicable.

However, due to the air and water in the pores among soil grains, which have viscous properties, the actual wave propagation in soil is very complex. As a result, in studying traffic induced ground vibrations, the atomic structure and molecular motion of soil particles and their internal pores are not considered, thus the variations of stress, deformation and displacement of soil can be described with continuous functions. The error caused by this assumption is acceptable as long as the particle size and the distance between adjacent particles are far smaller than the size of soil model. In addition, the initial stress of soil is neglected, and the stress, deformation and displacement of soil in infinity are deemed as zero. Moreover, since the actual soil is a kind of layered system, each layer of soil is assumed homogeneous and isotropic.

Base on the assumptions above, the viscoelastic layered half-space soil model is established in the cylindrical coordinate system, as shown in Figure 11, where *r*, θ and *z* represent the radial, ring and vertical coordinates, respectively.

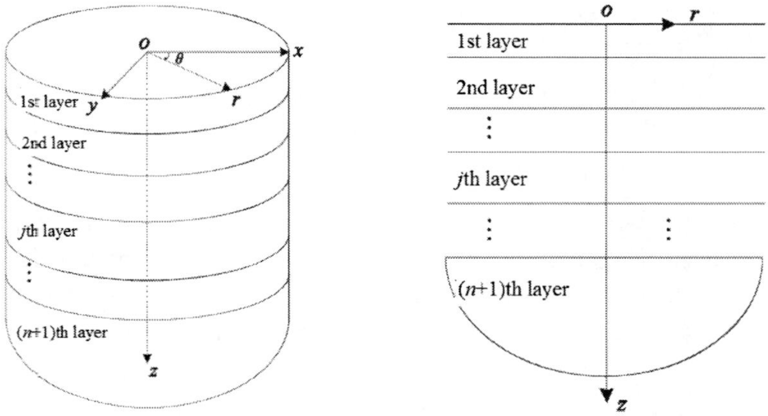

Figure 11. Viscoelastic layered half-space soil model in cylindrical coordinate.

Herein, the linear hysteretic damping theory (also called complex damping theory) [20] is adopted, by multipling the characteristic parameters of linear elastic soil with the hysteretic damping coefficients corresponding to damping ratio of soil. For example, the soil parameters under complex damping theory can be expressed as follows:

$$E^* = (1+2i\xi)E\,;\ \mu^* = (1+2i\xi)\mu\,;\ \lambda^* + \mu^* = (1+2i\xi)(\lambda+\mu) \tag{33}$$

$$c_{\mathrm{p}}^* = c_{\mathrm{p}}\sqrt{1+2i\xi} \approx (1+i\xi)c_{\mathrm{p}}\,;\ c_{\mathrm{s}}^* = c_{\mathrm{s}}\sqrt{1+2i\xi} \approx (1+i\xi)c_{\mathrm{s}} \tag{34}$$

$$k_{\mathrm{p}}^* = \frac{k_{\mathrm{p}}}{\sqrt{1+2i\xi}} \approx (1-i\xi)k_{\mathrm{p}}\,;\ k_{\mathrm{s}}^* = \frac{k_{\mathrm{s}}}{\sqrt{1+2i\xi}} \approx (1-i\xi)k_{\mathrm{s}} \tag{35}$$

where the superscript * denotes the parameters under complex damping theory; E represents the dynamic elasticity modulus, and ξ is the material damping ratio; c_{p}, c_{s}, k_{p} and k_{s} represent the velocity and wavenumber of P-wave and S-wave, respectively.

For convenient expression, the sign * for the complex parameters is omitted in the next sections.

4.2. Vibration Transmission in a Single Layer of Soil

For the j-th (j=1, 2, ···, n) layer soil with depth of h_j, the origin of z coordinate is on top surface, and its positive direction is vertically downward. Base on the elastic dynamics, the displacement and the stress of isotropic homogeneous elastic medium satisfy the following equations:

$$\frac{\partial \sigma_r}{\partial r} + \frac{\partial \tau_{zr}}{\partial z} + \frac{\sigma_r - \sigma_\theta}{r} = \rho_j \frac{\partial^2 u_r}{\partial t^2}\,;\ \frac{\partial \tau_{zr}}{\partial r} + \frac{\partial \sigma_z}{\partial z} + \frac{\tau_{zr}}{r} = \rho_j \frac{\partial^2 u_z}{\partial t^2} \tag{36}$$

$$\sigma_r = [(\lambda_j + 2\mu_j)\frac{\partial}{\partial r} + \frac{\lambda_j}{r}]u_r + \lambda_j \frac{\partial u_z}{\partial z};\ \sigma_\theta = (\lambda_j \frac{\partial}{\partial r} + \frac{\lambda_j + 2\mu_j}{r})u_r + \lambda_j \frac{\partial u_z}{\partial z};$$

$$\sigma_z = \lambda_j(\frac{\partial}{\partial r} + \frac{1}{r})u_r + (\lambda_j + 2\mu_j)\frac{\partial u_z}{\partial z};\ \tau_{zr} = \mu_j(\frac{\partial u_r}{\partial z} + \frac{\partial u_z}{\partial r}). \tag{37}$$

where the subscripts r, θ and z denote, the physical quantities in r, θ and z directions; u and σ represent displacement and stress of soil; ρ_j, λ_j and μ_j represent the density and Lamé constants of the j-th layer of soil, respectively.

Equations (36) and (37) are transformed into the frequency domain by the Fourier integral transform about time t as follows:

$$\frac{\partial \bar{\sigma}_r}{\partial r} + \frac{\partial \bar{\tau}_{zr}}{\partial z} + \frac{\bar{\sigma}_r - \bar{\sigma}_\theta}{r} = -\rho_j \omega^2 \bar{u}_r;\ \frac{\partial \bar{\tau}_{zr}}{\partial r} + \frac{\partial \bar{\sigma}_z}{\partial z} + \frac{\bar{\tau}_{zr}}{r} = -\rho_j \omega^2 \bar{u}_z \tag{38}$$

$$\bar{\sigma}_r = [(\lambda_j + 2\mu_j)\frac{\partial}{\partial r} + \frac{\lambda_j}{r}]\bar{u}_r + \lambda_j \frac{\partial \bar{u}_z}{\partial z};\ \bar{\sigma}_\theta = (\lambda_j \frac{\partial}{\partial r} + \frac{\lambda_j + 2\mu_j}{r})\bar{u}_r + \lambda_j \frac{\partial \bar{u}_z}{\partial z};$$

$$\bar{\sigma}_z = \lambda_j(\frac{\partial}{\partial r} + \frac{1}{r})\bar{u}_r + (\lambda_j + 2\mu_j)\frac{\partial \bar{u}_z}{\partial z};\ \bar{\tau}_{zr} = \mu_j(\frac{\partial \bar{u}_r}{\partial z} + \frac{\partial \bar{u}_z}{\partial r}). \tag{39}$$

Equation (39) can be changed to the following forms:

$$\frac{\partial \bar{u}_r}{\partial z} = -\frac{\partial \bar{u}_z}{\partial r} + \frac{1}{\mu_j}\bar{\tau}_{zr};\ \frac{\partial \bar{u}_z}{\partial z} = -\frac{\nu_j}{1-\nu_j}(\frac{\partial}{\partial r} + \frac{1}{r})\bar{u}_r + \frac{1}{2\mu_j}\frac{1-2\nu_j}{1-\nu_j}\bar{\sigma}_z;$$

$$\frac{\partial \bar{\sigma}_z}{\partial z} = -\rho_j \omega^2 \bar{u}_z - (\frac{\partial}{\partial r} + \frac{1}{r})\bar{\tau}_{zr}. \tag{40}$$

Combining Equations (38) and (39), the partial derivation of shear stress to z can be obtained as follows:

$$\frac{\partial \bar{\tau}_{zr}}{\partial z} = [-\rho_j \omega^2 - \frac{2\mu_j}{1-\nu_j}(\frac{\partial^2}{\partial r^2} + \frac{1}{r}\frac{\partial}{\partial r} - \frac{1}{r^2})]\bar{u}_r - \frac{\nu_j}{1-\nu_j}\frac{\partial \bar{\sigma}_z}{\partial r} \tag{41}$$

Equations (40) and (41) can be rewritten by combining as:

$$\frac{\partial}{\partial z}\begin{Bmatrix} \bar{u}_r \\ \bar{u}_z \\ \bar{\sigma}_z \\ \bar{\tau}_{zr} \end{Bmatrix} = \begin{bmatrix} 0 & -\frac{\partial}{\partial r} & 0 & \frac{1}{\mu_j} \\ a_1 & 0 & a_2 & 0 \\ 0 & -\rho_j \omega^2 & 0 & -(\frac{\partial}{\partial r} + \frac{1}{r}) \\ a_3 & 0 & a_4 & 0 \end{bmatrix} \begin{Bmatrix} \bar{u}_r \\ \bar{u}_z \\ \bar{\sigma}_z \\ \bar{\tau}_{zr} \end{Bmatrix} \tag{42}$$

where

$$a_1 = -\frac{\nu_j}{1-\nu_j}(\frac{\partial}{\partial r}+\frac{1}{r}), \qquad a_2 = \frac{1}{\mu_j}\frac{1-2\nu_j}{2(1-\nu_j)}, a_3 = -\rho_j\omega^2 - \frac{2\mu_j}{1-\nu_j}(\frac{\partial^2}{\partial r^2}+\frac{1}{r}\frac{\partial}{\partial r}-\frac{1}{r^2}),$$

$a_4 = -\frac{\nu_j}{1-\nu_j}\frac{\partial}{\partial r}$, and ν_j represents the Poisson ratio of the j-th layer of soil.

Then, the Hankel integral transform about radial coordinate r is carried to Equation (42), and the problem is transformed into the wave-number domain of k_r:

$$\frac{\mathrm{d}\{\tilde{\bar{\mathbf{Z}}}\}_j}{\mathrm{d}z} = \mathbf{B}_j \cdot \{\tilde{\bar{\mathbf{Z}}}\}_j \tag{43}$$

where the vector $\{\tilde{\bar{\mathbf{Z}}}\}_j = [\tilde{\bar{u}}_r(k_r,z,\omega), \tilde{\bar{u}}_z(k_r,z,\omega), \tilde{\bar{\sigma}}_z(k_r,z,\omega), \tilde{\bar{\tau}}_{zr}(k_r,z,\omega)]_j^{\mathrm{T}}$ represents the state vector of soil at the depth z in the j-th layer of soil.

$$\mathbf{B}_j = \begin{bmatrix} 0 & k_r & 0 & \frac{1}{\mu_j} \\ -\frac{\nu_j k_r}{1-\nu_j} & 0 & \frac{1}{\mu_j}\frac{1-2\nu_j}{2(1-\nu_j)} & 0 \\ 0 & -\rho_j\omega^2 & 0 & -k_r \\ \frac{2\mu_j k_r^2}{1-\nu_j} - \rho_j\omega^2 & 0 & \frac{\nu_j k_r}{1-\nu_j} & 0 \end{bmatrix}$$

Obviously, Equation (43) is the simple matrix differential equation with one unknown and the expression of its solution is:

$$\{\tilde{\bar{\mathbf{Z}}}\}_j = e^{\mathbf{B}_j z} \cdot \{\tilde{\bar{\mathbf{Z}}}\}_{j0} \tag{44}$$

where $\{\tilde{\bar{\mathbf{Z}}}\}_{j0} = [\{\tilde{\bar{u}}_r(k_r,0,\omega), \tilde{\bar{u}}_z(k_r,0,\omega), \tilde{\bar{\sigma}}_z(k_r,0,\omega), \tilde{\bar{\tau}}_{zr}(k_r,0,\omega)]_{j0}^{\mathrm{T}}$ represents the state vector on the surface (z=0) of the j-th layer of soil, and the subscript "$j0$" denotes top surface of the j-th layer.

When $z = h_j$, the relationship of vibration transmission from top surface to bottom surface of the j-th layer soil can be obtained:

$$\{\tilde{\bar{\mathbf{Z}}}\}_{j1} = \mathbf{Q}_j \cdot \{\tilde{\bar{\mathbf{Z}}}\}_{j0} \tag{45}$$

where the subscript "$j1$" denotes bottom surface of the j-th layer; $\mathbf{Q}_j = \mathrm{e}^{\mathbf{B}_j h_j}$ is the transmission matrix (dimension of 4×4) of vibrations between the top and bottom surfaces of the single soil layer.

4.3. Relationship of Displacement and Stress on Surface of Layered-soil with Rigid Foundation

The continuity of displacement and the equilibrium condition of stress on interface can be expressed as $\{\tilde{\bar{\mathbf{Z}}}\}_{11}=\{\tilde{\bar{\mathbf{Z}}}\}_{20}$, $\{\tilde{\bar{\mathbf{Z}}}\}_{21}=\{\tilde{\bar{\mathbf{Z}}}\}_{30}$, $\{\tilde{\bar{\mathbf{Z}}}\}_{31}=\{\tilde{\bar{\mathbf{Z}}}\}_{40}$, ... , and $\{\tilde{\bar{\mathbf{Z}}}\}_{n-1,1}=\{\tilde{\bar{\mathbf{Z}}}\}_{n,0}$, respectively. According to Equation (45), the following relationship can be drawn:

$$\{\tilde{\bar{\mathbf{Z}}}\}_{n,1}=\mathbf{Q}_n\cdot\{\tilde{\bar{\mathbf{Z}}}\}_{n,0}=\mathbf{Q}_n\cdot\{\tilde{\bar{\mathbf{Z}}}\}_{n-1,1}=\mathbf{Q}_n\cdot\mathbf{Q}_{n-1}\cdot\{\tilde{\bar{\mathbf{Z}}}\}_{n-1,0}=\mathbf{Q}_n\cdot\mathbf{Q}_{n-1}\cdots\mathbf{Q}_j\cdots\mathbf{Q}_1\cdot\{\tilde{\bar{\mathbf{Z}}}\}_{10} \quad (46)$$

Since the matrix $\mathbf{Q}_j$ is known, the product on right side of Equation (46) can be calculated. Let $\mathbf{T}=\mathbf{Q}_n\cdot\mathbf{Q}_{n-1}\cdots\mathbf{Q}_j\cdots\mathbf{Q}_1$, Equation (46) can be simplified as follows:

$$\{\tilde{\bar{\mathbf{Z}}}\}_{n,1}=\mathbf{T}\cdot\{\tilde{\bar{\mathbf{Z}}}\}_{10} \quad (47)$$

in which the size of matrix **T** is also 4×4 because the matrix **Q** has the dimension of 4×4.

Therefore, the transmission matrix of vibrations between the top surface and the bottom surface of all the *n* layers of soil is as follows:

$$\mathbf{T}=\exp\left(\sum_{j=1}^{n}\mathbf{A}_j h_j\right)=\begin{bmatrix} T_{11} & T_{12} & T_{13} & T_{14} \\ T_{21} & T_{22} & T_{23} & T_{24} \\ T_{31} & T_{32} & T_{33} & T_{34} \\ T_{41} & T_{42} & T_{43} & T_{44} \end{bmatrix} \quad (48)$$

For the bottom rigid foundation, i.e. the (n+1)-th layer of soil model, the displacement equals to zero, that is $\{\tilde{\bar{u}}_r \quad \tilde{\bar{u}}_z\}^{\mathrm{T}}{}_{n+1,0}=\{0 \quad 0\}^{\mathrm{T}}$ in the vector of $\{\tilde{\bar{\mathbf{Z}}}\}_{n+1,0}$. Meanwhile, the displacements between adjacent layers satisfy the continuity condition of $\{\tilde{\bar{u}}_r \quad \tilde{\bar{u}}_z\}^{\mathrm{T}}{}_{n+1,0}=\{\tilde{\bar{u}}_r \quad \tilde{\bar{u}}_z\}^{\mathrm{T}}{}_{n,1}$, hence $\{\tilde{\bar{u}}_r \quad \tilde{\bar{u}}_z\}^{\mathrm{T}}{}_{n,1}=\{0 \quad 0\}^{\mathrm{T}}$. Then Equation (47) can be written as:

$$\begin{bmatrix} T_{11} & T_{12} & T_{13} & T_{14} \\ T_{21} & T_{22} & T_{23} & T_{24} \\ T_{31} & T_{32} & T_{33} & T_{34} \\ T_{41} & T_{42} & T_{43} & T_{44} \end{bmatrix}\begin{Bmatrix} \tilde{\bar{u}}_r(k_r,0,\omega) \\ \tilde{\bar{u}}_z(k_r,0,\omega) \\ \tilde{\bar{\sigma}}_z(k_r,0,\omega) \\ \tilde{\bar{\tau}}_{zr}(k_r,0,\omega) \end{Bmatrix}_{10}=\begin{Bmatrix} 0 \\ 0 \\ \tilde{\bar{\sigma}}_z(k_r,z,\omega) \\ \tilde{\bar{\tau}}_{zr}(k_r,z,\omega) \end{Bmatrix}_{n1} \quad (49)$$

And thus,

$$\begin{bmatrix} T_{11} & T_{12} \\ T_{21} & T_{22} \end{bmatrix}\begin{Bmatrix} \tilde{\bar{u}}_r(k_r,0,\omega) \\ \tilde{\bar{u}}_z(k_r,0,\omega) \end{Bmatrix}_{10}+\begin{bmatrix} T_{13} & T_{14} \\ T_{23} & T_{24} \end{bmatrix}\begin{Bmatrix} \tilde{\bar{\sigma}}_z(k_r,0,\omega) \\ \tilde{\bar{\tau}}_{zr}(k_r,0,\omega) \end{Bmatrix}_{10}=\mathbf{0} \quad (50)$$

or

$$\begin{Bmatrix} \tilde{\bar{u}}_r(k_r,0,\omega) \\ \tilde{\bar{u}}_z(k_r,0,\omega) \end{Bmatrix}_{10} = -\mathbf{S} \begin{Bmatrix} \tilde{\bar{\sigma}}_z(k_r,0,\omega) \\ \tilde{\bar{\tau}}_{zr}(k_r,0,\omega) \end{Bmatrix}_{10} \tag{51}$$

where the matrix $\mathbf{S} = \begin{bmatrix} T_{11} & T_{12} \\ T_{21} & T_{22} \end{bmatrix}^{-1} \begin{bmatrix} T_{13} & T_{14} \\ T_{23} & T_{24} \end{bmatrix} = \begin{bmatrix} S_{11} & S_{12} \\ S_{21} & S_{22} \end{bmatrix}$ represents the transmission matrix of displacement and stress on the free surface of layered-soil with rigid foundation.

4.4. Relationship of Displacement and Stress on Surface of Half-space Soil

The local coordinate system in half-space is adopted in this section, in which the coordinate origin is on the top surface of half-space. The density, Lamé constants and Poisson ratio of the half-space soil are represented by ρ_{n+1}, λ_{n+1}, μ_{n+1} and ν_{n+1}, respectively.

Based on the Helmholtz theorem, the displacement vector in the cylindrical coordinate system (the case of axial symmetry) can be decomposed as follows:

$$u = \nabla\Phi + \nabla\times\Psi \tag{52}$$

where Φ and Ψ represent, respectively, scalar potential and vector potential, which meet $\nabla\times\nabla\Phi = 0$ and $\nabla\cdot\nabla\times\Psi = 0$. The specific expressions of Equation (52) can be rewritten as:

$$u_r = \frac{\partial\Phi}{\partial r} + \frac{\partial^2\Psi}{\partial z\partial r}; \qquad u_z = \frac{\partial\Phi}{\partial z} + \frac{\partial^2\Psi}{\partial z^2} - \nabla^2\Psi \tag{53}$$

Substituting Equation (53) into (39), the stress component in the z direction can be expressed in the form of potential functions as:

$$\begin{aligned} \sigma_z &= \lambda_{n+1}\nabla^2\Phi + 2\mu_{n+1}\frac{\partial}{\partial z}\left(\frac{\partial\Phi}{\partial z} + \frac{\partial^2\Psi}{\partial z^2} - \nabla^2\Psi\right) \\ \tau_{zr} &= \mu_{n+1}\frac{\partial}{\partial r}\left(2\frac{\partial\Phi}{\partial z} + 2\frac{\partial^2\Psi}{\partial z^2} - \nabla^2\Psi\right) \end{aligned} \tag{54}$$

According to the wave propagation in soil, two partial differential equations of displacement potential functions can be easily deduced as:

$$\nabla^2\Phi = \ddot{\Phi}/c_{\mathrm{p},n+1}^2; \qquad \nabla^2\Psi = \ddot{\Psi}/c_{\mathrm{s},n+1}^2 \tag{55}$$

where $c_{\mathrm{p},n+1}=\sqrt{(\lambda_{n+1}+2\mu_{n+1})/\rho_{n+1}}$ and $c_{\mathrm{s},,n+1}=\sqrt{\mu_{n+1}/\rho_{n+1}}$ represent P-wave speed and S-wave speed, respectively.

By applying the Fourier integral transform about t and the Hankel integral transform about r to Equation (55), the problem can be transformed from the time-space domain to the frequency-wavenumber domain as follows:

$$\frac{\mathrm{d}^2\bar{\tilde{\Phi}}(k_r,z,\omega)}{\mathrm{d}z^2}=\alpha_{n+1}^2\bar{\tilde{\Phi}};\qquad \frac{\mathrm{d}^2\bar{\tilde{\Psi}}(k_r,z,\omega)}{\mathrm{d}z^2}=\beta_{n+1}^2\bar{\tilde{\Psi}} \tag{56}$$

where $\alpha_{n+1}^2=k_r^2-k_{\mathrm{p},n+1}^2$; $\beta_{n+1}^2=k_r^2-k_{\mathrm{s},n+1}^2$; k_r represents the wavenumber of half-space vibration along the r direction; $k_{\mathrm{p},n+1}=\omega/c_{\mathrm{p},n+1}$ and $k_{\mathrm{s},n+1}=\omega/c_{\mathrm{s},n+1}$ represent the wavenumbers of P-wave and S-wave, respectively.

The solution of Equation (56) has the form:

$$\bar{\tilde{\Phi}}(k_r,z,\omega)=\varphi(k_r,\omega)\mathrm{e}^{-\alpha z};\qquad \bar{\tilde{\Psi}}(k_r,z,\omega)=\phi(k_r,\omega)\mathrm{e}^{-\beta z} \tag{57}$$

By applying the Fourier integral transform about t and the Hankel integral transform bout r to Equations (53) and (54), we obtain:

$$\bar{\tilde{u}}_r=-k_r\bar{\tilde{\Phi}}-k_r\frac{\partial\bar{\tilde{\Psi}}}{\partial z};\qquad \bar{\tilde{u}}_z=\frac{\partial\bar{\tilde{\Phi}}}{\partial z}+k_r^2\bar{\tilde{\Psi}} \tag{58}$$

$$\begin{aligned}\bar{\tilde{\sigma}}_z&=\lambda_{n+1}(-k_r^2\bar{\tilde{\Phi}}+\frac{\partial^2\bar{\tilde{\Phi}}}{\partial z^2})+2\mu_{n+1}\frac{\partial}{\partial z}(\frac{\partial\bar{\tilde{\Phi}}}{\partial z}+k_r^2\bar{\tilde{\Psi}})\\ \bar{\tilde{\tau}}_{zr}&=\mu_{n+1}(-2k_r\frac{\partial\bar{\tilde{\Phi}}}{\partial z}-2k_r\frac{\partial^2\bar{\tilde{\Psi}}}{\partial z^2}-k_{\mathrm{s},n+1}^2k_r\bar{\tilde{\Psi}})\end{aligned} \tag{59}$$

Substituting Equation (57) into (58) and (59) and let z=0, and the relationship of displacement and stress on the surface of half-space in the wavenumber-frequency domain can be obtained as follows:

$$\{\bar{\tilde{\mathbf{Z}}}\}_{n+1,0}=\{\mathbf{L}\} \tag{60}$$

in which $\{\bar{\tilde{\mathbf{Z}}}\}_{n+1,0}=\{\bar{\tilde{u}}_r\quad \bar{\tilde{u}}_z\quad \bar{\tilde{\sigma}}_z\quad \bar{\tilde{\tau}}_{zr}\}_{n+1,0}^{\mathrm{T}}$; $\{\mathbf{L}\}=\begin{Bmatrix}-k_r\varphi+k_r\beta_{n+1}\phi\\ -\alpha_{n+1}\varphi+k_r^2\phi\\ \mu_{n+1}(2k_r^2-k_{\mathrm{s},n+1}^2)\varphi-2\mu_{n+1}k_r^2\beta_{n+1}\phi\\ 2\mu_{n+1}k_r\alpha_{n+1}\varphi-\mu_{n+1}(2k_r^2-k_{\mathrm{s},n+1}^2)k_r\phi\end{Bmatrix}$.

4.5. Relationship of Displacement and Stress on Surface of Layered Half-space Soil

Since the transmission relationship of vibrations for layered-soil model has been known as $\{\tilde{\tilde{\mathbf{Z}}}\}_{n,1} = \mathbf{T} \cdot \{\tilde{\tilde{\mathbf{Z}}}\}_{10}$, and $\{\tilde{\tilde{\mathbf{Z}}}\}_{n,1} = \{\tilde{\tilde{\mathbf{Z}}}\}_{n+1,0}$ according to the continuity between adjacent soil layers, $\{\tilde{\tilde{\mathbf{Z}}}\}_{n+1,0} = \mathbf{T} \cdot \{\tilde{\tilde{\mathbf{Z}}}\}_{10}$ can be obtained, i.e.

$$\begin{Bmatrix} \tilde{\tilde{u}}_r(k_r, z, \omega) \\ \tilde{\tilde{u}}_z(k_r, z, \omega) \\ \tilde{\tilde{\sigma}}_z(k_r, z, \omega) \\ \tilde{\tilde{\tau}}_{zr}(k_r, z, \omega) \end{Bmatrix}_{n+1,0} = \mathbf{T} \cdot \begin{Bmatrix} \tilde{\tilde{u}}_r(k_r, 0, \omega) \\ \tilde{\tilde{u}}_z(k_r, 0, \omega) \\ \tilde{\tilde{\sigma}}_z(k_r, 0, \omega) \\ \tilde{\tilde{\tau}}_{zr}(k_r, 0, \omega) \end{Bmatrix}_{10} \tag{61}$$

By combining Equations (60) and (61), the relationship of displacement and stress on the top surface of layered half-space soil can be obtained:

$$\begin{bmatrix} T_{11} & T_{12} & T_{13} & T_{14} \\ T_{21} & T_{22} & T_{23} & T_{24} \\ T_{31} & T_{32} & T_{33} & T_{34} \\ T_{41} & T_{42} & T_{43} & T_{44} \end{bmatrix} \begin{Bmatrix} \tilde{\tilde{u}}_r(k_r, 0, \omega) \\ \tilde{\tilde{u}}_z(k_r, 0, \omega) \\ \tilde{\tilde{\sigma}}_z(k_r, 0, \omega) \\ \tilde{\tilde{\tau}}_{zr}(k_r, 0, \omega) \end{Bmatrix}_{10} = \{\mathbf{L}\} \tag{62}$$

4.6. Green's Functions of Ground Displacements in Wavenumber-Frequency Domain

Since the exciting force corresponding to the Green's functions of ground displacements is a unit impulse load $F(t)=\delta(t)$ vertically acting to the coordinate origin, the boundary conditions of stress are as follows:

$$\sigma_z(r,0,t) = \lim_{\varepsilon \to 0} \frac{\delta(t)}{\pi \varepsilon^2}; \qquad \tau_{zr}(r,0,t) = 0 \tag{63}$$

By applying the Fourier integral transform about t and the Hankel integral transform about r to Equation (63), we have:

$$\tilde{\tilde{\sigma}}_z(k_r, 0, \omega) = 1/2\pi; \qquad \tilde{\tilde{\tau}}_{zr}(k_r, 0, \omega) = 0 \tag{64}$$

Substituting Equation (64) into (51), (60) and (62), the Green's functions of ground displacements in the soil models of layered rigid foundation, half-space, and layered half-space can be obtained as:

$$\begin{Bmatrix} \tilde{\bar{G}}_{zr}(k_r,0,\omega) \\ \tilde{\bar{G}}_{zz}(k_r,0,\omega) \\ \varphi(k_r,\omega) \\ \phi(k_r,\omega) \end{Bmatrix} = -\frac{1}{2\pi}\begin{bmatrix} \mathbf{S}_{11} & \mathbf{S}_{12} \\ \mathbf{S}_{21} & \mathbf{S}_{22} \end{bmatrix}^{-1} \begin{Bmatrix} T_{13} \\ T_{23} \\ T_{33} \\ T_{43} \end{Bmatrix} \tag{65}$$

where $\mathbf{S}_{11} = \begin{bmatrix} T_{11} & T_{12} \\ T_{21} & T_{22} \end{bmatrix}$; $\mathbf{S}_{21} = \begin{bmatrix} T_{31} & T_{32} \\ T_{41} & T_{42} \end{bmatrix}$; $\mathbf{S}_{12} = \begin{bmatrix} k_r & -k_r\beta_{n+1} \\ \alpha_{n+1} & -k_r^2 \end{bmatrix}$;

$$\mathbf{S}_{22} = \begin{bmatrix} -\mu_{n+1}(2k_r^2 - k_{s,n+1}^2) & 2\mu_{n+1}k_r^2\beta_{n+1} \\ -2\mu_{n+1}k_r\alpha_{n+1} & \mu_{n+1}(2k_r^2 - k_{s,n+1}^2)k_r \end{bmatrix}.$$

4.7. Green's Functions of Ground Displacements in Time-Space Domain

The displacements in the Cartesian coordinate system and those in cylindrical coordinate system have the following relationships:

$$k_r^2 = k_x^2 + k_y^2 \tag{66a}$$

$$\tilde{\bar{G}}_{zx}(k_x,k_y,0,\omega) = 2\pi \cdot i \cdot \tilde{\bar{G}}_{zr}(\sqrt{k_x^2+k_y^2},0,\omega)\frac{k_x}{\sqrt{k_x^2+k_y^2}} \tag{66b}$$

$$\tilde{\bar{G}}_{zy}(k_x,k_y,0,\omega) = 2\pi \cdot i \cdot \tilde{\bar{G}}_{zr}(\sqrt{k_x^2+k_y^2},0,\omega)\frac{k_y}{\sqrt{k_x^2+k_y^2}} \tag{66c}$$

$$\tilde{\bar{G}}_{zz}(k_x,k_y,0,\omega) = 2\pi\tilde{\bar{G}}_{zz}(\sqrt{k_x^2+k_y^2},0,\omega) \tag{66d}$$

In the cylindrical coordinate system, the Green's functions of ground displacements in the space domain about r can be obtained by the inverse Hankel integral transform about k_r as follows:

$$\bar{G}_{zr}(r,0,\omega) = \int_0^\infty k_r\tilde{\bar{G}}_{zr}(k_r,0,\omega)J_1(k_r r)\mathrm{d}k_r;\quad \bar{G}_{zz}(r,0,\omega) = \int_0^\infty k_r\tilde{\bar{G}}_{zz}(k_r,0,\omega)J_0(k_r r)\mathrm{d}k_r \tag{67}$$

In the Cartesian coordinate system, the Green's functions of ground displacements in the space domain about x and y can be obtained by the inverse Fourier integral transform about k_x and k_y as follows:

$$\bar{G}_{zx}(x,y,0,\omega)=\frac{1}{(2\pi)^2}\int_{-\infty}^{\infty}\int_{-\infty}^{+\infty}\tilde{\bar{G}}_{zx}(k_x,k_y,0,\omega)e^{-ik_x x}e^{-ik_y y}dk_x dk_y$$

$$\bar{G}_{zy}(x,y,0,\omega)=\frac{1}{(2\pi)^2}\int_{-\infty}^{\infty}\int_{-\infty}^{+\infty}\tilde{\bar{G}}_{zy}(k_x,k_y,0,\omega)e^{-ik_x x}e^{-ik_y y}dk_x dk_y \tag{68}$$

$$\bar{G}_{zz}(x,y,0,\omega)=\frac{1}{(2\pi)^2}\int_{-\infty}^{\infty}\int_{-\infty}^{+\infty}\tilde{\bar{G}}_{zz}(k_x,k_y,0,\omega)e^{-ik_x x}e^{-ik_y y}dk_x dk_y$$

If only the inverse Fourier transform about k_y is carried, by substituting Equation (66) into (68), one obtains:

$$\tilde{\bar{G}}_{zx}(k_x,y,0,\omega)=\int_{-\infty}^{+\infty}i\cdot\tilde{\bar{G}}_{zr}(\sqrt{k_x^2+k_y^2},0,\omega)\frac{k_x}{\sqrt{k_x^2+k_y^2}}e^{-ik_y y}dk_y \tag{69a}$$

$$\tilde{\bar{G}}_{zy}(k_x,y,0,\omega)=\int_{-\infty}^{+\infty}i\cdot\tilde{\bar{G}}_{zr}(\sqrt{k_x^2+k_y^2},0,\omega)\frac{k_y}{\sqrt{k_x^2+k_y^2}}e^{-ik_y y}dk_y \tag{69b}$$

$$\tilde{\bar{G}}_{zz}(k_x,y,0,\omega)=\int_{-\infty}^{+\infty}\tilde{\bar{G}}_{zz}(\sqrt{k_x^2+k_y^2},0,\omega)e^{-ik_y y}dk_y \tag{69c}$$

Based on the property of Green's functions, the integrals in Equations (69a) and (69c) are even functions about y, so the exponential item in Fourier transform can be replaced by the cosine function; while the integral in Equation (69b) is an odd function about y, so the exponential item of Fourier transform can be replaced by the sine function, i.e.

$$\tilde{\bar{G}}_{zx}(k_x,y,0,\omega)=2\int_{0}^{+\infty}i\cdot\tilde{\bar{G}}_{zr}(\sqrt{k_x^2+k_y^2},0,\omega)\frac{k_x}{\sqrt{k_x^2+k_y^2}}\cos(k_y y)dk_y \tag{70a}$$

$$\tilde{\bar{G}}_{zy}(k_x,y,0,\omega)=2\int_{0}^{+\infty}\tilde{\bar{G}}_{zr}(\sqrt{k_x^2+k_y^2},0,\omega)\frac{k_y}{\sqrt{k_x^2+k_y^2}}\sin(k_y y)dk_y \tag{70b}$$

$$\tilde{\bar{G}}_{zz}(k_x,y,0,\omega)=2\int_{0}^{+\infty}\tilde{\bar{G}}_{zz}(\sqrt{k_x^2+k_y^2},0,\omega)\cos(k_y y)dk_y \tag{70c}$$

The integral expressions in Equations (67) and (70) can be rewritten uniformly as:

$$I(r)=\int_{0}^{+\infty}F(k_r)J_n(k_r r)dk_r \tag{71a}$$

$$I(y)=\int_{0}^{+\infty}F(k_y)\cos(k_y y)dk_y \tag{71b}$$

Obviously, the integrands in Equation (71) are products of two functions with strong oscillation, respectively corresponding to the Cartesian coordinate system and the cylindrical coordinate system. Both $F(k_r)$ and $F(k_y)$ are oscillatory kernel functions related to Green's functions $\tilde{\bar{G}}_{zr}$, $\tilde{\bar{G}}_{zx}$, $\tilde{\bar{G}}_{zy}$ or $\tilde{\bar{G}}_{zz}$ of ground vibrations and lead to a very slow convergence rate. $J_n(k_r, r)$ is Bessel function, a kind of periodic attenuation function of wave propagation, and is equal to zero when k_r tends to infinity, while the trigonometric function is a bounded periodic function. All of this makes solution of the integrals very difficult. Currently, the solution mainly concentrates on some numerical integral methods, such as discrete wave-number, self-adapting Filon, Simpson, Gauss, wavelet decomposition and so on. Cao et al. [21] proposed an effective method called modified Clenshaw-Curtis method to solve the integrals, which is based on the traditional Clenshaw-Curtis numerical integration method.

5. Dispersion Curve and Cutoff Frequency of Soil

Just like using the natural frequencies and mode-shapes to reflect the dynamic properties of a finite structure, the dispersion curves and cutoff frequencies are adopted to the infinite soil medium. When the frequency of external force equals to the cutoff frequency of the soil, resonance occurs. The dispersion curves of layered-soil model can be obtained by finding the relationship between wavenumber and frequency in wave propagation.

Since the relationship between displacement and stress on ground surface of layered-soil model has been achieved by Equation (51), the following formula can be obtained:

$$\mathbf{K}\cdot\begin{Bmatrix}\tilde{\tilde{u}}_r(k_r,0,\omega)\\ \tilde{\tilde{u}}_z(k_r,0,\omega)\end{Bmatrix}_{10}=\begin{Bmatrix}\tilde{\tilde{\sigma}}_z(k_r,0,\omega)\\ \tilde{\tilde{\tau}}_{zr}(k_r,0,\omega)\end{Bmatrix}_{10} \quad (72)$$

in which **K** is the stiffness matrix of layered-soil with rigid foundation, and is equal to the inverse of flexibility matrix **S**.

When the soil is not subjected to any external load, that is to say, the right side of Equation (72) equals to zero, the soil has a free vibration in form of modal waves. Moreover, it can be seen from Equation (72) that only when the determinant of coefficient matrix **K** equals to zero, the equation has nonzero solutions. Because the coefficient matrix **K** is a function of wavenumber and frequency, the equation $|K|=0$ corresponds to a group of curves about frequency and wavenumber, which are the dispersion curves of layered-soil model.

The equation $|K| = 0$ can be simplified as:

$$\begin{vmatrix}k_{11}(k_r,\omega) & 0 & 0\\ 0 & k_{22}(k_r,\omega) & k_{23}(k_r,\omega)\\ 0 & k_{32}(k_r,\omega) & k_{33}(k_r,\omega)\end{vmatrix}=0 \quad (73)$$

which can be decoupled as:

$$|k_{11}(k_r,\omega)|=0 \quad (74a)$$

$$|k_{22}(k_r,\omega)\times k_{33}(k_r,\omega)-k_{23}(k_r,\omega)\times k_{32}(k_r,\omega)|=0 \quad (74b)$$

where Equation (74a) is the dispersion curve equation of SH wave in soil, mainly corresponding to the horizontal displacement in horizontal plane of soil; Equation (74b) is the dispersion curve equation of P-SV wave, mainly corresponding to the vertical displacement and the transverse displacement in horizontal plane of soil.

There exist some critical frequencies in the dispersion curves, and the number of modal waves may change when the frequency of the soil crosses these critical frequencies from low to high. These critical frequencies are called the cutoff frequencies of the layered-soil. In actual calculation, the cutoff frequencies of layered-soil value are equal to the intersection frequencies of dispersion curves of elastic half-space at the bottom of soil and those of soil layers in upper half-space. The dispersion curve of half-space is a straight line expressed as:

$$k = \frac{\omega}{c_{\text{s-half}}} = \frac{2\pi}{c_{\text{s-half}}} f \tag{75}$$

By combining Equations (75) and (74), the cut-off frequency corresponding to SH-wave and P-SV wave in soil model can be solved, respectively. Since the contribution of SH wave to the vertical vibration of ground is very small, only the dispersion curves and cutoff frequencies of P-SV wave are studied herein.

Table 2. Parameters of elastic layered half-space soil model

Soil layer	Thickness /m	Elastic modulus /(MPa)	Poisson ratio	Damping ratio	Density /(kg/m^3)	c_p /(m/s)	c_s /(m/s)	c_R /(m/s)
1	3.0	95.59	0.37	0.05	1350	350	161	151
Halfspace	—	506.97	0.48	0.05	1898	1700	300	285

Take the soil parameters listed in Table 2 as example, and then the dispersion curves of P-SV wave can be obtained, as shown in Figure 12.

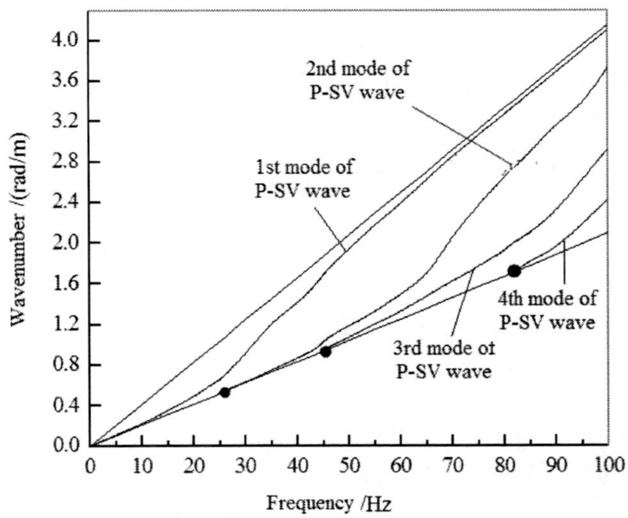

Figure 12. Dispersion curves of P-SV wave in foundation soil model.

It can be seen that all the dispersion curves of P-SV wave are between two boundary straight lines. The upper straight line is the dispersion curve of Rayleigh wave velocity c_R of the top soil layer, and the lower one is the dispersion curve of Shear wave velocity c_s of the bottom elastic half-space. Generally, the first mode of P-SV wave plays an important role, which closes to the dispersion curve of shear wave of half-space in lower frequencies while infinitely approaches to that of Rayleigh waves of the top soil in higher frequencies.

In Figure 12, the black dots at the intersections of dispersion curves of layered-soil and the shear wave of the bottom half-space indicate the cut-off frequencies of elastic layered half-space soil model. In this example, the cut-off frequencies of P-SV wave are 24.57 Hz, 43.99 Hz and 81.78 Hz, and the cutoff frequencies of SH-wave in the soil model are 31.8 Hz, 63.6 Hz and 95.4 Hz, respectively.

6. Soil Vibration Induced by Moving Loads

6.1. Betti-Rayleigh Dynamic Reciprocal Theorem

The Betti-Rayleigh Dynamic Reciprocal Theorem relates to two different elastodynamic states of an elastic body. It can be described briefly as: for the two states of an elastic body subjected to two different loads, the work on the displacement in the second state done by the external force in the first state equals to the work on the displacement in the first state done by the external force in the second state.

Considering a spatial region with an interior volume Ω and a boundary S, the vectors of body-force, surface-force, stress and displacement in the two states are denoted, respectively, by $\mathbf{F}_{b1}(\mathbf{X},t)$, $\mathbf{F}_{s1}(\mathbf{X},t)$, $\boldsymbol{\sigma}_1(\mathbf{X},t)$, $\mathbf{u}_1(\mathbf{X},t)$ and $\mathbf{F}_{b2}(\mathbf{X},t)$, $\mathbf{F}_{s2}(\mathbf{X},t)$, $\boldsymbol{\sigma}_2(\mathbf{X},t)$, $\mathbf{u}_2(\mathbf{X},t)$, where $\mathbf{X}=\{x,y,z\}$ represents the position vector of an arbitrary point in the spatial region and t represents the time. According to the Betti-Rayleigh reciprocal theorem, these vectors satisfy the following equation:

$$\begin{aligned}
&\int_S \int_0^t \mathbf{F}_{s1}(\mathbf{X},t-\tau)\cdot\mathbf{u}_2(\mathbf{X},\tau)\mathrm{d}\tau\mathrm{d}S + \int_\Omega \int_0^t \mathbf{F}_{b1}(\mathbf{X},t-\tau)\cdot\mathbf{u}_2(\mathbf{X},\tau)\mathrm{d}\tau\mathrm{d}V \\
&\quad + \rho\int_\Omega [\dot{\mathbf{u}}_1(\mathbf{X},0)\cdot\mathbf{u}_2(\mathbf{X},t) + \mathbf{u}_1(\mathbf{X},0)\cdot\dot{\mathbf{u}}_2(\mathbf{X},t)]\mathrm{d}V \\
&= \int_S \int_0^t \mathbf{F}_{s2}(\mathbf{X},t-\tau)\cdot\mathbf{u}_1(\mathbf{X},\tau)\mathrm{d}\tau\mathrm{d}S + \int_\Omega \int_0^t \mathbf{F}_{b2}(\mathbf{X},t-\tau)\cdot\mathbf{u}_1(\mathbf{X},\tau)\mathrm{d}\tau\mathrm{d}V \\
&+\rho\int_\Omega [\dot{\mathbf{u}}_2(\mathbf{X},0)\cdot\mathbf{u}_1(\mathbf{X},t) + \mathbf{u}_2(\mathbf{X},0)\cdot\dot{\mathbf{u}}_1(\mathbf{X},t)]\mathrm{d}V \qquad (76)
\end{aligned}$$

where ρ denotes the density of the medium material in the region.

Assuming zero initial speed and displacement, the parts including them in Equation (76) are equal to zeros. When the time approaches to infinite, i.e. $t \to -\infty$, the corresponding displacements and speeds in the two states are also equal to zeros. Therefore, Equation (76) can be simplified as:

$$\int_S \int_{-\infty}^{t} \mathbf{F}_{s1}(\mathbf{X},t-\tau)\cdot\mathbf{u}_2(\mathbf{X},\tau)\mathrm{d}\tau\mathrm{d}S + \int_\Omega \int_{-\infty}^{t} \mathbf{F}_{b1}(\mathbf{X},t-\tau)\cdot\mathbf{u}_2(\mathbf{X},\tau)\mathrm{d}\tau\mathrm{d}V$$

$$= \int_S \int_{-\infty}^{t} \mathbf{F}_{s2}(\mathbf{X},t-\tau)\cdot\mathbf{u}_1(\mathbf{X},\tau)\mathrm{d}\tau\mathrm{d}S + \int_\Omega \int_{-\infty}^{t} \mathbf{F}_{b2}(\mathbf{X},t-\tau)\cdot\mathbf{u}_1(\mathbf{X},\tau)\mathrm{d}\tau\mathrm{d}V \quad (77)$$

On the surface of the elastic body, the stress vector and the surface-force vector satisfy the following boundary conditions:

$$\mathbf{F}_{s1}(\mathbf{X},t)|_S=\boldsymbol{\sigma}_1(\mathbf{X},t)|_S \quad \text{and} \quad \mathbf{F}_{s2}(\mathbf{X},t)|_S=\boldsymbol{\sigma}_2(\mathbf{X},t)|_S \quad (78)$$

Substituting Equation (78) into (77) results in the following convolution integral form:

$$\int_S [\boldsymbol{\sigma}_1(\mathbf{X},t)*\mathbf{u}_2(\mathbf{X},t)]\mathrm{d}S + \int_\Omega [\mathbf{F}_{b1}(\mathbf{X},t)*\mathbf{u}_2(\mathbf{X},t)]\mathrm{d}V$$
$$= \int_S [\boldsymbol{\sigma}_2(\mathbf{X},t)*\mathbf{u}_1(\mathbf{X},t)]\mathrm{d}S + \int_\Omega [\mathbf{F}_{b2}(\mathbf{X},t)*\mathbf{u}_1(\mathbf{X},t)]\mathrm{d}V \quad (79)$$

where the sign * means Riemann convolution:

$$f(x,t)*g(x,t) = \int_0^t f(x,t-\tau)*g(x,\tau)\mathrm{d}\tau .$$

Equation (79) is the expression of the Betti-Rayleigh Dynamic Reciprocal Theorem in the space-time domain. Through the Fourier integral transform and utilizing $\overline{F}[f*g] = \overline{F}[f]\cdot\overline{F}[g]$, Equation (79) can be expressed in the space-frequency domain as:

$$\int_S \overline{\boldsymbol{\sigma}}_1(\mathbf{X},\omega)\cdot\overline{\mathbf{u}}_2(\mathbf{X},\omega)\mathrm{d}S + \int_\Omega \overline{\mathbf{F}}_{b1}(\mathbf{X},\omega)\cdot\overline{\mathbf{u}}_2(\mathbf{X},\omega)\mathrm{d}V$$
$$= \int_S \overline{\boldsymbol{\sigma}}_2(\mathbf{X},\omega)\cdot\overline{\mathbf{u}}_1(\mathbf{X},\omega)dS + \int_\Omega \overline{\mathbf{F}}_{b2}(\mathbf{X},\omega)\cdot\overline{\mathbf{u}}_1(\mathbf{X},\omega)\mathrm{d}V \quad (80)$$

6.2. Reciprocal Property of Green's Function of Subsoil

According to the definition of the Green's function of soil, when a unit-impulsive load $\delta(\mathbf{X}-\mathbf{X}_1)\delta(t)/\delta_{ij}$ is applied to the position $\mathbf{X}_1=\{x_1,y_1,z_1\}$ in a spatial region, the response at the location $\mathbf{X}=\{x,y,z\}$ along the $\mathbf{e}_j$-direction is named as the Green's function, denoted by $G_{ij}(\mathbf{X}_1,\mathbf{X},t)$, in which both i and j represent the directional vectors along three directions of x, y and z, and the Dirac tensor δ_{ij} with the value of $\delta_{ij}=1$ ($i=j$) or $\delta_{ij}=0$ ($i\neq j$) assures that the load acts along the $\mathbf{e}_i$-direction .

Consider the two elastodynamic states of the same soil body shown in Figure 13.

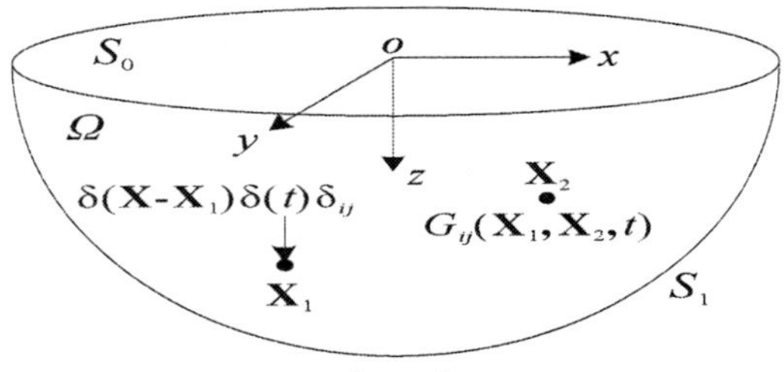

State 1

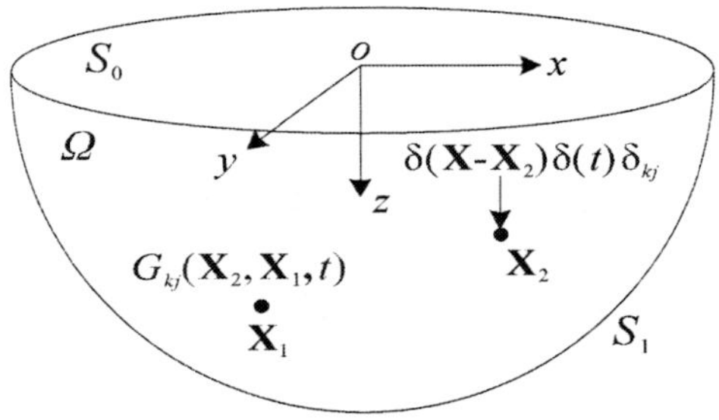

State 2

Figure 13. Two states demonstrating reciprocal property of Green's functions.

State 1

Body force: a unit-impulse load $\delta(\mathbf{X}-\mathbf{X}_1)\delta(t)/\delta_{ij}$ applied at the specified location of $\mathbf{X}_1$ along the $\mathbf{e}_i$ –direction.

Displacement: $G_{ij}(\mathbf{X}_1, \mathbf{X}, t)$ at an arbitrary position $\mathbf{X}$ in the $\mathbf{e}_j$-direction under the action of the body force $\delta(\mathbf{X}-\mathbf{X}_1)\delta(t)/\delta_{ij}$.

Stress: $\sigma_{ij}(\mathbf{X}_1, \mathbf{X}, t)$ at an arbitrary position $\mathbf{X}$ in the $\mathbf{e}_j$-direction under the action of the body force $\delta(\mathbf{X}-\mathbf{X}_1)\delta(t)/\delta_{ij}$.

State 2

Body force: unit-impulse load $\delta(\mathbf{X}-\mathbf{X}_2)\delta(t)/\delta_{kj}$ applied at the specified location of $\mathbf{X}_2$ along the $\mathbf{e}_k$-direction.

Displacement: $G_{kj}(\mathbf{X}_2, \mathbf{X}, t)$ at an arbitrary position $\mathbf{X}$ in the $\mathbf{e}_j$-direction under the action of the body force $\delta(\mathbf{X}-\mathbf{X}_2)\delta(t)/\delta_{kj}$.

Stress: $\sigma_{kj}(\mathbf{X}_2, \mathbf{X}, t)$ at an arbitrary position $\mathbf{X}$ in the $\mathbf{e}_j$-direction under the action of the body force $\delta(\mathbf{X}-\mathbf{X}_2)\delta(t)/\delta_{kj}$.

As the two states satisfy the Betti-Rayleigh Dynamic Reciprocal Theorem, they are substituted into Equation (79) to form a simplified expression as: x

$$\begin{aligned} &G_{ki}(\mathbf{X}_2, \mathbf{X}_1, t) + \iint_S \int_{-\infty}^{t} \sigma_{ij}(\mathbf{X}_1, \mathbf{X}, t-\tau) \cdot G_{kj}(\mathbf{X}_2, \mathbf{X}, \tau) \mathrm{d}\tau \mathrm{d}S \\ &= G_{ik}(\mathbf{X}_1, \mathbf{X}_2, t) + \iint_S \int_{-\infty}^{t} \sigma_{kj}(\mathbf{X}_2, \mathbf{X}, t-\tau) \cdot G_{ij}(\mathbf{X}_1, \mathbf{X}, \tau) \mathrm{d}\tau \mathrm{d}S \end{aligned} \tag{81}$$

For the soil model, the surface integral region S can be divided into two sub-regions: the free surface S_0 and the side surface S_1. Since the stress on the free surface S_0 equals zero, and only exists the integral on the side surface S_1 that locates at the infinite distance, i.e. $z \to \infty$, Equation (81) can be simplified as:

$$G_{ki}(\mathbf{X}_2,\mathbf{X}_1,t) = G_{ik}(\mathbf{X}_1,\mathbf{X}_2,t) + \lim_{z\to\infty} \iint_{S_1} \int_{-\infty}^{t} \sigma_{kj}(\mathbf{X}_2,\mathbf{X},t-\tau)\cdot G_{ij}(\mathbf{X}_1,\mathbf{X},\tau)\mathrm{d}\tau\mathrm{d}S$$

$$-\lim_{z\to\infty} \iint_{S_1} \int_{-\infty}^{t} \sigma_{ij}(\mathbf{X}_1,\mathbf{X},t-\tau)\cdot G_{kj}(\mathbf{X}_2,\mathbf{X},\tau)\mathrm{d}\tau\mathrm{d}S \tag{82}$$

According to the Sommerfeld Radiation Condition of propagating waves, only exits the outward energy flow in the infinite distance, which makes the last two integral terms in Equation (82) equal to zero. Therefore, Equation (82) can be finally simplified as:

$$G_{ki}(\mathbf{X}_2,\mathbf{X}_1,t) = G_{ik}(\mathbf{X}_1,\mathbf{X}_2,t) \tag{83}$$

which shows that the response at position $\mathbf{X}_1$ induced by a unit-impulsive load acting at location $\mathbf{X}_2$ is equal to that response at position $\mathbf{X}_2$ induced by a unit-impulsive load acting at location $\mathbf{X}_1$.

The expression in the frequency domain can be obtained by means of Fourier integral transform on Equation (83):

$$\overline{G}_{ki}(\mathbf{X}_2,\mathbf{X}_1,\omega) = \overline{G}_{ik}(\mathbf{X}_1,\mathbf{X}_2,\omega) \tag{84}$$

It can be seen from Equations (83) and (84) that for the Green's functions of the subsoil model, the source and the receiver can exchange their position, so the reciprocal property of the Green's function is deduced based on the Betti-Rayleigh Dynamic Reciprocal Theorem.

6.3. Soil Vibration Induced by Moving Loads

6.3.1. Integral Solution of Soil Vibration Induced by Moving Loads

Take the halfspace soil model shown in Figure 14 as an example. The load $P(t)$ applying toward the positive direction of z-axis moves in a constant speed along the path in positive direction of x-axis, and the acting point locates either on the surface of the spatial region or inside it.

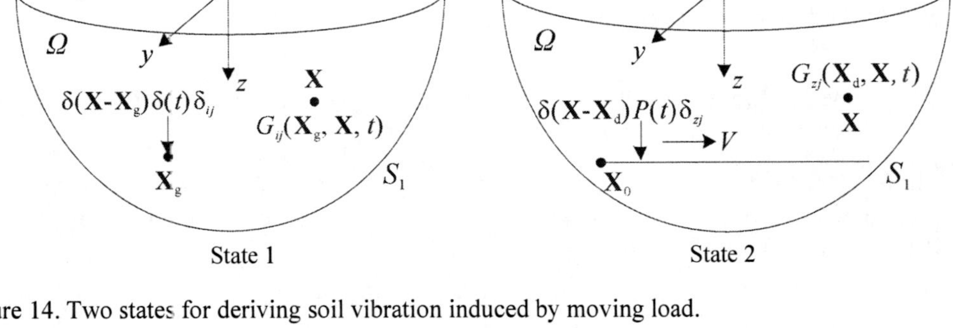

Figure 14. Two states for deriving soil vibration induced by moving load.

State 1

Body force: unit-impulse load $\delta(\mathbf{X}-\mathbf{X}_g)\delta(t)/\delta_{ij}$ applied at the specified location of $\mathbf{X}_g$ along the $\mathbf{e}_i$-direction, in which $\mathbf{X}_g=\{x_g,y_g,z_g\}$ and the subscript "g" means the load location is stationary.

Displacement: $G_{ij}(\mathbf{X}_g,\mathbf{X},t)$ at an arbitrary position $\mathbf{X}$ in the $\mathbf{e}_j$-direction under the action of the body force $\delta(\mathbf{X}-\mathbf{X}_g)\delta(t)/\delta_{ij}$.

Stress: $\sigma_{ij}(\mathbf{X}_g,\mathbf{X},t)$ at an arbitrary position $\mathbf{X}$ in the $\mathbf{e}_j$-direction under the action of the body force $\delta(\mathbf{X}-\mathbf{X}_g)\delta(t)/\delta_{ij}$.

State 2

Body force: the load $\delta[\mathbf{X}-\mathbf{X}_d(t)]P(t)/\delta_{zj}$ applied at the specified location of $\mathbf{X}_d(t)=\{x_0+Vt,y_0,z_0\}$ along the $\mathbf{e}_z$-direction, in which $P(t)$ is definite, $\mathbf{X}_0=\{x_0,y_0,z_0\}$ is the initial position of moving load and the subscript "d" means the load location is moving.

Displacement: $u_{zj}(\mathbf{X}_d,\mathbf{X},t)$ at an arbitrary position in the $\mathbf{e}_j$-direction under the action of the body force $\delta[\mathbf{X}-\mathbf{X}_d(t)]P(t)/\delta_{zj}$.

Stress: $\sigma_{zj}(\mathbf{X}_d,\mathbf{X},t)$ at an arbitrary position in the $\mathbf{e}_j$-direction under the action of the body force $\delta[\mathbf{X}-\mathbf{X}_d(t)]P(t)/\delta_{zj}$.

As the two states satisfy the Betti-Rayleigh Dynamic Reciprocal Theorem, Equation (79) with the two states substituted into becomes:

$$u_{zi}(\mathbf{X}_d,\mathbf{X}_g,t)=\int_{-\infty}^{t}P(\tau)\cdot G_{iz}(\mathbf{X}_g,\mathbf{X}_d(\tau),t-\tau)\mathrm{d}\tau \tag{85}$$

From the reciprocal property of the Green's function, we know $G_{iz}[\mathbf{X}_g,\mathbf{X}_d(\tau),t-\tau)]=G_{zi}[\mathbf{X}_d(\tau),\mathbf{X}_g,\ t-\tau)]$, thus Equation (85) becomes:

$$u_{zi}(x_0+Vt,y_0,z_0,x_g,y_g,z_g,t)=\int_{-\infty}^{t}P(\tau)\cdot G_{zi}(x_0+V\tau,y_0,z_0,x_g,y_g,z_g,t-\tau)\mathrm{d}\tau \tag{86}$$

where the components of vectors in Equation (85) are written out hereafter.

By assuming the soil parameters to be constant along the x-direction, the Green's functions of the soil are independent of the absolute coordinate of the source and the receiver point along the x-axis, but determined by the relative distance between them, and thus can be expressed as:

$$G_{zi}(x_0+V\tau,y_0,z_0,x_g,y_g,z_g,t-\tau)=G_{zi}(0,y_0,z_0,x_g-x_0-V\tau,y_g,z_g,t-\tau)$$

Substituting it into Equation (86) results in:

$$u_{zi}(x_0+Vt,y_0,z_0,x_g,y_g,z_g,t)=\int_{-\infty}^{t}P(\tau)\cdot G_{zi}(0,y_0,z_0,x_g-x_0-V\tau,y_g,z_g,t-\tau)\mathrm{d}\tau \tag{87}$$

in which the Green's function $G_{zi}(0,y_0,z_0,x_g-x_0-V\tau,y_g,z_g,t-\tau)$ denotes the soil displacement at the location of $\{x_g-x_0-V\tau,y_g,z_g\}$ in the $\mathbf{e}_j$-direction induced by a unit-impulsive load applied at the point of $\{0,y_0,z_0\}$ along the $\mathbf{e}_z$-direction.

Equation (87) is the integral solution of soil vibration induced by a moving load $\delta[\mathbf{X}-\mathbf{X}_d(t)]P(t)/\delta_{zj}$ acting either on the surface of the spatial region or inside it.

In the case of the loads moving on the ground surface along the x-axis from the origin ($z_0=0$ and $y_0=0$), Equation (87) can be written as:

$$u_{zi}(x_g, y_g, z_g, t) = \int_{-\infty}^{t} P(\tau)\cdot G_{zi}(x_g - x_0 - V\tau, y_g, z_g, t-\tau)\mathrm{d}\tau \tag{88}$$

Since the displacement function is meaningful only to positive values, the Green's function $G_{zi}(0, y_0, z_0, x_g-x_0-V\tau, y_g, z_g, t-\tau)$ becomes zero when $t-\tau<0$. Therefore, the upper integral limit in Equation (88) can be extended from t to $+\infty$, i.e.:

$$u_{zi}(x_g, y_g, z_g, t) = \int_{-\infty}^{+\infty} P(\tau)\cdot G_{zi}(x_g - x_0 - V\tau, y_g, z_g, t-\tau)\mathrm{d}\tau \tag{89}$$

Equation (89) is actually the integral solution of soil vibration induced by the load moving on the ground surface in the time-space domain. Since the improper integral in the right hand side of the equation is difficult, the integral transform method is herein adopted.

By means of double Fourier transforms to x and t, Equation (89) can be transformed into the wavenumber-frequency domain as follows:

$$\begin{aligned}\tilde{\bar{u}}_{zi}(k_{xg}, y_g, z_g, \omega) &= \int_{-\infty}^{+\infty} P(\tau)\cdot \tilde{\bar{G}}_{zi}(k_{xg}, y_g, z_g, \omega)\cdot \mathrm{e}^{ik_{xg}(x_0+V\tau)}\cdot \mathrm{e}^{-i\omega\tau}\mathrm{d}\tau \\ &= \bar{P}(\omega - k_{xg}V)\cdot \tilde{\bar{G}}_{zi}(k_{xg}, y_g, z_g, \omega)\cdot \mathrm{e}^{ik_{xg}x_0}\end{aligned} \tag{90}$$

where k_{xg} denotes the wavenumber of x-direction; $\{x_g, y_g, z_g\}$ is the coordinate of any specified point in the spatial region.

Since the location of the receiver point is an independent variable of displacement function, $\{x_g, y_g, z_g\}$ can be replaced by $\{x, y, z\}$, and k_{xg} by k_x, respectively. Thus Equation (90) can be rewritten in a more general form as:

$$\tilde{\bar{u}}_{zi}(k_x, y, z, \omega) = \bar{P}(\omega - k_x V)\cdot \tilde{\bar{G}}_{zi}(k_x, y, z, \omega)\cdot \mathrm{e}^{ik_x x_0} \tag{91}$$

Equation (91) illustrates in the wavenumber-frequency domain, the surface displacement of subsoil induced by moving loads can be simplified as the product of three factors as:

(1) The loading function in the frequency domain, representing the influence of load amplitude on the soil vibration;

(2) The Green's function $\tilde{\bar{G}}_{zi}(k_x, y, z, \omega)$ of soil displacement in the wavenumber-frequency domain determined by the dynamic characteristics of the soil, reflecting the transfer relation of vibration in the soil;

(3) The exponential function $\mathrm{e}^{ik_x x_0}$ in the wavenumber domain, representing the influence of initial position of action point on the soil vibration.

The solution in the space-frequency domain can be obtained by applying inverse Fourier transform to Equation (91) about k_x:

$$\bar{u}_{zi}(x, y, z, \omega) = \frac{1}{2\pi}\int_{-\infty}^{+\infty} \bar{P}(\omega - k_x V) \cdot \tilde{\bar{G}}_{zi}(k_x, y, z, \omega) \cdot \mathrm{e}^{-ik_x(x-x_0)} \mathrm{d}k_x \tag{92}$$

Introducing or into Equation (92) results in:

$$\bar{u}_{zi}(x, y, z, \omega) = \frac{1}{2\pi V}\int_{-\infty}^{+\infty} \bar{P}(\bar{\omega}) \cdot \tilde{\bar{G}}_{zi}\left(\frac{\omega - \bar{\omega}}{V}, y, z, \omega\right) \cdot \mathrm{e}^{-i(x-x_0)\frac{\omega-\bar{\omega}}{V}} \mathrm{d}\bar{\omega} \tag{93}$$

where is a new integral variable representing the loading frequency of the moving load (source). Through the transform, the integral to wavenumber of soil response is converted to the integral to frequency of the source, where the wavenumber parameter of the Green's function becomes $k_x = (\omega - \bar{\omega})/V$.

The soil vibration in the space-time domain can be further obtained from Equation (93) by using the inverse Fourier transform to ω.

6.3.2. Ground Vibration Induced by a Moving Constant/Harmonic Load

As shown in Figure 15, the excitation $P(t)$ moves along the path in the x-direction with the uniform speed of V, and the receiver point on the ground surface is located at $A(0, y, 0)$ with y denoting the perpendicular distance from the source path to the receiver point.

For the constant load $P(t)=P_0$ moving on the ground surface and along the path in the x-direction with speed V, the frequency content of loading function is $\bar{P}(\bar{\omega}) = 2\pi P_0 \delta(\bar{\omega})$. By substituting it into Equations (91) and (93), the ground displacement can be obtained respectively in the wavenumber-frequency domain and the frequency domain as:

$$\tilde{\bar{u}}_{zz}(k_x, y, 0, \omega) = 2\pi P_0 \tilde{\bar{G}}_{zz}(k_x, y, 0, \omega) \cdot \mathrm{e}^{ik_x x_0} \delta(\omega - k_x V) \tag{94}$$

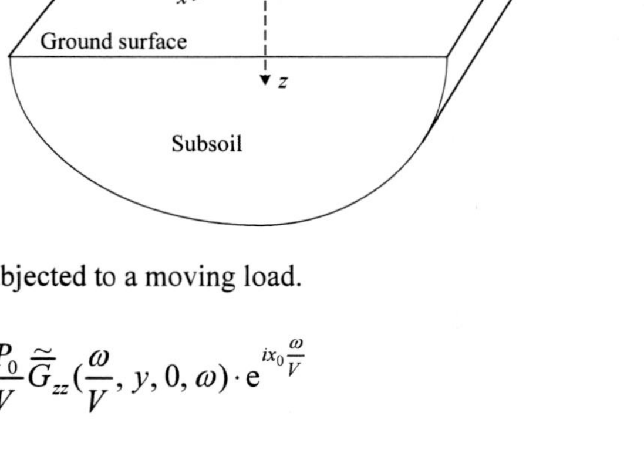

Figure 15. Soil model subjected to a moving load.

$$\bar{u}_{zz}(0, y, 0, \omega) = \frac{P_0}{V} \tilde{\bar{G}}_{zz}\left(\frac{\omega}{V}, y, 0, \omega\right) \cdot \mathrm{e}^{ix_0\frac{\omega}{V}} \tag{95}$$

For the harmonic load $P(t)=\overline{P}(\omega_0)\cdot e^{i\omega_0 t}$ moving on the ground surface and along the x-direction with speed V, the frequency content of $P(t)$ is $\overline{P}(\varpi)=2\pi\overline{P}(\omega_0)\delta(\varpi-\omega_0)$, in which denotes the loading frequency and represents the load amplitude. By substituting into Equations (91) and (93), the ground displacement can be obtained respectively in the wavenumber-frequency domain and the frequency domain as:

$$\tilde{\overline{u}}_{zz}(k_x, y, 0, \omega)=2\pi\overline{P}(\omega_0)\tilde{\overline{G}}_{zz}(k_x, y, 0, \omega)\cdot e^{ik_x x_0}\delta(\omega-\omega_0-k_x V) \quad (96)$$

$$\overline{u}_{zz}(0, y, 0, \omega)=\frac{\overline{P}(\omega_0)}{V}\tilde{\overline{G}}_{zz}\left(\frac{\omega-\omega_0}{V}, y, 0, \omega\right)\cdot e^{ix_0\frac{\omega-\omega_0}{V}} \quad (97)$$

It can be seen from Equations (94) to (97) that the wavenumbers in the x-direction become respectively and for the two types of unit-loads, and the ground displacements can be expressed directly by the Green's functions in the wavenumber-frequency domain. Consequently, the soil vibrations in the space-time domain induced by moving constant/harmonic loads can be obtained from Equations (95) and (96), respectively, by using the inverse Fourier transform about ω.

Therefore, it can be seen that by application of the Betti-Rayleigh Dynamic Reciprocal Theorem, the moving source problem is successfully converted to the problem of fixed source and moving receiver point, which avoids the complex integral calculation and makes the solution process much simpler.

In particular, if the directional vector i is chosen to be z-direction in Equations (94) to (97), the ground vibration becomes the vertical response $\overline{u}_{zz}(0, y, 0, \omega)$, and the corresponding Green's functions are and $\tilde{\overline{G}}_{zz}((\omega-\omega_0)/V, y, 0, \omega)$, respectively.

6.3.3. Verification of Present Approach

To verify the approach discussed above, the calculated ground displacements are compared with the results of Lombaert [19] by means of the Filon's numerical integral method, which can be seen in Figures 16 and 17, where the instant $t=0$ corresponds to the moment when the point load passes through the origin (0, 0, 0). Herein, the free-field responses at the location of (x, y, z)=(0, 8, 0) are illustrated, respectively for a constant moving unit load (V=110 m/s) and a harmonic moving unit load (V=60 m/s, $\omega_0=2\pi\cdot 15$ rad/s). The soil is modeled as a homogenous half-space with a density ρ=1800 kg/m^3, an S-wave velocity c_S=150 m/s, a P-wave velocity c_P=300 m/s, a hysteretic material damping ratio 0.025, and the Poisson ratio 0.333. The results of comparison show the ground responses computed by the present analytical methods agree well with the results of earlier literature.

The steady-state response for the same problem was also presented by Yang et al. [22] by the 2.5D finite/infinite element method. For the purpose of verification, the ground response at the location of (x, y, z)=(0, 10, 0) induced by a unit harmonic load $\delta(y)\delta(z)e^{-ikx}e^{-i\omega t}$ with $k=(\omega-\omega_0)/V$ are investigated. Shown in Figure 18 are the displacement spectra and the maximum velocity levels versus the distance to the source, respectively by the present approach and by Yang et al. [22], in which V=70 m/s, $\omega_0=2\pi\cdot 10$ rad/s, and the soil parameters are listed in Table 3.

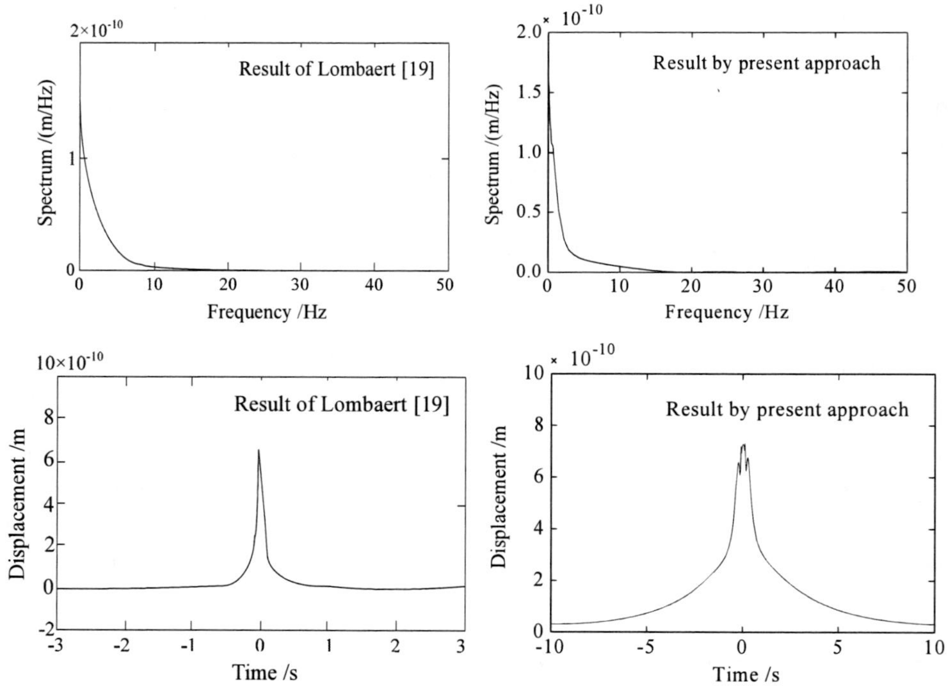

Figure 16. Comparison of ground vibrations induced by moving unit-constant-load: (top) displacement spectra, and (bottom) history curves.

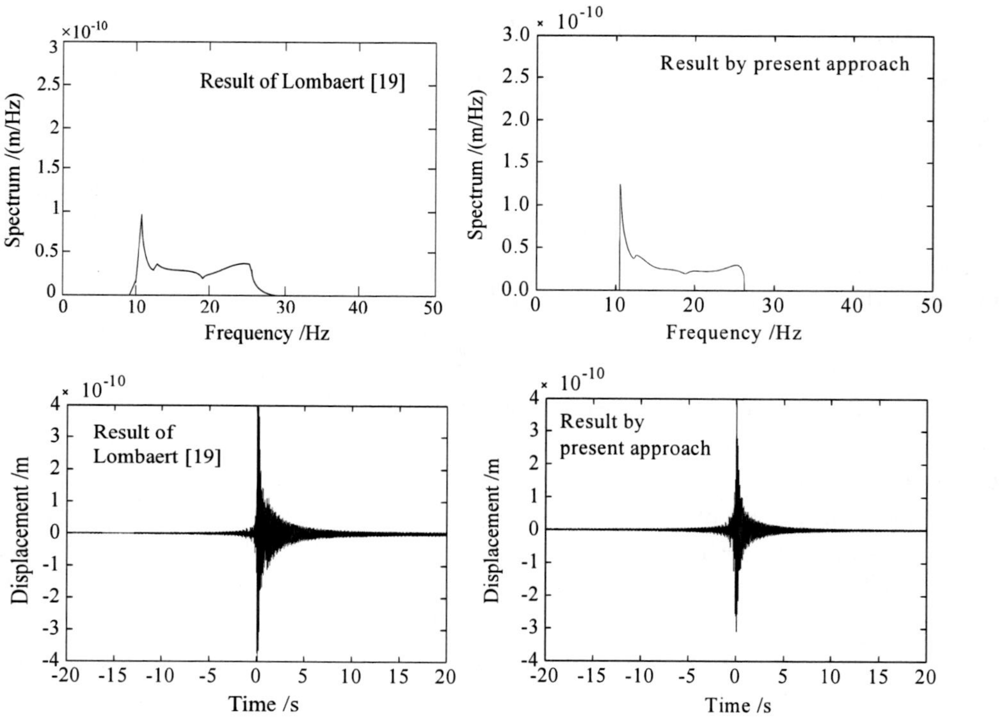

Figure 17. Comparison of ground vibrations induced by moving unit-harmonic-load: (top) displacement spectra, and (bottom) history curves.

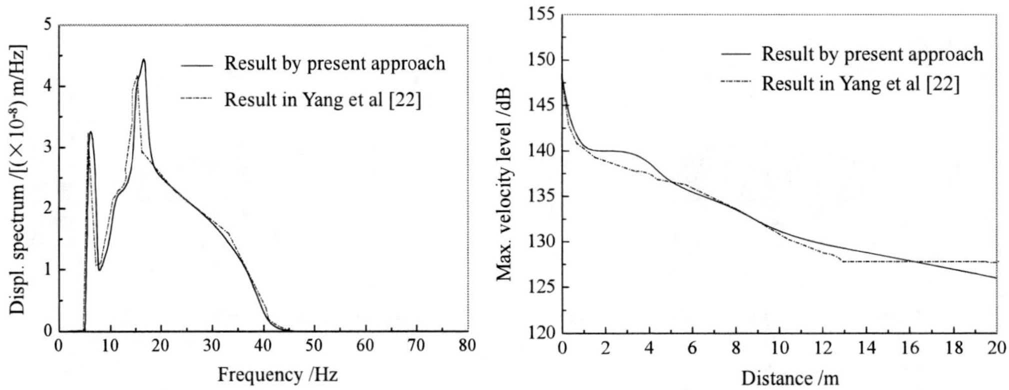

Figure 18. Verification of present approach by comparison with Yang [22]: (left) displacement spectra; and (right) maximum velocity levels vs distance to the source.

Table 3. Soil parameters adopted in Yang et al. [22]

Soil model	Elastic modulus /(MPa)	Poisson ratio	Damping ratio	Density /(kg/m^3)	c_s /(m/s)	c_R /(m/s)
3m	50	0.25	0.02	2000	100	92.1
Half-space	200	0.25	0.02	2000	200	184

In Figure 18, the velocity level is calculated by L(dB)=20×lg(v/v_0), where v is the computed ground velocity value and v_0 a reference value equal to 10^{-8}. As can be seen, the results obtained by the present approach are in good agreement with those of [22].

6.4. Case Study

The soil model with a single layer and a bottomed half-space is taken hereinafter again as an illustrative case study to analyze the vertical responses of subsoil subjected to a moving unit constant/harmonic load. The soil parameters are listed in Table 4.

Table 4. Parameters of layered elastic half-space soil

Soil model	Depth /m	Elastic modulus /(MPa)	Poisson ratio	Damping ratio	Density /(kg/m^3)	c_s /(m/s)	c_R /(m/s)
Layer 1	5.0	44	0.497	0.05	1475	100	95
Halfspace	—	507	0.48	0.05	1898	300	285

The Green's functions of ground displacements in the wavenumber-frequency domain are calculated and graphed in Figures 19 and 20, using a dimensionless wavenumber coordinate with $k_0=(\omega/c_s)_{min}$ as the reference value. It can be seen that there appear several peaks in the Green's functions of ground displacements corresponding to different frequencies and the distributions of the amplitude isograms are not so regular.

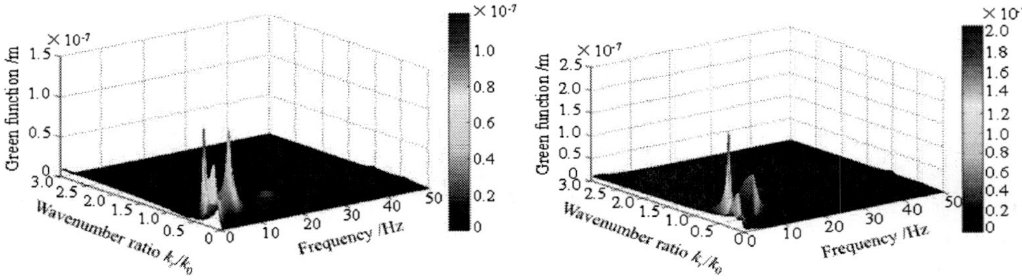

Figure 19. Green's functions of radial (left) and vertical (right) ground displacements.

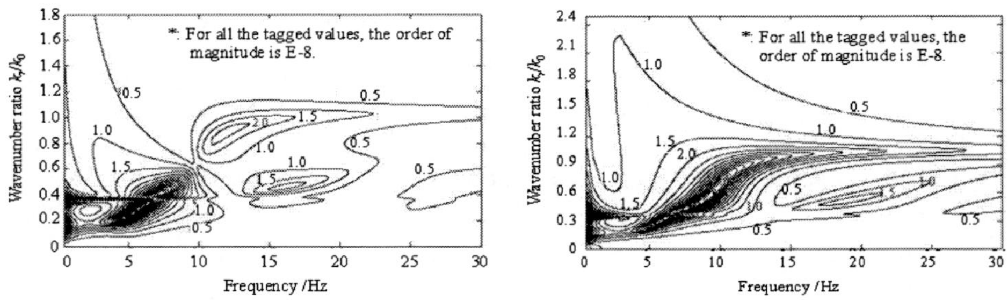

Figure 20. Isogram of Green's functions of radial (left) and vertical (right) ground displacements.

6.4.1. Ground Vibration Induced by a Moving Unit-constant-load

Illustrated in Figure 21 are the vertical ground displacement spectra and history curves at the locations of y=2 m and y=30 m from the load path, when a unit-constant-load moves along the path from a negative point on the x-axis at a speed of V=80 m/s.

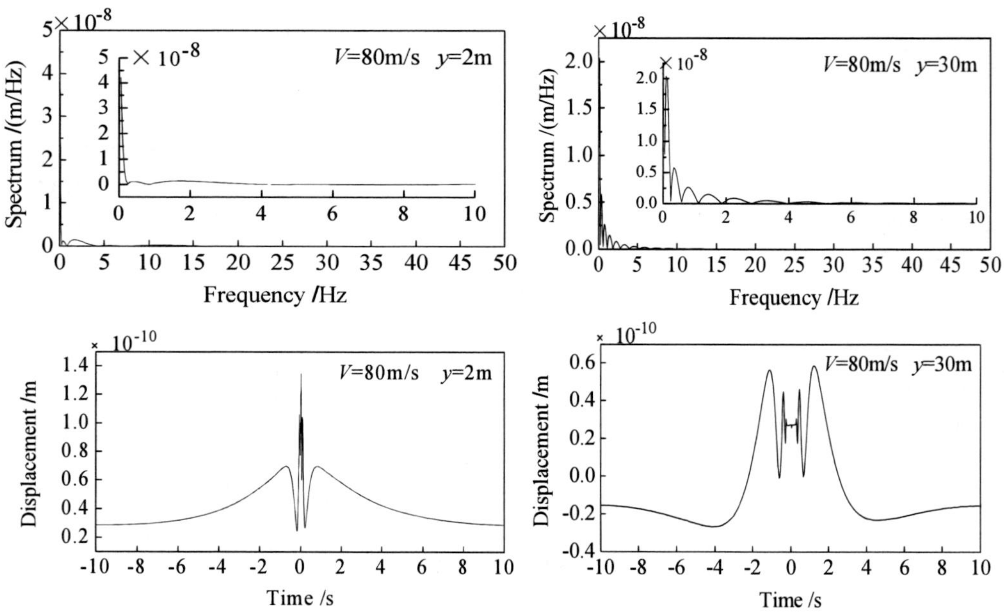

Figure 21. Vertical ground displacements induced by a moving unit-constant-load: (top) spectra, and (bottom) history curves.

The spectrum curves show that the vertical ground vibration subjected to a moving constant-load is featured of typical low frequency, which is also true even when the moving speed is very high (V=80 m/s). The history curves indicate that when the load moves along the path close to the position having the same x-coordinate with the receiver point, the ground vibration peak appears at t=0 for the receiver point at y=2 m from the load path, while they appear before and after t=0 for the receiver point at y=30 m. It can also be seen that the ground displacement at y=30 m is much smaller than that at y=2 m, indicating the vibration energy is dissipated by the material damping and the radiation damping in the soil during the wave propagation.

6.3.2. Ground Vibration Induced by a Moving Unit-harmonic-load

Illustrated in Figures 22 and 23 are, respectively, the vertical ground displacement spectra and history curves at the locations of y=2 m and y=30 m from the load path, when a unit-harmonic-load of 20 Hz moves along the path from a negative point on the x-axis. The moving speed of the harmonic-load is 20 m/s or 50 m/s, as indicated in the figures.

It is seen from Figure 22 that the frequencies of vertical ground vibration concentrate in a range around the loading frequency (20 Hz), and the frequency range corresponding to V=50 m/s is wider than that to V=20 m/s. The reason for this can be explained as follows. For the moving harmonic-load, the corresponding wavenumber in the x-direction is $k_x=(\omega-\omega_0)/V$. Note that the wavenumber $k_x=\omega/c_x$, in which c_x is the phase velocity of wave propagating along x-direction in the soil, one can easily obtain $\omega=\omega_0/(1-V/c_x)$, the correlation between the ground vibration frequency and the source frequency. Thus when the load moves towards the receiver along the path in x-direction, it will excite a response frequency $\omega=\omega_0/(1-V/c_x)$ that is higher than the source frequency ω_0; while when the load moves away from the receiver point, it will excite a response frequency $\omega=\omega_0/(1+V/c_x)$ that is lower than ω_0. Namely, the frequency of ground vibration is situated in the range of $[\omega_0/(1+V/c_x), \omega_0/(1-V/c_x)]$. As for the layered half-space soil model, the wave velocity c_x varies within the range $[c_{s\_down}, c_{R\_up}]$, where c_{s_down} is the shear wave velocity of the bottom half-space soil and c_{R_up} the Rayleigh's wave velocity of the top layer soil. Since generally $c_{s\_down}>c_{R\_up}$, the frequency range of receiver point becomes $[\omega_0/(1+V/c_{R\_up}), \omega_0/(1-V/c_{R\_up})]$, which shows that both the upper and the lower frequency limits of ground vibration are controlled by the Rayleigh's wave velocity of the top layer soil. In the layered half-space subsoil model of this example, the wave velocity range is $c_x \in [95, 300]$ m/s, so the frequency ranges are, respectively, [16.52, 25.33] Hz for the moving load speed V=20 m/s and [13.10, 42.22] Hz for V=50 m/s, which are in accordance with the frequency ranges illustrated in Figure 22.

On the other hand, It is well known that the three types of waves propagating in the soil have different velocities with $c_p>c_s>c_R$, so the P-wave, S-wave and R-wave have different effects on the ground vibration when the load moves towards and away from the source, which explains hereinafter the multi peaks appeared in Figure 22. According to the frequency range $[\omega_0/(1+V/c_x), \omega_0/(1-V/c_x)]$, the peaks induced by P-wave are located nearer to ω_0 while the contribution of the R-wave concentrates on the outermost peaks. For the load with speed of V=20 m/s, the outermost two peaks are more obvious at the location of y=2m while the middle peak is most obvious at the location of y=30m. This can be explained that when the receiver point is near the load path, the contribution of R-wave to ground vibration is bigger than that of P-wave or S-wave; while when it is farther from the path, the contribution of P-wave to ground vibration is the biggest.

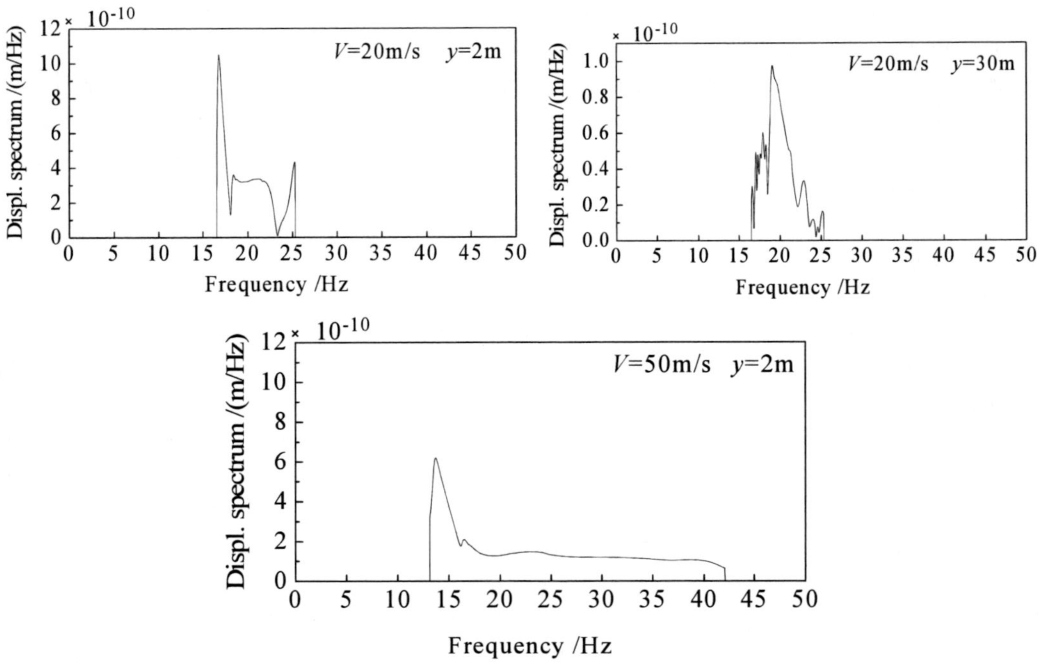

Figure 22. Ground displacement spectra induced by a moving unit-harmonic-load.

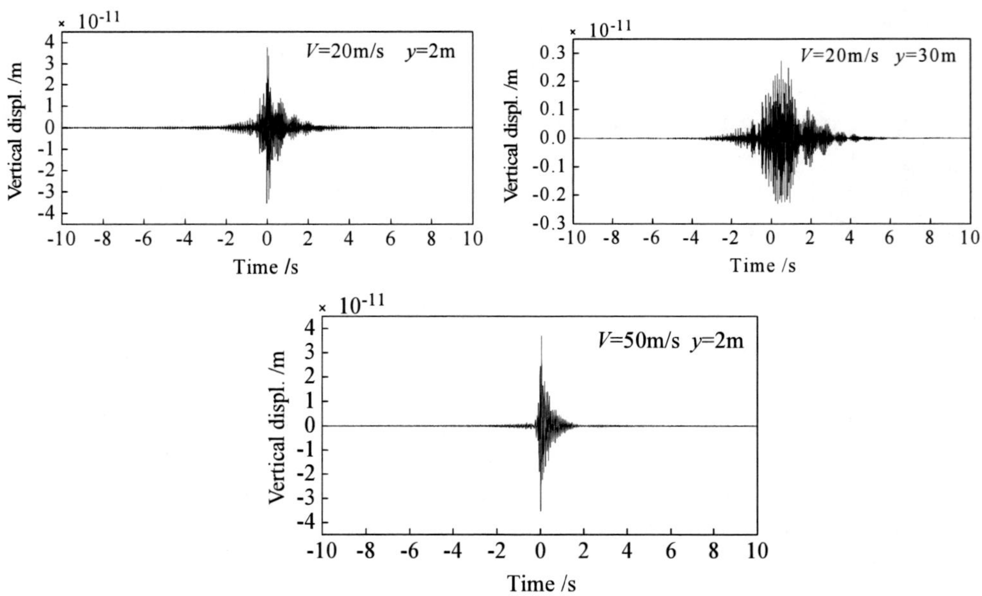

Figure 23. Ground displacement histories induced by a moving unit-harmonic-load.

In addition, from Figure 23 one can find the effect of wave propagation: when the source moves toward the receiver point along the x-axis, the ground vibration becomes bigger and bigger, and the peak appears at $t=0$ when it moves at the same x-coordinate with the receiver point; when the source moves outward the receiver point after $t=0$, the ground vibration becomes smaller and smaller. The frequency range of ground vibration $[\omega_0/(1+V/c_x), \omega_0/(1-V/c_x)]$, however, is only suitable for the load speed V smaller than the wave velocity c_x, namely, the sub-critical case. When the load speed approaches c_x or even exceeds c_x, namely, the critical case ($V=c_x$) or super-critical case ($V>c_x$), $[\omega_0/(1+V/c_x), \omega_0/(1-V/c_x)]$ is not applicable because $\omega_0/(1-V/c_x)$ becomes negative. It has been shown that the ground vibration characteristics induced by moving loads with a speed lower than or greater than the wave speed can be quite different [3]. Since this chapter does not focus in studying the effect of speed, only sub-critical case of load speed ($V<c_x$) is discussed.

CONCLUSION

The basic theory and analysis approaches in traffic-induced environmental vibrations are presented in this chapter, including vibration source, wave propagation, dynamic characteristics of soil, analytical model of train-track-ground, Green's functions of ground displacements, and so on. Moreover, the reciprocal property of Green's function and the solution of soil/ground vibration subjected to moving loads are specially derived. By application of the Betti-Rayleigh Dynamic Reciprocal Theorem, the moving source problem is converted to the fixed source problem with moving receiver point in opposite direction, which greatly simplifies the complex analytical solution of soil/ground vibration induced by moving load. The case study shows that when the receiver nears the source path, the R-wave contributes to ground vibration much more than the P-wave and the S-wave; while when the receiver is far from the source path, the contribution of the P-wave is more obvious. For the source of moving constant-load, the ground vibration is of typical low-frequency feature; while for the moving harmonic-load, the frequency range of ground vibration is controlled by the Rayleigh's wave velocity of the top layer soil.

ACKNOWLEDGMENTS

This study is sponsored by the National Natural Science Foundation of China (Grant No. 51108023), the Ph.D. Programs Foundation of Ministry of Education of China (20090009120021), and the Fundamental Research Fund of Beijing Jiaotong University (2011JBM088).

REFERENCES

[1] Lamb H. (1904). On the propagation of tremors over the surface of an elastic solid. *Philosophical Transactions of the Royal Society*, A203: 1–42.

[2] Grundmann H., Lieb M., Trommer E. (1999). The response of a layered half-space to traffic loads moving along its surface. *Archive Applied Mechanics*, 69(1): 55–67.

[3] Hung H. H., Yang Y. B. (2001). Elastic waves in visco-elastic half-space generated by various vehicle loads. *Soil Dynamics and Earthquake Engineering* , 21: 1–17.

[4] Bierer T., Bode C. (2007). A semi-analytical model in time domain for moving loads. *Soil Dynamic and Earthquake Engineering*, 27: 1073–1081.

[5] Martínez-Castro A. E., Museros P., Castillo-Linares A. (2006). Semi-analytic solution in the time domain for non-uniform multi-span Bernoulli-Euler beams traversed by moving loads. *Journal of Sound and Vibration*, 294: 278–297.

[6] Zhou H. F., Jiang J. Q. (2007). Dynamics responses of viscoelastic half-space under moving loads. *Chinese Journal of Theoretical and Applied Mechanics*, 39(4): 545–553.

[7] Galvin P., Francois S., Schevenels M., et al. (2010). A 2.5D coupled FE-BE model for the prediction of railway induced vibrations. *Soil Dynamics and Earthquake Engineering*, 30(12): 1500–1512.

[8] Hamid R. N., Morteza A., Hamid H. (2012). Numerical analysis of ground surface vibration induced by underground train movement. *Tunnelling and Underground Space Technology*, 29: 1–9.

[9] Motazedian D., Hunter J. A., Sivathayalan S., et al. (2012). Railway train induced ground vibrations in a low V_s soil layer overlying a high V_s bedrock in eastern Canada. *Soil Dynamics and Earthquake Engineering*, 36: 1–11.

[10] Xia H., De Roeck G., Goicolea J. M.. (2011). *Bridge Vibration and Controls: New Research*, New York: Nova Science Publishers Inc.

[11] Au F. T. K., Cheng Y. S., Cheung Y. K. (2001). Effects of random road surface roughness and long-term deflection of PC girder and cable-stayed bridges on impact due to moving vehicles. *Journal of Computers and Structures*, 79: 853–872.

[12] Majka M., Hartnett M. (2009) Dynamic response of bridges to moving trains: A study on effects of random track irregularities and bridge skewness. *Computers & Structures*, 87(19-20): 1233−1252.

[13] Gutowski T. G., Dym C. L. (1976). Propagation of ground vibration: a review. *Journal of Sound and Vibration*, 49(2): 179–193.

[14] Yang Y. B. (1995). *Soil Vibration Induced by High-speed Railway*. Taipei: Taiwan University.

[15] Wu T. X., Thompson D. J. (2001). Vibration analysis of railway track with multiple wheels on the rail. *Journal of Sound and Vibration*, 239(1): 69–97.

[16] Yoshioka O. (1976). Some considerations on generating mechanism of vibration due to running trains. *Geophysical Exploration*, 29(2): 23–33.

[17] Krylov V. V., Ferguson C. (1994). Calculation of low-frequency ground vibrations from railway trains. *Applied Acoustics*, 42: 199–213.

[18] Jones C. J. C., Block J. R. (1996). Prediction of ground vibration from freight trains. *Journal of Sound and Vibration*, 193(1): 205–213.

[19] Lombaert G. (2001). *Development and Experimental Validation of a Numerical Model for the Free Vibrations Induced by Road Traffic*. Belgium: Leuven University.

[20] Liao Z. P. (2002). *Introductory Theory of Engineering Wave* (2nd Edit). Beijing: Science Publishing House.

[21] Cao Y. M. (2006). *Theoretical Analysis and Experimental Study of Free field and Building vibrations Induced by Railway Trains*. Beijing: Beijing Jiaotong University.

[22] Yang Y. B., Hung H. H., Chang D. W. (2003). Train-induced wave propagation in layered soils using finite/infinite element simulation. *Soil Dynamics and Earthquake Engineering*, 2003, 23: 263–278.

In: Traffic Induced Environmental Vibrations and Controls ISBN: 978-1-62618-240-0
Editors: H. Xia and R. Calçada

Chapter 4

NUMERICAL MODELING OF VIBRATIONS INDUCED IN TUNNELS: A 2.5D FEM-PML APPROACH

Patrícia Lopes[1*], Pedro Alves Costa[2], Rui Calçada[3] and António Silva Cardoso[4]
[1,2,3,4]Faculty of Engineering, University of Porto,
Porto, Portugal
[1]Polytecnical Institute of Engineering of Porto,
Porto, Portugal

ABSTRACT

The simulation and study of problems related with vibrations induced by railway traffic in tunnels is a difficult and complex task. The semi-infinitive character of the domain surrounding the tunnel associated to the 3-dimensional characteristics of the problem is one of the major causes of that complexity. An efficient and accurate solution for the simulation of the dynamic behavior of very long structures, i.e., railways, roads or tunnel, is proposed by using 2.5D techniques in the context of the finite elements method, and adopting perfect matched layers (PML's) for the treatment of the boundaries due to the truncation of the finite elements mesh. Since the resource to 2.5D FEM-PML approach is not very usual, the equations of the 2.5D PML are derived, highlighting the compatibility with the 2.5D FEM. After that brief presentation of the model, a range of validation examples are depicted, showing the accuracy of the model. By last, a parametric study is presented in order to evaluate the influence of some tunnel and ground properties in the vibrations assessed at the ground surface.

1. INTRODUCTION

During the last decade, a considerable research effort has been devoted to the development of numerical and semi-analytical methods for the modelling of unbounded

* E-mail address: cpl@isep.ipp.pt.

structures submitted to dynamic load conditions. The increasing significance of this issue is related to the current expansion of modern transportation infrastructures as, for instance, high-speed railway and metro lines. This kind of structure is difficult to model due to several reasons, namely the infinite character of its geometry subjected to moving loads. However, this drawback has been contently overcome through some assumptions often made: the medium is usually assumed as periodic or invariant along the development direction, which can drastically reduce the numerical effort associated to the numerical solution of the three-dimensional problem.

The periodic models for the dynamic simulation of tunnels have been extensively applied by Gupta et al. [1–3] using a coupled FEM-BEM model previously developed by Clouteau et al. [4]. This numerical approach takes advantage of the FEM for the simulation of complex geometries, such as a tunnel, and of the BEM for the simulation of the unbounded domain represented by the ground. In spite of the high accuracy of that kind of model, the numerical effort is quite high, discouraging its intensive application.

Alternatively, for longitudinally invariant structures, a 2.5D approach can be applied, being the 3D solution obtained by numerical discretization of the cross-section combined with the Fourier transformation of the domain along the orthogonal direction. This method has been employed for the study of several railway lines applications both in surface and in tunnels [5–10]. This methodology can be applied in both finite [7, 10–11] and boundary elements formulations [8], as well as for coupled FEM-BEM approaches [6, 9, 12]. Due to the unbounded nature of the ground, the 2.5D BEM formulation is particularly adequate for dealing with this kind of problems.

Nevertheless, the versatility of the FEM turns this method attractive for dealing with complex geometries, although the drawbacks inherent to the simulation of unbounded domains require special procedures that are not always accurate or simple to implement. To overturn these limitations of both methods, the FEM-BEM coupling has been adopted by several authors. Following that approach, Alves Costa et al. [12] recently presented the experimental validation of a 2.5D FEM-BEM model, which has also been applied for the study of the dynamics of railway tracks with ballast mats [13–14].

In spite of the potential virtues of the 2.5D FEM-BEM approach, its application requires complex numerical procedures [15] and, generally, it computational efficiency trend decreases in the case of embedded structures. In such cases, the option by a 2.5D FEM approach for whole domain is somewhat more simple and efficient. Regarding this aspect, it should be mentioned that the finite element approach is by excellence suitable for dealing with very complex geometries or materials. In fact, the resource to other numerical or analytical approaches, although possible, requires strong theoretical knowledge, sometimes limited to the academia.

Actually, the consideration of complex restrictions to the problem, as for instance the soil anisotropic behavior can be dealt by an easier way by a FEM approach when compared with a BEM formulation [16], since the knowledge of specific Green's functions is not necessary. Other examples can also be presented, for instance complex geometries represented by twin tunnels as analyzed by Kuo et al. [17], can be simulated by a FEM approach without any particular change of the original equations. In the same way, material heterogeneities, which can present an important role on the prediction of vibrations induced by traffic in tunnels [18–19], can be introduced easily in a FEM approach.

Despite of the vantages of the FEM approach presented above, an important drawback cannot be neglected: the FEM approach requires a complete definition of the domain which creates relevant difficulties when dealing with unbounded domains. Indeed, a topic of particular interest is the formulation of special procedures to treat the boundary effects that are inherent to the truncation of the domain associated with the finite element discretization. For static problems, the contribution of the ground is reflected in terms of stiffness, so it is possible to truncate the domain at a sufficiently remote location where the ground deformation is so small that it can be neglected. However, in dynamic analyses the ground model should fulfill the requirements of representing not only the dynamic ground stiffness but also the radiation condition. The latter requirement demands a special treatment of the boundary conditions, since spurious reflection of the waves at the mesh boundary should not occur. For achieving this requirement several strategies have been proposed and applied in the context of tunnel-ground interaction problems. Viscous boundaries [20] are one of the simpler procedures, which was adopted by Bian et al. [21] for the study of the ground motions induced by railway traffic. The same objective can also be reached by the coupling of 2.5D infinite elements along the artificial boundary, as proposed by Yang et al. [10–11] and applied by Lopes et al. [5]. Comparing both methodologies, Alves Costa et al. [7] found that the accuracy achieved by the latter is quite higher than by the former, in spite of the impossibility of guarantee a perfect absorption condition in both approaches. Alternatively to the referred procedures, Bian et al. [22] proposed a 2.5D formulation of the gradually damped artificial boundaries previously presented by Liu & Jerry [23].

Despite the proposals referred above, other approach is becoming the method of choice in several domains leading with propagating waves – the PML (Perfect Matched Layers) formulation. A PML is a layer that respects the two main requirements: is not reflective and is absorbing [24]. The first mention to the PML formulation is due to Bérenger [25], in the framework of electromagnetics, suggesting that the stretching of the coordinates for a complex domain allows achieving the perfect absorbing condition. After that, elastodynamic applications start to arise, mainly due to the work developed by Chew & Liu [26] and, more recently, by Basu & Chopra [27].

However, in spite of the potential virtues of this method, which can supplant the methods previously presented, its divulgation and application for the simulation of transport infrastructures is not widespread yet. Thus far, only one paper has been published about this topic in the context of 2.5D formulation [28], at least in the authors' knowledge.

The goal of the present chapter is to show the potentialities and accuracy that can be reached by a 2.5D FEM-PML framework developed by the authors for the simulation of the tunnel-ground interaction and wave propagation. Following a similar formalism to the 2.5D FEM, the 2.5D PML formulation is presented with emphasis to the similarities and differences in relationship to the formal finite elements approach. The validation of the model by comparison with the analytical solution for a full-space [29] follows the presentation of the methodology. Without losing the focus on the paper goal, the solution obtained for a tunnel embedded in a full-space is compared with the solution provided by a semi-analytical code, called PiP, which was developed by Hussein and Hunt [30].

Finally, a parametric study is performed in order to discuss the influence of several aspects of the tunnel and of the ground on the propagation of vibrations.

2. 2.5D FEM-PML: CONCEPT AND FORMULATION

2.1. Generalities

Consider a structure with longitudinally invariant cross-section as shown in Figure 1. The whole domain, i.e., the cross section, is discretized into finite elements, being the external layers that bound the "box domain" formed by PML's. The truncation of the domain is applied after the PML's. So, this numerical device has the function of absorbing, without spurious reflection, the arbitrary direction waves that impinge the boundary between the domains described by the 2.5D FEM and by the 2.5D PML.

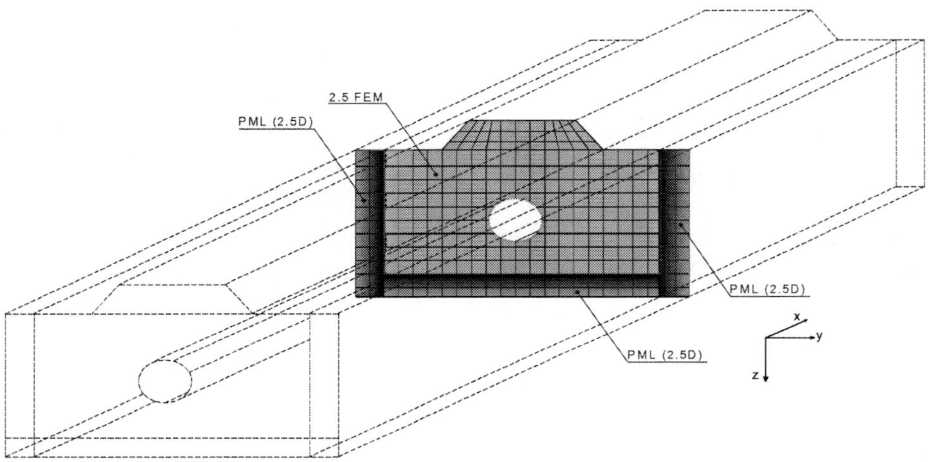

Figure 1. Infinite and invariant structure in one direction.

The absorbing condition is achieved by the stretching of the coordinates (in the present case y and z, since the coordinate x will be transformed to the wavenumber domain) for a complex domain, which induces an artificially increase of the wave attenuation that propagates along the PML, as is highlighted in Figure 2.

The trick of this method resides on the stretching functions adopted, since the PML must act not only like an absorbing material but also like a non-reflective absorbing material. Only fulfilling these two requirements, it is possible to obtain an accurate solution inside of the "box of interest".

2.2. Formulation in the Context of the 2.5D FEM-PML

2.2.1. Formulation of 2.5D FEM

In order to avoid extremely high costs of computation intrinsic to a full 3 D simulation, several attempts have been developed in order to respect the tri-dimensionally of the domain by one hand, but imposing some restrictions by the other hand in order to minimize computational effort. Tri-dimensional structures with infinite development and invariable properties (geometrical and mechanical) can be dealt by a 2.5D formulation. In these cases, the structure main properties can be represented by its cross-section, being, by that way, faced as a two-dimensional medium. However, despite of that condition, it should not be forget that the loading is not two-dimensional, which demands the solution of a 3D problem. The main concept behind the proposed solution to the problem is the use of a method which is between the two and the three dimensional domain. This method was firstly proposed by Hwang & Lysmer [31], in the FEM context, for the study of underground structures under travelling seismic waves. Subsequently, the method has been applied by a few researchers to the study of vibrations induced by traffic. In this field, special attention should be dedicated to the following works: [9, 11–12, 32–34].

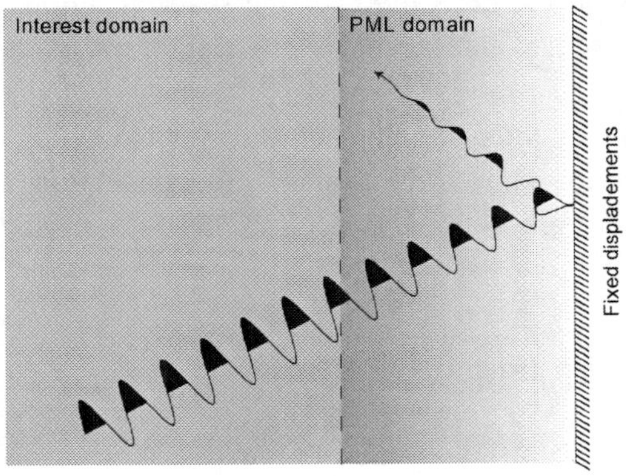

Figure 2. Schematic representation of the wave attenuation inside a PML layer.

Assuming that the response of the structure is linear, the analysis can be carried out in the wavenumber/frequency domain. All the variables, i.e., loads (action) and displacements (response), must be transformed to the wavenumber/frequency domain by means of a double Fourier transform, related with the direction along the track (x direction) and with time. Transformed quantities are functions of the Fourier images of x and t, defined as wavenumber and frequency and are represented by k_1 and ω, respectively.

Following the usual steps of the finite element procedure, namely the strong and weak formulations, the following equilibrium equation can be derived for any point of a three dimensional domain:

$$\int_V \delta\varepsilon\,\sigma\,dV + \int_V \delta u\,\rho\ddot{u}dV = \int_S \delta u\,p\,dS \tag{1}$$

where $\delta\varepsilon$ is the virtual strain field, σ represents the stress field, δu is the virtual displacement field, u is the displacement field, ρ is the mass density and p represents the applied loads.

After the transformation, the cross-section of the domain remains on the untransformed domain and is discretized into finite elements. This approach enables to rewrite Equation (1) in terms of nodal variables.

The concept of virtual work can be applied in the transformed domain by recourse to Parserval's theorem [35]:

$$\int \delta f(x)\, p\, dx = \int \delta f(-k_1)\, p(k_1)\, dk_1$$

Considering Equation (1), the virtual work of the internal stresses and inertial forces in the transformed domain is given by, respectively:

$$\int_V \delta\varepsilon\, \sigma\, dV = \int_{k_1} \delta u_n^T(-k_1,\omega) \int_z \int_y B^T(-k_1) D B(k_1)\, dy\, dz\, u_n(k_1,\omega)\, dk_1 \tag{2}$$

$$\int_V \delta u\, \rho \ddot{u}\, dV = -\omega^2 \int_{k_1} \delta u_n^T(-k_1,\omega) \int_z \int_y N^T \rho N\, dy\, dz\, u_n(k_1,\omega)\, dk_1 \tag{3}$$

where: B is the matrix with the derivatives of the shape functions; N is the shape function matrix; D is the strain-stress matrix; u_n is the vector of nodal displacements (in the transformed domain).

The virtual work done by the external loads is computed taking advantage of the fact that the geometry is only discretized on the ZY plane. So, considering a coordinate s, parallel to the edge of the element where traction is applied, the virtual work developed by the load system is given by,

$$\int_S \delta u\, p\, dS = \int_{k_1} \delta u_n^T(-k_1)\, p_n(k_1,\omega)\, dk_1 \tag{4}$$

Replacing and rearranging Equations (3) to (5) in Equation (1) yields,

$$\left(\int_z \int_y B^T(-k_1) D B(k_1)\, dy\, dz - \omega^2 \int_z \int_y N^T \rho N\, dy dz \right) u_n(k_1,\omega) = p_n(k_1,\omega) \tag{5}$$

Adopting the classic finite element notation leads to,

$$[K] = \int_z \int_y B^T(-k_1)\, D\, B(k_1)\, dy\, dz \tag{6}$$

and

$$[M] = \int_z \int_y N^T \rho N\, dy\, dz \tag{7}$$

where [K] and [M] are the stiffness and mass matrices, respectively.

As usual, matrix [B] is derived from the product of the differential operator matrix [L] (in the transformed domain) and matrix [N]. Since the direction x is transformed to the wavenumber domain, the derivatives with respect to k_1 are analytically computed, as presented in the following expression.

$$[L] = \begin{bmatrix} ik_1 & 0 & 0 & \frac{\partial}{\partial y} & 0 & \frac{\partial}{\partial z} \\ 0 & \frac{\partial}{\partial y} & 0 & ik_1 & \frac{\partial}{\partial z} & 0 \\ 0 & 0 & \frac{\partial}{\partial z} & 0 & \frac{\partial}{\partial y} & ik_1 \end{bmatrix}^T \tag{8}$$

A hysteretic damping model is considered, i.e., with complex stiffness parameters.

The computational efficiency can be improved by dividing matrix [K] into sub-matrices, independent of the wavenumber and frequency. This is achieved by considering the matrix [B] as the sum of two matrices, whereby the numerical and analytical derivatives are separated. Equation 6 can then be replaced by:

$$\left([K]_1 + ik_1[K]_2 + k_1^2[K]_3 - \omega^2[M]\right)u_n(k_1) = p_n(k_1) \tag{9}$$

2.2.2. Formulation of 2.5D PML

The finite element approach is not the most suitable way for simulating unbounded domains. Actually, the finite character of the domain of interest and the need of definition of a truncation boundary do not allow the reaching of the Sommerfeld condition by a FEM approach. Therefore, PML layers must be applied along the interest domain in order to avoid the spurious reflection of waves that impinge the border of the interest domain.

As mentioned before, the PML layer is discretized into 2.5D finite elements with special properties that allows for the absorption of the wave field. It should be remembered that the analytically continued solution outside the interest domain satisfies the same differential equations. So, in the 2.5D context, the coordinates y and z are stretched into the complex domain. The natural and modified coordinates are related by the following relationships:

$$\tilde{y} = \int_0^y \lambda_y(y)dy \tag{10}$$

$$\tilde{z} = \int_0^z \lambda_z(z)dz \tag{11}$$

where λy and λz are the stretching function in y and z directions, respectively.

Concerning the derivatives, it is also possible to find the following relationships:

$$\frac{\partial}{\partial \tilde{y}} = \frac{1}{\lambda_y(y)}\frac{\partial}{\partial y} \tag{12}$$

$$\frac{\partial}{\partial \tilde{z}} = \frac{1}{\lambda_z(z)}\frac{\partial}{\partial z} \tag{13}$$

Since the solution inside the PML domain satisfies the same differential equation as in the interest domain, the change of coordinates in Equation (1) is enough to create a 2.5D PML approach. Introducing Equations (11) to (14) to the Garlekin formalism, the following stiffness, $[K^*]$ and mass, $[M^*]$, matrices can be derived for the PML region:

$$[K^*] = \int_z \int_y B^{*T}(-k_1)\, D\, B^*(k_1)\, \lambda_y \lambda_z dy\, dz \tag{14}$$

and

$$[M^*] = \int_z \int_y N^T \rho\, N\, \lambda_y \lambda_z dy\, dz \tag{15}$$

As usual, the matrix B^* is derived by the product of the differential operator matrix $[L^*]$ (in the transformed and stretched domain) and matrix [N]. In order to respect Equations (13) and (14), this new differential operator matrix is given by :

$$[L^*] = \begin{bmatrix} ik_1 & 0 & 0 & \frac{1}{\lambda_y(y)}\frac{\partial}{\partial y} & 0 & \frac{1}{\lambda_z(z)}\frac{\partial}{\partial z} \\ 0 & \frac{1}{\lambda_y(y)}\frac{\partial}{\partial y} & 0 & ik_1 & \frac{1}{\lambda_z(z)}\frac{\partial}{\partial z} & 0 \\ 0 & 0 & \frac{1}{\lambda_z(z)}\frac{\partial}{\partial z} & 0 & \frac{1}{\lambda_y(y)}\frac{\partial}{\partial y} & ik_1 \end{bmatrix}^T \tag{16}$$

It should be stressed that the PML region can be absorbing in both directions, in one direction or in neither direction. The latter situation corresponds to $\lambda y = \lambda z = 1$, i.e., to a 2.5D finite element formulation. Kausel and Barbosa [36] have recently proposed an alternative formulation for 2D PMLs where the matrices are directly derived in stretched space. However, at the end, the result is the same.

Assembling the 2.5D FEM and 2.5 PML matrices, the solution is obtained in the transformed wavenumber - frequency domain, requiring a double Fourier inverse operation to change for the untransformed domain (space-time).

2.2.3. Stretching Functions

From the presented in the last two sections, one can conclude that the incorporation of a 2.5D PML procedure into a 2.5D FEM numerical application is quite simple. However, the stretching functions must be defined. These functions are derived from the continuation of the solution of the wave field beyond the domain of interest. Moreover, careful must be taken regarding the increase of the attenuation along the thickness of the PML in order to avoid spurious reflections. The evolution of the attenuation must be smooth, which is generally achieved by stretching functions of the following form:

$$\lambda_y(y) = \frac{2\pi}{\text{abs}(k)}\frac{y}{H_y} - i\frac{k_0}{k}\left(\frac{y}{H_y}\right)^2 \tag{17}$$

$$\lambda_z(z) = \frac{2\pi}{\text{abs}(k)} \frac{y}{H_y} - i\frac{k_0}{k}\left(\frac{z}{H_z}\right)^2 \tag{18}$$

where k_0 is an appropriate constant (in the present study it was found that k_0=20 gives rise to good results); H_y and H_z are the thickness of the PML in y and z directions, respectively, k is the effective wavenumber for waves that propagate along the cross-section, being given by:

$$k = \sqrt{\left(\frac{\omega}{C_s}\right)^2 - k_1^2} \tag{19}$$

where C_s is the shear wave velocity.

The real part of the stretching functions is introduced to proceed to a stretching of the mesh in order to obtain a thickness of the PML domain of, at least one wavelength. So, in the proposed approach, the PML mesh is always generated with 1 meter of thickness and divided into 5 or 6 sub-layers. Actually, an automatic expansion of the refereed domain is achieved by the stretching functions. The proposed stretching function were validated by comparison between the results provided by the proposed model and by theoretical solutions,

It should be noticed that if the radicand of Equation (20) is negative, the stretching functions don't have imaginary part, i.e., the propagate wave along the cross-section of the problem is evanescent and, consequently, the stretching functions (18) and (19) are pure real values, or, in other point of view, there isn't attenuation of the wave but there is a stretch of the mesh in order to better simulate the infinitive character of the domain.

3. Theoretical Validation

3.1. Vertical Load in a Full-space

In order to check the reliability and accuracy of the implemented model, a simple example was simulated and the results were compared with the closed-form solution given by Tadeu and Kausel [29]. Figure 3 shows the main characteristics of the problem as well as the geometry of the 2.5D finite element mesh adopted. For the numerical analysis, it was taken advantage of the symmetric and anti-symmetric conditions of the problem.

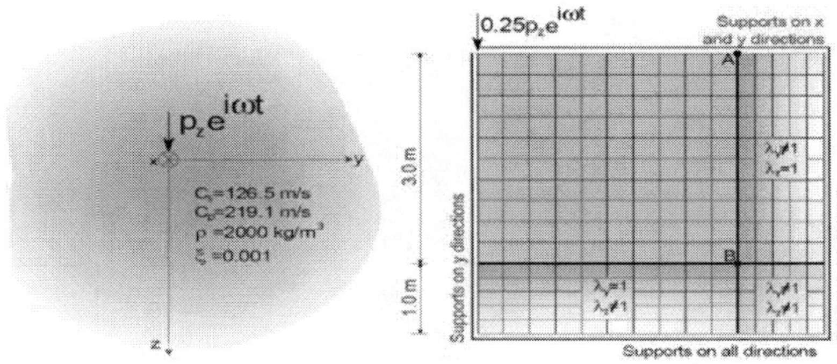

(a) (b)

Figure 3. Description of the validation problem: (a) Fullspace properties; (b) 2.5D finite element mesh.

The interest field has 3m×3m of cross section and is represented by 2.5D finite elements with element size l_e=0.3 m. By other hand, 1m thickness PML layer is attached along the artificial boundaries, being its dimension expanded as function of the wavelength that propagates in the yoz plane.

The fullspace is excited by a vertical load that is harmonic with respect to the longitudinal coordinate x with wavenumber k_1. In the frequency-space domain, the load is given by:

$$p_j(x, y, z, \omega) = \delta_{ij}\delta(y)\delta(z)\cos(k_1 x) \qquad (20)$$

Since the load is harmonic, the resulting displacements are also harmonics:

$$u_j(x, y, z, \omega) = u_j(k_1, y, z, \omega)\cos(k_1 x) \qquad (21)$$

Figures 4 and 5 show the vertical displacements along the y axis for dimensionless wave numbers k_1, defined as $k_1 = k_1 C_s/\omega$, and for frequencies of 10 Hz and 75 Hz, respectively.

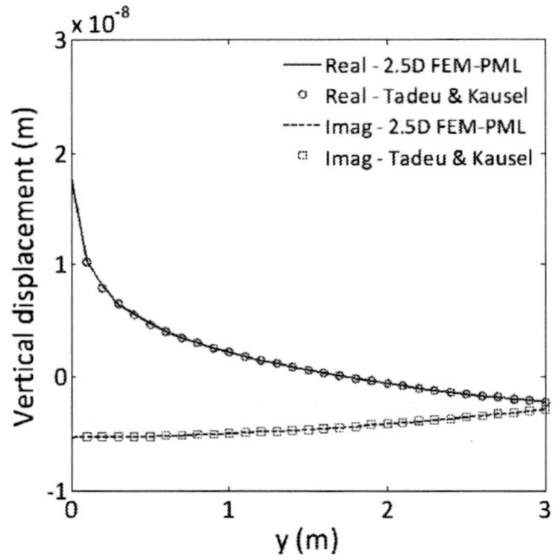

(a)

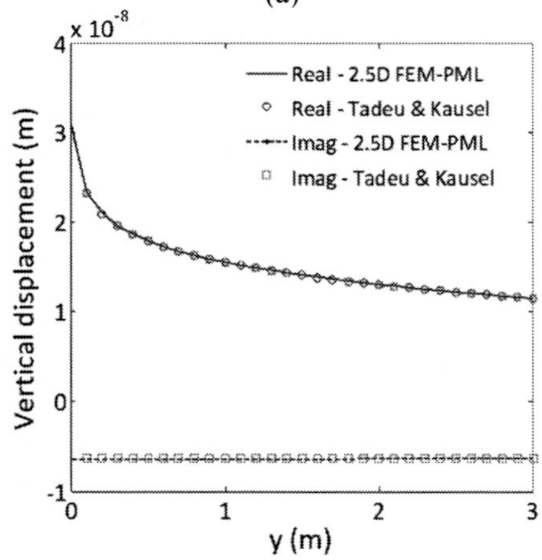

(b)

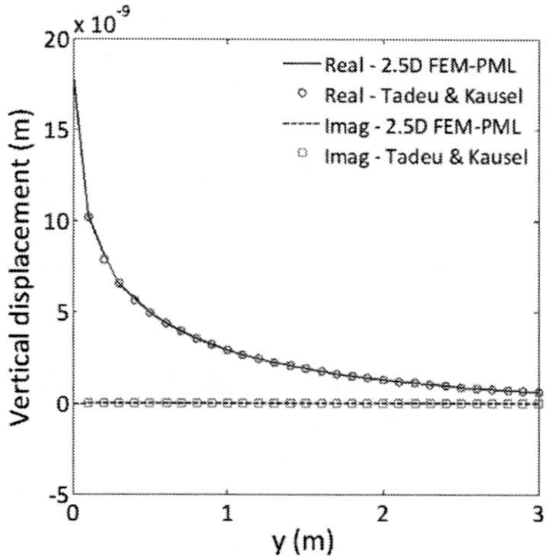

(c)

Figure 4. Real and imaginary parts of vertical displacement uz(k_1,y,z,ω) along y-axis of the fullspace excited at 10 Hz by a vertical load at dimensionless wavenumber: (a) k_1=0.5; (b) k_1=1.0; (c) k_1=1.5.

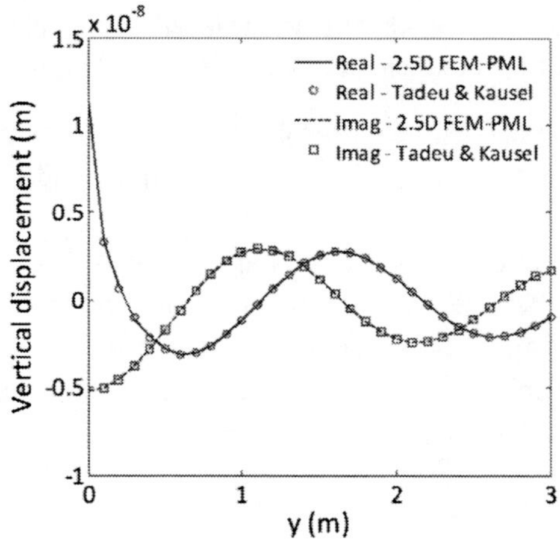

(a)

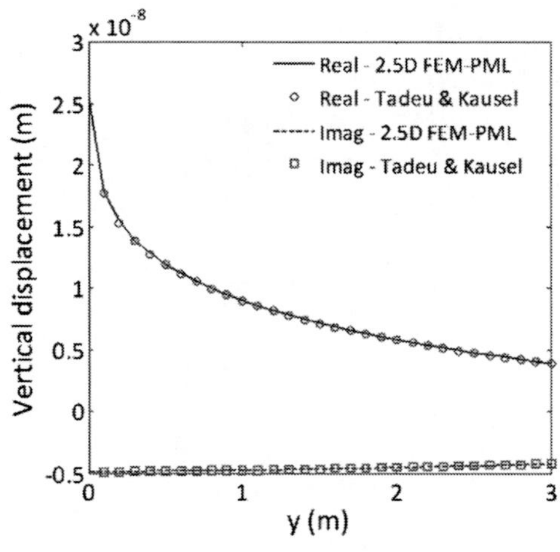

(b)

Figure 5. (Continued)

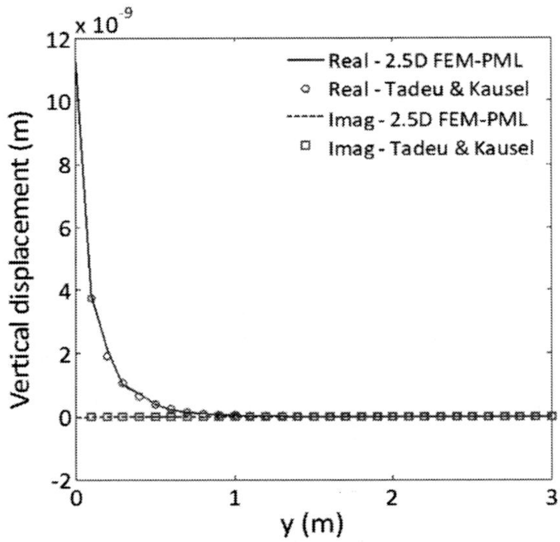

(c)

Figure 5. Real and imaginary parts of vertical displacement uz(k_1,y,z,ω) along y-axis of the fullspace excited at 75Hz by a vertical load at dimensionless wavenumber: (a) k_1=0.5; (b) k_1=1.0; (c) k_1=1.5.

Analyzing the results depicted in Figures 4 and 5, one can conclude that the agreement between numerical and theoretical solutions is splendid, independently of the wavenumber or frequency. It should be pointed out that for k_1 higher than 1 only evanescent waves can propagate along the cross-section. This effect is also very well reproduced by the numerical model, with a perfect match between numerical and theoretical results.

In order to extend the validation procedure, a 2D analysis was also performed. In fact, imposing k_1=0, the 2.5D problem redounds on a 2D plane strain analysis. For that situation and assuming the properties depicted in Figure 3, the excitation induced by the Ricker pulse (Figure 6) was imposed to the full-space.

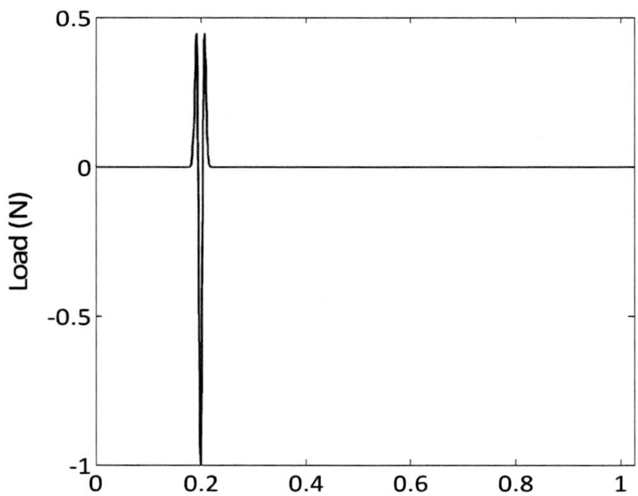

(a)

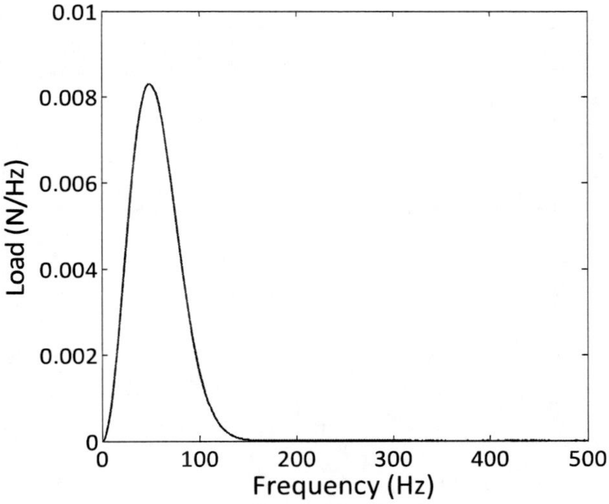

(b)

Figure 6. Ricker pulse adopted: (a) time representation; (b) Frequency content.

Figure 7 shows the time-space response of the points A and B indicated in Figure 3, which belong to the boundary of the interest domain. Overlapped to the numerical results, the theoretical ones (computed by the 2.5D Green's functions proposed by Tadeu and Kausel) are also plotted. Once again, the adherence between numerical and theoretical solution is perfect, having no distinction between the results even for points located along the boundary of the interest domain.

A more comprehensive comparison is given by Figure 8, where the field of vertical displacements is depicted for few time instants. Only the 2nd quadrant of the figure was computed by the numerical method, being the remaining quadrants obtained by the theoretical solution.

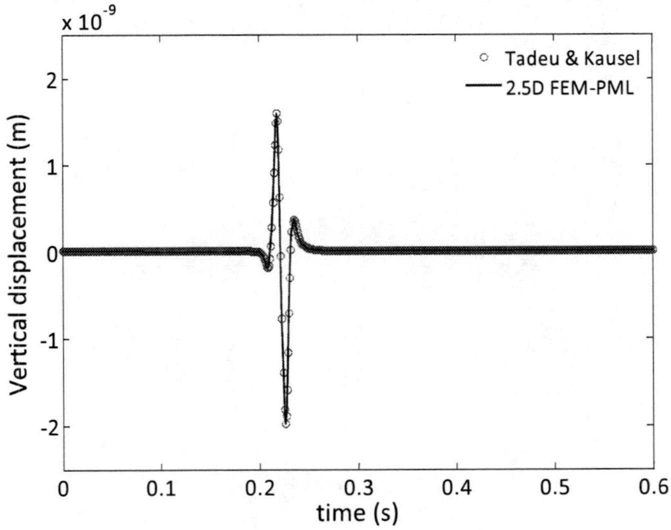

(a)

Figure 7. (Continued)

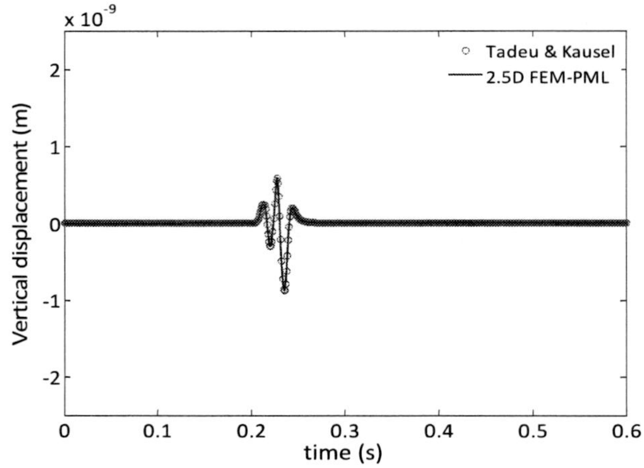

(b)

Figure 7. Time domain vertical displacements due to the excitation induced by Ricker pulse: (a) Point A; (b) Point B.

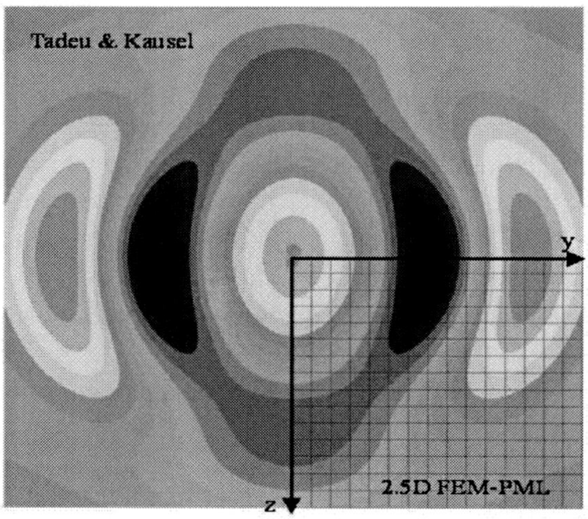

(a)

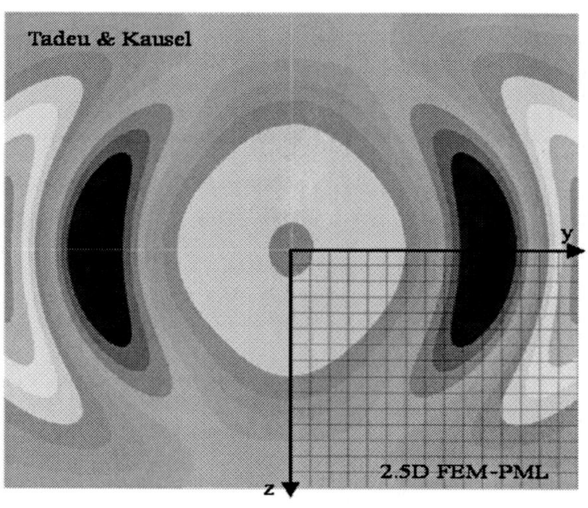

(b)

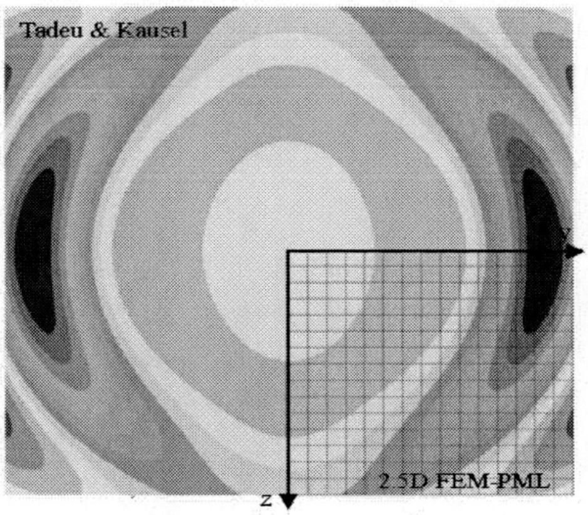

(c)

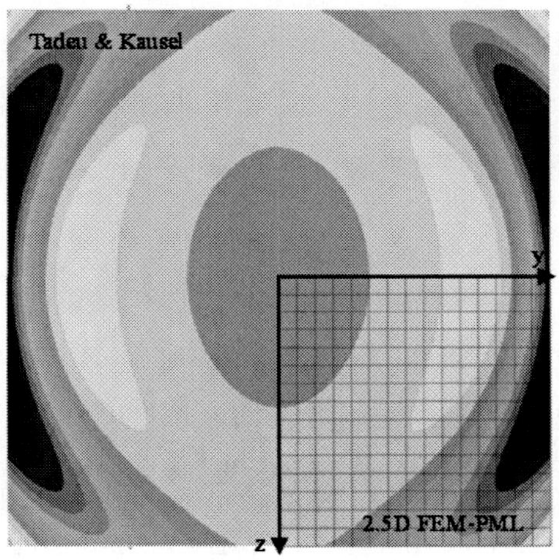

(d)

Figure 8. Vertical displacement field of the full-space due to the Ricker pulse excitation at distinct time instants: (a) $t=0.214$s, (b) $t=0.219$s, (c) $t=0.224$s, and (d) $t=0.229$s.

3.2. Tunnel Embedded on a Full-Space

Despite of the interest of the results presented in above section, which highlight the accuracy of the developed tool, it should not be forgotten that the goal is the development of a numerical tool for the analysis of the dynamic response of tunnels. Having that goal in mind, a case study previously presented by Gupta et al. [2] is here revised. This case study was used for the validation of a numerical model developed by those authors, comparing the numerical solution with the semi-analytical results provided by the model PiP developed by Hussein and Hunt [30]. Actually, the following presented study compares three distinct methodologies for the study of the dynamic response of a tunnel embedded in a full-space: i) the 2.5D FEM-PML; ii) the periodic FEM-BEM model presented by Gupta et al. [2]; iii) the semi-analytical model PiP, which is here assumed as reference solution. Figure 9 summarizes the properties of the tunnel and evolving fullspace. The tunnel, with inner radius of 2.75 m and a concrete liner with 0.25 m of thickness, is subjected to a vertical harmonic point load applied in its invert. The observation points are defined by polar coordinates in the cross-section of the problem and by rectangular coordinates in the orthogonal direction, as follows: x=0, r, θ. The load is applied in the point with coordinates (0,2.75,0).

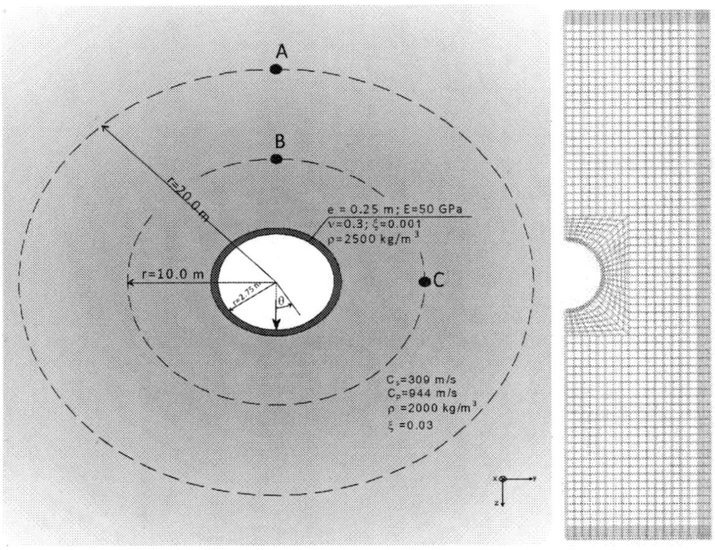

(a) (b)

Figure 9. Tunnel embedded on a fullspace: (a) Problem description; (b) 2.5D finite elements mesh.

The finite elements mesh has 10 m of width and 40 m of height. The domain was discretized into 2.5D finite elements with edge size smaller than 0.6 m in order to obtain accurate results for frequencies up to 80 Hz. To avoid spurious reflections of waves that can hit the artificial boundaries, 1 m thickness PML's, comprising 5 layers of 2.5D finite elements, were coupled around the artificial boundaries. Figure 10 shows the real part of the norm of the displacement field for the plane yoz, i.e., the plane that comprises the load and the observation points, and for distinct frequencies.

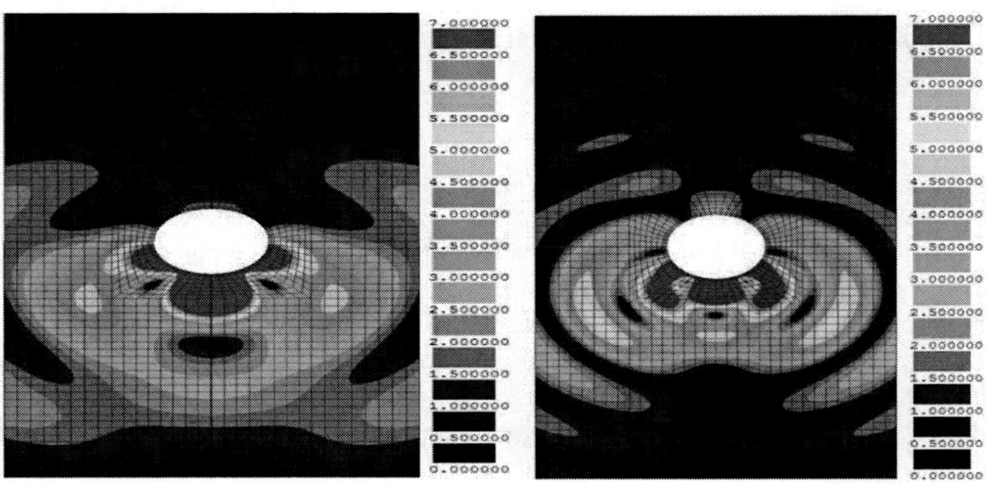

(a) (b)

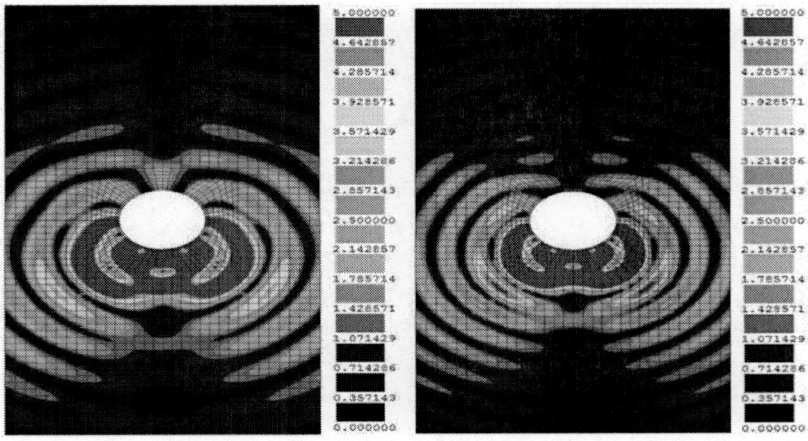

(c) (d)

Figure 10. Real part of the field of displacements along the yoz plane induced by a harmonic point load, applied at the tunnel invert, with distinct frequencies ($\times 10^{-11}$): (a) 20 Hz; (b) 40 Hz; (c) 60Hz; (d) 80 Hz.

The comparison between the results provided by the distinct models is given by Figure 11, for the three observation points.

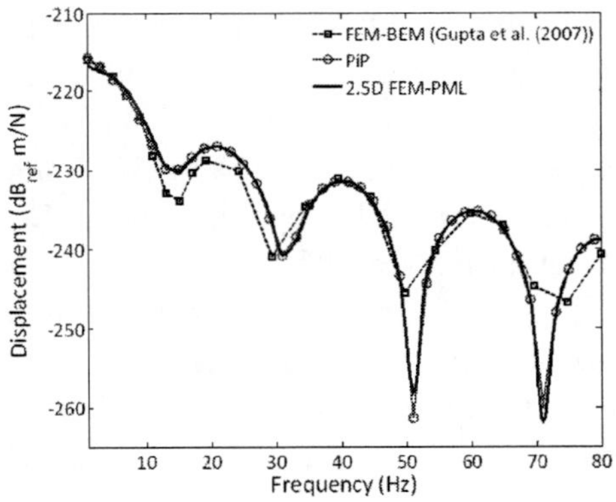

(a)

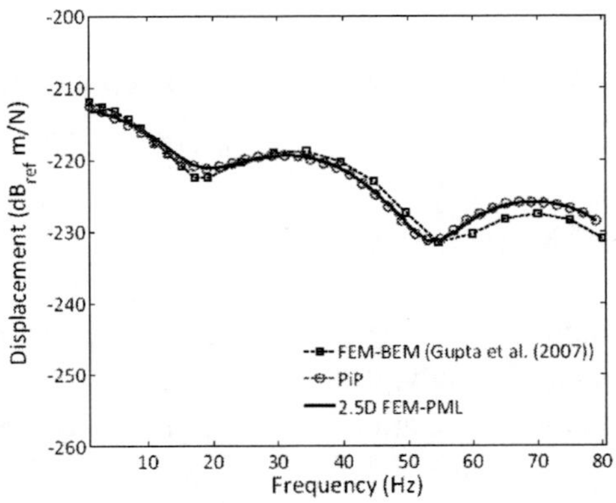

(b)

Figure 11. (Continued)

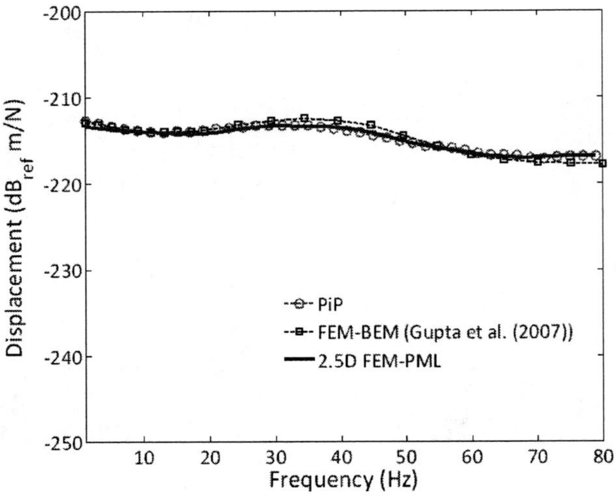

(c)

Figure 11. Transfer function of vertical displacements at different locations: (a) Point A; (b) Point B; (c) Point C.

As can be seen, a very good accuracy is achieved with the proposed model. Actually, there is an almost perfect agreement between the numerical results provided by the proposed model and the results obtained using PiP. Moreover, the numerical model is quite versatile, allowing complex geometries that are not possible to include in a semi-analytical model as the PiP model.

Concerning the comparison with the periodic FEM-BEM model proposed by Gupta et al. [2], it is clear, from the results expressed above, that the 2.5D FEM-PML can supplant its accuracy. In fact, the undulation between troughs with a step of 22 Hz that is present in Figure 11(a) is very well reproduced. As discussed by Gupta et al. [2], those undulations are due to the interference between propagating shear and compression waves in the full-space.

It should be also stressed the high accuracy reached even for points located in the boundary between the interest domain, i.e., the domain described by 2.5D FEM, and the PML domain.

4. Dynamic Response of a Tunnel Embedded in a Half Space

4.1. Description of the Reference Scenario

Since the numerical model is adequately validated, at least from the theoretical point of view, in this section a parametric study, developed in order to better understand the influence of soil-tunnel properties on the propagation of vibrations induced by dynamic loads applied in the interior of the tunnel, is presented.

Figure 12 depicts the geometry and mechanical properties adopted for the reference scenario.

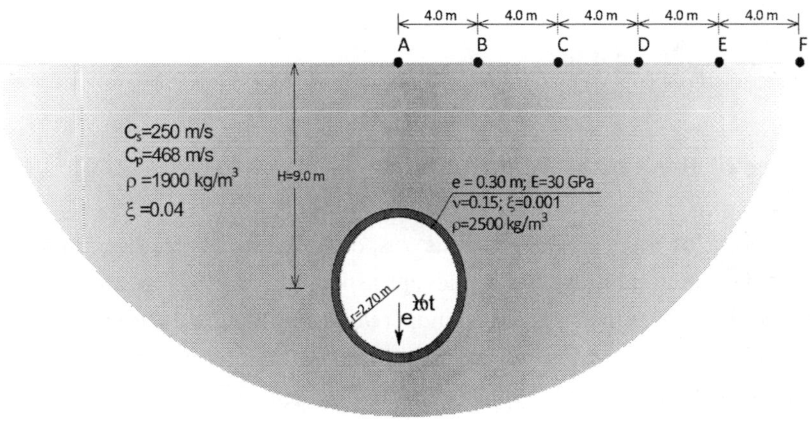

Figure 12. Geometry and material properties adopted for the reference scenario.

As can be seen, the study now proposed is quite simple, being a shallow tunnel considered embedded in a homogeneous half-space. The dynamic excitation is provided by a vertical harmonic load, with a variable frequency, applied on the tunnel floor. Concerning the observation points, it was considered the 6 locations at the ground surface indicated in the figure as well as the point of the tunnel beneath the load application. The observation points belong to the same cross-section of the point where the load is applied (stand-still loading conditions). However, the problem remains tri-dimensional since a point load is considered.

Regarding the numerical simulation, a 2.5D FEM-PML model was developed taking advantage of the symmetrical conditions of the problem. The domain of interest was considered with a width of 20 m, i.e., the vertical alignment that includes point F of Figure 12 corresponds to boundary of the interest domain. The finite elements dimension were designed in order to analyze a frequency range up to 80 Hz. Coupled to the interest domain, a layer of 2.5D PML's with 1 m thickness and divided into 6 sub-layers, was considered. Figure 13 shows the finite elements mesh adopted for the reference scenario.

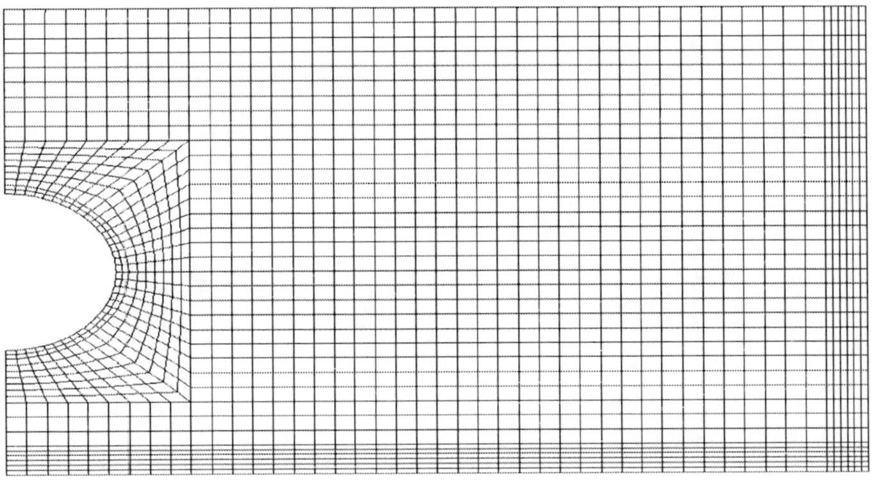

Figure 13. Finite elements mesh adopted for the reference scenario.

From the reference scenario, a parametric study was performed and the main conclusions are presented in the following subsections.

4.2. Influence of the Tunnel Depth

From the point of view of the dynamic response induced into the ground surface, it is quite obvious that the tunnel depth is a relevant aspect that must be addressed in the analysis of present problem. Having this in mind, two additional depths were considered, in correspondence to H=12 m and H=15 m, in addition to the reference scenario where H=9 m.

Figure 14 shows the vertical displacement induced in the reference points located at the ground surface and for the three distinct scenarios under analysis.

A general overview of Figure 14 allows to conclude that the trend of the dynamic response of the ground surface is quite complex. Several aspects contribute for that complexity, namely the interaction of shear, compression and Rayleigh waves, being the consequence of this interaction denoted by the troughs present in the curves. The frequency step of these troughs varies as function of the tunnel depth because its occurrence is also dependent of the distance between the source and the receiver.

However, although the complexity of the trend, it is indisputable that as deeper is the tunnel lower are the vibration levels. Actually, this effect is quite obvious for the points closer to the tunnel alignment, as can be seen in Figures 14a and b, or for the points located far away from the tunnel alignment (Figures 14 e and f). In latter case, the effect is more pronounced for frequencies above 20 Hz. This attenuation effect as function of the tunnel depth can be attributed to two main causes: i) to the damping, thus as the distance source-receiver increases with the tunnel depth; ii) to the decrease of the frequency range where Rayleigh waves generated by the excitation. [3] showed that the frequency content of the Rayleigh waves is confined to a frequency range up to C_R/d, where d is the source depth and C_R is the velocity of Rayleigh waves. Thus, the increase of the tunnel depth leads to a reduction of the frequency range where Rayleigh waves are important.

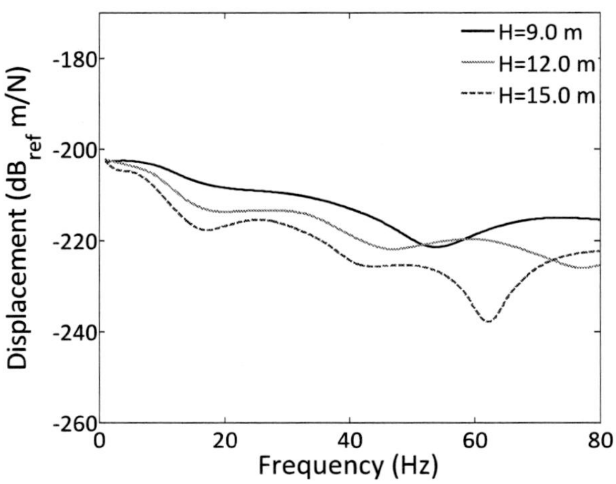

(a)

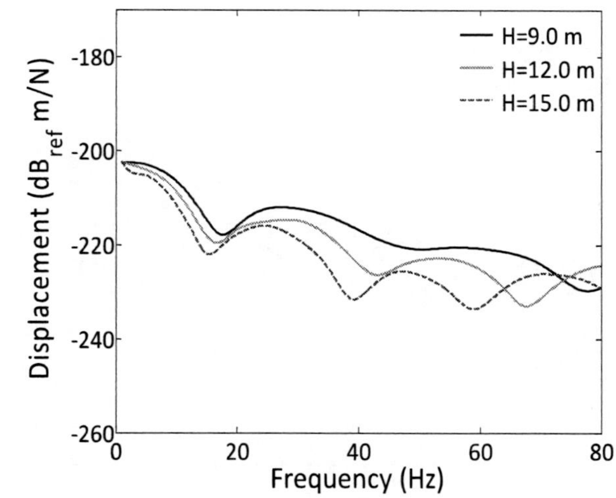

(b)

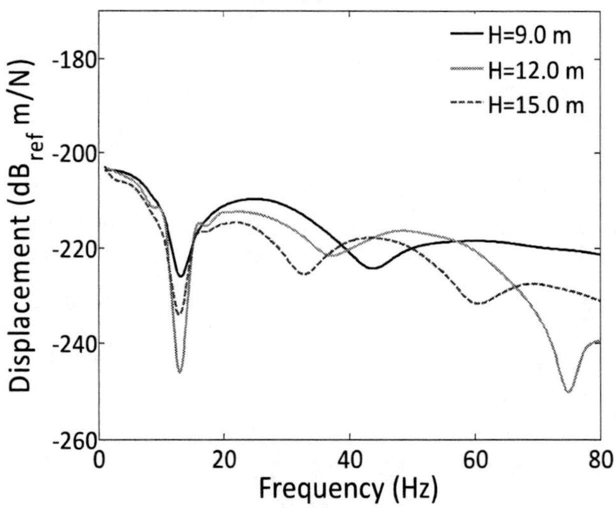

(c)

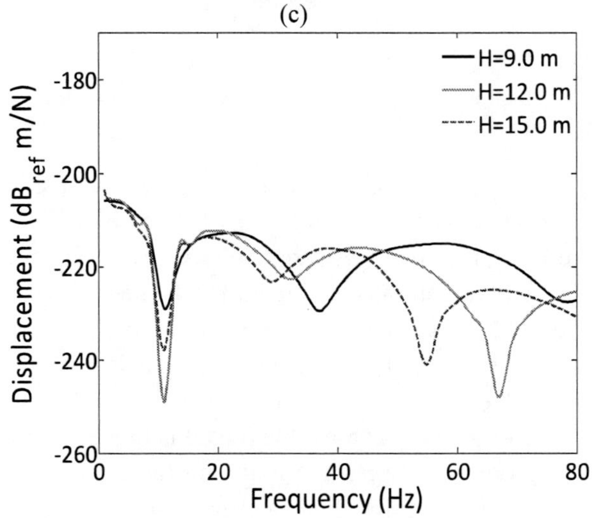

(d)

Figure 14. (Continued)

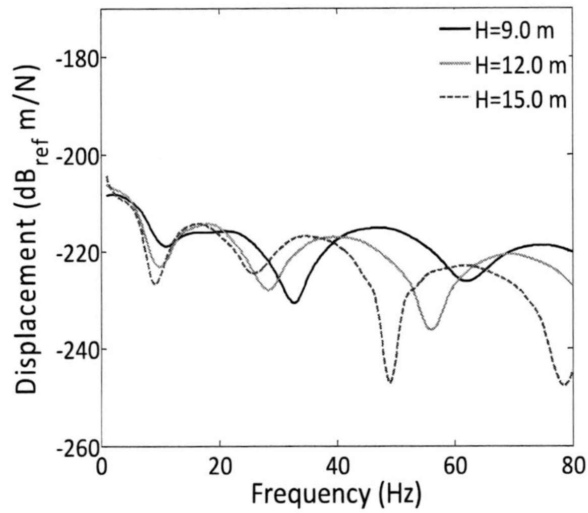

(e)

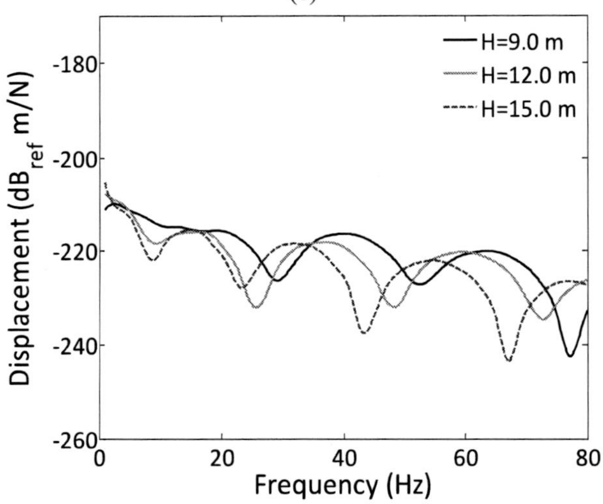

(f)

Figure 14. Vertical displacement at different locations for distinct tunnel depths: (a) point A; (b) point B; (c) point C; (d) point D; (e) point E; (f) Point F.

Regarding the dynamic response of the point located beneath the load application, it was found that the tunnel depth was a negligible effect, as can be observed in Figure 15.

As can be seen, only slight differences can be observed in the results presented in Figure 15, mainly when the focus of attention is appointed for the deeper tunnels. This conclusion is quite important for the point of view of vibrations induced by railway traffic. Indeed, the dynamic loads generated during the train passage are dependent of the dynamic behavior of the track support [8, 12]. Thus, since the dynamic response of the tunnel looks to be very similar, mainly if the ground thickness above the tunnel is higher than 1.5 times its diameter, the dynamic interaction problem between the train and the remaining system (track-tunnel-ground) should be independent of the tunnel depth. So, the dynamic interaction loads generated in a tunnel with a certain depth can be used as input excitation for the study of vibrations generated in a tunnel with a different depth.

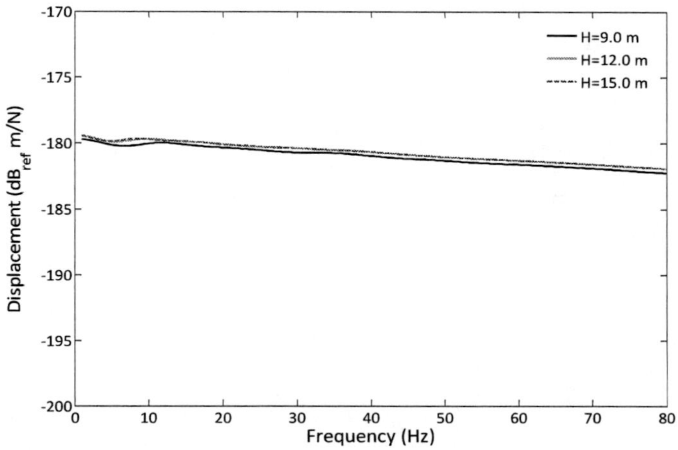

Figure 15. Displacement of the point beneath the load application for distinct frequencies and for different tunnel depths.

4.3. Influence of the Thickness of the Tunnel Liner

The thickness of the liner tunnel is one of the parameters that affect the in-plane dynamic behaviour of the tunnel, as well as the longitudinal flexure bending stiffness, and, consequently, the tunnel-ground interaction mechanism. In order investigate the effect of this parameter on the global dynamic behaviour of the system, two additional scenarios are now considered for the liner thickness: d=0.5 m and d=0.7 m. The external radius of the tunnel is assumed constant with the value depicted in Figure 12. Although the values assumed could be faced as exaggerated in comparison with the tunnel diameter, from the theoretical point of view, it is important to consider a high thickness contrast in order to highlight the influence of this parameter. Figure 16 depicts the vertical displacements of the observation points located at the ground surface as function of the frequency of excitation and for the three scenarios under consideration.

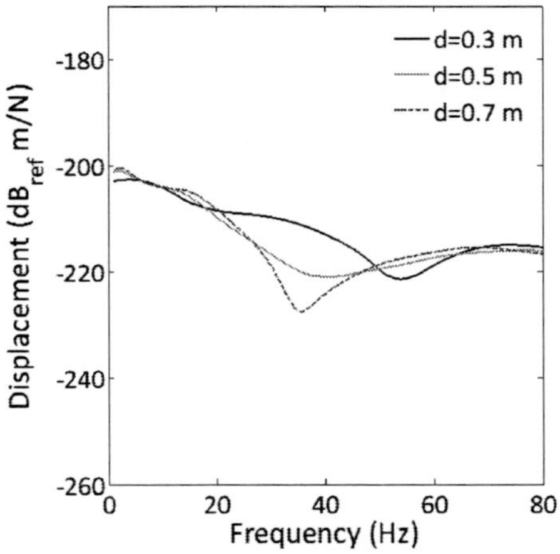

(a)

Figure 16. (Continued)

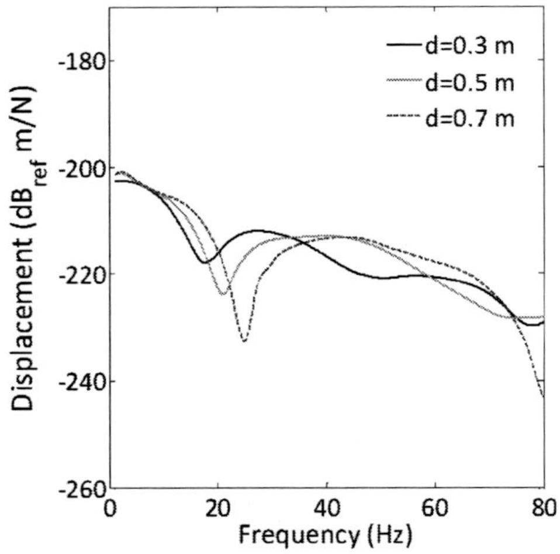
d=0.3 m
d=0.5 m
d=0.7 m
Displacement (dB$_{ref}$ m/N)
-180
-200
-220
-240
-260
0
20
40
60
80
Frequency (Hz)

(b)

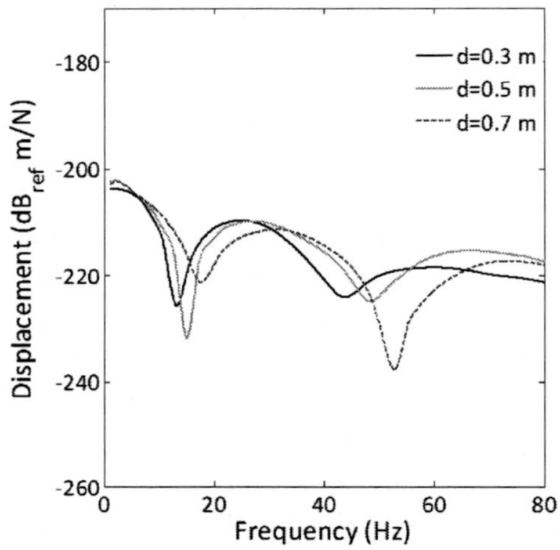
d=0.3 m
d=0.5 m
d=0.7 m
Displacement (dB$_{ref}$ m/N)
-180
-200
-220
-240
-260
0
20
40
60
80
Frequency (Hz)

(c)

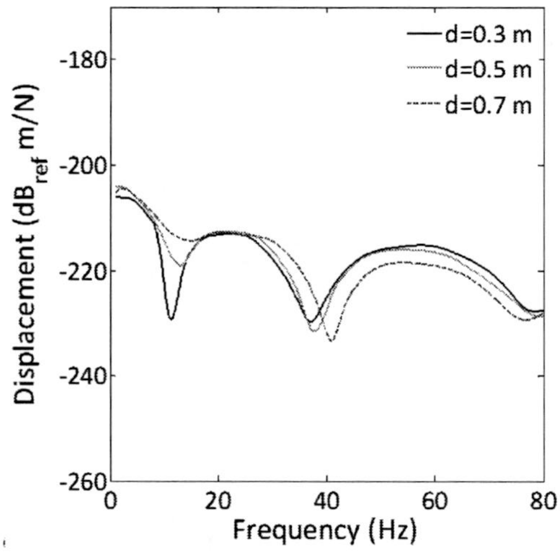
d=0.3 m
d=0.5 m
d=0.7 m
Displacement (dB$_{ref}$ m/N)
-180
-200
-220
-240
-260
0
20
40
60
80
Frequency (Hz)

(d)

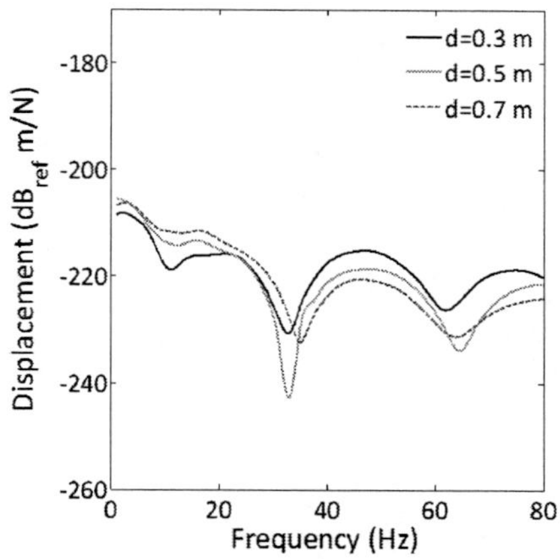

(e)

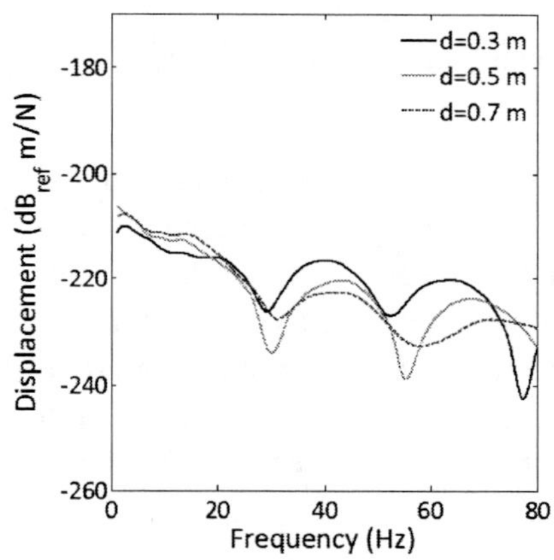

(f)

Figure 16. Vertical displacement at different locations for distinct thicknesses of the tunnel liner: (a) point A; (b) point B; (c) point C; (d) point D; (e) point E; (f) Point F.

A first observation of Figure 16 suggests the division of the results into two main groups: i) dynamic response for distances up to 8 m (Figures 16a, b and c); ii) dynamic response for distances above 8 m (Figures 16d, e and f). Regarding the first group, it is denoted that the stiffening of the tunnel by increasing its liner thickness do not affect substantially the vertical displacements of the ground surface. Nevertheless, in Figures 16b and c, a shift for higher frequencies of the first trough is observed with the increasing of the tunnel stiffness. This effect can be due to the increase of the eigenfrequencies of the in-plane flexural modes of the free tunnel induced by the increase of its liner thickness. Comprising the dynamic response for distances above 8 m, the general trend follows for the reduction of the vertical displacements of the surface with the increasing of the liner thickness, mainly for frequencies above 30 Hz, approximately. Thus, the increase of the tunnel liner stiffness can be faced as a mitigation measure for the abatement of vibrations induced by traffic when the distance between the source and the receiver is higher than 1.5 times the tunnel diameter.

Regarding the influence of the thickness of the liner on the dynamic response of the tunnel itself, Figure 17 allows a comparison between the vertical displacements of the point of load application for the three scenarios considered.

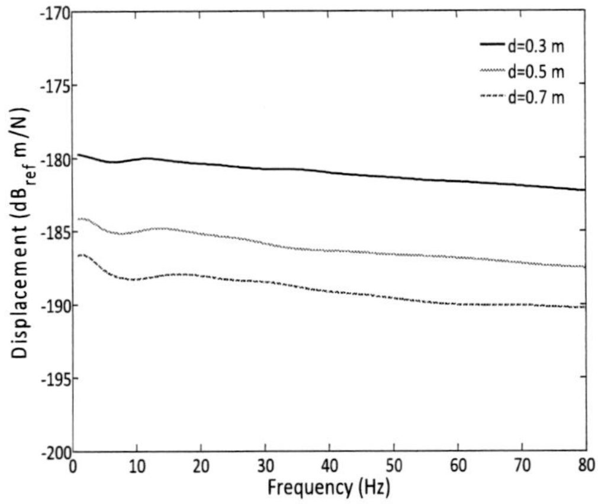

Figure 17. Displacement of the point beneath the load application for distinct frequencies and for different tunnel liner thicknesses.

The general trend of the amplitude of the vertical displacement of the point of the tunnel where the load is applied is similar for the three scenarios now considered. In spite of that, a considerable reduction of the amplitude of the displacements is achieved by the increase of the tunnel liner. Thus, on contrary to what observed in the latter sub-section, the thickness of the tunnel liner is a parameter that can play a non-negligible role on the train-track-tunnel interaction mechanism. Surely, it should not be forgotten that the mechanical properties of the track, namely the resilient materials introduced beneath the rail, can minimize the relevance of this effect.

4.4. Influence of Ground Material Damping

The problem of vibrations induced inside tunnels is quite complex, evolving not only the mechanical properties of the tunnel but also properties of the ground as for instance its stiffness, layering and damping. In order to illustrate the influence of the material damping of the ground a parametric study is now presented where, in addition to the reference scenario where $\xi=0.04$, two other situations are analyzed for damping ratios of 0.02 and 0.06.

The attenuation of vibration levels is due to two distinct mechanisms: the geometrical damping and the material damping. The geometrical damping refers to a spread of energy due to the wave propagation and it is dependent of the wave type. On the other hand, the material damping comprises the energy dissipation, being more evident its effect with the increase of frequency and distance receiver-source. This effect is well patent in Figure 18, where the amplitude of the vertical displacements of the points located at the ground surface are depicted.

From Figure 18 several remarks can be pointed out: i) as expected, the increase of damping is associated to a decrease of the vibration levels; ii) the effect of damping is as much denoted as higher is the frequency of excitation, independently of the distance source-receiver; iii) the abatement effect due to material damping is quite relevant when the distance source-receiver is large (see Figure 18f).

The remarks pointed above are justified by the dependence of the damping effect of the number of vibration cycles the material has undergone [3]. For higher frequencies, i.e., smaller wavelengths, a considerable abatement of vibration levels can be achieved when the distance source-receiver implies a significant number of cycles that the material has undergone.

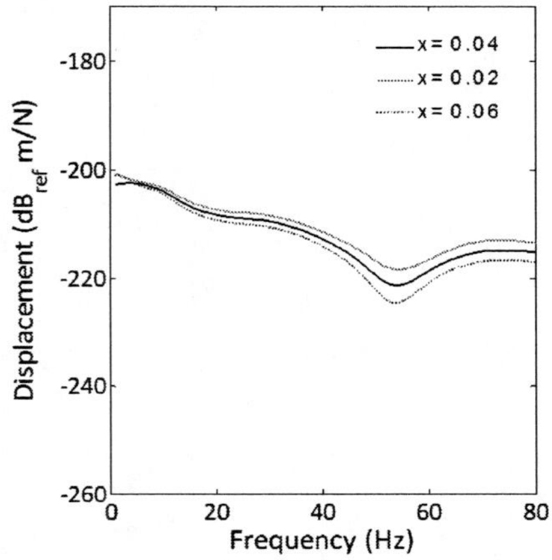

(a)

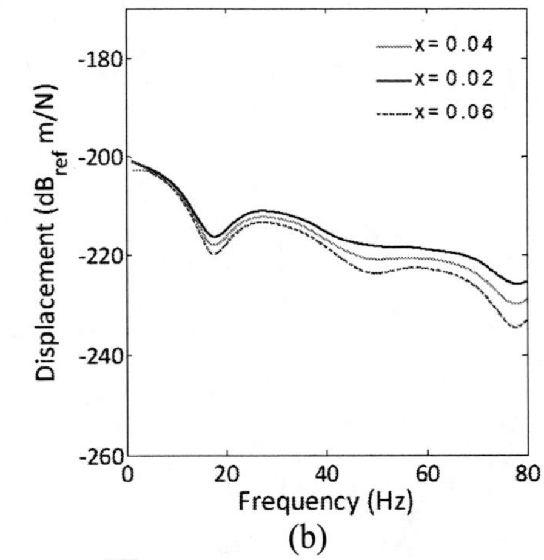

(b)

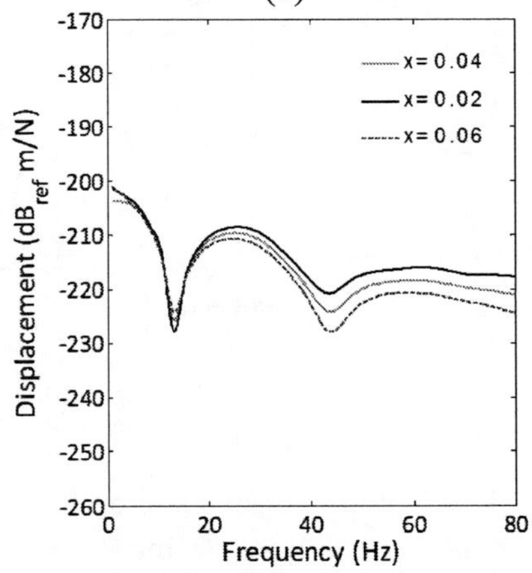

(c)

Figure 18. (Continued)

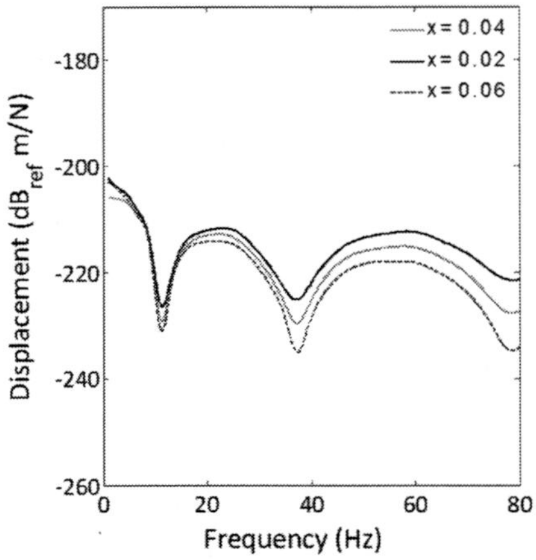

(d)

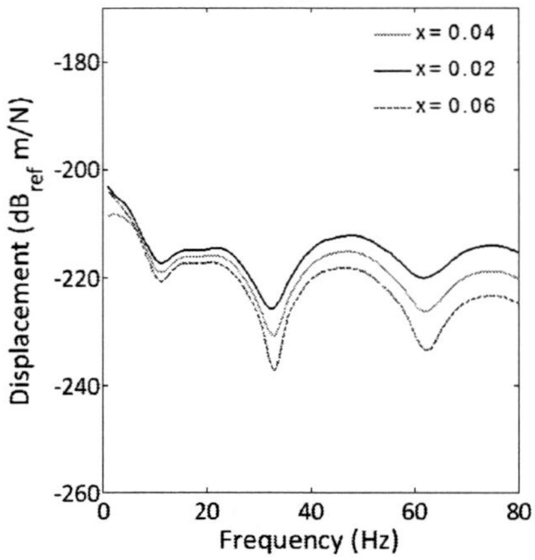

(e)

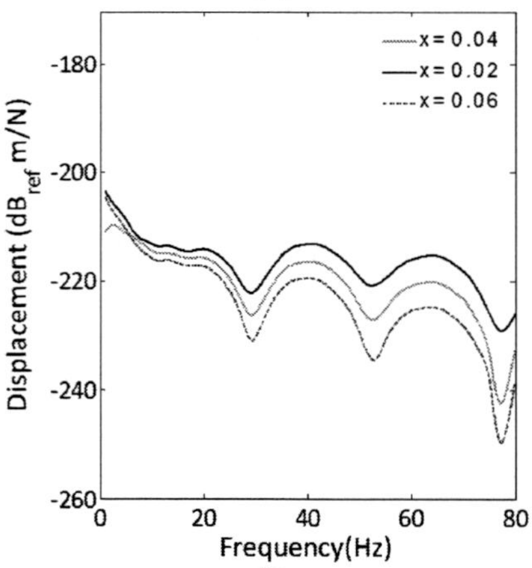

(f)

Figure 18. Vertical displacement at different locations for distinct values of the material damping: (a) point A; (b) point B; (c) point C; (d) point D; (e) point E; (f) Point F.

Regarding the influence of the damping of the soil on the dynamic response of the tunnel floor, the obtained results are depicted in Figure 19. Indeed, the material damping is a more or less negligible parameter for the assessment of the displacements of the tunnel itself. Although not analyzed in the present chapter, the conclusions here found allow advancing that the material damping has a small influence on the train-track dynamic mechanism.

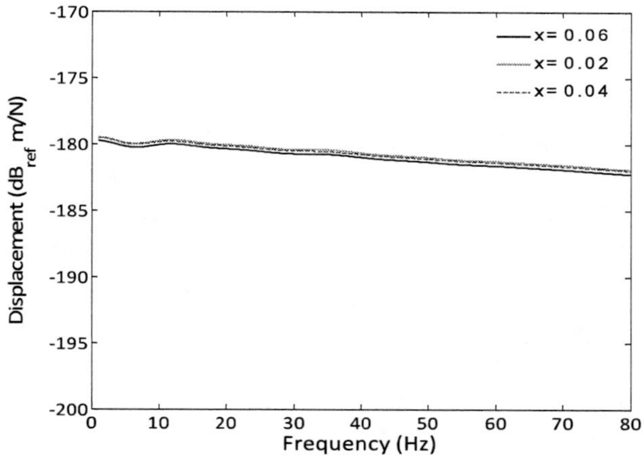

Figure 19. Displacement of the point beneath the load application for distinct frequencies and different values of the material damping.

CONCLUSION

In this chapter, a 2.5D FEM-PML model has been derived and validated. This kind of model is especially adequate for the study of infinite structures, as for instance tunnels and other transportation infrastructures.

Since the PML equations are quite similar to the equations often used in a finite element analysis, the compatibility of both methods is fulfilled without the need of special numerical procedures. The stretching functions are the key aspect of the method; it was suggested and adopted a kind of stretching function that is able for leading with evanescent and propagating waves along the cross-section of the domain.

The validation of the method was done by a two step approach. Firstly, a comparison between the numerical results and 2.5D Green's functions was performed. From those simple examples it was demonstrated the high accuracy that can be reached by the proposed model. In fact the adherence between theoretical and numerical results was perfect, allowing inferring the supremacy of the PML technique when compared with conventional local procedures for wave absorption, as for instance the viscous boundaries or the infinite elements. Moreover, it was also shown that unbounded domains can be accurately simulated even adopting finite element meshes with reduced dimension.

After the preliminary theoretical validation, a second example, comprising the tunnel-ground dynamic interaction, was simulated and the obtained solution was compared with the previously ones depicted by other authors [2]. Although the simplicity of the proposed example, it has the advantage of has been adopted by other authors to validate their proposals for dealing with this complex problem. The comparison between the 2.5D FEM-PML's solution with the semi-analytical ones revealed an almost perfect result, supplanting the results previously achieved by the usage of a periodic FEM-BEM model. Actually, even using reduced finite element mesh, with dimension equal to the interest domain, it was possible to achieve results with very high accuracy.

From the theoretical validation analysis it is possible to conclude, or at least to suggest, that the 2.5D FEM-PML procedure can be faced as an interesting alternative for more complex numerical procedures as for instance the 2.5D FEM-BEM. In fact, the proposed approach looks like to overcome one of the main drawbacks of the finite elements procedure: its inability for dealing with unbounded domains.

Finally, in order to highlight the potentialities of the proposed approach, a parametric study, comprising the dynamic interaction between a tunnel and the halfspace where the former is embedded, was presented. The influence of three variables (tunnel depth, thickness of tunnel liner and soil damping) on the vertical displacements of the ground surface due to a stand-still vertical harmonic load applied inside the tunnel was investigated. The study performed allowed to illustrate the influence of each variable at the ground surface, namely: it was shown that the vibration levels at the ground surface decrease with the increasing of the tunnel depth; the reduction of vibration levels induced by an increase of the damping material of the ground is so much more evident as higher is the frequency of excitation and the distance source-receiver; the influence of the thickness of the tunnel liner on the abatement of vibrations is only appreciable for larger distances source-receiver. Moreover, it was also possible to suggest that the damping of the ground and the tunnel depth are parameters of negligible influence into the train-track-tunnel dynamic interaction mechanism.

A final remark must be done involving the potentialities of the presented approach on the prediction of vibrations induced by traffic. Actually, although the analysis of moving loads was not presented in the present chapter, it should be stressed that the present numerical model was developed from a previous 2.5D FEM-BEM tool that allows the study of train-track-ground interaction [12, 37], being the same potentialities also available for the present proposal, i.e., the numerical model is able for dealing with moving loads as well as with the train-track interaction, being, from that point of view, an comprehensive numerical tool for dealing with problems induced by railway traffic in tunnels.

ACKNOWLEDGMENTS

This paper reports research developed under the financial support of "FCT - Fundação para a Ciência e Tecnologia", Portugal, namely from the research project - PTDC/ECM/114505/2009.

REFERENCES

[1] Gupta S., Berghe H., Lombaert G., Degrande G. (2010).Numerical modelling of vibrations from a Thalys high speed train in the Groene Hart tunnel. *Soil Dynamics and Earthquake Engineering*, 30(3): 82–97.

[2] Gupta S., Hussein M., Degrande G., Hunt H., Clouteau D. (2007). A comparison of two numerical models for the prediction of vibrations from underground railway traffic. *Soil Dynamics and Earthquake Engineering*, 27(7): 608–624.

[3] Gupta S., Stanus Y., Lombaert G., Degrande G. (2009). Influence of tunnel and soil parameters on vibrations from underground railways. *Journal of Sound and Vibration*, 327: 70–91.

[4] Clouteau D., Arnst M., Al-Hussaini T., Degrande G. (2005). Free field vibrations due to dynamic loading on a tunnel embedded in a stratified medium. *Journal of Sound and Vibration*, 283(1–2): 173–199.

[5] Lopes P., Alves Costa P., Calçada R., Silva Cardoso A. (2010). Análise numérica de vibrações induzidas por tráfego ferroviário em túneis baseada em modelos 2.5D, *in 12° Congresso Nacional de Geotecnia*, G. Correia, Editor: Guimarães. CD-ROM publication.

[6] Galvín P., François S., Schevenels M., Bongini E., Degrande G., Lombaert G. (2010). A 2.5D coupled FE-BE model for the prediction of railway induced vibrations. *Soil Dynamics and Earthquake Engineering*, 30(12): 1500–1512.

[7] Alves Costa P., Calçada R., Couto Marques J., Cardoso A. (2010). A 2.5D finite element model for simulation of unbounded domains under dynamic loading, *in Numerical Methods in Geotechnical Engineering - Proceedings of the 7th European Conference on Numerical Methods in Geotechnical Engineering* T. Benz and S. Nordal, Editors: Trondheim. 397–404

[8] Lombaert., G.. G. Degrande. (2009). Ground-borne vibration due to static and dynamic axle loads of InterCity and high-speed trains. *Journal of Sound and Vibration*, 319(3–5): 1036–1066.

[9] Sheng X., Jones C., Thompson D. (2006). Prediction of ground vibration from trains using wavenumber finite and boundary element method. *Journal of Sound and Vibration*, 293: 575-586.

[10] Yang Y., Hung H. (2008). Soil Vibrations Caused by Underground Moving Trains. *Journal of Geotechnical and Geoenvironmental Engineering*, 134(11): 1633–1644.

[11] Yang Y.B., Hung H.H. (2001). A 2.5D finite/infinite element approach for modelling visco-elastic body subjected to moving loads. *International Journal for Numerical Methods in Engineering*, 51: 1317–1336.

[12] Alves Costa P., Calçada R., Cardoso A. (2012). Track–ground vibrations induced by railway traffic: In-situ measurements and validation of a 2.5D FEM-BEM model. *Soil Dynamics and Earthquake Engineering*, 32: 111–128.

[13] Alves Costa P., Calçada R. (2011). Influence of ballast mats on the reduction of track-ground vibrations induced by railway traffic, in *Proceedings of 5th International Symposium on Environmental Vibration (ISEV2011)*: Chendu, China. 793–800.

[14] Alves Costa P., Calçada R., Cardoso A. (2012). Ballast mats for the reduction of railway traffic vibrations. Numerical study. *Soil Dynamics and Earthquake Engineering*, 42: 137–150.

[15] François S., Schevenels M., Galvín P., Lombaert G., Degrande G. (2010). A 2.5D coupled FE–BE methodology for the dynamic interaction between longitudinally invariant structures and a layered halfspace. *Computer Methods in Applied Mechanics and Engineering*, 199(23–24): 1536–1548.

[16] Rieckh G., Kreuzer W., Waubke H., Balazs P. (2012). A 2.5D-Fourier-BEM model for vibrations in a tunnel running through layered anisotropic soil. *Engineering Analysis With Boundary Elements*, 36: 960–967.

[17] Kuo K.A., Hunt H., Hussein M. (2011). The effect of a twin tunnel on the propagation of ground vibration from an underground railway. *Journal of Sound and Vibration*, 330: 6203–6222.

[18] Jones S., Hunt H. (2011). Voids at the tunnel–soil interface for calculation of ground vibration from underground railways. *Journal of Sound and Vibration*, 330: 245–270.
[19] Jones S., Hunt H. (2012). Predicting surface vibration from underground railways through inhomogeneous soil. *Journal of Sound and Vibration*, 331: 2055–2069.
[20] Lysmer J., R.L. (1969). Kuhlemeyer, Finite dynamic model for infinite media. *Journal of Engineering Mechanics Division*, 95: 859–877.
[21] Bian X., Zeng E., Chen Y., (2007). Ground motions generated by harmonic loads moving in subway tunnel, in *Proceedings of the Third International Symposium on Environmental Vibrations: Prediction, Monitoring, Mitigation and Evaluation. ISEV 2007*. Taipei, Taiwan.
[22] Bian X., Jin W., Jiang H. (2012). Ground-borne vibrations due to dynamic loading from moving trains in subway tunnels. *Journal of Zhejiang University - Science A (Applied Physics & Engineering)*, 13(11): 870–876.
[23] Liu G., Jerry S. (2003). A non-reflecting boundary for analyzing wave propagation using the finite element method. *Finite Elements in Analysis and Design*, 39(5–6): 403–417.
[24] Johnson S. (2010). *Notes on Perfectly Matched Layers (PMLs)*, L. notes, Editor. 2010, Massachusetts Institute of Technology.
[25] Berenger J. (1994). A perfectly matched layer for absorption of electromagnetic waves. *Journal of computational physics*, 1114: 185–200.
[26] Chew W., Liu Q. (1996). Perfectly matched layers for elastodynamics: a new absorbing boundary condition. *Journal of Computational Acoustics*, 4: 341–359.
[27] Basu U., Chopra A.K. (2003). Perfectly matched layers for time-harmonic elastodynamics of unbounded domains: theory and finite-element implementation. *Computer Methods in Applied Mechanics and Engineering*, 192: 1337–1375.
[28] François S., Schevenels M., Lombaert G., Degrande G. (2010). A 2.5D displacement-based PML for elastodynamic wave propagation, in *IV European Conference on Computational Mechanics* 2010: Paris, France.
[29] Tadeu A., Kausel E. (2000). Green's functions for two-and-a-half-dimensional elastodynamic problems. *Journal of Engineering Mechanics*, 126(10): 1093–1096.
[30] Hussein M., Hunt H. (2007). A numerical model for calculating vibration from a railway tunnel embedded in a full-space. *Journal of Sound and Vibration*, 305(3): 401–431.
[31] Hwang R., Lysmer J. (1981). Response of buried structures to travelling waves. *Journal of Geotechnical Engineering Division*, 107(2): 183–200.
[32] Muller,K. (2007). Dreidimensionale dynamische Tunnel-Halbraum-Interaktion, in *Lehrstuhl fur Baumechanik*. 2007, Technische Universitat Munchen: Munchen.
[33] Alves Costa P. (2008). Moving loads on the ground: a 2.5D transformed finite element code for train-track-soil interaction, in *(Research Report - FEUP)*. FEUP: Porto.
[34] Alves Costa P., Calçada R., Silva Cardoso A., Bodare A. (2010). Influence of soil non-linearity on the dynamic response of high-speed railway tracks. *Soil Dynamics and Earthquake Engineering*, 30(4): 221–235.
[35] Grundmann H., Dinkel J. (2000). Moving oscillating loads acting on a dam over a layered half space, in *Wave 2000*, C. Schmid, Editor., Balkema: Bochum. p. 53–70.
[36] Kausel E., Barbosa J. (2012). PMLs: A direct approach. *International Journal for Numerical Methods in Engineering*, 90(3): 343–352.

[37] Alves Costa P., Calçada R., Silva Cardoso A. (2012). Track-ground vibrations induced by railway traffic, in *Applications of Computational Mechanics in Geotechnical Engineering,*, L. Sousa, et al., Editors. Balkema. p. 125–159.

In: Traffic Induced Environmental Vibrations and Controls ISBN: 978-1-62618-240-0
Editors: H. Xia and R. Calçada

Chapter 5

INVESTIGATION OF TRAIN-INDUCED GROUND VIBRATIONS USING FEM AND FIELD EXPERIMENTS

Shen-Haw Ju and Hsin-Chang Li
Department of Civil Engineering, National Cheng-Kung University, Taiwan

ABSTRACT

The main purpose of this chapter is to compare the ground vibration induced by trains moving on bridges using finite element analyses and field experiments. The finite element meshes include bridge superstructures, bridge foundations, moving trains, soils, and absorbing boundaries. The comparison indicates that the predicted numerical and measured experimental results are in good agreement, and the train-induced vibration will be large at the dominant frequencies of nV/L, where n is a positive integer, V is the train speed, and L is the distance between two carriage centers. Different from the behavior of trains moving on embankments, the train-bridge-induced ground vibrations are not greatly influenced by rail irregularities, and only in the high frequency range (over 15 Hz), the rail irregularity effect may not be ignored. However, those high frequency vibrations will be damped fast at a far field, such as 50 m from the railway.

1. INTRODUCTION

Moving high-speed trains often produce significant ground vibration, which transmits to the near-by building and produces environmental problems. Especially, when the hospitals or high-tech factories are adjacent to the railroad, a small vibration transmitted from the high-speed train may harm the surgery operation or high-tech product. Moreover, large vibration also makes uncomfortable for human. Therefore, correctly measuring and simulating the noise and vibration induced by MRT or moving trains has been an important issue.

Madshus and Kaynia [1] presented in situ measurement for a high-speed train running on a soft soil site in Sweden, and they further developed computer code to predict ground response to high-speed trains. Kaynia et al. [2] performed experiments in southern Sweden to measure the vibration induced by high-speed trains, and developed a rigorous numerical model to predict embankment/ground response with a good agreement with experimental results. Hall [3] used finite element analyses to simulate train-induced ground vibrations, and performed field measurements of train-induced ground vibrations at a site. In his study, the measured and calculated responses from the three-dimensional analyses were in good agreement. Lombaert and Degrande [4] completed the experimental validation of a numerical model for the prediction of traffic-induced vibrations. The experimental validation shows that the prediction model describes the essential physical phenomena with very good accuracy. Takemiya [5] presented the available field measurements to demonstrate the author's simulation model for predicting the train-track and nearby ground-borne vibrations by the Swedish high-speed train X-2000 at Ledsgard. Kwark et al. [6] performed a large number of field tests to investigate the dynamic safety of bridges crossed by the Korean high-speed train (KHST) running at high-speed and to verify the numerical analysis. Auersch [7] presented an integrated model to compute vehicle-track interaction and the ground vibrations of passing trains, and the theoretical methods and results have been proven by field experiments. Xia and Zhang [8] established a computational model of the dynamic interaction between high-speed trains and bridges. In their study, the proposed analysis model and the solution method were verified by comparing the calculated results and the data measured in situ. Ju and Lin [9] compared the dynamic responses of finite element analyses and field experiments. Their finite element result showed that vibrations in the bridge's longitudinal direction are twice larger than those from experiments if the continuous rail between two simply supported beams is neglected. Zhai et al. [10] measured the ground vibration induced by high-speed trains, and then they used the experimental results to validate their numerical prediction with the proposed train-track-ground model. Ju et al. [11] investigated the behavior of ground vibrations induced by trains moving on embankments using theoretical formulations, finite element analyses, and field experiments. They found that the train-induced vibrations are large at the train dominant frequencies, even though the rail is very rough. Costa el al. [12] performed field experiments to measure the train-induced vibration waves induced by traffic both on the railway track and free-field. Then, 2.5D finite element results were compared with the measurements.

In general, the accurate estimation of ground vibrations induced by moving trains is particularly complicated, and any mistake in the numerical modeling may cause imprecision in the analyses. This study aims to generate a correct finite element model to simulate the ground vibration induced by high-speed trains passing over multi-span simply supported bridges and embankments. The dynamic responses of the finite element model are compared with field measurements to investigate the accuracy of this model.

2. Finite Element Models for the Interaction of Trains and Rails

Moving trains are modeled as the combination of moving wheel elements, spring-damper elements, lumped mass elements, and rigid links. The train is assumed to move in the X direction (railway direction) and the Y direction is perpendicular to the railway.

2.1. Rail Irregularities

Rail irregularities are a major source of train vibrations, which may enlarge the train derailment effect during seismic loading. Thus, this paper also investigated train derailment due to rail irregularities. A sample function $r_v(X)$ of the rail irregularity [13] is used in this study.

$$r_v(X) = \sum_{k=1}^{N} a_k \cos(\omega_k X + \phi_k) \tag{1}$$

where a_k is the amplitude, ω_k is a frequency (rad/s) within the upper and lower limits of the frequency $[\omega_l, \omega_u]$, ϕ_k is a random phase angle in the interval $[0, 2\pi]$, X is the global coordinate in the rail direction, and N is the total number of terms. The parameters a_k and ω_k are computed by:

$$a_k = 2\sqrt{G_{rr}(\omega_k)\Delta\omega}\ ;\ \omega_k = \omega_l + (k-1/2)\Delta\omega\ (k=1,2,\ldots,N) \tag{2}$$

$$G_{rr}(\omega) = \frac{A_r \omega_2^2 (\omega^2 + \omega_1^2)}{\omega^4 (\omega^2 + \omega_2^2)} \tag{3}$$

where $\Delta\omega = (\omega_u - \omega_l)/N$, $G_{rr}(\omega)$ is a power spectral density (PSD) function, A_r is the roughness coefficient, and ω_1 and ω_2 are frequencies that change the shape of $G_{rr}(\omega)$. We set ω_1 to zero in this study, so Equation (3) is the same as the PSD function in Lei and Noda [14]. The rail irregularity has an amplitude of 2 mm per 20 m of the rail in the vertical and transverse directions of the rail, and the irregularity parameters are shown in Table 1 for the both directions. It is noted that the irregularity magnitudes of the two directions are similar, but the angle phases of them (ϕ_k) are different due to using a different random number.

Table 1. Rail irregularity parameters used in the numerical analyses

A_r /(m^2 rad/m)	ω_1 /(rad/m)	ω_2 /(rad/m)	ω_l /(rad/m)	ω_u /(rad/m)	N
0.8E-7	0	1	0.08	20	1000

2.2. Moving Wheel Element

As shown in Figure 1(left), the wheel element contains a stiffness k_r between the rail and wheel, where the direction of the stiffness k_r can be in the global X, Y or Z directions. The element includes a wheel node and a number of target nodes, while the current wheel position equals the initial wheel position plus the duration time multiplied by the train speed. The two target nodes between which the wheel node is located can then be found. If the two target nodes and the wheel node are nodes 1, 3 and 2 respectively, the three-node element stiffness with five degrees of freedom for the nodal displacements (d_1, θ_1, d_2, d_3, θ_3), as shown in Figure 1(right), is:

$$\mathbf{S}=\mathbf{T}^{\mathrm{T}}\begin{bmatrix} k_r & -k_r \\ -k_r & k_r \end{bmatrix} T\ ;\ \mathbf{T}=\begin{bmatrix} 0 & 0 & 1 & 0 & 0 \\ N_1 & N_2 & 0 & N_3 & N_4 \end{bmatrix} \tag{4}$$

where d_1, θ_1, d_3 and θ_3 are the translations and rotations at target nodes 1 and 3, d_2 is the translation of the wheel node, and N_i = the cubic Hermitian interpolation functions, as follows:

$$N_1=1-\frac{3X_2^2}{L^2}+\frac{2X_2^3}{L^3};\ N_2=-X_2\left(1-\frac{2X_2}{L}+\frac{X_2{}^2}{L^2}\right) \tag{5a,b}$$

$$N_3=\frac{3X_2^2}{L^2}-\frac{2X_2^3}{L^3};\ N_4=\frac{X_2^3}{L^2}-\frac{X_2^2}{L} \tag{5c,d}$$

where X_2 is the X coordinate at node 2 while the origin is at node 1. The irregularity between the rail and wheel can be easily included. Only the element forces $\mathbf{f_r}$ need to be added into the global force vector, as follows:

$$\mathbf{f}_r{}^{\mathrm{T}}=[\ N_1\quad N_2\quad -1\quad N_3\quad N_4\ ]\,k_r r_v(X) \tag{6}$$

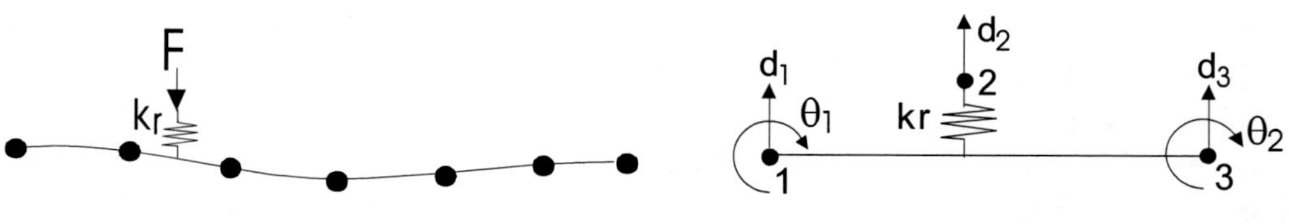

Figure 1. Moving wheel element with a spring: element includes a wheel node and a number of target nodes (left); and nodal number and degrees-of-freedom for the wheel element (right).

The internal force vector of the wheel element is:

$$\begin{aligned}\mathbf{F}&=[f_1\quad m_1\quad f_2\quad f_3\quad m_3]^{\mathrm{T}}\\ &=\mathbf{S}[X_1\quad \theta_1\quad X_2\quad X_3\quad \theta_3]^{\mathrm{T}}-\mathbf{f_r}\end{aligned} \tag{7}$$

where the nodal forces should exclude the terms of rail irregularities, and (f_1, m_1, f_2, f_3, m_3) are internal forces and moments at nodes 1, 2 and 3. The value of f_2 is the contact force between the wheel and rail. If it is a tensile force (positive value), the wheel and rail are separated, and k_r will be set to zero for the next Newton-Raphson iteration. Otherwise, they are in contact together, and we set k_r as follows:

$$k_r=a+bf_2^c \tag{8}$$

where a, b and c are constant, which can be determined using the Hertz contact theory, such as a smooth sphere with the radius of R in contact with a smooth flat surface under a load f_2, with the stiffness as follows [8]:

$$k_r = \frac{3}{2}\left(\frac{16RE_*^{\ 2}}{9}\right)^{1/3} f_2^{\ 1/3} \tag{9}$$

where $\frac{1}{E_*} = \frac{1-\nu_1^2}{E_1} + \frac{1-\nu_2^2}{E_2}$, E_1, E_2, ν_1 , and ν_2 are the Young's modulus and Poisson's ratio for the sphere and plate, respectively. It is clear that Equation (9) can be used to model the above stiffness.

However, the Hertz contact theory is generally used for regular disks or plates, and may produce errors for the actual rails and wheels. For this reason, one can find these constants, a, b and c, by fitting the results of a static contact finite element analysis between the rail and wheel [15].

If b equals zero, $k_r=a$ is simply changed to a constant stiffness during the contact period. When one only considers the contact mode between wheels and rails, a constant k_r produces a linear elastic finite element analysis of the interaction between trains and rails. In this study, since only the train-induced vibration without derailment was simulated, we set a constant $k_r=5\times10^5$ kN/m. in the finite element analysis.

2.3. Spring-Damper and Lumped Mass Elements

The spring-damper element contains stiffness k_v and damping c_v in parallel (Kelvin model), and can be in the X, Y or Z direction. The stiffness and damping matrices of this element are

$$[S] = \begin{bmatrix} S_1 & S_2 \\ S_2 & S_3 \end{bmatrix} \tag{10}$$

where S_1, S_2, and S_3 are obtained from the input data of the damping c_v or stiffness matrix k_v of the element. For the damping matrix, $S_1=S_3=c_v$ and $S_2=-c_v$, and for stiffness matrix, $S_1=S_3=k_v$ and $S_2=-k_v$.

The lumped mass matrix is a diagonal matrix as follows:

$$[S] = \lfloor\backslash \quad m \quad m \quad m \quad I_x \quad I_y \quad I_z \quad \backslash\rfloor \tag{11}$$

where m is the mass and I_x , I_y, and I_z are the moment of inertia in three global directions.

2.4. Rigid Link Effect

Rigid links can connect the above finite elements for rigid bodies. If the global force $\mathrm{F}_i^{\mathrm{S}}$ (i=X, Y or Z) at a slave node is known, the force and moments at the mass center, called the master node, can be calculated as follows [16]:

$$\mathrm{F}_i^{\mathrm{M}} = \mathrm{F}_i^{\mathrm{S}} \tag{12}$$

$$\begin{Bmatrix} \mathrm{M}_\mathrm{Y}^\mathrm{M} \\ \mathrm{M}_\mathrm{Z}^\mathrm{M} \end{Bmatrix} = \begin{bmatrix} -\Delta\mathrm{Z} \\ \Delta\mathrm{Y} \end{bmatrix} \mathrm{F}_\mathrm{X}^\mathrm{S}, \quad \begin{Bmatrix} \mathrm{M}_\mathrm{X}^\mathrm{M} \\ \mathrm{M}_\mathrm{Z}^\mathrm{M} \end{Bmatrix} = \begin{bmatrix} \Delta\mathrm{Z} \\ -\Delta\mathrm{X} \end{bmatrix} \mathrm{F}_\mathrm{Y}^\mathrm{S} \text{ and } \begin{Bmatrix} \mathrm{M}_\mathrm{X}^\mathrm{M} \\ \mathrm{M}_\mathrm{Y}^\mathrm{M} \end{Bmatrix} = \begin{bmatrix} -\Delta\mathrm{Y} \\ \Delta\mathrm{X} \end{bmatrix} \mathrm{F}_\mathrm{Z}^\mathrm{S}$$

$$\text{or } \{\mathrm{M}^\mathrm{M}\} = \begin{bmatrix} D_1 \\ D_2 \end{bmatrix} \mathrm{F}_i^\mathrm{S} \tag{13}$$

where $(\Delta X, \Delta Y, \Delta Z)$ is the coordinate difference between the master and slave nodes, and $\mathrm{M}_\mathrm{X}^\mathrm{M}$, $\mathrm{M}_\mathrm{Y}^\mathrm{M}$, and $\mathrm{M}_\mathrm{Z}^\mathrm{M}$ are the global moments at the master node.

For an n-degree-of-freedom mass, damping or stiffness matrix ($[\mathrm{S}^\mathrm{S}]$) and the force vector ($\{\mathrm{F}^\mathrm{S}\}$) with a slave degree-of-freedom (assume at k^{th} degree-of-freedom), the stiffness or damping matrix ($[\mathrm{S}^\mathrm{M}]$) and the force vector ($\{\mathrm{F}^\mathrm{M}\}$) at the master node are:

$$[\mathrm{S}^\mathrm{M}] = \begin{bmatrix} & [\mathrm{I}] & \\ [0]_1 & [\mathrm{D}] & [0]_2 \end{bmatrix} [\mathrm{S}^\mathrm{S}] \begin{bmatrix} & [\mathrm{I}] & \\ [0]_1 & [\mathrm{D}] & [0]_2 \end{bmatrix}^\mathrm{T} \text{ or } [\mathrm{S}^\mathrm{M}] = [\mathrm{T}][\mathrm{S}^\mathrm{S}][\mathrm{T}]^\mathrm{T} \tag{14}$$

$$\{\mathrm{F}^\mathrm{M}\} = [\mathrm{T}]\{\mathrm{F}^\mathrm{S}\} \tag{15}$$

where n by n matrix [I] is the unit matrix, 2 by k−1 matrix $[0]_1$ and 2 by n−k matrix $[0]_2$ are zero matrices and $[\mathrm{S}^\mathrm{S}]$ is obtained from Equation (4) or (10). Usually, an element contains slave parts and parts with normal degrees of freedom. Equation (14) can be used to do the transformation for each slave degree-of-freedom successfully, and at each transformation the matrix will increase by two rotation degrees-of-freedom. One can successfully use Equation (14) to do the transformation for each slave degree of freedom.

For a spring or damper connected to two master nodes using the above theory, the stiffness or damping matrix is:

$$\mathbf{S} = s\mathbf{B}\mathbf{B}^\mathrm{T} \tag{16}$$

where s is the spring constant for stiffness matrix **S** or the damping constant for damping matrix **S**, $\mathbf{B}^\mathrm{T} = [1 \quad -1 \quad D_1^R \quad D_2^R \quad -D_1^L \quad -D_2^L]$. D_1^R and D_2^R (or D_1^L and D_2^L) are the D_1 and D_2 mentioned in Equation (13).

A vehicle or train containing car bodies, bogies, wheel-axis sets, springs and dampers can be modeled as a combination of nonlinear moving wheel elements, spring-damper elements with the rigid link effect, and lumped mass. In this study, the Newmark direct integration method and the Newton-Raphson method are used to solve this time-dependent nonlinear problem. If a constant k_r is set, the equation changes to linear elastic, and the standard linear Newmark direct integration method without iterations can then be used. The SKS-700 high-speed train was modeled as follows:

As shown in Figure 2, the wheel contact nodes are set from nodes 1 to 4 and 101 to 104 for the 4 wheels on the left and right rails, respectively. Nodes 301 to 304 and nodes 401 to 404 are located at the 4 wheel axes, where springs and dampers are connected between these points and train bogies. The above nodes are slave nodes controlled by master nodes 1201 to 1204 on the mass centers of the 4 wheel sets.

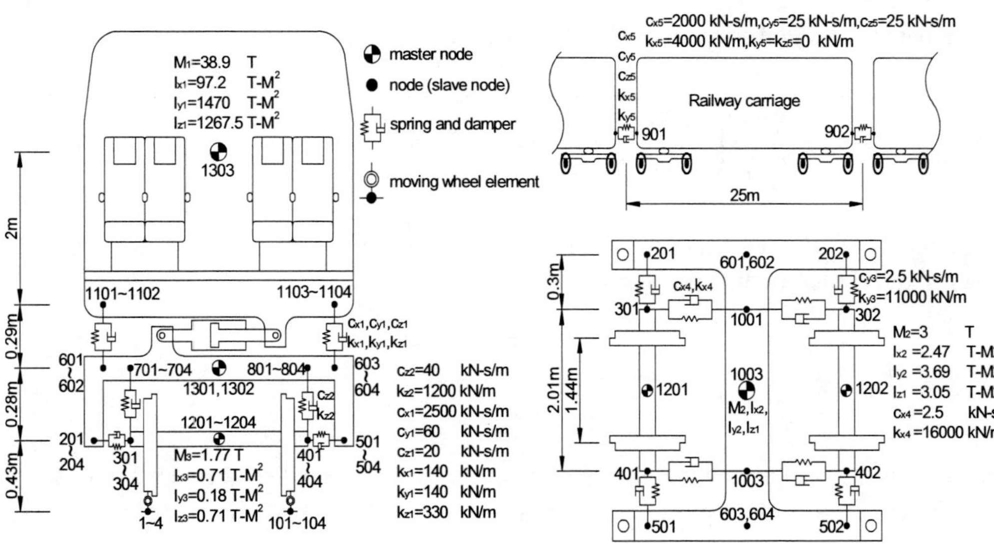

Figure 2. Three-dimensional model of the high-speed train.

1) Nodes 201 to 204, 501 to 504, 701 to 704, 801 to 804 and 1001 to 1004 are located at the 2 bogies to generate the string-damper elements between wheel sets and bogies. Nodes 601 to 604 at the 2 bogies are used to mesh the string-damper elements in three global directions between the two bogies and the carriage. The above nodes are slave nodes controlled by master nodes 1301 and 1302 on the mass centers of the two bogies.
2) Nodes 1101 to 1104 at the carriage are used to mesh the string-damper elements in three global directions between bogies and carriages. Nodes 901 and 902 at the edges of the carriage are used to mesh the string-damper elements in three global directions between two adjacent carriages. The above nodes are slave nodes controlled by master node 1303 at the mass center of the carriage.

The above arrangement gives the mesh of the first carriage. The nodal numbers of the nth carriage can be simply shifted by adding 4(n-1) to those of the first carriage, and the nodal X-coordinates are shifted by adding the carriage interval times (n-1).

2.4 Absorbing Boundary Condition

In this study, absorbing boundary conditions are used in the finite element analysis to reduce fake reflections from artificial boundaries. For $\{(x_1, x_2, x_3): x_1>0\}$, x_1 is positive into the domain while $x_1=0$ at the absorbing boundary, and u_i is the x_i-displacement. The absorbing boundary condition is [18]:

$$(\frac{\partial}{\partial t} - c_i \frac{\partial}{\partial x_1})u_i = 0 \quad \text{for } i=1,2 \text{ and } 3 \tag{17}$$

where c_i is the i-direction velocity divided by the cosine of the incidence angle. Ju and Wang [18] evaluated c_i using the square-root-of-the-sum-of-the-squares method as:

$$c_i = \sqrt{\left(\sum_{k=1}^{N} \frac{\partial u_{ik}}{\partial t} \frac{\partial u_{ik}}{\partial t}\right) \Big/ \left(\sum_{k=1}^{N} \frac{\partial u_{ik}}{\partial x_1} \frac{\partial u_{ik}}{\partial x_1}\right)} \quad \text{for } i=1,2 \text{ and } 3 \tag{18}$$

where N is the number of nodes to obtain the average c_i, and u_{ik} means displacement at node k. The forward Euler method derived from Equation (17) is:

$$u_{i,n+1} = u_{i,n} + \Delta t c_i \frac{\partial u_{i,n}}{\partial x_1} \tag{19}$$

where $u_{i,n+1}$ is the displacement at the current time step, $u_{i,n}$ is the displacement at the last time step, and Δt is the time step length. In Equation (19), all the values on the right side are known, so the current controlled displacements are obtained. Thus, the time-history simulation of the infinite-domain wave can be achieved.

3. Dominant Frequencies of Train-Induced Vibrations

The trainload is an important source of induced vibrations, so its dominant frequencies should be known first. To evaluate them, the wheel loading of a train passing a point is assumed to be $P(t)$ as follows:

$$P(t) = \sum_{j=1}^{N_c} \sum_{k=1}^{N_w} P_{\text{axle}} \left[\delta(t - t_k - jt_c)\right] \tag{20}$$

where the train has N_c carriages, a carriage has N_w pairs of wheels, P_{axle} is the load of a pair of wheels, δ is the unit impulse function, $t_k=s_k/V$, in which V is the train speed and s_k is the distance between the kth wheel and the beginning of the carriage, and $t_c=L/V$, in which L is the distance between two carriage centers. The Fourier transform can be applied to Equation (20) to convert the trainload function into the frequency domain as [19]:

$$P(f) = \int_{-\infty}^{\infty} P(t)e^{-i2\pi f}\,\mathrm{d}t = \sum_{j=1}^{N_c}\sum_{k=1}^{N_w} P_{\mathrm{axle}}\mathrm{e}^{-i2\pi f(t_k + jt_c)} = \sum_{j=1}^{N_c}\sum_{k=1}^{N_w} P_{\mathrm{axle}}\mathrm{e}^{-i2\pi f(s_k + jL)/V} \tag{21}$$

and $P(t) = \int_{-\infty}^{\infty} P(f)e^{2\pi f t i}\,\mathrm{d}f$

The influence factors R_f can be determined as follows:

$$R_f = |P(f)| = \left|\sum_{j=1}^{N_c}\sum_{k=1}^{N_w} P_{\mathrm{axle}}\mathrm{e}^{-i2\pi f(s_k + jL)/V}\right| \tag{22}$$

The dominant frequencies are defined as the relatively largest values of R_f, but they cannot be derived directly from Equation (22). If $f \times t_c$ of Equation (21) is a positive integer (n), or

$$f = n/t_c = nV/L, \tag{23}$$

Equation (22) is changed to the sum of N_c carriage loads as follows:

$$R_{f=nV/L} = N_c P_{\mathrm{axle}}\left|\sum_{k=1}^{N_w} \mathrm{e}^{-i2n\pi S_k/L}\right| \tag{24}$$

Otherwise, R_f can be cancelled by each other due to the positive or negative sign of the load in the exponential function. Thus, if N_c is large enough, the frequencies of Equation (23) should be the dominant frequencies of the trainload. Figure 3(a) shows R_f using Equation (22) for the SKS-700 high-speed train as illustrated in Figure 2, in which the number of carriages is set from 1 to 12, and Figure 3(b) is the R_f of 12 carriages.

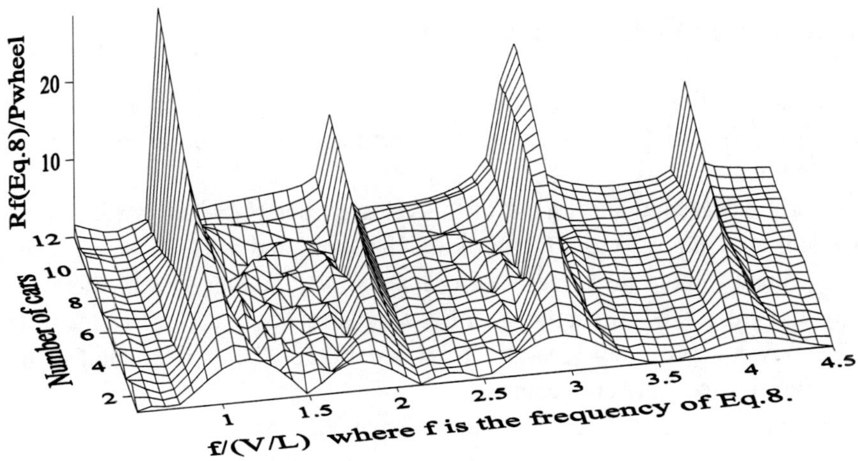

(a)

Figure 3. (Continued)

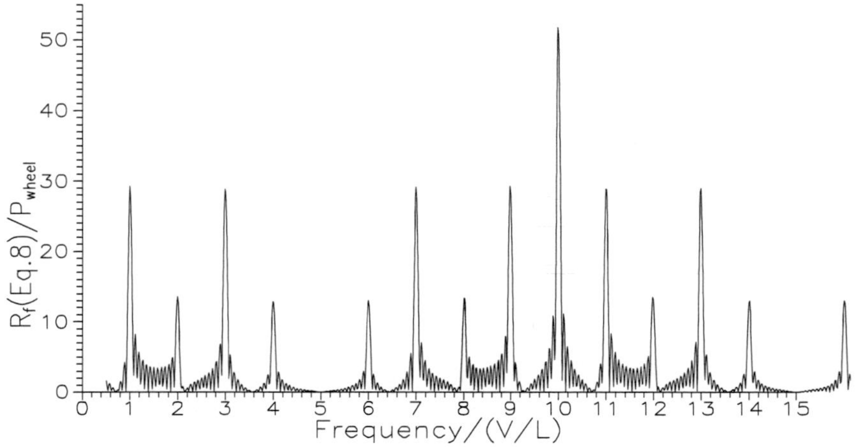

(b)

Figure 3. Influence factors R_f using Equation (22) for SKS-700 high-speed train: (a) number of cars from 1 to 12, and (b) number of cars equal to 12.

This figure indicates that the dominant frequencies can be precisely predicted as shown in Equation (23) when the number of carriages is larger than 3. It is noted that due to the arrangement of the wheels of the Japan SKS-700 high-speed train, the influence factors at n=5, 15, 25 etc. are zero. Equations (23) and (24) can be used to evaluate dominant frequencies and influence factors.

The dominant frequencies are only dependent on the ratio of the train speed to the carriage length, and the influence factors are only dependent on the ratio of wheel locations to the carriage length, but independent of the train speed.

4. ILLUSTRATION OF FIELD EXPERIMENTS

4.1. Explanation of Field Measurements

There are two stations with the distances of 3 m and 50 m, respectively, from the railroad centerline, and each station with three velocity sensors were set up to measure X, Y, and Z vibration velocities.

The experimental setup and layout of the velocity meters are shown in Figure 4, where the surface soil was removed to a depth of about 40 cm, and a 21×21×1-cm steel plate was fixed to the ground to hold the three velocity meters. A computer system was used to record the velocity histories of the two stations with a sample rate of 512 Hz (512 measurements per second). The power supply was generated from batteries, so the experimental setup, as shown in Figure 4 (right), is considerably compact.

The field measurements were continuous for about eight hours per day, and computer software was used to pick up the 40-second data for trains passing over this region. Two groups of the 40-second ambient vibration data were also obtained 200 seconds before and after each passing train.

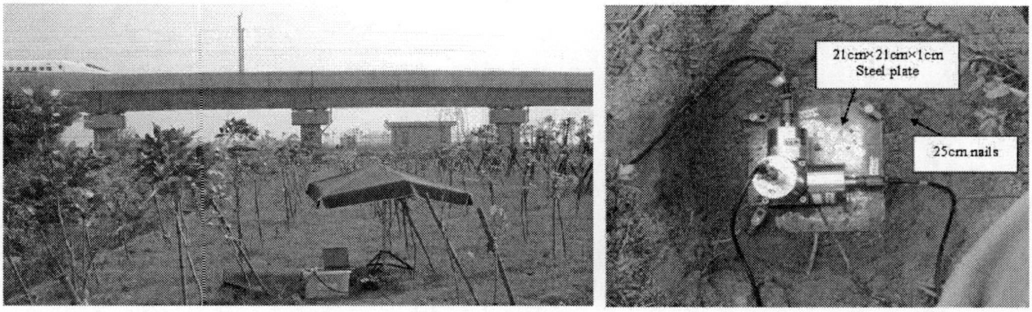

Figure 4. Field experiments for measuring train-induced vibration (left) and Installation of velocity meters (right).

Figure 5 shows a 40-second vibration data ($y_t(t)$) taken at the 3-m and 50-m observation stations when the train was passing the experimental region at 300 km/h. The duration of the trainload at the 3-m and 50-m stations was about 5 and 10 seconds, respectively. The ambient vibrations taken at the two sides of this period of time are much smaller than those taken during the train passing, which indicate that the measured vibration during the train passing is real. Figure 6 displays the FFT results of the above data, and this figure clearly shows that the vibration is greatest at the frequencies of nV/L except at n=5 and 15, precisely as illustrated in Figure 3 using Equation (22). This figure also indicates that the high-frequency vibration is damped faster than the low-frequency vibration. For trains moving on bridges, the resonant vibrations are serious when the natural frequencies of the bridge are close to the trainload dominant frequencies. The trainload dominant frequency and its influence factor can be computed using the two simple equations as listed in Equations (23) and (22).

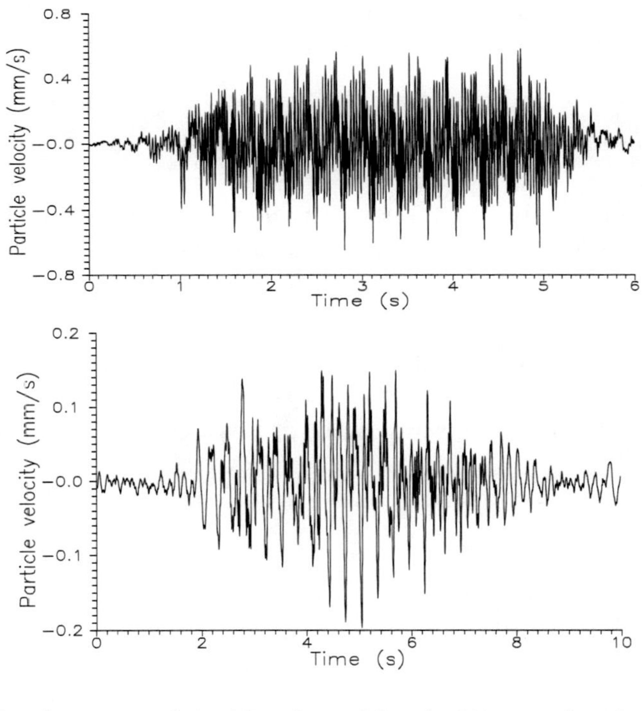

Figure 5. Vertical-direction measured time-domain particle velocities at stations 3-m (top) and 50-m (bottom) under a train speed of 300 km/h.

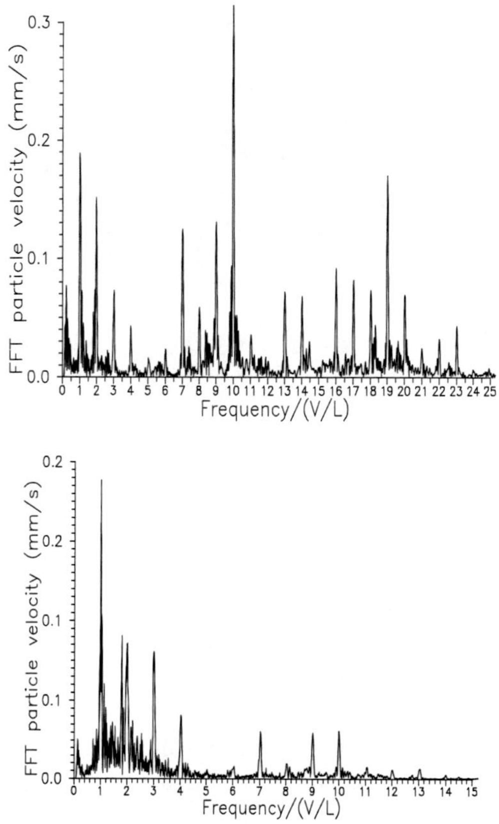

Figure 6. Vertical-direction measured frequency-domain particle velocities at stations 3-m (left) and 50-m (right) under a train speed of 300 km/h.

4.2. Calculation of 1/3 Octave Band in the Frequency Domain

Analyzing vibrations using the 1/3 octave band in the frequency domain has become the international standard in the semi-conductor industry [20]. Moreover, this method can be applied to vibration guidelines for human living. Thus, this method, with an output in decibels (dB), is utilized in this research, and Table 2 briefly presents the recommended vibration guidelines. The calculation of this method includes the following steps:

(1) Select a section of the velocity record, $y(t)$, to analyze by the Fast Fourier Transform (FFT) method, and calculate the power spectrum density function (PSDF), $S_y(f)$:

$$S_y(f) = \frac{2|Y(f)|^2}{T} \tag{25}$$

where $|Y(f)|$ is the FFT amplitude, T is the period of $y(t)$, and f is the frequency (Hz).

(2) To accumulate PSDF, $E_y(f_c)$:

$$E_y(f_c) = \int_{f_l}^{f_u} S_y(f)\,\mathrm{d}f \tag{26}$$

where f_l, f_u, and f_c are the lower band, upper band, and center frequencies, respectively, obtained from Gordon [20].

(3) Calculate the root mean square(RMS), $\sigma_y(f_c)$:

$$\sigma_y(f_c) = \sqrt{E_y(f_c)} - \sigma_{\text{ambient}}(f_c) \tag{27}$$

where $\sigma_{\text{ambient}}(f_c)$ is the average ambient RMS vibration 300 seconds before and after the moving train in field measurements. It is zero for finite element analyses.

(4) Calculate RMS velocity level ($L_y(f_c)$) in decibels (dB) :

$$L_y(f_c) = 20\log_{10}\frac{\sigma_y(f_c)}{\sigma_0} \tag{28}$$

where the referred velocity $\sigma_0=10^{-6}$ in/sec (2.54×10^{-8} m/sec).

For finite element results, all the nodal displacements on the certain locations were obtained first. The central difference method was then used to transform the nodal displacements into nodal velocities. For field experiments, vibration velocities were measured directly. The average ambient vibration dB of 40 seconds before and after the moving train was removed from the data using Equation (27). At each location, a time period T in Equation (25) equal to 8 seconds was selected, and shifted 1 second to select another 8 seconds until the last time period was less than 8 seconds. Computer software was then used to find all the 1/3 octave band results for all the time periods, finally showing the extreme dB changes in the frequency domain at each location.

Table 2. Recommended vibration guidelines

Description of use	Max. level of vibration	Description
Workshop	90 dB	Distinctly perceptible vibration. Appropriate for workshops and non-sensitive areas.
Office	84 dB	Perceptible vibration. Appropriate for offices and non-sensitive areas.
Residential day	78 dB	Barely perceptible vibration. Appropriate for sleeping areas in most instances.
Operation room	72 dB	Vibration not perceptible. Adequate in most instance for optical microscopes to 100X.
VC-A standard in semi-conductor industries	66 dB	Adequate in most instances for optical microscopes to 400X, microbalances, optical balances, proximity and projection aligners, etc.
VC-D standard in semi-conductor industries	48 dB	Adequate in most instances for optical microscopes (TEM and SEM), microbalances, optical balances, proximity and projection aligners, etc.

5. Comparison of FEM and Experimental Results for Train-Induced Vibrations

5.1. Bridge Dimensions and Soil Properties

The field measurements were performed in southern Taiwan. The X axis is defined along the bridge direction and the negative Z axis is the soil depth direction. The bridge is a series of simply supported beams with a span of 30 m. Figure 7 shows the dimensions of the bridge and foundation, in which the pile diameter is 2 m and the pile length is 62.4 m. The top surfaces of the piles are connected by a reinforced concrete (RC) cap with a thickness of 2.8 m, which is buried underground with a soil layer of 2 m. The bridge girder, with a cross section as shown in Figure 7, is supported on a 2.8 m × 3.2 m rectangular pier 9.5 m high, using four bearing plates. Other than the beam mass itself, the bridge beam supports an extra-mass per unit length of 18 t/m for the railway, parapet, and devices. The Young's modulus of the bridge beams, bridge piers, and foundation caps and piles is 3.02×10^7 kN/m^2, 2.7×10^7 kN/m^2, and 2×10^7 kN/m^2, respectively. The Poisson's ratio and mass density of the complete bridge are 0.15 and 2.4 t/m^3, respectively.

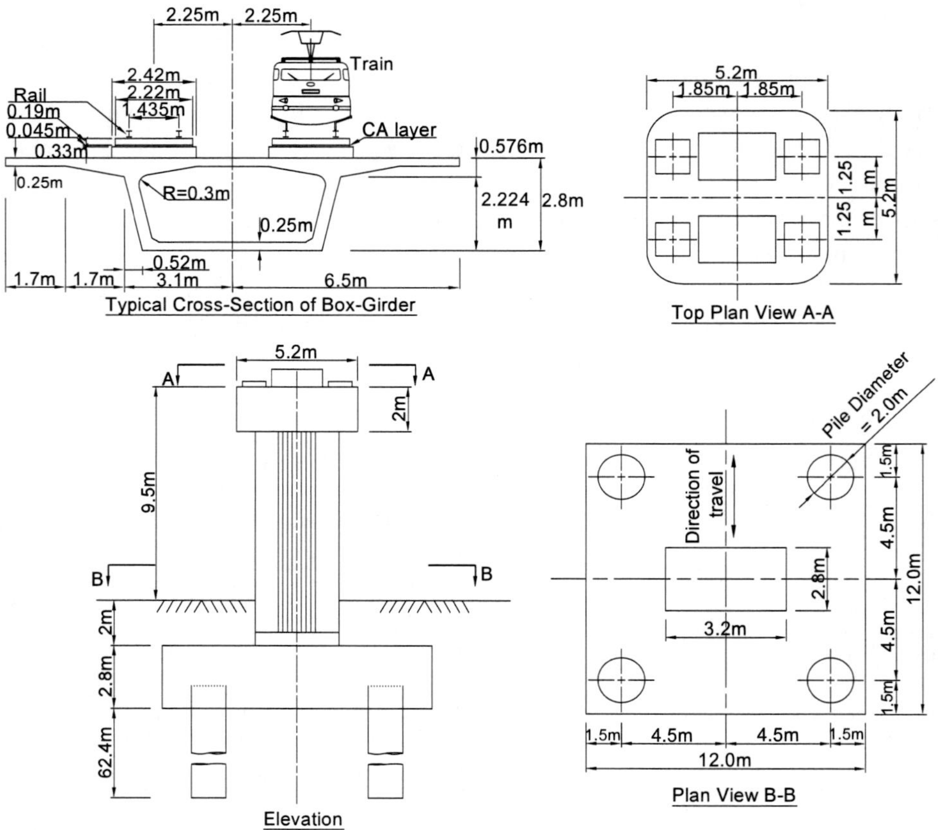

Figure 7. Bridge pier, piles, and beam dimensions.

The rail is the JIS-60-kg type supported on the upper concrete bed with a support interval of 0.625 m in the rail direction (X-direction). The concrete slab system shown in Figure 7 includes a 19-cm-thick upper concrete bed for sustaining the rail, a 4.5-cm-thick cement-asphalt layer (CA layer) for cushion, and a 33-cm-thick bottom concrete bed for sustaining the track structure on the soil foundation.

The Young's modulus, Poisson's ratio, and mass density for the concrete bed are 2×10^7 kN/m^2, 0.15, and 2.4 t/m^3, those for the cement-asphalt layer are 0.6×10^6 kN/m^2, 0.25, and 1.6 t/m^3, and those for the rail are 2×10^8 kN/m^2, 0.3, and 7.85 t/m^3, respectively.

The soil profile contains 10-m inorganic clays of low to medium plasticity, 40-m silty sands, and the rest is very hard sand. Soil wave velocity experiments indicate that the Young's modulus of the surface soil is 7.5×10^4 kN/m^2, and that of more than 50 m under the ground is 8×10^5 kN/m^2. Linear interpolation was applied to determine it between these two depths. The mass density and Poisson's ratio of the soil are 2 t/m^3 and 0.49, respectively. The two factors of Rayleigh damping ([Damping]=α [Mass]+ β[Stiffness]), α and β, for the soil equal 0.78/s and 3.7×10^{-4} s, respectively, which gives an approximately 2% damping ratio at frequencies of 4 Hz and 15 Hz. The field experiments used to obtain soil damping data are described in [15]. The two factors of Rayleigh damping, α and β, for the pre-stressed bridge system equal 1.7/s and 0.23×10^{-4} s, respectively, which gives an approximately 2% damping ratio at a frequency of 7 Hz and a 1% damping ratio at a frequency of 16 Hz. The bridge damping was calculated from a free vibration experiment just after a train passed over the bridge.

5.3. Finite Element Model

The Newmark direct integration method and the consistent mass scheme were used to solve this problem with the solution scheme of the SSOR (Symmetric Successive Over-Relaxation) preconditioned conjugated gradient method [19]. The finite element model is 1048 m long, 370 m wide, and 150 m deep with a maximum solid element size of 2.5 m. The major part of the mesh is modeled by 8-node 3D solid elements for soil, rail concrete slabs, and the bridge foundation. As shown in Figure 8, rails are modeled by 2-node 3D beam elements connected to the concrete slabs modeled by 8-node solid elements every 0.625 m in the X-direction using two spring elements (K_{rail}=spring constant=2.5×10^5 kN/m in X, Y and Z directions) and the nodes along rails are the wheel target nodes of the train model. The nodes along the bottom surface of rail concrete slabs are slave nodes controlled by the nodes of the bridge beams. Thus, the bridge superstructure including rails, concrete slabs, AC layers, and bridge main beams are coupled together.

Figure 8 shows the finite element mesh of the bridge and pile foundation. Within the 2.8 m by 3.2 m area on the center of the pile cap surface, a column modeled by two layers of 3D solid elements is used to simulate the 2-m covered soil layer. A master-slave node scheme is used to model the connection between the column simulated by the 3D solid and beam elements. The master node is arranged at the column center, and other nodes within this 2.8 m by 3.2 m area are set to be slave nodes, whose three degrees of freedom are dominated by the 6-degree-of-freedom master node. A 35-bay bridge system modeled by 3D beam elements, as shown in Figure 8, was adopted in this study. The bearing plates of the simply supported beam are not located on the pier center, so the rigid beam system shown in Figure 8(c) was used to model this eccentricity equal to 1.25m wide. Finally, the five surfaces, except for the top surface of the mesh, are modeled by the absorbing boundary condition of Ju and Wang, [18].

Figure 9 shows a typical 3D finite element mesh, which contains 8,776,152 degrees-of-freedom. The time step length was 0.005 seconds with 2,400 time steps simulated. The computer memory and time requirements are about 5.6 GB and 3 days, respectively, using an Intel I7-2600K personal computer.

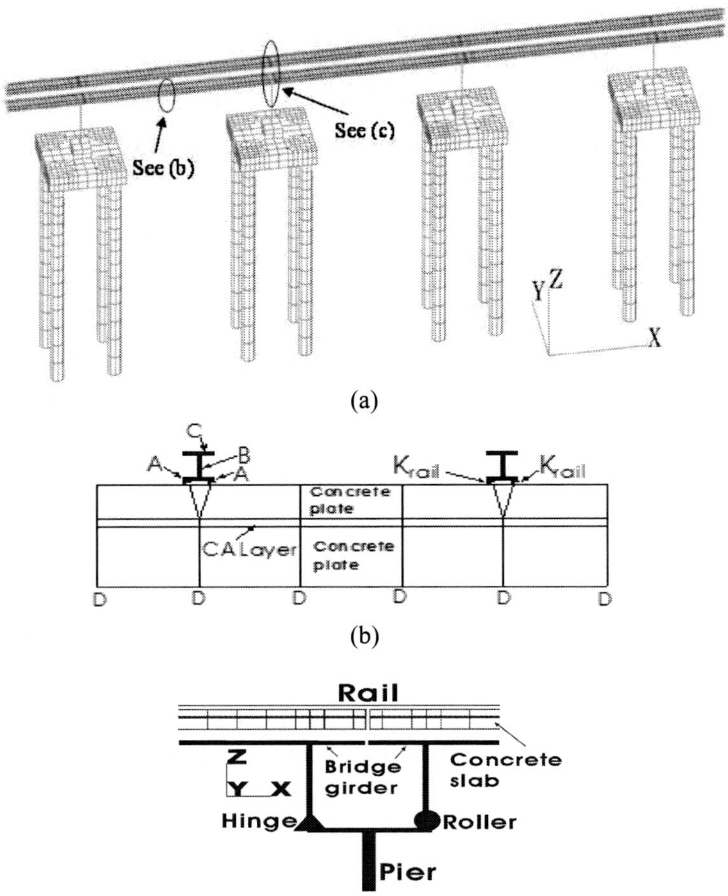

(c)

Figure 8. Finite element mesh and illustration: (a) part of mesh for piles and bridges, (b) mesh near the rail, and (c) Simulation of beam supports. (A=slave nodes that connect the beam to the concrete plate, C=slave nodes that are the target nodes of train wheels, B=rail center nodes that are the master nodes of nodes A and C, D=slave nodes that connect the concrete plate and the bridge beam center (master node), and K_{rail}=three-direction springs).

5.4. Comparisons between Finite Element and Experimental Results

Finite element analyses with and without rail irregularities were performed. The first case had A_r=0 that means a perfectly smooth rail, and the second case was the regular case (A_r $=0.8\times10^{-7}$ m^2 rad/m as shown in Table 1). Figure 10 shows the experimental and finite element results at the two stations under a train speed of 300 km/h. The figure indicates that the accuracy of the finite element analysis is acceptable for vibrations in the free field.

Figure 10 also shows that the train-induced ground vibrations are not greatly influenced by rail irregularities, and only vibrations higher than 10 Hz were obviously different in simulations with and without rail irregularities. The reason is because the low frequency vibrations of the bridge highly depend on its natural frequencies, especially the first one in each direction. Normally, these first natural frequencies in the X and Y directions of the bridge are smaller than 3 Hz, and that in the Z direction is smaller than 7 Hz, which may cause resonance of the trainload dominant frequencies (nV/L) and the bridge natural frequencies at low frequency vibrations. For the high frequency range, such as frequencies larger than 10 Hz, the rail irregularity effect cannot be ignored.

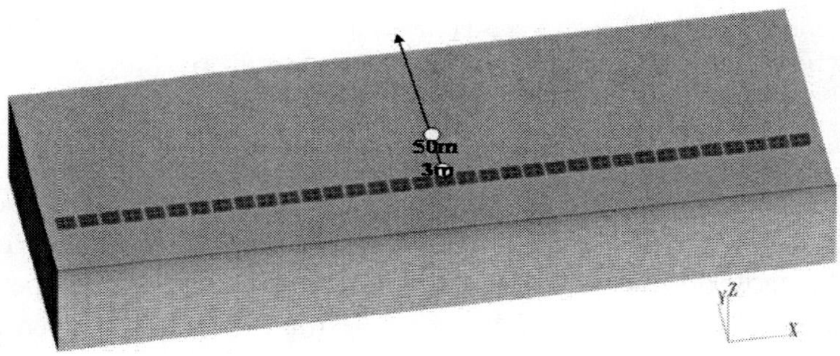

Figure 9. The 3D finite element mesh (the bridge superstructure is not shown).

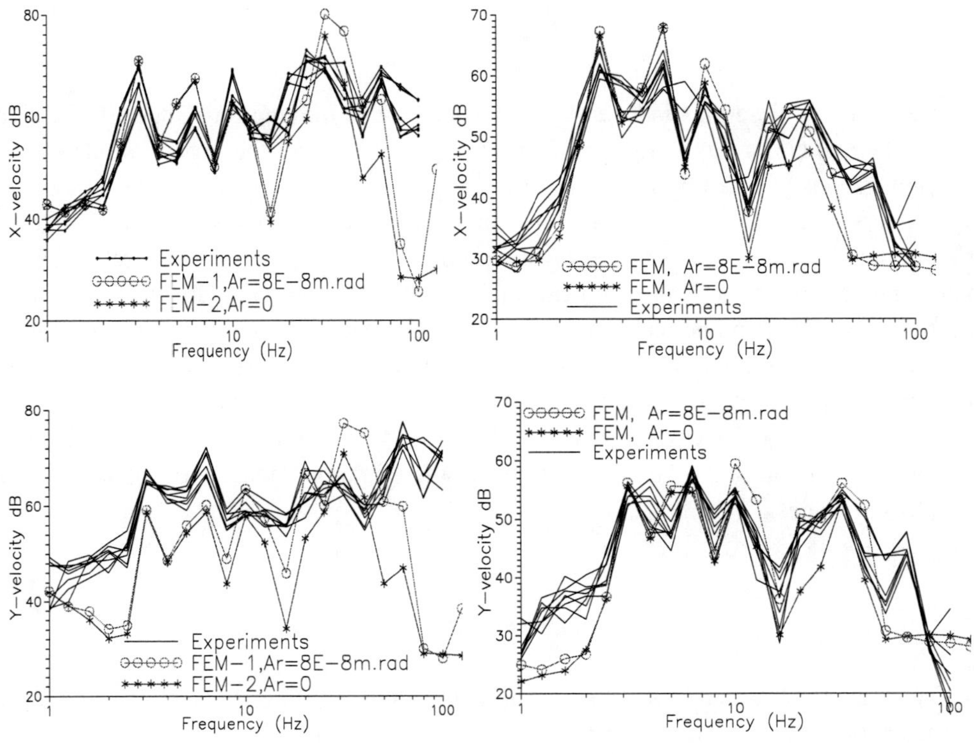

Figure 10. (Continued)

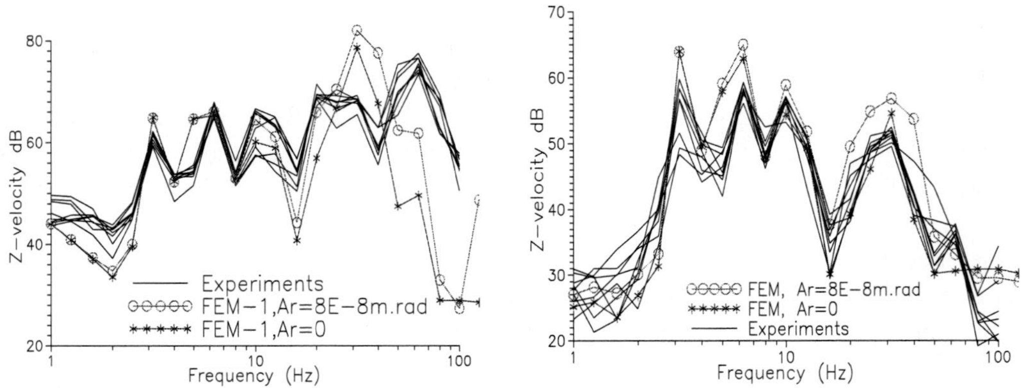

Figure 10. Comparison of finite element and experimental results at 3-m (left) and 50-m (right) from the railroad centerline under a train speed of 300 km/h (Experiments=eight field experiments, FEM-1=FEM with rail irregularity, FEM-2=FEM without rail irregularity).

For trains moving on embankments, Ju and Liao [11] indicated that for subsonic train speeds, the train–induced ground vibration is major come from the rail irregularities; for supersonic train speeds, the vibration of the first dominant frequency having the longest wavelength is the most obvious wave. This condition is different from the behavior of trains moving on bridges that the resonance between the train dominant frequencies and bridge natural frequencies is the major source to generate ground vibration.

Conclusion

To model soil vibrations induced by high-speed trains moving on bridges, finite element meshes often contain a large number of degrees of freedom, so the proper validation of the finite element analysis is important. This paper generated finite element meshes, including bridge superstructures, bridge foundations, moving trains, and soils. Moreover, along the mesh boundaries, numerical schemes were used to avoid fictitious wave reflections. The dynamic responses of the finite element analyses were then compared with those of field measurements. Both numerical and experimental results show similar locations and magnitudes of the vibration peaks caused by the trainload whose frequency equals an integer times the train speed divided by the carriage length. Since the comparison indicates that the predicted numerical and measured experimental results are in good agreement, the proposed finite element model can be used to simulate soil vibration induced by high-speed trains moving over bridges.

References

[1] Madshus C., Kaynia A.M. (2000). High-speed railway lines on soft ground: Dynamic behaviour at critical train speed. *Journal of Sound and Vibration*; 231(3): 689–701.

[2] Kaynia A.M., Madshus C., Zackrisson P. (2000). Ground vibration from high-speed trains: Prediction and countermeasure. *Journal of Geotechnical and Geoenvironmental Engineering-ASCE*; 126(6): 531−537.

[3] Hall L. (2003). Simulations and analyses of train-induced ground vibrations in finite element models. *Soil Dynamics and Earthquake Engineering*. 23(5): 403−413.

[4] Lombaert G., Degrande G. (2003). The experimental validation of a numerical model for the prediction of the vibrations in the free field produced by road traffic. *Journal of Sound and Vibration*. 262(2):309−331.

[5] Takemiya H. (2003). Simulation of track-ground vibrations due to a high-speed train: the case of X-2000 at Ledsgard. *Journal of Sound and Vibration*. 261(3):503−526.

[6] Kwark J.W., Choi E.S., Kim Y.J., Kim B.S., Kim S.I. (2004). Dynamic behavior of two-span continuous concrete bridges under moving high-speed train. *Computers & Structures*. 82(4−5): 463−474.

[7] Auersch L. (2005). The excitation of ground vibration by rail traffic: theory of vehicle-track-soil interaction and measurements on high-speed lines. *Journal of Sound and Vibration*. 284 (1−2): 103−132.

[8] Xia H., Zhang N. (2005). Dynamic analysis of railway bridge under high-speed trains. *Computers & Structures*. 83(23−24): 1891−1901.

[9] Ju S.H., Lin H.T. (2007). Experimentally investigating finite element accuracy for ground vibrations induced by high-speed trains. *Engineering Structures*. 30(3): 733−746.

[10] Zhai W.M., He Z.X., Song X.L. (2010). Prediction of high-speed train induced ground vibration based on train-track-ground system model. *Earthquake Engineering and Engineering Vibration*. 9(4):545−554.

[11] Ju S.H., Liao J.R., Ye Y.L. (2010). Behavior of ground vibrations induced by trains moving on embankments with rail roughness. Soil Dynamics and Earthquake Engineering. 30: 1237–1249.

[12] Costa P.A., Calcada R., Cardoso A.S. (2012). Track-ground vibrations induced by railway traffic: In-situ measurements and validation of a 2.5D FEM-BEM model. *Soil Dynamics and Earthquake Engineering*. *32*(1):111−128.

[13] Au F.T.K., Wang J.J., Cheung Y.K. (2002). Impact study of cable-stayed railway bridges with random rail irregularities. *Engineering Structures*. 24(5): 529−541.

[14] Lei X., Noda N.A. (2002). Analyses of dynamic response of vehicle and track coupling system with random irregularity of track vertical profile. *Journal of Sound and Vibration*. 258(1): 147−165.

[15] Ju S.H. (2012). A simple finite element for nonlinear wheel/rail contact and separation simulations. Accepted by *Journal of Vibration and Control*, DOI: 10.1177/1077546312463753.

[16] Ju S.H., Lin H.D., Hsueh C.C., Wang S.L. (2006). A simple finite element model for vibration analyses induced by moving vehicles. *International Journal for Numerical Method in Engineering*. 68(12): 1232−1256.

[17] Higdon R.L. (1992). Absorbing boundary conditions for acoustic and elastic waves in stratified media. *Journal of Computational Physics*. 101:386−418.

[18] Ju S.H., Wang Y.M. (2001). Time-dependent absorbing boundary conditions for elastic wave propagation. *International Journal for Numerical Methods in Engineering*. 50(9): 2159−2174.

[19] Ju S.H., Lin H.T., Huang J.Y. (2009). Dominant frequencies of train-induced vibrations. *Journal of Sound and Vibration*. 319: 247–259.

[20] Gordon C.G. (1991). Generic criteria for vibration sensitive equipment. *Optics and* Metrology. 1619:71–75.

[21] Ju S.H. (2010). A simple OpenMP scheme for parallel iteration solvers in finite element analysis. *CMES-Computer Modeling in Engineering & Sciences*. 64(1): 91–108.

In: Traffic Induced Environmental Vibrations and Controls ISBN: 978-1-62618-240-0
Editors: H. Xia and R. Calçada

Chapter 6

HUMAN INDUCED VIBRATIONS OF STEEL AND ALUMINUM BRIDGES

Horst Werkle *

Department of Civil Engineering, University of Applied Sciences (HTWG)
Konstanz, Germany

ABSTRACT

This paper reviews the modeling and numerical simulation of human induced vertical vibrations of pedestrian bridges. In particular the load models of Bachmann and Seiler/Hüttner are discussed and compared to others. Simplified formula to assess the maximum accelerations of bridges with and without TMD's are presented. Measurements of human induced vibrations were performed on an aluminum bridge, which was excited in the second harmonic of the load function. Furthermore a steel bridge being excited by the first harmonic of the load is investigated. The results obtained by a numerical simulation with a finite element analysis are in close agreement with the measured values. However, some models in literature are found to possess very large values of the Fourier coefficient of the second harmonic. In cases where the second harmonic of the load function is predominant, a realistic modeling of its size is important. Recommendations are given for the second Fourier term of the load, in order to obtain results which are in accordance with the measurements on real bridges. The measurements also indicate that higher Fourier coefficients may be significant for the response of the bridge at several measuring points.

1. INTRODUCTION

Slender, lightweight bridges of steel and aluminum have found an increasing interest in the last decades. They were made possible by new structural materials with increasing strength and production technologies. Steel bridges with cable stayed suspensions allow for greater spans and new architectural forms. Aluminum bridges are beneficial due to their light

* E-mail address: werkle@htwg-konstanz.de.

weight and their low maintenance requirements. They can be fabricated in a plant and transported economically over large distances.

In order to assure the serviceability of pedestrian bridges, a dynamic analysis should be part of the design process at an early stage. The fundamentals of the excitation of bridges by pedestrians and appropriate analytical methods have been known since the 80's of the 20th century. However in many countries, codes including serviceability requirements for human induced vibrations in the design of bridges have been introduced only in the last decade. The guidelines and requirements among different national and international codes are not comprehensive and indicate the need for further research. In some cases even new bridges suffered excessive vibrations under dynamic loads induced by pedestrians, requiring later major retrofitting measurements to ensure their serviceability. In cases where limit values of accelerations are exceeded and the serviceability requirements cannot be fulfilled, Tuned Mass Dampers (TMD's) are appropriate means to reduce the vibrations considerably. This paper gives a review of the methods in the design of pedestrian bridges of human induced vibrations for vertical vibrations with some emphasis on methods used in Germany and Europe. For two bridges, experimental investigations are presented and compared with numerical simulations based in the models given in the paper in order to assess their accuracy.

2. Load Models

2.1. General

Humans can excite structures to vibrations by different ways of motion. These are walking, running, jumping and deliberate excitation of vibrations also denoted as vandalism. Their description by complex biomechanical models of the human body which include the mechanisms of motion and the resulting mass forces (see e.g. [1]) is in general not required when the vibrations of structures are considered. Instead of that the interaction between the structure and the excitation is ignored and a predefined load time history is assumed as excitation of the structure. Only in the case of the so-called "lock-in" effect the interaction between the loading and the vibrational response of the structure has to be taken into account. When the lock-in effect occurs the pedestrians adapt their step frequency unintentionally to the eigenfrequency of the bridge. This causes self-excited vibrations being limited only by damping combined with nonlinear effects. The "lock-in" effect occurs in bridges with insufficient stiffness in lateral direction and large vibration amplitudes. It is characteristic for tensile structures and suspension bridges. However, also bridges with some lateral stiffness where the lock-in effect can be excluded, may exhibit disturbing lateral vibrations. In this paper vertical excitations are considered. This is the most significant type of excitation for the majority of footbridges, having a first horizontal eigenfrequency below 5 Hz.

Load functions for the movement of a single human have been described in literature. A literature review is given e.g. in [2] and [3]. Most models describe the load as Fourier series. But there are also deterministic models using polynomials as well as probabilistic models.

2.2. Walking

Humans walking generate forces in vertical as well as in horizontal direction acting on a structure. Measurements of the load time-history show that these forces depend not only on the weight of the human and the step frequency but also on other parameters as the type of the underground and of the footwear.

Figure 1 shows a typical vertical force time history of a single footstep. The first maximum of the time history corresponds to the contact with the heel, whereas the second peak occurs when rolling off with the ball of the foot. The steps of the left and the right foot can be added up to a load time history for walking. This results in a single load $F(t)$ varying in time and propagating with the velocity c as

$$c = f_s \cdot l_s \tag{1}$$

where f_s denotes the step frequency and l_s the step length. The time duration of a single step equals

$$T_s = \frac{1}{f_s} \tag{2}$$

A distribution of step frequencies based on a statistical evaluation of different humans in [5] is given in Figure 2 for walking, running (jogging) and sprinting (fast running). It shows that typical step frequencies for walking are between 1.6 and 2.2 Hz. However, in a dynamic analysis the step frequency has to be assumed in an unfavourable fashion, i.e. in resonance with the structure, if the structure possesses an eigenfrequency in that range.

The step length depends on the step frequency f_s. Typical values are given in Table 1.

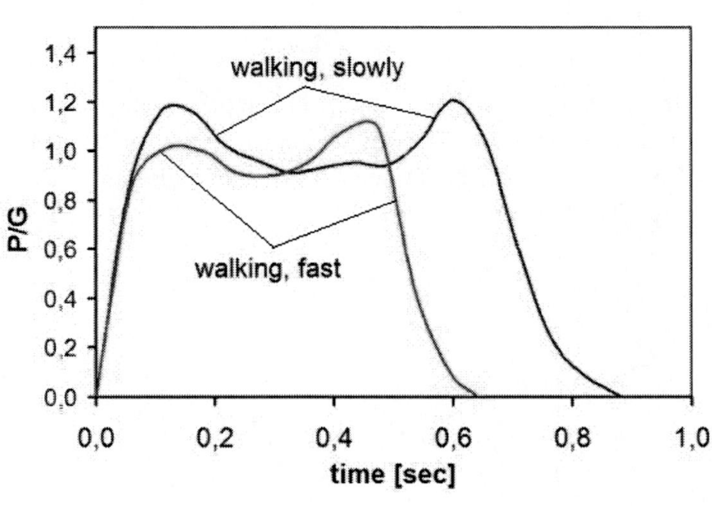

Figure 1. Time histories of the vertical load for a single footstep (after Wheeler [4]).

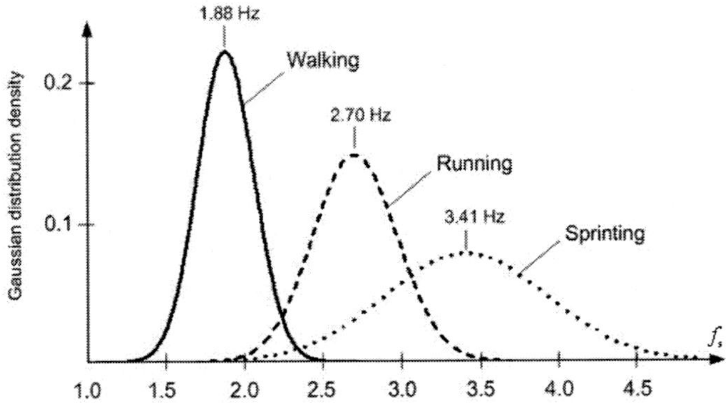

Figure 2. Distribution of step frequencies for walking, running and sprinting acc. to [5].

Table 1. Step frequency and step length acc. [7, 8]*

Type of motion		f_s [Hz]	L_s [m]
Walking	slowly	1.6 (1.4 – 1.7)*	0.60
	normally	2.0 (1.7 – 2.2)*	0.75
	fast	2.3 (2.2 – 2.4)*	1.00
Running	slowly	2.1 (1.9 – 2.2)*	-
	normally	2.5 (2.2 – 2.7)*	1.30
	fast	>3.2 (2.7 – 3.3)*	1.75
Jumping	slowly	1.6 (1.3 – 1.9)*	0
	normally	2.5 (1.9 – 3.0)*	
	fast	3.2 (3.0 – 3.4)*	

The load time history can be expressed by a Fourier series as

$$F(t) = G \cdot \left[1 + \sum_{j=1}^{4} \alpha_j \cdot \sin(2\pi \cdot j \cdot f_s \cdot t - \varphi_j \right] \tag{3}$$

where G denotes the weight of the human and f_s the step frequency. The weight force of a human is often assumed to be 0.8 kN. The Fourier coefficients α_j, φ_j have been determined by different authors. Table 2 shows the coefficients and phase angles as given by Bachmann [9] and Seiler-Hüttner [6]. The load time histories of both models are shown in Figure 3.

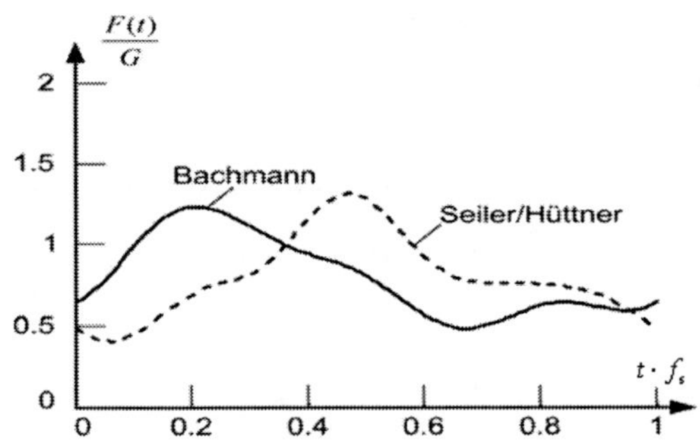

Figure 3. Load time histories of the vertical forces for “walking”.

Table 2. Parameters of the vertical forces for "walking" acc. to [9] and [6]

		Bachmann [9]	Seiler-Hü. [6]
j	f_s	1.5–2.5 [Hz] (mean value 2 Hz)	1.3–3.4 [Hz]
1	α_1	0.4 for $f_s \leq 2$ Hz $0.4 + 0.1 \cdot (f_s - 2) / 0.4$ for 2.0 Hz $\leq f_s \leq$ 2.4 Hz 0.5 for $f_s \geq 2.4$ Hz	0.4
2	α_2	0.1	0.15
3	α_3	0.1	0.10
4	α_4	0	0.05
1	φ_1	0	$\pi / 2$
2	φ_2	$\pi / 2$	$-\pi \cdot 5/6$
3	φ_3	$\pi / 2$	$\pi /2$
4	φ_4	0	$-\pi \cdot 5/6$

Frequency dependent coefficients have been given in [10] and [11]. An overview on models developed in the UK, Canada and Japan is given in [2]. Due to the large number of parameters having an influence the coefficients show a large scattering. Generally only the variation of the coefficients with the step frequency is taken into account. Design values for the Fourier coefficients as recommended by ARUP with a probability to be exceeded of 25% and 50%, respectively, according to [12] (after [2]) are given in Table 3. A comparison of different coefficients given in literature is shown in Figure 4.

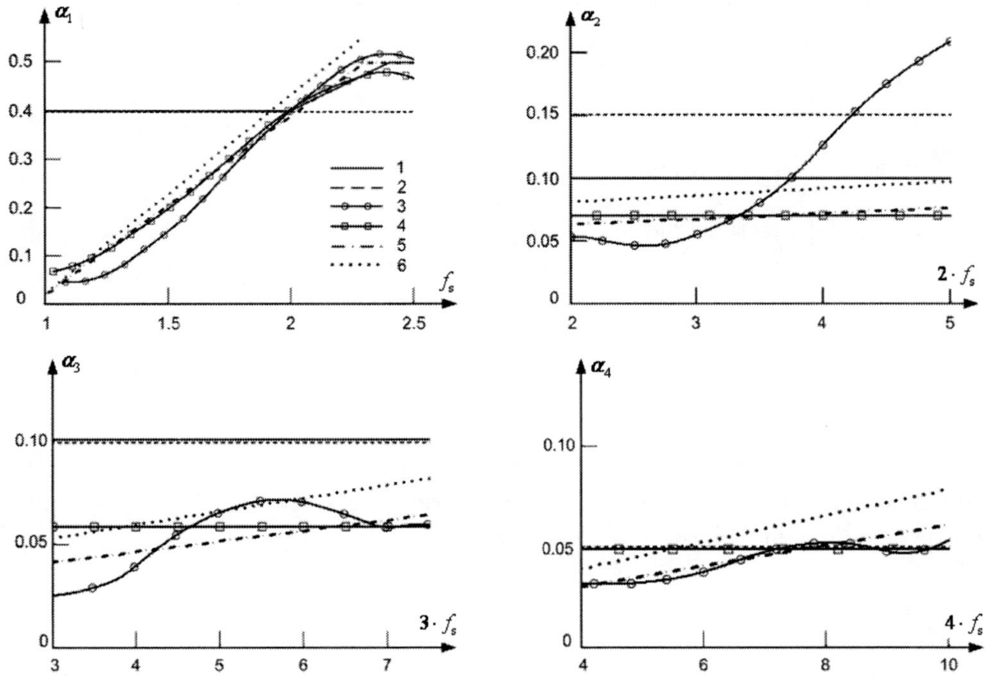

Figure 4. Frequency dependent coefficients α_1–α_4 for walking (according to: 1-Bachmann [7]; 2-Seiler/Hüttner [6]; 3-Rainer [10]; 4-Kerr [11]; 5-Young, 50% exceedance probability; 6-Young, 25% exceedance probability) [12].

Table 3. Parameters of the vertical forces for "walking" acc. to [12] (after [2])

j		Exceedance probability		$f = j \cdot f_s$
		50%	25%	
1	α_1	$0.37 \cdot (f - 0.95) \leq 0.5$	$0.41 \cdot (f - 0.95) \leq 0.56$	$f = f_s = 1 - 2.8$ Hz
2	α_2	$0.054 + 0.0044 \cdot f$	$0.069 + 0.0056 \cdot f$	$f = 2 \cdot f_s = 2 - 5.6$ Hz
3	α_3	$0.026 + 0.0050 \cdot f$	$0.033 + 0.0064 \cdot f$	$f = 3 \cdot f_s = 3 - 8.4$ Hz
4	α_4	$0.010 + 0.0051 \cdot f$	$0.013 + 0.0065 \cdot f$	$f = 4 \cdot f_s = 4 - 11.2$ Hz

2.3. Running

A walking human has a permanent contact with both feet or even with a single foot with the ground. When running or jumping a human loses temporarily this contact. This must be reflected by the load time history. According to Figure 5 the motion sequence consists of contact phase in the time $0 \leq t \leq t_c$ and a get-off phase between $t_c < t < T_s$.

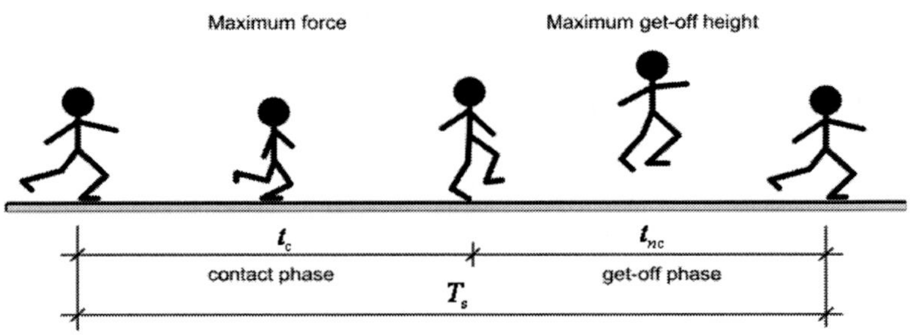

Figure 5. Motion-sequence during running.

Bachmann [9] describes the load time history during the contact phase by a half sine function. For this so called half sine model $F(t)$ is obtained with T_s and t_c acc. to Figure 5 as

$$F(t) = \begin{cases} G \cdot k_p \cdot \sin(\pi \cdot t / t_c) & \text{for } 0 \leq t \leq t_c \\ 0 & \text{for } t_c \leq t \leq T_s \end{cases} \tag{4}$$

Here, k_p is a dynamic impact factor which depends on the ratio t_c / t_s. By equating the integral of the half sine function over the time duration of a single step T_s to the integral over a static load in the same time one obtains

$$k_p = \frac{\pi}{2} \cdot \frac{1}{t_c / T_s}. \tag{5}$$

The contact time t_c as given by Wheeler [4] is given in Figure 6. It may be approximated by

$$t_c = \frac{0.4}{f_s - 1} + 0.044 \cdot (2.5 - f_s) \qquad \text{with } f_s > 1.74 \tag{6}$$

With Equations (2), (5) and (6) for a given step frequency e.g. acc. to [7] in the range of $2\text{Hz} < f_s < 3\text{Hz}$, the load time history acc. to Equation (4) can be determined.

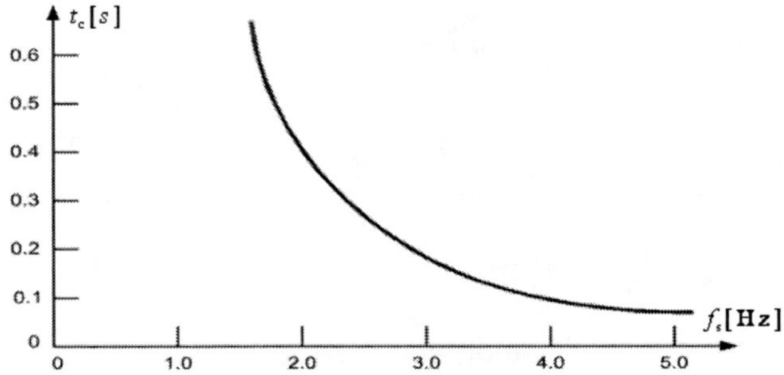

Figure 6. Contact time t_c [4].

The model of Seiler-Hüttner [6] describes also a contact phase and a get-off phase during running. Whereas the model of Bachmann is justified heuristically, the model of Seiler-Hüttner is based on a simple physical model of a running human. The model consists of a moving spring-mass-system which allows only the transfer of compression forces between the spring and the ground. For a continuous contact between the human and the ground which characterizes "walking", the force time history obtained by the model is a sine function acc. to Equation (3) with only one Fourier term ($j = 1$). For "running" the following force time history is obtained as

$$F(t) = \begin{cases} G \cdot \left[1 + s \cdot \sin(\pi \cdot t / t_G + \arcsin(-t/s)) \right] & \text{for } t \le t_c \\ 0 & \text{for } t_c < t \le T_s \end{cases} \tag{7}$$

with

$$t_G = \frac{\pi}{2} \cdot \frac{t_c}{\arccos(-1/s)} \tag{8}$$

$$\frac{t_c}{T_s} = \left(1 + \frac{\sqrt{s^2 - 1}}{\arccos(-1/s)} \right)^{-1} \approx \frac{2 + \pi \cdot s}{2 + \pi \cdot s + 2 \cdot s \cdot \sqrt{s^2 - 1}} \ . \tag{9}$$

The impact factor s in Equation (7) is shown in Figure 7. It may be approximated as

$$s \approx k_p - 0.77 = \frac{\pi}{2} \cdot \frac{1}{t_c / T_s} - 0.77 \ . \tag{10}$$

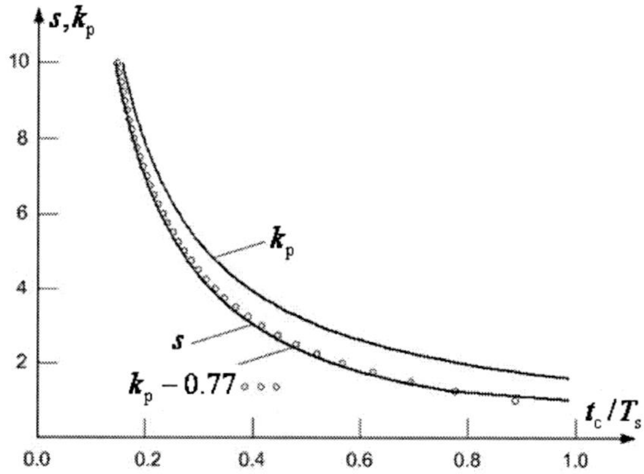

Figure 7. Impulse factors acc. to [9] and [6].

In the range $0.2 \le t_c / T \le 0.88$ the difference to the exact solution in Equation (9) is less than 3%.

The load time history for a running human can now be determined with Equation (7) for a given step frequency $f_s = 1/T_s$ and contact time t_c.

Due to its mechanical base the model of Seiler-Hüttner has a significant benefit. For long contact times $t_c \rightarrow T_s$ corresponding to walking, a harmonic vibration (with $j=1$) is obtained whereas the half sine model gives a series of half sine functions. For three step frequencies the load time histories of a single footstep obtained by both models are shown in Figure 8.

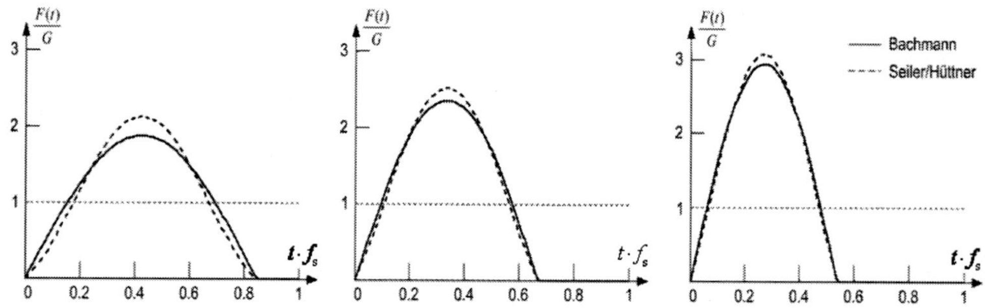

(a) slowly (b) normal (c) fast

Figure 8. Load-time histories of vertical forces for running due to Bachmann and Seiler/Hüttner (one single footstep).

For a running human the steps of the left and the right foot can again be added up to a time dependent single load propagating with the velocity c. The time history of the load is again described by the first terms of a Fourier series acc. to Equation (3). Table 4 gives the amplitudes and phase angles of the Fourier terms as given by Bachmann [9] and Seiler-Hüttner [6]. The resulting time histories are given in Figure 9 where the phase angles given by Seiler-Hüttner are also used with the coefficients of Bachmann. The maxima show a remarkable difference indicating that the coefficient of Bachmann is quite huge. Figure 10 compares these coefficients with the frequency dependent coefficients given by Rainer et.al. [10].

Table 4. Fourier coefficients of vertical forces for "running" acc. to [9] and [6]

j		Bachmann [9]	Seiler-Hüttner [6]
1	α_1	1.6	1.38
2	α_2	0.7	0.31
3	α_3	0.2	−0.10
4	α_4	0	0
1	φ_1	N/A	$j\cdot\pi\cdot\left(1+\frac{\sqrt{8}}{\arccos(-1/3)}\right)^{-1}-\frac{\pi}{2}=\pi\cdot\left(0.40\cdot j-0.5\right)$
2	φ_2	N/A	
3	φ_3	N/A	
4	φ_4	N/A	N/A

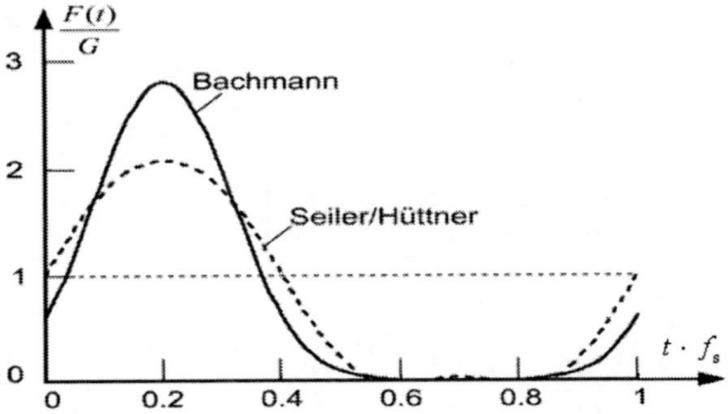

Figure 9. Time histories of vertical forces for running due to [6] and [9] (with phase angles φ_i acc. to [6]).

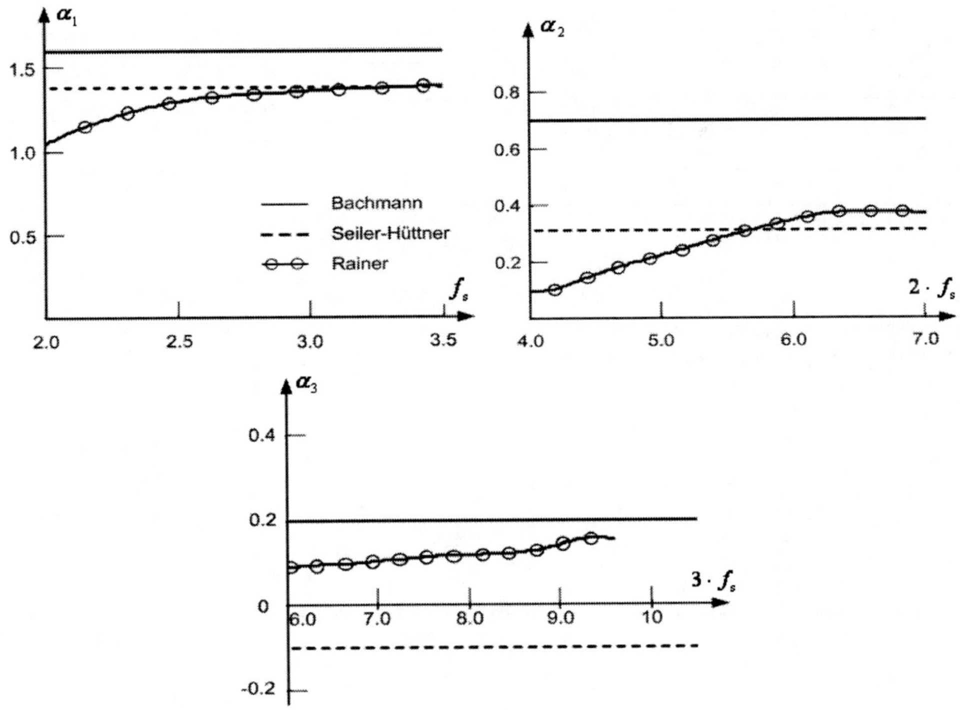

Figure 10. Fourier coefficients α_1–α_3 depending on the pacing frequency for "running".

2.4. Jumping

Jumping of a human is represented by a stationary time dependent load. The time history is again represented by the first terms of a Fourier series acc. to Equation (3).

Table 5 gives the Fourier coefficients given by Bachmann [7] and Seiler-Hüttner [6]. Figure 11 compares the frequency dependent coefficients of Bachmann [7], Seiler-Hüttner [6] and Rainer et. al. [10].

Table 5. Fourier coefficients of vertical forces for "jumping" due to [7] and [6]

j		Bachmann [7]		Seiler-Hüttner [6]
1	α_1	1.8/1.9* $-0.1\cdot f_s+2$ $-0.1\cdot f_s+2.1$* 1.7/1.8*	for $f_s \le 2$ Hz for 2.0 Hz $< f_s \le 3.0$ Hz for $f_s > 3$ Hz	1.72
2	α_2	1.3/1.6* $-0.2\cdot f_s+1.7$ $-0.3\cdot f_s+2.2$* 1.1/1.3*	for $f_s \le 2$ Hz for 2.0 Hz $< f_s \le 3.0$ Hz for $f_s > 3$ Hz	1.05
3	α_3	0.7/1.1* $-0.2\cdot f_s+1.1$ $-0.3\cdot f_s+1.7$* 0.5/0.8*	for $f_s \le 2$ Hz for 2.0 Hz $< f_s \le 3.0$ Hz for $f_s > 3$ Hz	0.36
4	α_4	0		−0.04
1	φ_1	0		$j\cdot\pi\cdot\left(1+\frac{\sqrt{1.56}}{\arccos(-1/1.6)}\right)^{-1}-\frac{\pi}{2}=\pi\cdot(0.64\cdot j-0.5)$
2	φ_2	$\pi\cdot(1-f_s\cdot t_c)$		
3	φ_3			
4	φ_4	-		-

*value for high jumping.

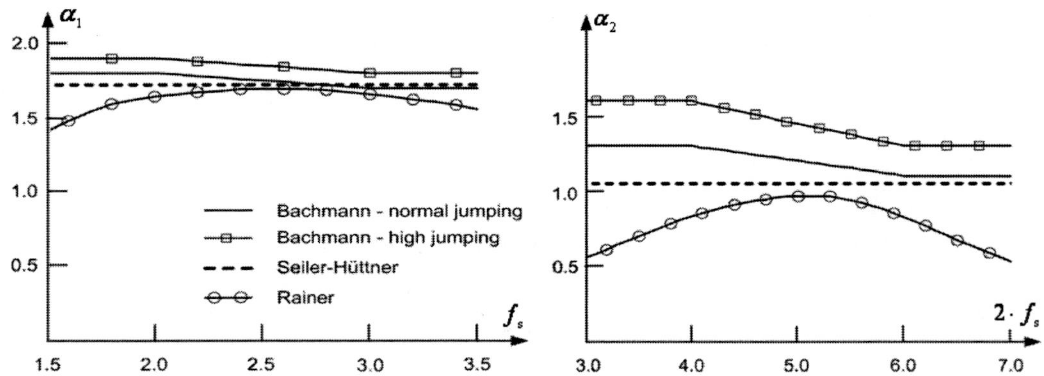

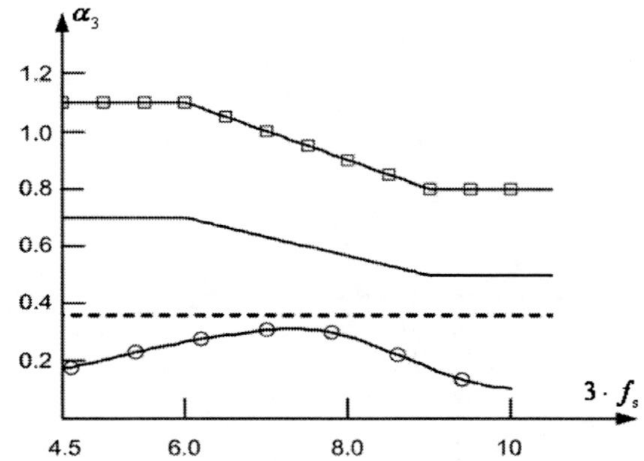

Figure 11. Fourier coefficients α_1–α_3 depending on the step frequency for "jumping".

2.5. Loading by Several Individual Persons, Groups of Persons and Crowds

For the design of bridges in general the vibrations due to several persons have to be considered. Since people almost never walk with the same step frequency and phase – except may be military units in lockstep -, an increase of the acceleration proportional to the number of persons on the bridge can be excluded in practice. According to Matsumoto [13], for n persons walking on the bridge the structural response obtained for one person has to be augmented by the factor

$$n_{\text{eq}} = \sqrt{n} \tag{11}$$

where n_{eq} is the equivalent number of fully synchronized persons. Equation (11) is based on the assumption that all persons are walking with the same frequency but with random phase distribution. The number of persons on the bridge may be determined based on the arrival rate λ [persons/s] and the time T_0 required traversing the bridge as

$$n = \lambda \cdot T_0 = \lambda \cdot L \cdot c \tag{12}$$

where the bridge length is L. In order to reflect the fact that step frequencies outside the range of 1.6 to 2.4 Hz are rare, Bachmann suggested to use n_{eq} acc. to Equation (11) for step frequencies between 1.8 and 2.1 Hz and a limit of 2 for step frequencies less than 1.6 Hz and greater than 2.4 Hz [9] (Figure 12).

The number of persons to be considered depends on the use of the bridge. The following loading scenarios may be defined:

- Traffic with low density (e.g. 0.2 persons/m^2) where individuals can move freely.
- Groups of persons.
- Crowd loading (e.g. 0.5−1.0 persons/m^2) where free movement of individuals is restricted.

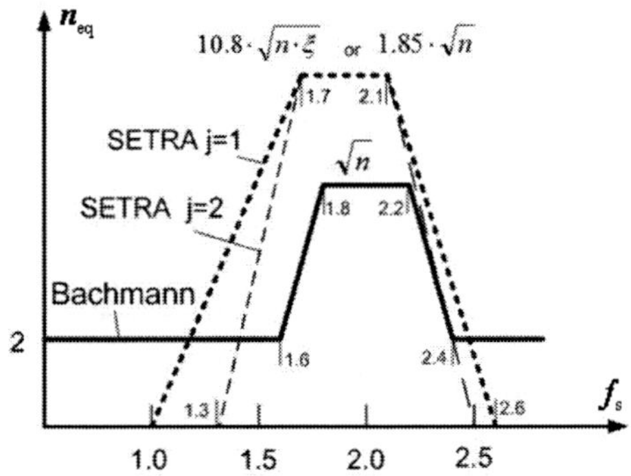

Figure 12. Factor *m* when considering more than one person acc. to [8], [13] and [14].

These scenarios are defined in literature in more detail (e.g. in [15, 16]). The restriction of the movement in crowds also influences the correlation of the movement in the crowd and therewith the number n_{eq} of equivalent fully synchronized walking persons. Extended probabilistic response simulations of pedestrian streams with random phase and frequency distribution done by SETRA [14] lead to two relationships, one for sparse and dense crowds and another for very dense crowds:

$$n_{eq} = \begin{cases} 10.8 \cdot \sqrt{n \cdot \xi} & \text{for } d \leq 0.8 \text{ persons/m}^2 \\ 1.85 \cdot \sqrt{n} & \text{for } d \geq 1.0 \text{ persons/m}^2 \end{cases} \tag{13}$$

Herein ξ is the damping ratio of the structure. The frequency dependence of the load function is considered by multiplying n_{eq} with the distribution in Figure 12. A similar formula is given in the UK National Annex (NA) to Eurocode 1 [17, 18]. However the percentile value of peak accelerations which can be exceed is chosen larger than in the SETRA model. According to [19] this gives $n_{eq} = 6.1 \cdot \sqrt{n \cdot \xi}$.

The frequency distribution to be considered in addition is different from [15] but not given here (see [17]). A comparison of the Matsumoto model with the SETRA model is shown in Figure 13. It is interesting to note that the number n_{eq} of equivalent persons walking in phase increases considerably when the damping increases.

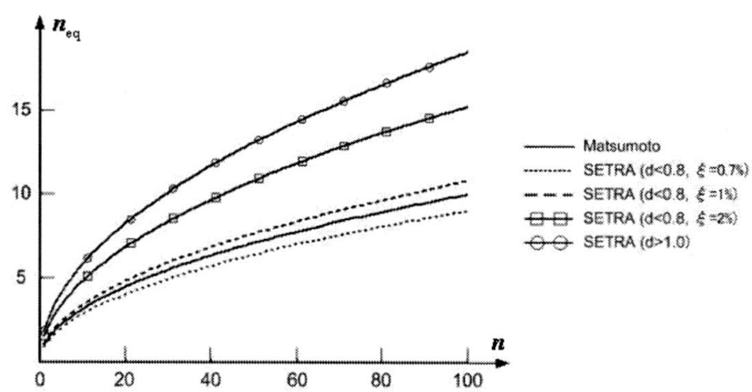

Figure 13. Factor n_{eq} for different models.

Eurocode 5 [20] defines $n_{eq}=0.23 \cdot n$ with $n=13$ for a group of persons and $n=0.64 \cdot A$ for a stream of persons where A is the bridge area in m². Comparative computations by Pavic [19] showed that this formula may give much larger values than other approaches e.g. acc. to Equation (13) and is not recommended for use.

A more sophisticated model taking into account not only the probability distribution of step frequencies, but also of arrival times, human mass and force amplitudes has been developed by Butz [21, 22]. The model gives a formula for the 95th percentile of the maximum acceleration as function of the crowd density d (in persons/m²), the modal mass m_i^* (in tons) and damping ratio ξ as well as the eigenfrequency f_i of the structure as

$$a_{\max\_95} = k_{a,95\%} \cdot \frac{\sqrt{k_1 \cdot C \cdot k_F \cdot n \cdot \xi^{k_2}}}{m_i^*} \tag{14}$$

with

$$k_1 = a_1 \cdot f_i^2 + a_2 \cdot f_i + a_3 \tag{14a}$$

$$k_2 = b_1 \cdot f_i^2 + b_2 \cdot f_i + b_3 \tag{14b}$$

All constants are given in Table 6. The model is applicable only for structures where the mean step frequency of the pedestrian stream coincides with the i-the eigenfrequency of the structure and the first Fourier term of the load is excited. It has been implemented into the HIVOSS Guidelines [15] and is extended in [23].

Table 6. Constants for the probabilistic model of Butz [21, 15]

d [P/m²]	k_F	C	a_1	a_2	a_3	b_1	b_2	b_3	$k_{a,95\%}$
≤ 0.5	1.20×10^{-2}	2.95	−0.07	0.60	0.075	0.003	−0.040	−1.000	3.92
1.0	7.00×10^{-3}	3.70	−0.07	0.56	0.084	0.004	−0.045	−1.000	3.80
1.5	3.34×10^{-3}	5.10	−0.08	0.50	0.085	0.005	−0.060	−1.005	3.74

3. Structural Models

The numerical simulation of human induced vibrations of bridges is generally performed by a dynamic finite element analysis of the structure. When solving the equations of motion for the time dependent displacements $\underline{u}(t)$ of the system

$$\underline{K} \cdot \underline{u}(t) + \underline{C} \cdot \dot{\underline{u}}(t) + \underline{M} \cdot \ddot{\underline{u}}(t) = \underline{F}(t) \tag{15}$$

with $\underline{K}$, $\underline{C}$ and $\underline{M}$ as stiffness, damping and mass matrix, respectively, the load vector $\underline{F}(t)$ varies in time and space. The solution may be obtained by direct integration and also by modal analysis as long as no special damping devices as TMD's are to be implemented in the model.

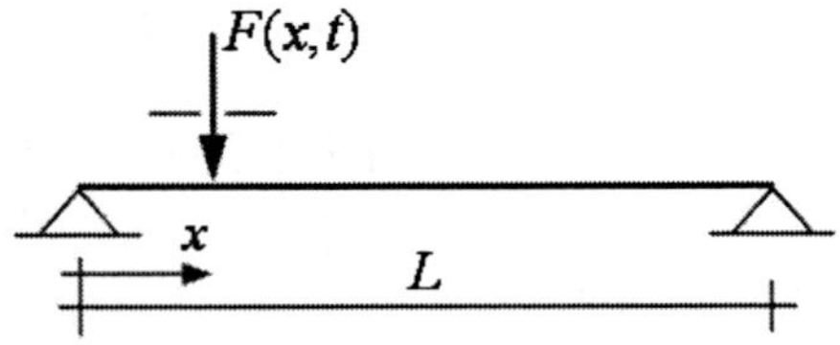

Figure 14. Simply supported beam with a moving point load.

For simple structures, analytical solutions can be formulated. In the following a moving point load on a simply supported beam is considered (Figure 14). The eigenfrequencies f_i and mode shapes $\phi_i(x)$ of the system are given by

$$f_i = \frac{i^2 \cdot \pi}{2 \cdot L^2} \sqrt{\frac{E \cdot I}{\mu}} \tag{16}$$

$$\phi_i(\overline{x}) = \sin(\frac{i \cdot \pi \cdot \overline{x}}{L}) \tag{17}$$

where $E{\cdot}I$, μ are the bending stiffness and the mass per length, respectively, of the beam. The modal displacements $\alpha(t)$ and accelerations $\ddot{\alpha}(t)$ are obtained by solving the modal equations

$$k_i^* \cdot \alpha(t) + c_i^* \cdot \dot{\alpha}(t) + m_i^* \cdot \ddot{\alpha}(t) = \phi_i(c \cdot t) \cdot F(t) \tag{18}$$

with the modal mass, the modal stiffness and the modal damping

$$m_i^* = \mu \cdot L/2; \quad k_i^* = (2 \cdot \pi \cdot f_i)^2 \cdot m_i^*; \quad c_i^* = 2 \cdot \xi \cdot \sqrt{k_i^* \cdot m_i^*} \tag{19}$$

and $F(t)$ acc. to Equation (3) in time domain. The time histories of the displacements $u(\overline{x},t)$ and accelerations $\ddot{u}(\overline{x},t)$ at the coordinate $\overline{x}$ are obtained as

$$u(\overline{x},t) = \sum_i \alpha(t) \cdot \phi(\overline{x}) \; ; \; \ddot{u}(\overline{x},t) = \sum_i \ddot{\alpha}(t) \cdot \phi(\overline{x}) \tag{20}$$

More simplified formulas have been derived in order to easily assess the maximum vertical acceleration of beam-like bridges. For a 1-DOF system with mass $\overline{m}$, spring constant k and damping ratio ξ excited by a harmonic force $F(t) = F_0 \cdot \sin(\Omega \cdot t)$ caused by a single person, the stationary response for the maximum acceleration at resonance is

$$a_{\text{vert},1} = \omega^2 \cdot u_{\text{vert},1} = \frac{k}{\overline{m}} \cdot \frac{1}{2 \cdot \xi} \cdot \frac{F_0}{k} = \frac{F_0}{2 \cdot \overline{m} \cdot \xi} \tag{21}$$

with $\omega = \sqrt{k/\bar{m}}$ being the circular eigenfrequency and $1/(2 \cdot \xi)$ the magnification factor of the static displacement $u_{\text{vert},1} = F_0 / k$ at resonance. With $F_0 = \alpha_j \cdot G$ one obtains

$$a_{\text{vert},1} = \frac{\alpha_j \cdot G}{2 \cdot \bar{m} \cdot \xi} \tag{22}$$

where $\bar{m}$ being the modal mass of the system. It has to be evaluated with the mode shape set to 1 at the place where the force $F(t)$ is acting.

For a simply supported beam with the mass per length m and the span width L the modal mass is $\bar{m} = m \cdot L / 2$ when $F(t)$ is acting in the middle of the bridge. In order to take into account that the load is transient and moving on the beam with a velocity c instead of being a stationary harmonic load on a 1-DOF system, a reduction factor α_{red} is introduced [24]. Hamm [25] suggests a factor $\alpha_{\text{red}} = 0.75$. With $M = m \cdot L$ one obtains

$$a_{\text{vert},1} = \alpha_{\text{red}} \frac{\alpha_j \cdot G}{M \cdot \xi} \tag{23}$$

With $\alpha_{\text{red}} = 0.71$, $G = 700$ N and $\alpha_1 = 0.4$ corresponding to the first Fourier term, the formula of the maximum vertical acceleration for a single person given in Eurocode 5 [20] is obtained as

$$a_{\text{vert},1} = \frac{200}{M \cdot \xi} \quad \text{for } f_{\text{vert}} \leq 2.5 \text{ Hz} \tag{24a}$$

where M denotes the total mass of the bridge in kg, ξ the damping ratio and f_{vert} the first eigenfrequency of the (one-span) bridge in vertical direction. It should be noted that the formula gives the maximum acceleration at resonance of a 1-DOF system with the modal mass of the first mode, a harmonic load $F(t) = \alpha_1 \cdot G \cdot \sin(\Omega \cdot t)$ and a reduction factor α_{red} without any further assumptions.

For the second Fourier term with $G = 700$ N, $\alpha_2 = 0.14$ and $\alpha_{\text{red}} = 1$ one obtains the formula given in Eurocode 5 [20]

$$a_{\text{vert},1} = \frac{100}{M \cdot \xi} \quad \text{for } 2.5 \text{Hz} \leq f_{\text{vert}} \leq 5.0 \text{Hz} \tag{24b}$$

4. Threshold Values of Accelerations

The level of acceptance of vertical vibrations by pedestrians depends on their individual perception, attitude and expectation of vibrations. Another important factor is the frequency of occurrence of a peak acceleration value. In order to consider the sensitivity to the frequency content of the vibration, Blanchard [26] suggested a frequency dependent threshold value for the vertical acceleration as

$$a_{\mathrm{v,lim}} = 0.5 \cdot \sqrt{f_{1,\mathrm{v}}} \ [\mathrm{m/s^2}] \qquad (25)$$

where $f_{1,v}$ is the lowest vertical eigenfrequency of the bridge in [Hz].

In his study, Matsumoto [13] took a vertical acceleration of 1.0 m/s^2 as threshold value to avoid discomfort and insecurity. Bachmann [9] suggested a range of vertical accelerations between 0.5−1.0 m/s^2, where the lower value is considered as safe against any objections. Eurocode 5 requires a threshold value 0.7 m/s^2 in vertical direction [20].

A more detailed examination associates acceleration ranges with comfort levels (Table 7). In [14] the following comfort levels are defined:

- *Maximum comfort:* Accelerations undergone by the structure are practically imperceptible to the users.
- *Average comfort:* Accelerations undergone by the structure are merely perceptible to the users.
- *Minimum comfort:* under loading configurations that seldom occur, accelerations undergone by the structure are perceived by the user, but do not become intolerable.

The comfort level required is chosen according to the design situation depending on the traffic situation and its frequency of occurrence. In practice several different probable design situations should be considered.

Table 7. Comfort classes with acceleration ranges [14] and [15]

Comfort class	Degree of comfort	$a_{v,limit}$
CL 1	Maximum	< 0.50 m/s^2
CL 2	Medium	0.50 – 1.00 m/s^2
CL 3	Minimum	1.00 – 2.50 m/s^2
CL 4	Unacceptable discomfort	> 2.50 m/s^2

5. Codes and Guidelines

In the last decade a number of codes and design guidelines concerning pedestrian-induced dynamic loading of footbridges have been published. They describe loading scenarios and loading functions, give simplified formula for the maximum acceleration and threshold values for accelerations. Most important among them are the Eurocodes 1 and 5 which may also be applied to other than wooden structures due to its general mechanical basis [17, 20], the ISO 19137 standard [27] and the HIVOSS [15, 16] and SETRA [14] guidelines. A comparison of these codes and guidelines with special emphasis on vertical crowd dynamic action is presented in [19].

6. Tuned Mass Dampers

In cases where the required acceleration limits cannot be kept, Tuned Mass Dampers (TMD's) are an appropriate means to reduce the vibrations. They consist of a mass m_D, a spring k_D and a damper c_D attached at the main system e.g. the footbridge. The main system with a TMD can be simplified as a 2-DOF system in which the bridge (or its modal mass) is represented as a 1-DOF system with the parameters k_H, c_H and m_H, as in Figure 15.

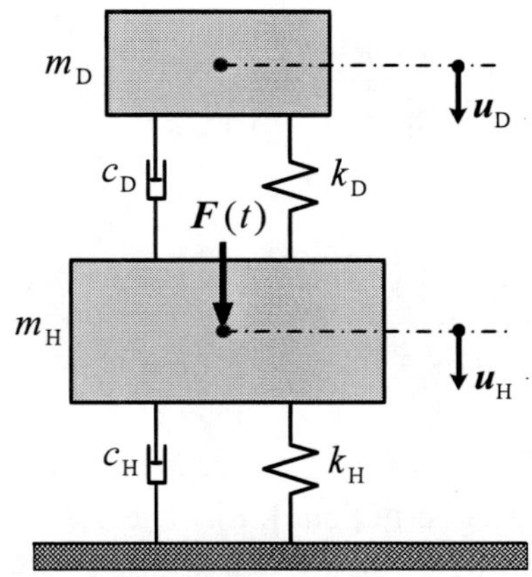

Figure 15. Simplified model for a main system with a TMD.

The solution of its equation of motion for stationary harmonic motion due to a harmonic force

$$F(t) = F_0 \cdot \sin(\Omega \cdot t) \tag{26}$$

is:

$$\bar{u}_H(t) = u_H \cdot \sin(\Omega \cdot t + \varphi) \tag{26a}$$

$$\ddot{\bar{u}}_H(t) = -\Omega^2 \cdot u_H \cdot \sin(\Omega \cdot t + \varphi) = -\Omega^2 \cdot \bar{u}_H(t) \tag{26b}$$

The amplitudes of the vibrations of the main system depend on the frequency ratio $\eta = \Omega / \omega_H$ and may also be written as

$$u_H(\eta) = V_H(\eta) \cdot u_{H,stat} \tag{27}$$

with the dynamic amplification ratio [3]

$$V_H(\eta) = \frac{u_H}{u_{H,stat}} = \sqrt{\frac{A^2 + B^2}{C^2 + D^2}} \tag{28}$$

and

$$A(\eta) = -\eta^2 + \kappa^2 \,;\; B(\eta) = 2 \cdot \kappa \cdot \xi_D \cdot \eta \tag{29a}$$

$$C(\eta) = \eta^4 - [1 + (1+\mu) \cdot \kappa^2 + 4 \cdot \kappa \cdot \xi_H \cdot \xi_D] \cdot \eta^2 + \kappa^2 \tag{29b}$$

$$D(\eta) = [-2 \cdot [\xi_H + (1+\mu) \cdot \kappa \cdot \xi_D] \cdot \eta^3] + 2 \cdot \kappa \cdot (\kappa \cdot \xi_H + \xi_D) \cdot \eta \tag{29c}$$

$$\eta = \frac{\Omega}{\omega_H} \,;\; \mu = \frac{m_D}{m_H} \,;\; \kappa = \frac{f_D}{f_H} = \sqrt{\frac{k_D \cdot m_H}{k_H \cdot m_D}} \tag{29d}$$

$$\omega_H = 2 \cdot \pi \cdot f_H = \sqrt{\frac{k_H}{m_H}} \tag{29e}$$

$$\xi_H = \frac{c_H}{2 \cdot \sqrt{m_H \cdot k_H}} \,;\; \xi_D = \frac{c_D}{2 \cdot \sqrt{m_D \cdot k_D}} \,;\; u_{H,stat} = \frac{F_0}{k_H} \tag{29f}$$

The properties of the TMD are generally based on an optimization criterion for steady state motion as the Den Hartog criterion [28]. The target of this optimization is the minimization of the displacements of the system with $\xi_H = 0$. According to Den Hartog the frequency ratio should be chosen to be

$$\kappa_{opt} = \frac{f_{D,opt}}{f_H} = \frac{1}{1+\mu} \tag{30}$$

and the damping ratio as

$$\xi_{D,opt} = \sqrt{\frac{3 \cdot \mu}{8 \cdot (1+\mu)^3}} \tag{31}$$

The corresponding parameters of the damper are

$$m_D = \mu \cdot m_H \tag{32a}$$

$$f_{D,opt} = \kappa_{opt} \cdot f_H \tag{32b}$$

$$k_{D,opt} = 4 \cdot \pi^2 \cdot m_D \cdot f_{D,opt}^2 \tag{32c}$$

$$c_{D,opt} = 4 \cdot \pi \cdot m_D \cdot f_{D,opt} \cdot \xi_{D,opt} \tag{32d}$$

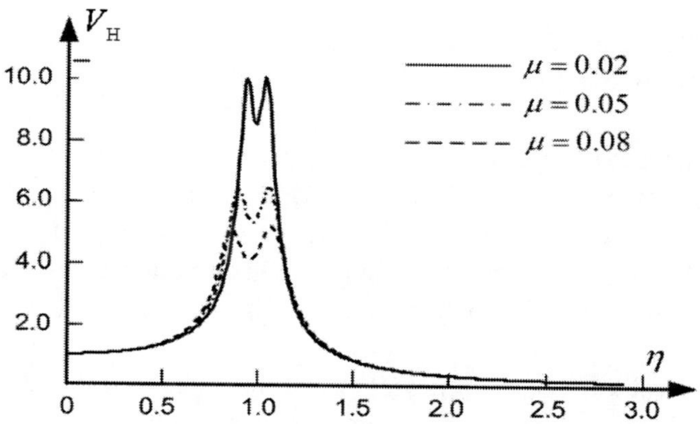

Figure 16. Amplitudes of main system with a TMD, μ=0.02, 0.05, 0.08 and $\xi_H = 0$.

The main parameter for the design of the TMD is the mass ratio μ. In addition the mass and the eigenfrequency of the main structure have to be known. In practice, values of μ=0.02 – 0.08 are chosen. Figure 16 shows the magnification function of the amplitudes of the main system for mass ratios μ=0.02, 0.05 and 0.08.

With these parameters of the TMD the maximum dynamic amplification ratio for $\xi_H = 0$ is given by Den Hartog as

$$V_{H,max} = \sqrt{1 + \frac{2}{\mu}} \tag{33}$$

Using the maximum dynamic amplification factor of a 1-DOF-system which is $V_{H,max} = 1/(2 \cdot \xi)$ an equivalent damping of the TMD can be determined as

$$\xi_{H,equival} = \frac{1}{2 \cdot \sqrt{1 + \frac{2}{\mu}}} \tag{34}$$

As this equation has been derived for $\xi_H = 0$ some portion of the damping of the main system e.g. $0.5 \cdot \xi_H$ may be added.

The maximum acceleration of the main mass of the 2-DOF system acc. to Equation (26b) is

$$a_{H,opt} = \max\left|\ddot{u}_H(t)\right| = \Omega^2 \cdot \left|\max u_H(t)\right| = \omega_H^2 \cdot \max\left|\eta^2 \cdot V_H(\eta)\right| \cdot \frac{F_0}{k_H} \tag{35}$$

In the case of a beam-like bridge the equivalent damping may be used to determine the maximum acceleration system similarly as in Equation (24a) for the system without TMD. As comparative computations show, due to the large damping of the system the reduction factor should be chosen to be $\alpha_{red} = 1$. By introducing $\alpha_{red} = 1$, $G = 700$ N and $\alpha_1 = 0.4$ in Equation (23) one obtains for a system with a TMD designed according the Den Hartog

$$a_{\text{vert,1}} = \frac{280}{M \cdot \xi_{\text{H,equival}}} = \frac{560}{M} \cdot \sqrt{1 + \frac{2}{\mu}} \quad \text{for } f_{\text{vert}} \leq 2{,}5\,\text{Hz} \tag{36}$$

where M again denotes the total mass of the bridge in kg.

Equation (36) presumes that the first mode shape is still relevant after the installation of the TMD. This must be checked carefully before applying Equation (36). If after the installation of a TMD the motion is governed by a higher mode e.g. a torsional mode, which is not damped by the TMD, more sophisticated models including all relevant modes must be investigated.

TMD's show a robust behavior in cases where the spring, mass and damper of a TMD possess slight fabrication tolerances or the damper element of the TMD changes its properties due to temperature [29, 30].

7. Experimental Investigations

7.1. Aluminum Footbridge

7.1.1. Generals

Aluminum bridges are beneficial due to their light weight and their low maintenance requirements. Aluminum is especially suited for prefabricated pedestrian bridges. They are fabricated with spans of 12 m to 45 m in a plant and can be economically transported over distances up to several hundred kilometers at the construction site. Whereas aluminum has a bearing capacity similar to steel, its stiffness is considerably lower. The Young's modulus of aluminum is only 1/3 of that of steel. Therefore aluminum pedestrian bridges are prone to the excitation of vibrations induced by humans passing over the bridge.

The truss bridge investigated has a span of 29.60 m, a width of 2.6 m and a mass of 6.7 tons (Figure 17). It possesses a slight curvature in the vertical direction. In the horizontal direction the bridge is stiffened by a horizontal truss under the paving slab and the slab itself. Figure 18 (a, b) shows the bridge in the plant and after placement on site.

In order to assess the vibrations of aluminum bridges, extensive numerical and experimental investigations have been performed by Püschel on a pedestrian bridge [31–33]. Tests have been undertaken on the bridge at the plant as well as on the site in Leingarten, Germany, where the bridge was later built in.

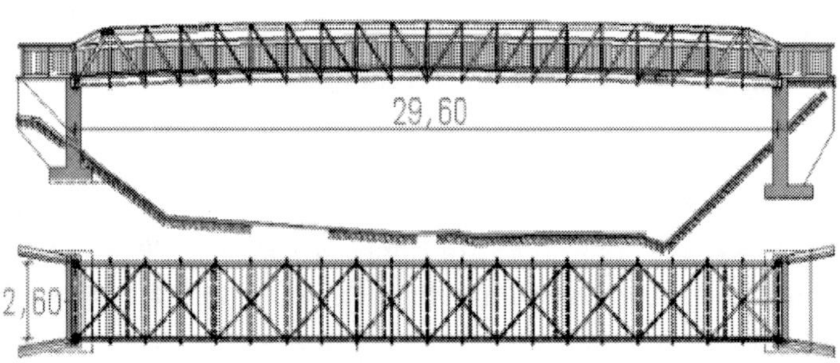

(a) bridge construction

(b) Finite Element model

Figure 17. Aluminum bridge, Leingarten, Germany.

(a) in the plant

(b) on site

Figure 18. Aluminum footbridge.

7.1.2. Computational Simulation

The computational simulation of the bridge has been performed using a very detailed finite element model with 2060 nodes and 1750 elements, Figure 17b. The first eigenfrequency has been determined to be f_1=4.0 Hz. The corresponding mode shape is shown in Figure 19 (a). All other frequencies are above 5 Hz and hence not relevant for the vibration excitation by pedestrians.

The computational simulation of the vibrations induced by pedestrians has been performed using different load models:

1. single footsteps with load time histories acc. to Wheeler [4], (Figure 1)
2. load model of Bachmann [7, 9]
3. load model of Seiler/Hüttner [6]

Time histories of the accelerations have been obtained by a modal analysis with 25 modes and a modal damping of 1% in the first mode and 0.5% in the higher modes. The damping values have been derived from field tests of the bridge with impulse loading.

(a) $f_1 = 4.0$ Hz

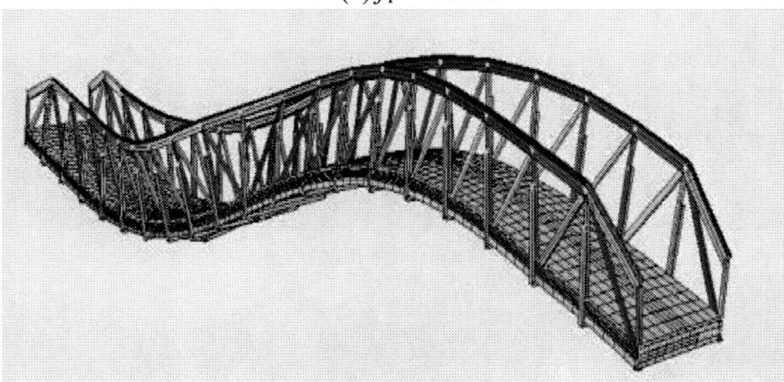

(c) $f_2 = 13.0$ Hz

Figure 19. Mode shapes for vertical vibrations.

Figure 20 shows the accelerations in the middle of the bridge for a step frequency of 2.0 Hz when the load is applied by single footsteps. A maximum vertical acceleration of 0.5 m/s² is obtained.

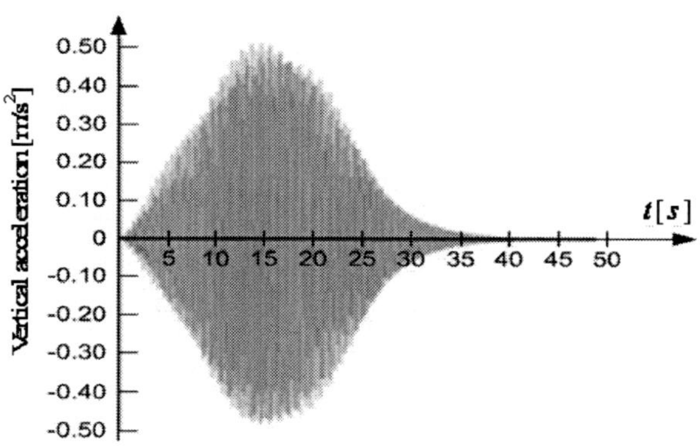

Figure 20. Computed vertical acceleration time history, f_s=2.0 Hz, G=0.8 kN.

For the load models with a continuously propagating load, the maximum accelerations in the middle of the bridge are plotted in Figure 21 over the step frequency. They show a pronounced resonance peak at half of the first eigenfrequency of the structure. At the same frequency the footstep model with time histories according to Wheeler [4] shows a considerably lower acceleration. The approximate formula, Equation (23), gives with $\alpha_2 = 0.10$ $G = 800$ N and $\alpha_{red} = 1$ for the second Fourier coefficient

$$a_{\text{vert},1} = 1.0 \cdot \frac{0.10 \cdot 800}{6700 \cdot 0.01} = 1.2 \ \text{m/s}^2 \tag{37}$$

According to Equation (24b) given in Eurocode 5 a peak acceleration of

$$a_{\text{vert},1} = \frac{100}{M \cdot \xi} = \frac{100}{6700 \cdot 0.01} = 1.49 \ \text{m/s}^2 \tag{38}$$

is obtained. The great differences in the peak accelerations reflect the differences in the Fourier term α_2 given by various authors (see Tables 2, 3 and Figure 4).

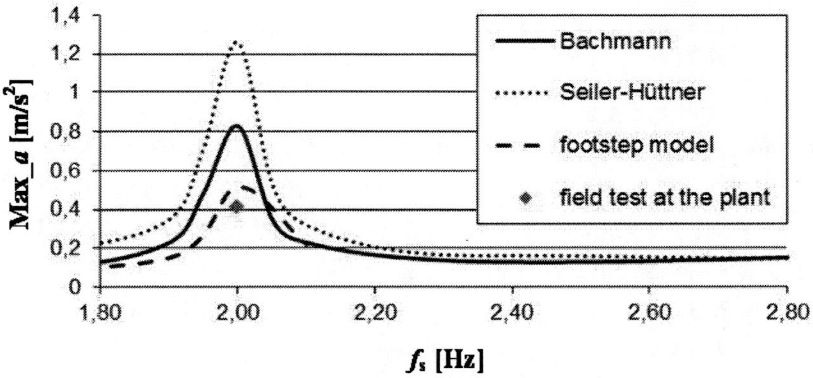

Figure 21. Maximum acceleration vs. step frequency at the plant, $G = 0.80$ kN.

7.1.3. Field Tests

Tests of the bridge have been performed at the plant. A second series of tests has been undertaken at the finished bridge in its final location.

The natural frequencies of the bridge have been derived from the FFT spectrum of the bridge response due to impulse excitation as well as by ambient vibration measurements. Values of the first eigenfrequency are given in Table 8. The eigenfrequency of the finite element model agrees closely with the frequency measured in the plant. However there is a significant difference between the eigenfrequencies measured at the plant and at the site. It originates from an arc effect, resulting from the horizontal fixing of the bearings of the bridge when the bridge was built in at its final location.

Field tests with a person passing the bridge have been performed with different step frequencies. Figure 22 shows a typical acceleration time history as measured at the plant with a step frequency f_s=2.0 Hz. The maximum acceleration compares closely with the single-footstep model, (see Figure 20).

Table 8. First and second natural frequencies of the bridge

	First eigenfrequency	Second eigenfrequency
Computational simulation	3.98 Hz	13.0 Hz
Measurement at the plant	3.85 Hz	11.8 Hz
Measurement on site	4.50 Hz	12.1 Hz

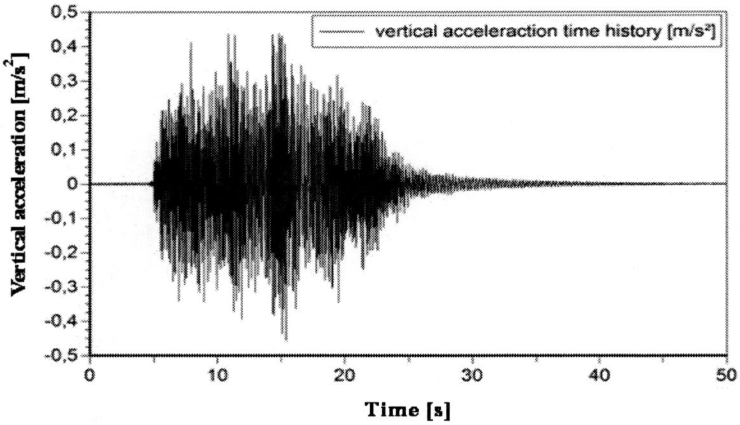

Figure 22. Measured vertical acceleration time history, f_s=2.0 Hz, G=0.95 kN.

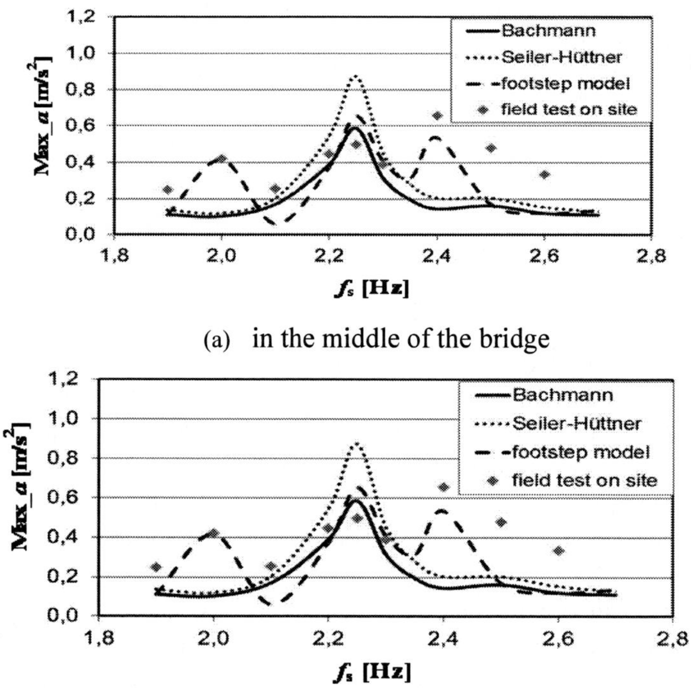

(a) in the middle of the bridge

(b) at the quarter point of the bridge

Figure 23. Maximum acceleration vs. step frequency on site, G=0.80 kN.

Field tests with different step frequencies have been performed at the bridge at the final location (Figure 23). A shift of the acceleration peak from 2.00 to 2.25 Hz can be observed in

Figures 21 and 23. It results from the different natural frequencies in the plant and at the site, due to the different bearing conditions of the structure. In the middle of the bridge only the first eigenfrequency at 4.5 Hz is excited whereas at the quarter point of the bridge also the second eigenfrequency at ~12.1 Hz has a significant influence. This can be seen in the frequency spectrum of the bridge vibration (Figure 24). Whereas often the first Fourier term seems to be the only one which is considered, the measurements indicate that the second and even higher order Fourier terms may be of significant importance especially if places outside the middle of the bridge are considered.

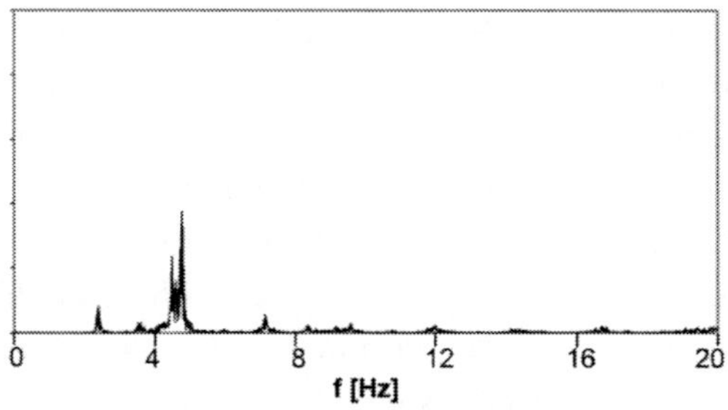

(a) in the middle of the bridge

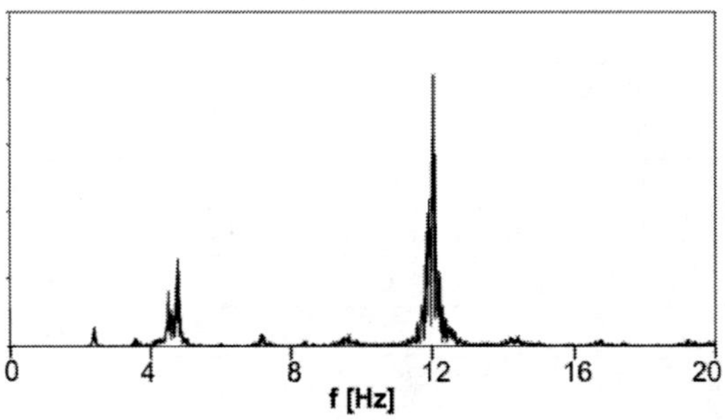

(b) at the quarter point of the bridge

Figure 24. Frequency spectrum of the bridge vibration for a person passing the bridge with f_s=2.4 Hz.

A computational simulation of the bridge vibrations has been performed after adapting the bridge model to the measured eigenfrequencies of 4.5 Hz and 12.1 Hz, respectively. This has been achieved by introducing a spring in order to model the horizontal bearing conditions and a slight modification of the overall stiffness. Especially the load models of Bachmann and Seiler-Hüttner as well as the single-footstep model have been applied. The maximum acceleration is strongly dependent on the step velocity, see Figure 23. The peak accelerations differ considerably between the load models. Again the footstep model based on Wheeler [4] shows the closest agreement with the measurement. The measurements and numerical simulation at the quarter point of the bridge reflect the influence of higher order Fourier terms and the excitation of higher mode shapes caused by them.

When considering the maximum acceleration in Figure 23(a) it should be noted that the resonance peak is very sharp. The difference between two measuring points of the step frequencies at the peak is 2.25−2.20=0.05 Hz, which corresponds to a difference in the time period between two steps of 0.01 s. To get the peak acceleration value, the bridge has to be passed exactly with one half of the eigenfrequency. A slight change of e.g. 0.01s in the time period between two steps reduces the acceleration by half. In the tests, special measurements like the use of a metronome were necessary in order to keep the step frequency constant at a predefined value. In practice these conditions are not given and hence it is very unlikely to obtain the peak values of the accelerations measured.

7.1.4. Assessment of the Results

In the present study the first eigenfrequency of the structure is about twice the typical step frequency of 2.0 Hz or 2.25 Hz, respectively. Therefore, in Equation (3) the second Fourier term with j=2 is significant for the structural response. Table 9 summarizes the Fourier coefficients α_2 obtained by different methods for f_s=2.0 and 2.25 Hz, respectively.

The Fourier coefficient varies between 0.06 and 0.15 according to the model chosen. The results of the tests presented here indicate that a Fourier coefficient $\alpha_2 \approx 0.07$ will be an appropriate choice. In addition, one may take into account that the probability of reaching the resonance peak is low for naturally passing persons. The sharpness of the resonance peak (Figures 21, 23) is related to the damping of the system. Therefore also a small damping ratio as in the case of aluminium structures is presumed in this conclusion. Additional results for running and tests with several persons walking on the same bridge are given in [31] and [32].

Table 9. Second Fourier coefficient

Numerical Simulation	
- Bachmann [7, 9]	0.10
- Seiler [6]	0.15
- Young [2, 12] - 50% exceedance probability	0.064
- Young [2, 12] - 25% exceedance probability	0.082
- Rainer [10]	0.065
- Eurocode 5, [20] Equation (24b)	0.14
Field tests	
- Measurement in the plant	0.06
- Measurement at the site	0.07

Figure 25. Stückisteg footbridge, Basel, Switzerland.

7.2. Steel Bridge

A footbridge being excited by the first Fourier coefficient of the load is studied. It is a steel bridge in Basel, Switzerland with a span width of 45 m, a total mass of 160 tons and a first eigenfrequency of 1.75 Hz (Figure 25). The damping of the steel bridge has been measured by the Logarithmic decrement method with impact excitation tests to be 2% [34].

A finite element analysis has been performed on a simplified beam model with constant cross section having the same first eigenfrequency as the original bridge. The bridge is loaded by a time-dependent point load as given in Equation (2) propagating with the velocity c acc. to Equation (1) on the bridge [35]. The weight of the person is assumed to be 0.7 kN, the step length 0.80 m and the step frequency equals the first eigenfrequency of 1.75 Hz of the bridge. Figure 26 shows the time history of the vertical accelerations in the middle of the bridge. A maximum acceleration of 0.08 m/s² is obtained. In a test with 13 persons a maximum of 0.25 m/s² was measured when walking over the bridge with arbitrary phase and step frequencies and of 0.45 m/s² when trying to walk in phase in the first eigenfrequency [34]. Applying Equations (11) and (13) to a group of 13 persons $\sqrt{13}\cdot 0.08\,\mathrm{m/s^2} = 0.29\ \mathrm{m/s^2}$ and $10.8\cdot\sqrt{13\cdot 0.02}\cdot 0.08\,\mathrm{m/s^2} = 0.45\ \mathrm{m/s^2}$, respectively, are obtained. The HIVOSS [15] formula with m^*=80 tons, f_1=1.75 Hz and ξ=2% for a density d<0.5 Persons/m² gives $a_{max,95}$=0.25 m/s². The measured accelerations compare well to the theoretical calculations for randomly distributed step frequencies and phases for practical purposes.

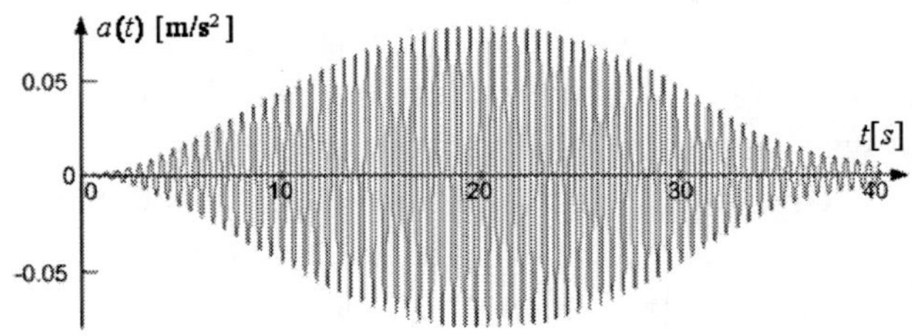

Figure 26. Acceleration time history at the middle of the bridge without TMD, a_max=0.08m/s².

Now a TMD is added to the finite element model in the middle of the bridge. Its parameters are determined according to the Den Hartog optimization criteria. The lowest eigenfrequencies are now obtained as 1.61 Hz and 1.86 Hz. With a TMD with μ=0.02 (referring to a modal mass of 80 t) and a step frequency of 1.86 Hz the vertical accelerations in Figure 27 are obtained.

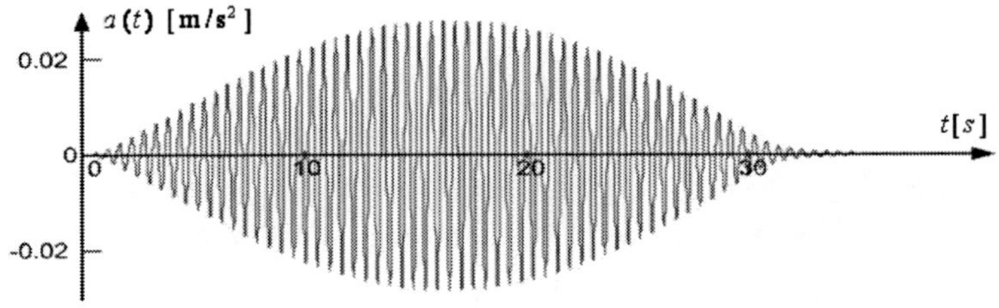

Figure 27. Acceleration time history at the middle of the bridge with TMD, μ=0.02, _max=0.028m/s².

Table 10. Maximum vertical acceleration in m/s² of the bridge

Method	Without TDM	With TDM		
		μ=0.02	μ=0.05	μ=0.08
FEM	0.08	0.03	0.02	0.02
Equations (24a) and (36) resp.	0.06	0.04	0.02	0.02

The maximal vertical accelerations show a good agreement with the approximate relationships Equations (24a) and (36), respectively, in Table 10.

Conclusion

Load models and computational methods for vibrations induced by a single walking human are confirmed well by experimental investigations for structures being excited by the first harmonic of the load. For this case, comprehensive investigations have also been performed for excitation by a pedestrian stream. For structures excited by the second Fourier coefficient of the load, however, the Fourier coefficients given in the literature differ in a wide range. When comparing measured accelerations on pedestrian bridges and values obtained by numerical modeling often remarkable differences are reported. Accelerations measured in tests are often lower than those predicted by the computational model. This is true in the bridge investigated here too. One reason can be found in the difference between the natural walking load varying slightly in frequency and the loading in the model which possesses a very constant frequency, causing pronounced sharp resonance peaks especially for structures with low damping. Therefore in a numerical model the Fourier coefficient of the loading should be chosen as $a_2 \leq 0.1$ in order not to overestimate the bridge vibrations. Further research is required for a pedestrian stream on bridges excited by the second Fourier term of the load.

Acknowledgments

The author would like to thank Dipl.-Ing. (FH) Wolfram Püschel, M.Eng., and Dr.-Ing. Christiane Butz for providing some information on the experimental investigations in this paper as well as Andrei Firus, B.Eng., for the drawing work.

References

[1] Qin J., Law S. S., Yang Q., Yang N. (2011). Smart Modeling of the Human-Structure Interaction with Human Intervention, *Asian-Pacific-Vibration Conference (APVC)*, Hong Kong (SAR), China, 2011.

[2] Zivanovic S., Pavic A., Reynolds P. (2005). Vibration serviceability of footbridges under human-induced excitation: a literature review, *Journal of Sound and Vibration*,

279: 1–74.

[3] Butz C., Distl J. (2008). Personen-induzierte Schwingungen von Fußgängerbrücken, *Stahlbau-Kalender 2008, Ernst&Sohn*, Berlin, Germany (in German).

[4] Wheeler J.E. (1982), Prediction and Control of Pedestrian Induced vibration in Footbridges, *J. Struct. Division, ASCE* 108 (9): 2045–2065.

[5] Hawryszkow P. (2011). Analysis of dynamical sensitivity and comfort of footbridges, *4th International Conference Footbridge 2011*, Wroclaw, Poland.

[6] Seiler Ch., Hüttner S. (2004). *Ein einheitliches Modell zur Beschreibung von Fußgängerlasten für verschiedene Bewegungsarten – Theorie, experimentelle Messungen und praktische Anwendung*, Bauingenieur 79, Springer, Berlin, p. 483–496 (in German).

[7] Bachmann et.al (1997). *Vibration Problems in Structures*, Birkhäuser Verlag, Basel, Switzerland.

[8] Bachmann H. (2004). *"Lebendige" Fußgängerbrücken – eine Herausforderung*, Bautechnik 81, Ernst&Sohn, Berlin, p. 227-236 (in German).

[9] Bachmann H. (1988). *Schwingungsprobleme bei Fußgängerbauwerken*, Bauingenieur 63, Springer, Berlin, p. 67–75 (in German).

[10] Rainer J. H., Pernica G., Allen, D. E. (1988). Dynamic loading and response of footbridges. *Canadian Journal of Civil Engineering*, 15: 66–71.

[11] Kerr S.C. (1998). *Human Induced Loading on Staircases*, PhD Thesis, University College London, UK.

[12] Young P. (2001). Improved floor vibration prediction methodologies, *ARUP Vibration Seminar*, October 4, 2001.

[13] Matsumoto Y., Nishioka T., Shiojiri H., Matsuzaki K. (1978). Dynamic Design of Footbridges, *IABSE Proceedings* No. P.17/78, p. 1–15.

[14] SETRA (2006). Footbridges, assessment of vibrational behavior of footbridges under pedestrian loading, *Technicla guide. Service d'études techniques des routes et autoroutes*, Association Française de Génie Civil, Paris, France.

[15] Hivoss (2008). Human induced vibration of steel structures, *Design of Footbridges Guideline, RFS2-CT-2007-00033*, www.stb.rwth-aachen.de/projekte/2007/HIVOSS/-download.php

[16] Hivoss (2008). Human induced vibration of steel structures, *Background documentation*, www.stb.rwth-aachen.de/projekte/2007/HIVOSS/download.php

[17] BSI (2008). *UK National Annex to Eurocode 1: Actions on Structures - Part 2: Traffic Loads on Bridges*, NA to BS EN 1991-2:2003, British Standards Institution.

[18] Barker C., Mackenzie D. (2008). Calibration of the UK National Annex. *Proceedings of the Third International Conference Footbridge 2008*, Porto.

[19] Pavic A. (2011). Vertical Crowd Dynamic Action on Footbridges: Review of Design Guidelines and their Application, *4th International Conference Footbridge 2011*, Wroclaw, Poland.

[20] Eurocode 5 (2004). *Design of timber structures – Part 2: Bridges*; German version EN 1995-2.

[21] Butz C. (2006). *Beitrag zur Berechnung fußgängerinduzierter Brückenschwingungen*, Shaker Verlag, Aachen, Germany. (in German).

[22] Butz C. (2008). A Probabilistic Engineering Load Model for Pedestrian Streams, *Proceedings of the International Conference Footbridge 2008*, Porto, Portugal.

[23] Butz C., Feldmann M., Heinemeyer C. et. al. (2008). *Advanced load models for*

synchronous pedestrian excitation and optimized design guidelines for steel footbridges, European Commission, final report, Research Fund for Coal and Steel, ISBN: 978-92-79-08303-7

[24] Kreuzinger H. (2002). Dynamic design strategies for pedestrian and wind actions, *Footbridge 2002, Proceedings of the International Conference on the Design and dynamic behaviour of footbridges*. Paris.

[25] Hamm P. (2003). *Ein Beitrag zum Schwingungs- und Dämpfungsverhalten von Fußgängerbrücken aus Holz*, PhD Dissertation, Technical University München, (in German).

[26] Blanchard J., Davies B.L., Smith J.W. (1977). Design criteria and analysis for dynamic loading of Footbridges, *Proceedings of the DOE and DOT TRRL Symposium on Dynamic behavior of bridges*, UK, p. 90−106.

[27] ISO (2007). *Bases for design of structures – Serviceability of buildings and walkways against vibration*, ISO 10137:2007, International Organization for standardization.

[28] Den Hartog J. P. (1984). *Mechanical Vibrations*, Dover Publications, New York.

[29] Werkle H., Butz C., Tatar R. (2011). Effectiveness of "Detuned" TMD's for Beam-like Footbridges, *14th Asia-Pacific Vibration Conference (APVC'2011)*, Hong Kong (SAR), China

[30] Werkle H., Butz C., Tatar R. (2012). Effectiveness of "Detuned" TMD's for Beam-like Footbridges, *Advances in Structural Engineering*. ISSN 1369-4332, Multi-Science Publishing Company, Brentwood, Essex CM15 9TB, UK, (to be published)

[31] Püschel W. (2006). *Menschenerregte Schwingungen von Fußgängerbrücken aus Aluminium – Messungen und numerische Simulation*, Master Thesis, HTWG Konstanz, Konstanz (in German).

[32] Werkle H., Püschel W. (2009). *Durch Menschen induzierte Schwingungen von Fußgängerbrücken aus Aluminium*, Bauingenieur, Springer, Berlin (in German).

[33] Werkle H., Püschel W. (2009), Human induced Vibrations of Aluminium Pedestrian Bridges, *ISEV2009 – 4th International Symposium on Environmental Vibration: Prediction, Monitoring and Evaluation*, Beijing, China.

[34] Butz C. (2009), *Ermittlung der Schwingungsanfälligkeit des Stückistegs in Basel und der Wirkung des Massendämpfers*, Technischer Bericht, MAURER SÖHNE – Bauwerkschutzsysteme, München, Germany (unpublished, in German).

[35] Fritsche D. (2011). *Effektivität von Massendämpfern für Fußgängerbrücken bei transienter Belastung*, Master Thesis, HTWG Konstanz (in German).

In: Traffic Induced Environmental Vibrations and Controls ISBN: 978-1-62618-240-0
Editors: H. Xia and R. Calçada

Chapter 7

PREDICTION OF ENVIRONMENTAL VIBRATIONS FROM UNDERGROUND TRAINS

Weining Liu[1*], Weifeng Liu[2], Deyun Ding[3], Yingxuan Jia[4], Meng Ma[5] and Kefei Li[6]

[1,2]School of Civil Engineering, Beijing Jiaotong University. Beijing, China
[3]Beijing Urban Engineering Design and Research Institute Co., td. Beijing, China
[4]Beijing Institute of Economic Management. Beijing, China
[5]China Academy of Railway Sciences. Beijing, China
[6]Beijing MTR Construction Administration Corporation, Beijing,China

ABSTRACT

In general, the vibration from underground trains is regarded in the prediction as an integrated system of vehicle-track-tunnel-stratum, but a composed system by the subsystems of vehicles, tracks, tunnel linings and soil strata is practical for analysis. In this model, the wheel-rail contact force is the main excitation source of the vibration, while tunnel linings and soil strata are the main mediums for vibration propagation. In this chapter, the simulation of train loads, the propagation of metro-induced ground-borne vibration and the prediction of environmental vibration from underground trains are studied by combination of theoretic analysis, numerical simulation, in situ tests and empirical estimations.

1. INTRODUCTION

The problem of environmental vibration from underground trains can be studied with an integrated vehicle-track-tunnel-stratum system, but by convenience, the composition of the sub-systems of vehicle, track, tunnel lining and soil stratum is practical in the analysis. Thereinto, vehicle-unsprung mass directly affects the contact forces between the wheel and rail, which is the main excitation of vibration; while tunnel lining and soil stratum are the

* E-mail address: weiningliu@126.com.

main mediums for vibration propagation. These sub-systems interact and intercoupled with each other, which makes this problem extremely complicated. Great efforts have been made in the research for the prediction of environmental vibration induced by underground trains.

The train loads simulation, also called vehicle-track dynamic load analysis, has been studied since 1867, beginning with the elastic fundamental beam established by Winkler [1], and the theory was quickly used in track modeling. However, there were few new track calculation models to solve the practical problem until 1950's. More detailed track models were established to study the random vibration of the track structure [2]. Vehicle-track interaction was also studied with single-degree vehicle-track model, combining with in site measurement [3]. Analytical research of the wheel-rail dynamic interaction on high-speed track, heavy track and speed-up track were carried out by Zhai et al. [4] with detailed models, in which vertical and horizontal vehicle-track interaction are all taken into account.

For numerical modeling, 2-D, 2.5-D and 3-D models have been developed [5–7], based on the finite element (FE) method, the boundary element (BE) method or the coupled FE-BE method, to study the propagation of metro-induced ground vibration. Since the 3-D models were expensive for vibration analysis at higher frequencies, a 3-D coupled periodic FE-BE method was introduced using the Floquet transform to exploit the longitudinal periodicity of the track-tunnel-soil system. The 3-D dynamic track-tunnel-soil interaction problem was solved with a subdomain formulation where a FE method was used for the track and the tunnel, while a BE method for the soil, modelled as a horizontally layered half-space.

In-situ measurement may provide useful data for evaluating vibration induced by metros. As early in 1977, the metro-induced vibration was conducted in the Berlin Metro [8]. As part of the CONVURT project, the vibration measurement was carried out in the Regent's Park on the Bakerloo Line in London during 35 passages of a test train at speed 20–50 km/h [9−10]. In China, the dynamic response at the tunnel lining were measured in a Beijing metro, and the characteristics of vibration were measured in the running tunnel and on the ground surface at the Dongdan-Jianguomen Section of Line 1 of the Beijing Metro [11].

In practice, it is often difficult to evaluate and predict the environmental vibration induced by underground trains with only one method. The current studies on the simulation of train loads, the propagation of metro-induced ground vibration and the prediction of environmental vibration from underground trains, are introduced in this chapter, by combination of theoretic analysis, numerical simulation, in-situ tests and empirical estimation.

2. SIMULATION OF TRAIN LOADS

2.1. Empirical Estimation

Empirical method is dedicated to calculate the train loads for new lines which are not in operation and lack of measured data. There into the excitation load generated by moving trains on the uneven rail is simulated by an excitation load function composed of static and three harmonic forces, expressed as

$$F(t) = P_0 + P_1 \sin \omega_1 t + P_2 \sin \omega_2 t + P_3 \sin \omega_3 t \tag{1}$$

where P_0 is static load; P_1, P_2 and P_3 are vibration loads accordance to the unevenness vector height α_i , $P_i = M_0\alpha_i\omega_i^2$, where M_0 is unsprung mass, ω_i is radian frequency, $\omega_i = 2\pi v / L_i$, where v is train velocity and L_i is unevenness wavelength. Figure 1 (top) illustrates the empirical train load, with the parameters $P_0 = 125\,\text{kN}$, $M_0 = 1510$ N·s²/m , v=90 km/h.

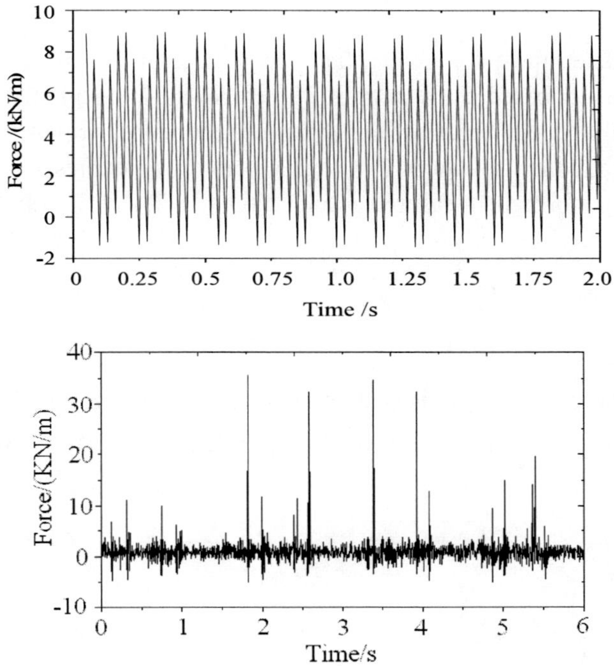

Figure 1. Time histories of emperial (top) and tested (bottom) train loads.

2.2. In Situ Test Measurement

The vertical train loads can be obtained using the shear-stress method, via conversion of the rail stress measured with the strain gauge sticked on the rail web when the train passes through, with the quasi-static calibration of the measuring point.,

The dynamic train load is supposed to be the uniformly distributed load along the track centerline, and the calculation formula for the train load is as follows:

$$F(t) = K \cdot \frac{2P(t) \cdot N \cdot n}{L} \tag{2}$$

where $F(t)$ is the distributed train load along the track centerline, in unit of kN/m; $P(t)$ is the vertical train load obtained through the conversion of the in situ recorded data; N is the number of vehicle bogies for each car (N=2); L is the length of the car; K is the dispersion coefficient, considering the dispersion of the wheel loads during transmission through the rail, the sleeper and the track bed. K is generally 0.6 to 0.9.

The time history of the obtained distributed train load is shown in Figure 1 (bottom). The six-carriage vehicle (the axle load is 14 t) and 60 kg/m rail (m=60.64 kg/m) are applied in the calculation. The train speed is v=53.6 km/h and the sleeper spacing is 0.6m.

2.3. Simulated Deterministic Method

The simulated deterministic method is adopted to calculate the dynamic wheel-rail forces, which can be obtained based on a simplified vehicle-track coupled model, in which the vehicle is simplified as a combination of a series of secondary spring-mass model uniformly distributed along the tunnel, shown in Figure 2, a conversion expression are presented using FFT algorithm to transform the measured acceleration of rail, as follows:

$$\ddot{z}_0(t) = \sum_{n=0}^{N/2-1} \left(A_n \cos n\omega t + B_n \sin n\omega t \right) \tag{3}$$

where $\ddot{z}_0$ is the rail acceleration; N is sampled data number; A_n, B_n are FFT coefficient; ω is fundamental frequency, $\omega = 2\pi/T = 2\pi/N\Delta t$, T is recorded periodic time, Δt is sample interval.

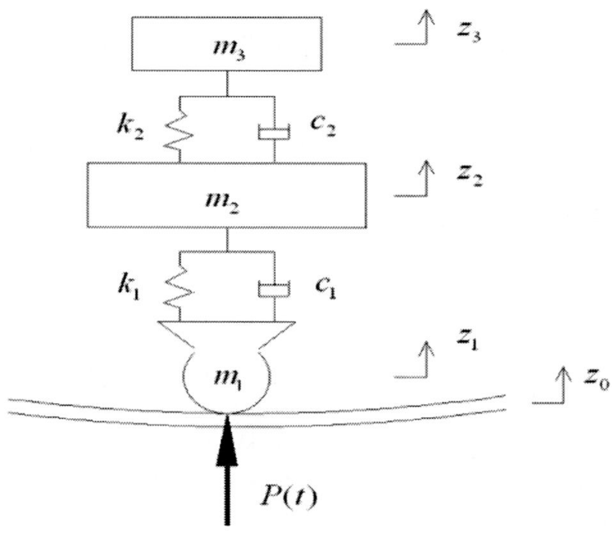

Figure 2. Simplified vehicle model.

The vertical dynamic equations of the vehicle can be obtained as follows:

$$\begin{cases} m_3\ddot{z}_3 + c_2\left(\dot{z}_3 - \dot{z}_2\right) + k_2\left(z_3 - z_2\right) = 0 \\ m_2\ddot{z}_2 + c_1\left(\dot{z}_2 - \dot{z}_1\right) + k_1\left(z_2 - z_1\right) - k_2\left(z_3 - z_2\right) - c_2\left(\dot{z}_3 - \dot{z}_2\right) = 0 \end{cases} \tag{4}$$

Assuming the moving wheel always contacted with the rail, and set the relative displacement as $\xi_2 = z_2 - z_1$, $\xi_3 = z_3 - z_2$, respectively. the equations can be written as:

$$\begin{cases} m_3\left(\ddot{\xi}_2+\ddot{\xi}_3\right)+c_2\dot{\xi}_3+k_2\xi_3=-m_3\sum_{n=0}^{N/2-1}\left(A_n\cos n\omega t+B_n\sin n\omega t\right) \\ m_2\left(\ddot{\xi}_2+\ddot{\xi}_3\right)+k_1\xi_2-k_2\xi_3-c_2\dot{\xi}_3=-m_2\sum_{n=0}^{N/2-1}\left(A_n\cos n\omega t+B_n\sin n\omega t\right) \end{cases} \tag{5}$$

According to the D'Alembert Principle and the vehicle parameters, the interaction force between wheel and rail can be obtained as:

$$\begin{aligned} P(t) &= (m_1+m_2+m_3)g+m_1\ddot{z}_1+m_2\ddot{z}_2+m_3\ddot{z}_3 \\ &= (m_1+m_2+m_3)g+\begin{bmatrix} m_1 & m_2 & m_3 \end{bmatrix}\left(\begin{Bmatrix}1\\1\\1\end{Bmatrix}\ddot{z}_0+\begin{bmatrix}1&0&0\\1&1&0\\1&1&1\end{bmatrix}\begin{Bmatrix}\ddot{\xi}_1\\\ddot{\xi}_2\\\ddot{\xi}_3\end{Bmatrix}\right) \end{aligned} \tag{6}$$

As the train loads are assumed to be uniformly distributed along the tunnel, the reduction factor can be neglected in Equation (2). According to the recorded acceleration of the in situ tests in Beijing Metro Line1, the simulated dynamic wheel-rail forces is shown in Figure 3.

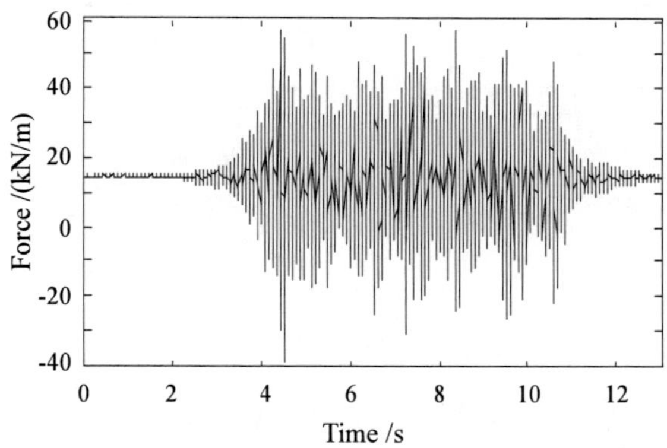

Figure 3. The simulated dynamic wheel-rail forces.

2.4. Analytical Model of Coupled Vehicle and Track

There are three key problems in implementation of the vehicle-track coupled dynamic analysis: establishment of a rational mathematical model to reflect the physical essence of the vehicle-track coupling system, programming of numerical solution, and determination of the basic parameters of vehicle-track system.

2.4.1. General Idea of the Analytical Model

A novel analytical model based on "coupled vehicle and track basic cell" is established, as shown in Figure 4, in which the factors of whole vehicle system, infinite track structure, track irregularity and wheel-rail Hertz non-linear contact are taken into account. The dynamic equations and the solving process are in the wave number-frequency domain, and the displacement influence matrices of the track system and the vehicle system are obtained with the "coupled vehicle-track basic cell". The dynamic responses of the vehicle-track structure is solved with the wheel-rail contact force added to the infinitely long track system; the force under the sleeper is obtained though the composite stiffness. The dynamic response of the entire system in the frequency domain is then transformed into the time domain by the Inverse Fourier Transformation, which re-simplified the differential equation and also facilitate the dual-domain analysis.

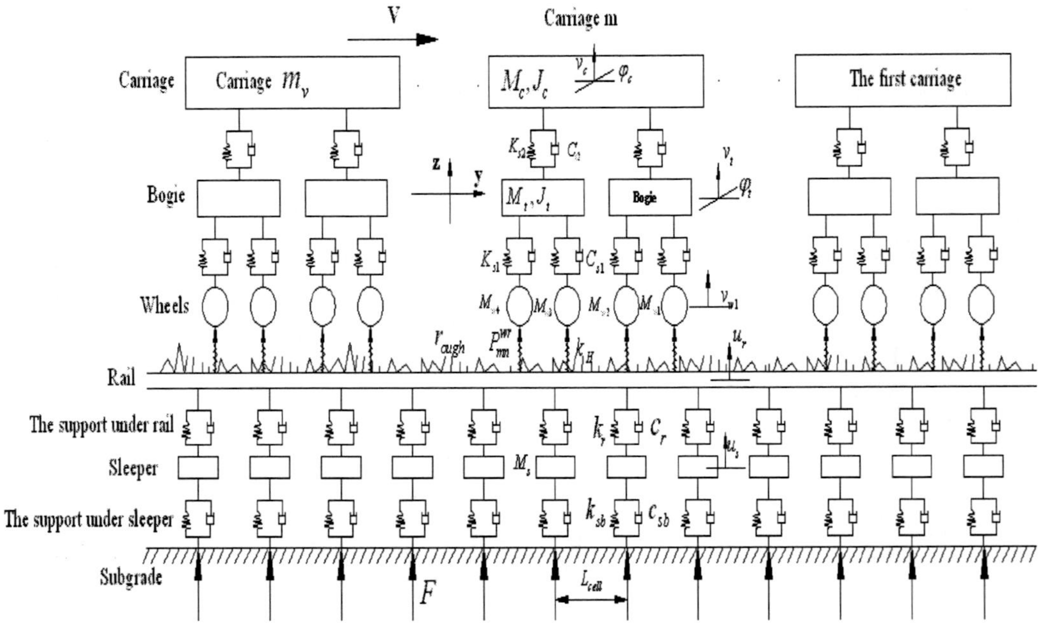

Figure 4. Dynamic model of coupled vehicle and track.

The whole system is decomposed into two subsystems, and the two subsystems coupled with each other through the wheel-rail interaction force. The basic assumptions of the model include:

(1) Only the wheel-rail dynamic interaction in vertical direction is taken into account.
(2) The track and vehicle systems are supposed to be symmetrical along the track.
(3) The vehicle subsystem is considered as a vehicle model with secondary suspension system, and 10 degrees-of-freedom are taken into account as follows: the nodding, up-and-down of the whole vehicle body and rear bogies, and up-and-down of the wheels.
(4) The wheel-rail contact force obeys Hertz Contact Theory.
(5) The rail is considered to be an infinite double-layer Timoshenko Beam, which is periodically supported.
(6) The elasticity and damping of the support under the rail are equivalent to elastic coefficient k_r and damping coefficient c_r (composite parameter of parallel plate and fasteners under the rail); The elasticity and damping under the sleeper are equivalent to elastic coefficient k_{sb} and damping coefficient c_{sb}, which are tandem complex

parameter of sleeper cushion parameters (k_s, c_s) and the track bed parameters(k_b, c_b). The track bed between the sleeper and foundation is simulated as spring-damper.

(7) The rail is supposed to be discretely supported on the elastic sleeper, the mass of which is the integration of the sleeper and bed is applied on each bearings node.

2.4.2. Vehicle System Dynamic Equations

Obviously, the vehicle can be considered as a multi-rigid-body system moving in the vertical longitudinal plane. The number of the freedom for the four-axis vehicle model with secondary suspension system is 10, as shown in Figure 4.

The system equation for the *m*th car can be expressed as

$$M_m \ddot{A}_m(t) + C_m \dot{A}_m(t) + K_m A_m(t) = Q_m(t) \tag{7}$$

where M_m, C_m and K_m are the mass, stiffness and damping matrices; $K_m(t)$ is the equivalent load vector; $A_m(t)$, $\dot{A}_m(t)$ and $\ddot{A}_m(t)$ are the the displacement, velocity and acceleration vectors, respectively, the explicit expression of which are as follows

$$A_m = \{v_{cm}, \varphi_{cm}, v_{tm1}, v_{tm2}, \varphi_{tm1}, \varphi_{tm2}, v_{wm1}, v_{wm2}, v_{wm3}, v_{wm4}\}^{\mathrm{T}} \tag{8}$$

The wheel displacement of the whole vehicle with m_v carriages can be defined as:

$$\begin{Bmatrix} \hat{A}_{w1}(\omega) \\ \hat{A}_{w2}(\omega) \\ \hat{A}_{w3}(\omega) \\ \vdots \\ \hat{A}_{wm_v}(\omega) \end{Bmatrix} = \begin{bmatrix} \hat{V}_{w1} & 0 & 0 & \cdots & 0 \\ 0 & \hat{V}_{w2} & 0 & \cdots & 0 \\ 0 & \cdots & \hat{V}_{w2} & \cdots & 0 \\ 0 & \cdots & 0 & \ddots & 0 \\ 0 & \cdots & 0 & 0 & \hat{V}_{wm_v} \end{bmatrix} \begin{Bmatrix} \hat{Q}_1^{wr}(\omega) \\ \hat{Q}_2^{wr}(\omega) \\ \hat{Q}_3^{wr}(\omega) \\ \vdots \\ \hat{Q}_{m_v}^{wr}(\omega) \end{Bmatrix} \tag{9}$$

where $\hat{\mathbf{V}}_w = \mathrm{diag}[\hat{V}_{w1}, \ \hat{V}_{w2}, \ \hat{V}_{w3}, \ \cdots, \ \hat{V}_{wm_v}]$ is defined as the displacement influence matrix. $\hat{Q}_m^{wr}(\omega)$ is defined as: $\hat{Q}_m^{wr}(\omega) = \{\hat{P}_{m1}^{wr}(\omega), \hat{P}_{m2}^{wr}(\omega), \hat{P}_{m3}^{wr}(\omega), \hat{P}_{m4}^{wr}(\omega)\}^{\mathrm{T}}$ is the equivalent load vector, $P_{mi}^{wr}(t)$ is wheel-rail contact forces obtained by the Hertz formula based on the vertical wheel-rail contact relative displacement, (i=1,2,3,4).

2.4.3. Dynamic Equations of Track System

The vibration induced by train running on the track can be decomposed into the coupling vibration of wheel-axle dynamic load and track structure. The problem can be described with a basic mechanical model, a time-varying amplitude load of coupled vehicle-wheel system moving on the infinite track at the speed *v*. As the time is infinite, the above time-domain equation can be transformed into the frequency domain solution by means of the Fourier Transform, the load position along the track in the space domain can also be converted to wave number space equation. With this two transformations, the train moving on the infinite track can be converted to the wave number-frequency domain, that is, the coupled integral equations in the time and space domain are mapped into an domain independent of time and space to be solved, and a complex space-time integral equation is transformed into the simple combination in the wave number-frequency domain. Through the inverse transform, the results in the space-time domain can be obtained finally.

In case of the periodic track structure, the general response in the frequency domain is simplified in a form of summation within the track sleeper period instead of integral, which is obtained by transferring the moving loads on the rail to the moving of receiver backward within periodic sleeper distance L_{cell}. The advantage of this method is that all the dynamic analysis of the track system is carded out within sleeper distance L_{cell}. In this way, the track response to moving train can be obtained by a very simple calculation program, see Figure 5.

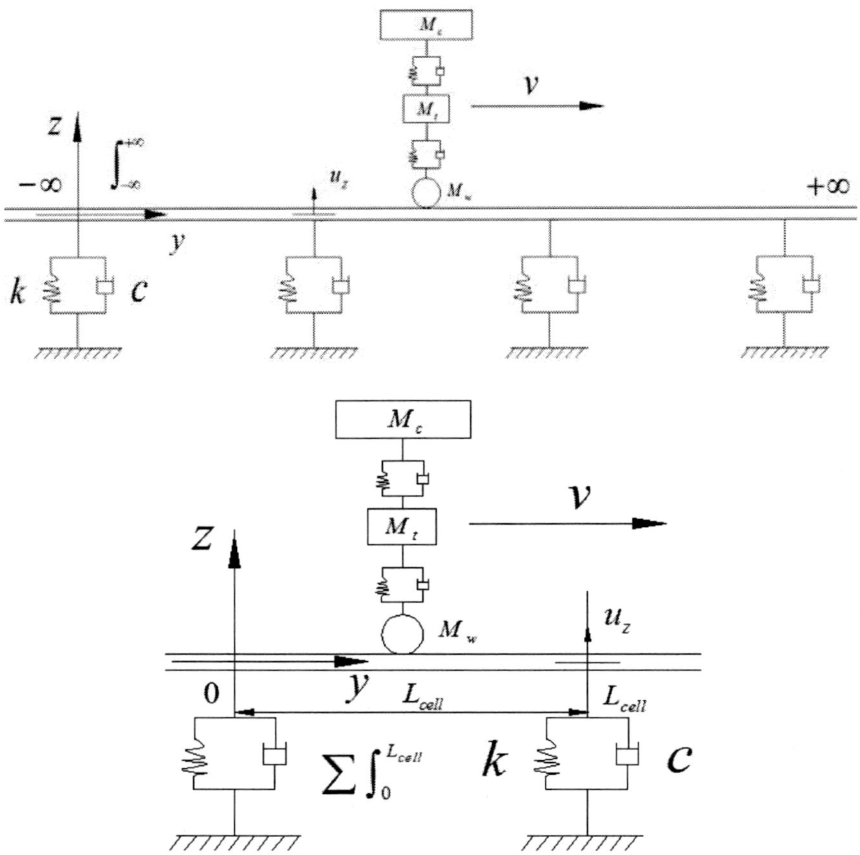

Figure 5. Infinite track subjected to running train (top) and coupled vehicle-track basic unit (bottom).

The periodic analytical solution for the receiver point on the track, with the moving load on the track, can be expressed as

$$\hat{u}_{i\text{-single}}(\xi_y,\omega)=\sum_{n_y=-\infty}^{+\infty}\frac{1}{v}\cdot\int_{\tilde{y}_{k0}}^{\tilde{y}_{k0}+L_{\text{cell}}} g_k\left[\frac{\tilde{y}-\tilde{y}_{k0}+(n_y-n_{k0})L_{\text{cell}}}{v}\right]\cdot\hat{h}_{zi}(\tilde{y},\tilde{\xi}_y$$
$$+(n_{\xi_y}-n_y)L_{\text{cell}},\omega)\cdot\exp\left[-i\omega\left(\frac{\tilde{y}-\tilde{y}_{k0}+(n_y-n_{k0})L_{\text{cell}}}{v}\right)\right]d\tilde{y} \tag{10}$$

2.4.4. Transfer Function of the Periodic Supported Timoshenko Beam

$\hat{h}_{zi}(\tilde{y},\tilde{\xi}_y+(n_{\xi_y}-n_y)L_{\text{cell}},\omega)$ in Equation (10) is expanded as the unit impulse force at the point $\tilde{y}$, the state variables at $\tilde{\xi}_y+(n_{\xi_y}-n_y)L_{\text{cell}}$ is as follows:

$$\begin{aligned}\hat{h}_{zi}(\tilde{y},\tilde{\xi}_y+(n_{\xi}-n_y)L_{\text{cell}},\omega)&=S(\tilde{\xi}_y+(n_{\xi}-n_y)L_{\text{cell}},\omega)\\&=T_{\text{track}}(\tilde{\xi}_y-\tilde{y},\tilde{y},\omega)\cdot T_{\text{track}}([(n_{\xi_y}-n_y)L_{\text{cell}}],\tilde{y},\omega)\cdot S(\tilde{y},\omega)\end{aligned} \tag{11}$$

Equation (11) indicates that as the loads moving on the Timoshenko beam periodically supported, the state variables of ξ_y and transfer functions $\hat{h}_{zi}(\tilde{y},\tilde{\xi}_y+(n_{\xi_y}-n_y)L_{cell},\omega)$ between the observation point ξ_y and the excitation point y can be solved using the transfer matrix method as the product of the state vector of the load excitation point y and the transfer matrix of $(n_{\xi y}-n_y)+1$ cells. Then the dynamic response solution of any receive point in the beam can be obtained. Figure 6 shows the track basic units between the excitation point y and the observation point ξ_y. $T_{\text{track}}(L_{\text{cell}},\tilde{y},\omega)$ and $S(\tilde{y},\omega)$ have been settled in some papers [12].

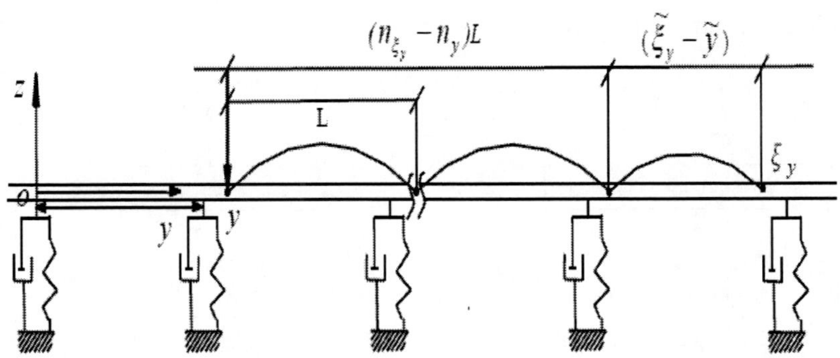

Figure 6. Timoshenko beam element between excitation and receiver points.

2.4.5. Equations of the Vehicle-track System

Above all, the track is supposed to be infinite Timoshenko beam system with multi-layer discrete supports, and the vehicle is supposed to be two-spring system vehicle systems containing 10 degrees-of-freedom: The equation of the system is as follows:

$$\hat{\mathbf{Q}}^{\text{wr}}(\omega)=[k_{\text{H}}\cdot(\mathbf{V}_{\text{w}}-\hat{\mathbf{D}}_{\text{rail}})-\mathbf{I}]^{-1}\cdot k_{\text{H}}\cdot\hat{R}^{\text{ough}}(\omega) \tag{12}$$

where : $\hat{\mathbf{Q}}^{\text{wr}}(\omega)=[\hat{Q}_1^{\text{wr}},\ \hat{Q}_2^{\text{wr}},\ \hat{Q}_3^{\text{wr}},\ \cdots,\ \hat{Q}_{m_v}^{\text{wr}}]^{\text{T}}$ is wheel-rail contact force in the frequency domain; k_{H} is the wheel-rail contact spring stiffness coefficient. The influence matrix of the vehicle system is $\hat{\mathbf{V}}_{\text{w}}=\text{diag}[\hat{V}_{\text{w1}},\ \hat{V}_{\text{w2}},\ \hat{V}_{\text{w3}},\ \cdots,\ \hat{V}_{\text{w}m_v}]$; $\hat{\mathbf{D}}_{\text{rail}}$ is wheel-wheel displacement influence function matrix. **I** is the unit matrix.

$$\hat{\mathbf{D}}_{\mathrm{rail}}=\begin{bmatrix} \hat{D}_{\mathrm{rail}}^{11} & \hat{D}_{\mathrm{rail}}^{12} & \hat{D}_{\mathrm{rail}}^{13} & \cdots & \hat{D}_{\mathrm{rail}}^{1m_v} \\ \hat{D}_{\mathrm{rail}}^{21} & \hat{D}_{\mathrm{rail}}^{22} & \hat{D}_{\mathrm{rail}}^{23} & \cdots & \hat{D}_{\mathrm{rail}}^{2m_v} \\ \hat{D}_{\mathrm{rail}}^{31} & \hat{D}_{\mathrm{rail}}^{32} & \hat{D}_{\mathrm{rail}}^{33} & \cdots & \hat{D}_{\mathrm{rail}}^{3m_v} \\ \vdots & \vdots & \vdots & \vdots & \vdots \\ \hat{D}_{\mathrm{rail}}^{\mathrm{m}_v1} & \hat{D}_{\mathrm{rail}}^{\mathrm{m}_v2} & \hat{D}_{\mathrm{rail}}^{\mathrm{m}_v3} & \cdots & \hat{D}_{\mathrm{rail}}^{m_vm_v} \end{bmatrix}$$

$\hat{R}^{\mathrm{ough}}(\omega)=[\hat{R}_{\mathrm{r}1},\ \hat{R}_{\mathrm{r}2},\ \hat{R}_{\mathrm{r}3},\ \cdots,\ \hat{R}_{\mathrm{r}m_v}]^{\mathrm{T}}$ is the track irregularity vector in the frequency domain.

The wheel-rail contact force solved in Equation (12) is substituted into the dynamic analytical expression Equation (10) of the infinite track system, that is, the analytical solution of the response of any point on the rails, under the loads of the whole vehicle. And then substituted into Equation (10) and thus the counter-force under the sleeper can be further obtained.

According to the analysis above, there are two key dynamic interaction forces in the basic vehicle-track-basis model: the force on the rail induced from the running trains and the force applied on the basis from the sleeper. The analytical vehicle-track coupling system can be obtained with the solution of the two forces above.

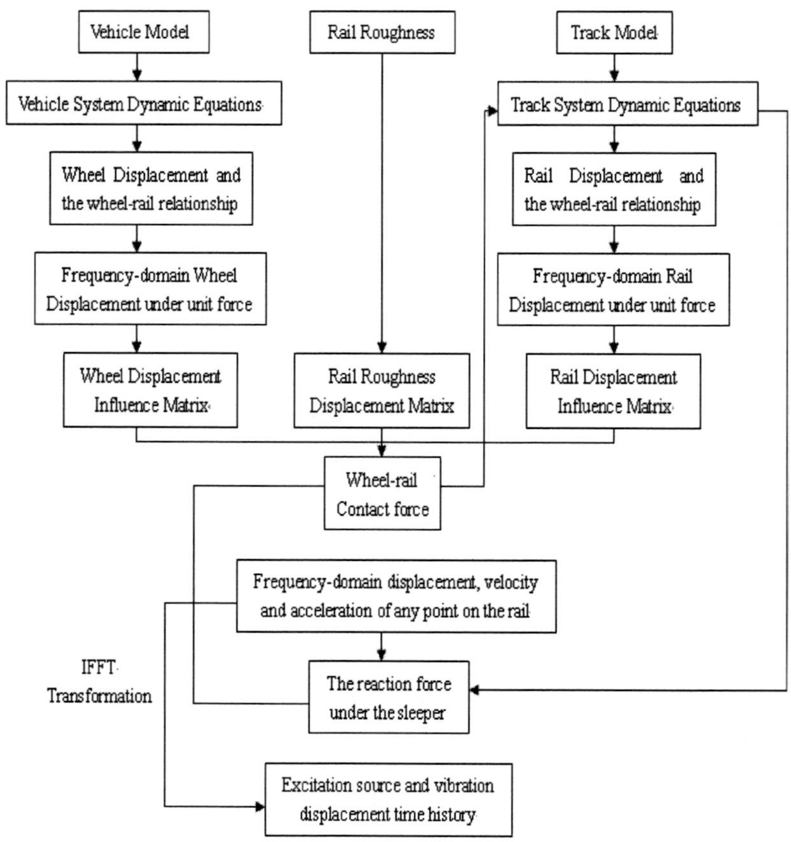

Figure 7. Program structure of TMCVT.

2.4.6. Calculation Examples

Based on the theory of the analytical vehicle-track coupling system, the vibration analysis program TMCVT (Theoretical Model of Coupled Vehicle and Track) is achieved with MATLAB, the program structure of TMCVT is shown in Figure 7.

The vehicle-track coupling system in this paper is a model different from the traditional dynamic vehicle-track model, the calculated results of the model is obtained with the analysis program TMCVT.

Table 1. Parameter of metro train

Body mass M_c (t)	35.0	Bogie spring coefficient K_t (kN/m)	2080
Body mass moment of inertia J_c (tm^2)	1700	Bogie damping C_t (kN.s/m)	240
Bogie mass M_t (t)	4.60	Wheel spring coefficient K_w (kN/m)	2450
Bogie mass moment of inertia J_t (tm^2)	9.62	Wheel damping coefficient C_w (kN.s/m)	240
Vehicle length L (m)	19	Fixed distance l_2 (m)	12.6
Axis Distance l_1 (m)	2.20	Wheel mass M_w (t)	1.42

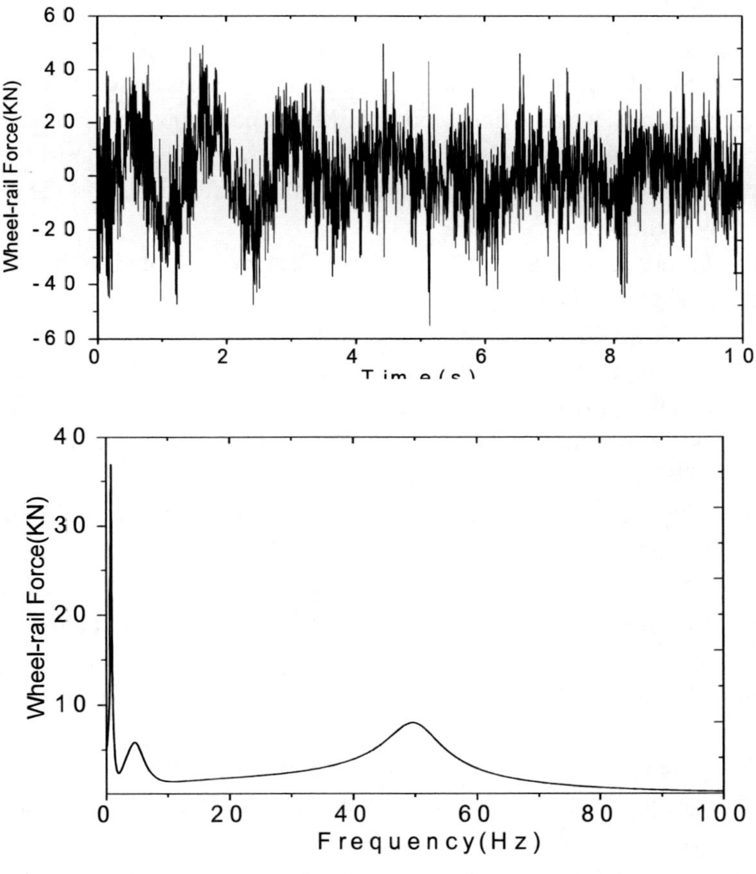

Figure 8. Contact force coupled with wheel & rail: time history (top) and spectrum (bottom).

The Parameters of metro train vehicles are shown in Table 1. The parameters of the track system are as follows: rail mass of unit length $\overline{m}_r$=60 kg/m; elastic modulus E=2.10×10^{11} Pa; cross-sectional area A=7.60×$10^{-3}$$m^2$; sectional moment of inertia I=3.04 ×$10^{-5}$$m^4$; damping ratio ξ_r=0.01. The sleeper mass of unit length m_s=50 kg; spacing d=0.60 m. Track bed: mass m_b=260 kg (combined mass of sleeper and track bed). $DTVI_2$ fastener is used: spring/damping coefficients under rail k_r=78 MN/m, c_r=50 kN·s/m, and under sleeper k_{sb}=100 MN/m, c_{sb}=50 kN·s/m. The vehicle-track contact force of the three fasteners is calculated under the train speed of v, as shown in Figure 8.

3. Metro-Induced Ground-Borne Vibration

Ground-borne vibration induced by metro train propagates through the tunnels and the surrounding soil which interacts with the nearby buildings, where malfunction of sensitive instruments, discomfort to people and re-radiated noise in buildings may be caused.

3.1. *In Situ* Measurements

As Metro Line 4 in Beijing ran in close proximity of the physics laboratory of a University, environmental vibration should be examined to ensure installation of some high-precision Transmission Electron Microscopy (TEM). A non-stop in situ measurement of 92 hours was performed to estimate the vibration of three orthogonal directions in three rooms of the laboratory, which was induced by the subway traffic, road traffic and the combination of the two previously mentioned.

As shown in Figure 9, the distances from the laboratory is located to the ground road and the central line of the east tunnel are 53m and 70m, respectively.

The figure also shows the positions of the accelerometers in three rooms of the laboratory, where some sensitive electron microscopes were to be installed. At each measurement point, three accelerometers with three orthogonal directions were installed: vertical direction, north-south direction and east-west direction. The distances from the measurement points to the tunnel center were 76.8m, 89.7m and 124.3m, respectively.

The experimental configuration for the vibration measurements consists of 8 Lance-AS130 piezoelectric accelerometers and a FOCUS-II dynamic data acquisition system. The accelerometer has the sampling frequency between 0.2 and 600 Hz, the sensitivity of 40 V/g and the maximum measuring acceleration of 0.125 g. For FOCUS-II dynamic data acquisition system, the maximum sampling frequency of the experimental setup can be tuned to 42 kHz. An anti-aliasing filter and a 160 dB/octave linear phase digital filter have been employed to avoid aliasing and phase distortion.

These measurement equipments ran 92 hours continuously from 0:00 Oct. 17th to 20:00 Oct. 20th 2009, in which time frame subway Line 4 was still running as normal with the road traffic varying from morning till night.

In order to see the differences between how the subway train runs under varied road traffic conditions, 6 different measurement cases were studied, details see Table 2.

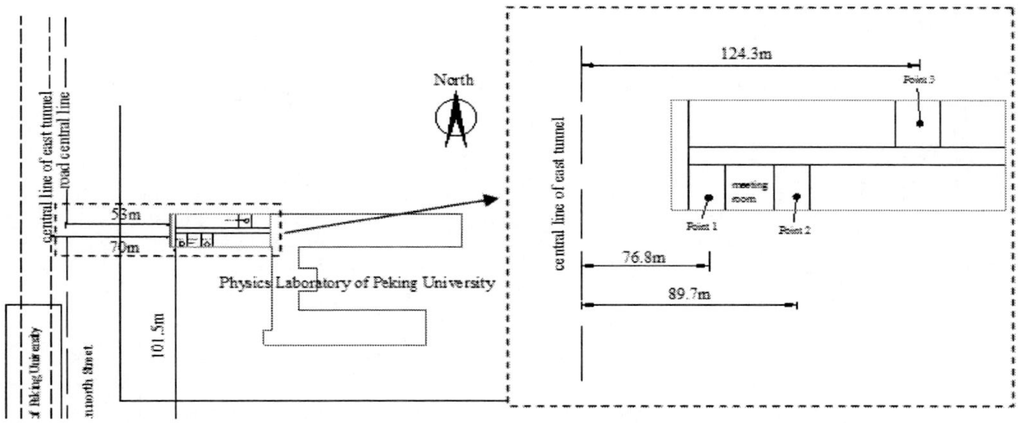

Figure 9. Test site in the Physics Laboratory.

Table 2. Measurement cases

Vibration induced by	Case	Metro	Bus/ truck
Background	1	No	No
Metro trains	2	Train only in the eastern tunnel	No
	3	Trains in both tunnels	No
Road traffic	4	No	Yes
Metro trains and road traffic	5	Train only in the eastern tunnel	Yes
	6	Trains in both tunnels	Yes

The data were analyzed in one-third octave band and were compared with the vibration criterion of Tecnai 30 TEM. For each case, results include three orthogonal directions at Point 1, 2 and 3. The final average curves were based on at least 10 useful records.

Figure 10 shows the background vibration measured at the midnight when there was no subway train and little road traffic. It can be seen that vibration in vertical direction is larger than those in longitudinal and transverse directions in most frequency bands. Vibration at Point 1 is larger, for it is closer to the street. Accelerations over 50Hz attenuate with distance. The two curves in the figure are used for evaluating the results. If all frequency components are below curve I, the high resolution performance will not be affected. If some components lie between curves I and II, the results must evaluated for the acceptance of the site. If any components above curve II appeared, the site is unsuitable for installation.

Shown in Figure 11 are the measured vibrations in laboratory generated by metro trains. Figure 11 (left) describes the accelerations induced by metro trains when trains run in the east tunnel only, where three averaged curves remain below the criterion curve I, which means no effect on the lab equipement. In Figure 11 (right), three averaged curves show vibration induced by trains running in both tunnels at the same time. Due to the eigen frequency of SSFST, vibration around 10Hz goes up and almost reach Curve I, but still below Curve I.

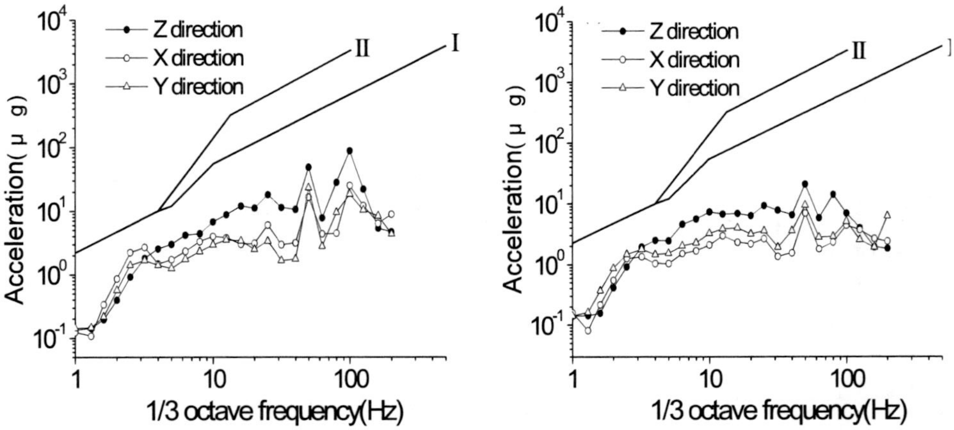

Figure 10. Background vibration measured at Point 1(left) and Point 3 (right).

Shows in Figure 12 are the averaged vibrations in three rooms induced by road traffic only. The curves at Points 1 and 2 exceed Curve I around 10Hz, which probably affect the reading veracity of TEM. Even at Point 3, with a distance more than 100m, the curve approaches the criterion. Compared with subway traffic, road traffic has larger influences on the equipment in the laboratory. In fact, the heavy traffic in rush hours has caused a big problem so that those instruments have to be operated at midnight.

Abundant measurement data provide some complex combinations of different sources for analysis. For example, shown in Figure 13 are the vibrations at Point 1 induced by combination of two traffic sources in Cases 5 and 6 in Table 2. It is obvious that vertical the acceleration exceeds Curve I. Compared with Figures 11 and 12, it can be concluded that road traffic contributes more.

As mentioned above, vibration in vertical direction plays a bigger role than those in the other two directions. A vibration envelope in Z direction could be drawn for all cases, which represents the largest vibration in certain measured room of the laboratory. Figure 14 shows the vibration envelope for Point 1. The others can be drawn accordingly in the same way.

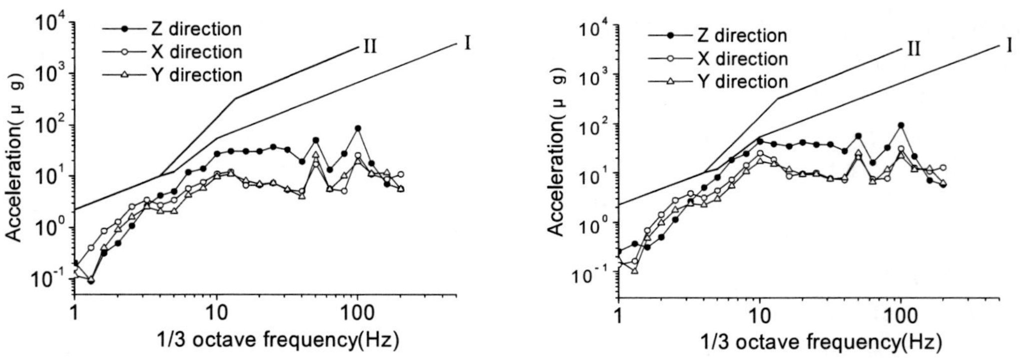

Figure 11. Vibration at Point 1 induced by trains in east tunnel only (left) and in both tunnels (right).

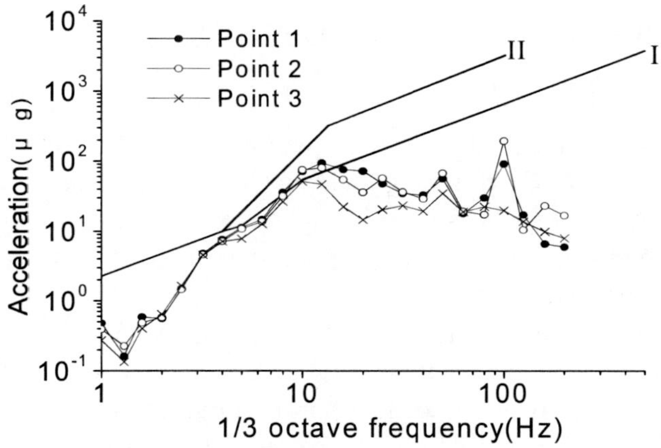

Figure 12. Vibration induced by road traffic at three different measured points.

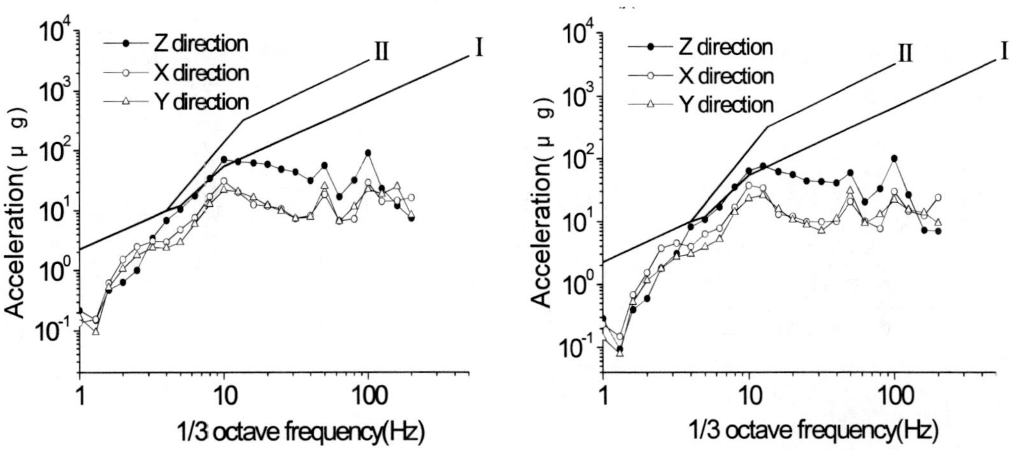

Figure 13. Vibration at Point 1 induced by the combination of road traffic and metro trains: trains in east tunnel only (left) and in both tunnels (right).

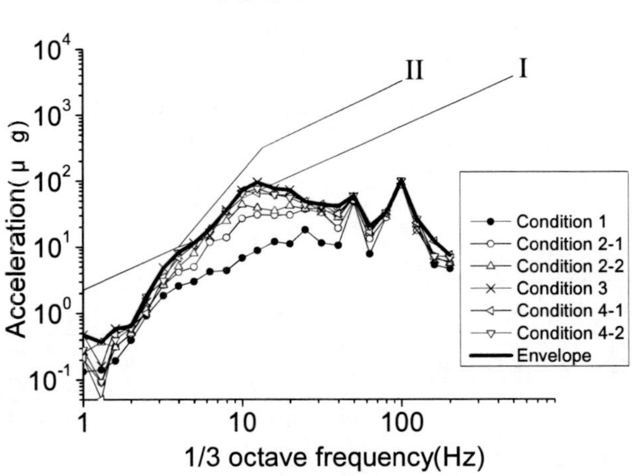

Figure 14. Acceleration curves and their envelope in Z direction of all cases at Point 1.

3.2. Low Frequency Vibration Tests in Laboratory

To obtain the characteristics of low frequency vibration and the isolation performance of a floating slab track (FST), full-scale tests were carried out in the Lab of Track Vibration Abatement and Control (LTVAC) in Beijing Jiaotong University, as shown in Figure 15.

The LTVAC has two storeys, as shown in Figure 15. Floor 1 is a curved tunnel with a radius of 12.5 m, embedded in sandy silt to a depth of 6 m underground. Floor 2 is an L-type tunnel, inserted in gravel to a depth of 14 m.

The FST was mounted on the invert of Floor 2 (Figure 16). The length, width, and thickness of the slab were 6, 3.5, and 0.45 m, respectively, and the mass of the slab M_{slab} was 22680 kg. It was supported by two rows of steel springs installed in holes. The stiffness of the springs was either 6.9 or 5.3 MN/m, and they were spaced at intervals of 1.2, 1.8 or 2.4 m.

T-60 rails with a mass per unit length of 60.64 kg/m were used, and rubber rail pads with a 50 MN/m stiffness discretely supported the rails at intervals of 0.6 m on the slab.

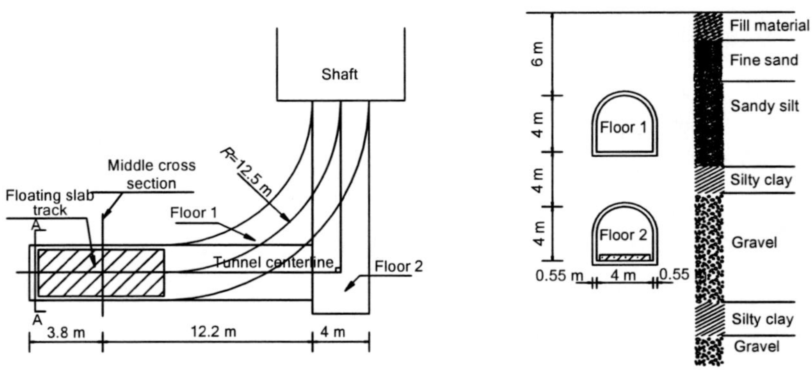

Figure 15. Laboratory of Track Vibration Abatement and Control in Beijing Jiaotong University.

An unbalanced shaker SBZ30 (Figure 17) was made to simulate train induced vibration, with a total mass M_{shaker}=18 400 kg. It consisted of a vibration generator and a bogie. The vibration generator had a static eccentric moment in the range of 0 to 250 N·m, and the rotation speed of the eccentric axis between 0 and 1800 rpm. The SBZ30 shaker could produce harmonic loads with amplitudes ranging from 0 to 250 kN within the frequency range between 5 and 30 Hz. The wheelbase of the bogie was 2.2 m, which was similar to that of the metro train.

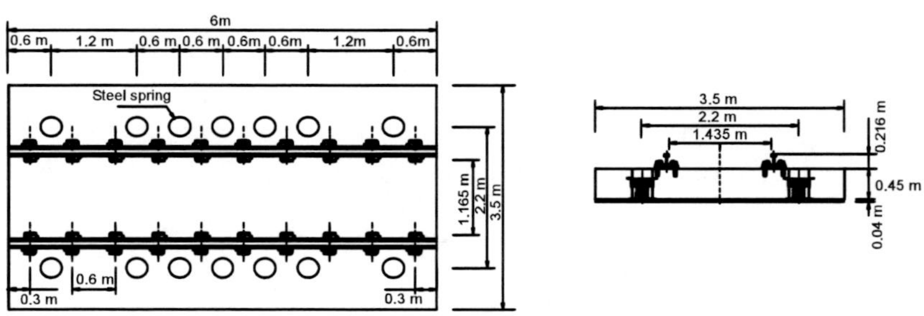

Figure 16. The floating slab track: Plan (left); cross section (right).

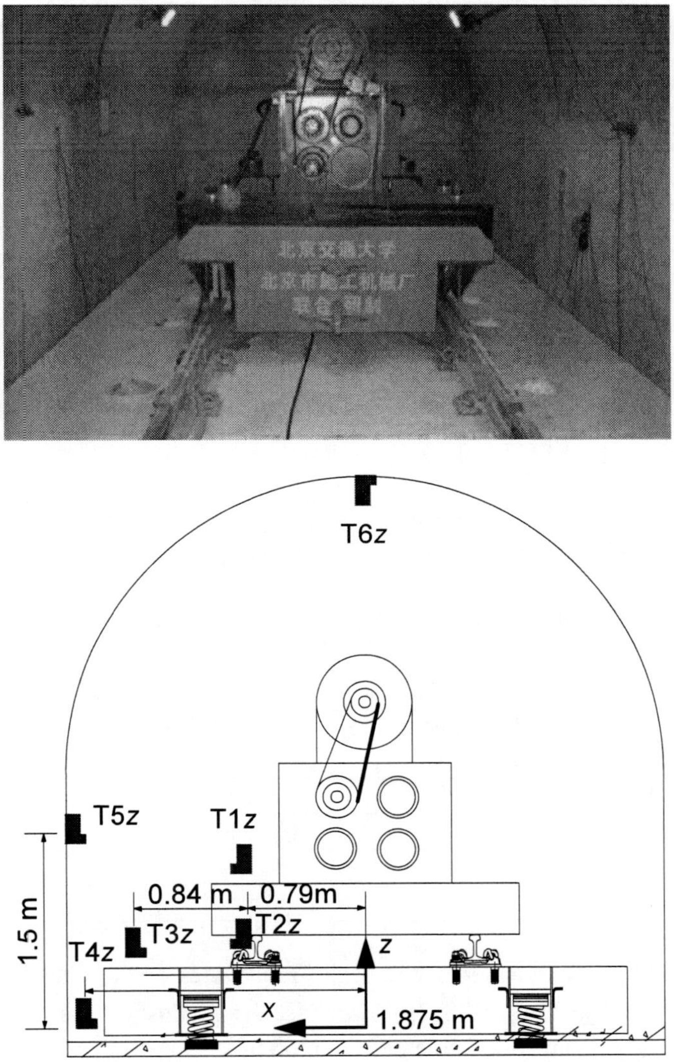

Figure 17. Unbalanced shaker SBZ30 used for the dynamic simulation.

According to the stiffness, spacing, and allowable deformation of the steel springs, some test cases were designed (Table 3). Case 1 was regarded as a non-isolation measure without the springs installed between the slab and the tunnel invert; Cases 2, 3, 4, and 5 were considered as the isolation measures.

Vibration measurements were performed on the shaker bogie, rail, slab, tunnel invert, wall and apex, and also on the ground surface and several floors of an adjacent building. The attenuation of low frequency vibration with the distance from the track is presented, as well as the responses of different floors of the building were reported.

For Cases 2−5, four kinds of steel spring combinations were used with the total effective stiffness K_{case_i} shown in Table 4. The basic frequencies of the track with isolation measures were calculated approximately using the formula $f_{case\_i}=[K_{case\_i}/(M_{slab}+M_{rail})]^{1/2}/(2\pi)$, the results were slightly different from those calculated using the 3-D finite element method based on the theory of MDOF system [13].

Table 3. Test cases and measured deflection of springs

Case	Stiffness /(MN/m)	Spacing*	Spring number	Deflection	Allowance D_{all} /mm	Spring Location (black filled circle)
1	Non-isolation					
2	6.9	1.2 m×4	10	7.4	15	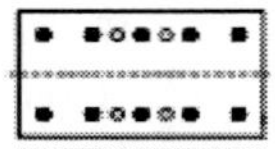
3	5.3	1.2 m×4	10	9.6	10	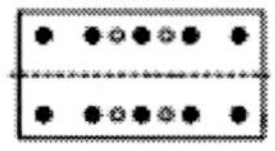
4	6.9	1.2, 1.8, 1.8 m	8	9.2	15	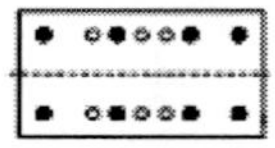
5	6.9	2.4 m×2	6	12.3	15	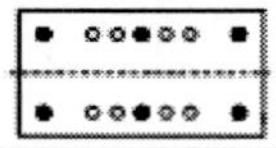

* Spacings are vertically symmetrical distributed, and only a half presented, i.e., 1.2 m×4 means four spacings of 1.2 m each between every two springs in the upper half.

Table 4. Basic frequencies in different cases

Case	K_{case_i} /(MN/m)	basic frequencies of the track f_{case_i} /Hz		
		Without shaker mass	Ding et al.	With shaker mass
2	69	8.65	8.12	6.47
3	53	7.58	7.12	5.67
4	55.2	7.73	7.31	5.79
5	41.1	6.70	6.50	5.01

With the shaker mass considered, the basic frequencies of the track were calculated by $f_{case\_i}=[K_{case\_i}/(M_{slab}+M_{rail}+M_{shaker})]^{1/2}/(2\pi)$. For Case 3 and Case 4, the basic frequencies are similar, while the spacing and allowable deformation of steel springs are different.

For all cases, the harmonic excitation forces 11−100 kN were produced by the shaker between 5.2 and 16 Hz. The maximum deformations of the steel springs were obtained by $D_{case\_i}=[(M_{slab}+M_{rail}+M_{shaker})g+F_{max}]/K_{case\_i}$, as shown in Table 3, which indicated that the deformations of the steel springs were within the allowable range.

In the measurement, vertical accelerations in the tunnel were measured at six different points, T1−T6, along the tunnel centerline. Six Lance accelerometers were glued to the bogie, rail, slab, tunnel invert, tunnel wall, and tunnel apex, respectively.

the x-axis perpendicular to the tunnel centerline, the longitudinal y-axis in the direction of the tunnel centerline, and the z-axis pointing upwards.

Vertical and horizontal (along the tunnel centerline) accelerations were measured with ten Lance accelerometers at five points, S1−S5, located on the ground surface, at horizontal distances of 0, 20, 40, 60, and 80 m from the middle cross section, as shown in Figure 18. A FOCUS II dynamic data acquisition analysis apparatus was used to record the vibration signals at a sampling frequency of 512 Hz.

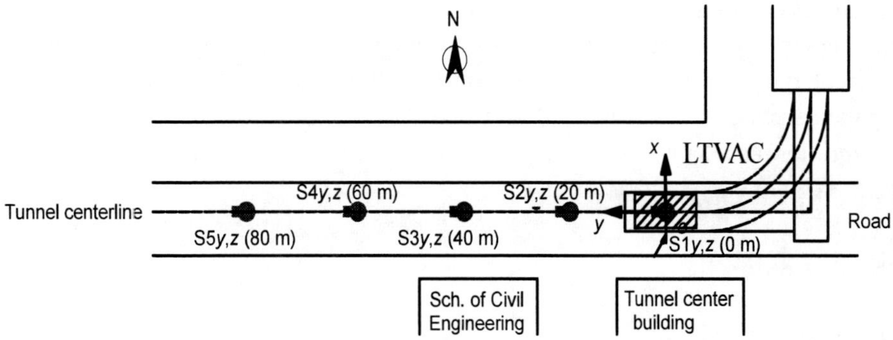

Figure 18. Measurement setup on the ground surface.

Figure 19 shows the location of the measurement points B1, B2, B3, and B4 in the Tunnel Center Building and accelerometers placed on each floors, vertical and horizontal (in the *x*-direction) accelerations were measured in the corridor at a horizontal distance of 23 m from the tunnel centerline. The same accelerometers and data acquisition analysis apparatus were used as on the ground surface.

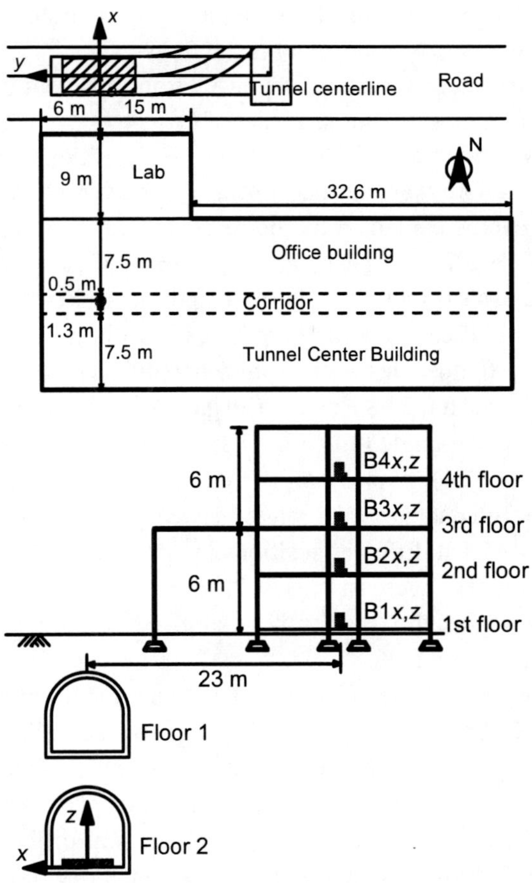

Figure 19. Measurement setup in the building: plan (top) and Cross section (bottom).

To obtain comparable results, it was assumed that the excitation force varied directly with vibration acceleration, based on the small deformation of elasticity theory. Therefore, the recorded accelerations could be normalized into unit-force accelerations, namely, 1 kN. In the following data analysis, the normalized accelerations were adopted.

The vibration acceleration level (VAL), vibration transfer loss (VTL) and transmission coefficient (TC) are expressed as follows:

$$\mathrm{VAL} = 20\log_{10}(a / a_0) \ \text{(dB)} \tag{13}$$

$$\mathrm{VTL} = \mathrm{VAL}_i - \mathrm{VAL}_k \ \text{(dB)} \tag{14}$$

$$\mathrm{TC} = \frac{\frac{1}{2\pi}\int_{-\infty}^{+\infty} Y(t)\exp(-j2\pi ft)}{\frac{1}{2\pi}\int_{-\infty}^{+\infty} X(t)\exp(-j2\pi ft)} = \frac{Y(f)}{X(f)} \tag{15}$$

where a is the root mean square (RMS) of acceleration, and a_0=10^{-6} m/s^2 is the reference vibration acceleration; subscripts 'i' and 'k' refer to different positions; $X(t)$ and $Y(t)$ are the input and output of the system, $X(f)$ and $Y(f)$ are the input and output Fourier transformations of the system, respectively; t is the time, and f is the frequency.

In the following analysis of vibration responses from different cases, Case 3 is omitted, because the response was almost the same as that of Case 4.

Figure 20 summarizes the vertical VALs at different positions in the tunnel. In all cases, the trends at different positions were similar. In Case 1, the vertical VAL reached a maximum at about 12 Hz, which can be explained by the resonance of the shaker. There was a marked peak at about 7 Hz in Case 2, and peaks around 6 Hz in Cases 4 and 5, explained by the resonance of the whole system including the FST and the shaker. It can be clearly seen that the resonance frequency shifted from 12 Hz to 6−7 Hz with the installation of steel springs. Note that the resonance frequencies seem somewhat different from the basic frequencies, because the interfaces between the FST and the shaker not rigid during the tests.

The vertical VALs on the bogie (T1), rail (T2), and slab (T3) were obviously larger than those on the tunnel invert (T4), tunnel wall (T5), and tunnel apex (T6). The VTL from the slab to the tunnel invert attenuated in the range between 29.7 and 46.2 dB, according to the statistic of the vertical VALs at different positions in different cases. The maxima of the VTL from the slab to the tunnel invert in Cases 2, 4, and 5 were 42.4, 44.2, and 46.2 dB respectively, which shows that the smaller the total effective spring stiffness, the greater the VTL in the frequency range from 5 to 16 Hz.

Figure 21 gives the vertical TCs from the slab to the tunnel invert and the tunnel apex. The vertical TC values in Case 1 were obviously the highest of all the cases, ranged between 0.012 and 0.028, and they became smaller in Cases 2, 4, and 5, ranged between 0 and 0.01, showing moderate variation between 5 and 16 Hz in all cases with isolation measures.

Figure 22 shows the ground VALs for different cases at different positions. Generally, the vertical response tended to be greater than the horizontal one; both the maximum vertical and horizontal responses in Cases 1, 2, 4, and 5 appeared at about 12, 7, 6, and 6 Hz, respectively. Due to the steel springs, the ground responses declined markedly at above 9 Hz.

In general, at frequencies greater than 7 Hz, Case 2 showed the biggest vertical response, and Case 5 the smallest; while the horizontal responses showed a different variation pattern at each position.

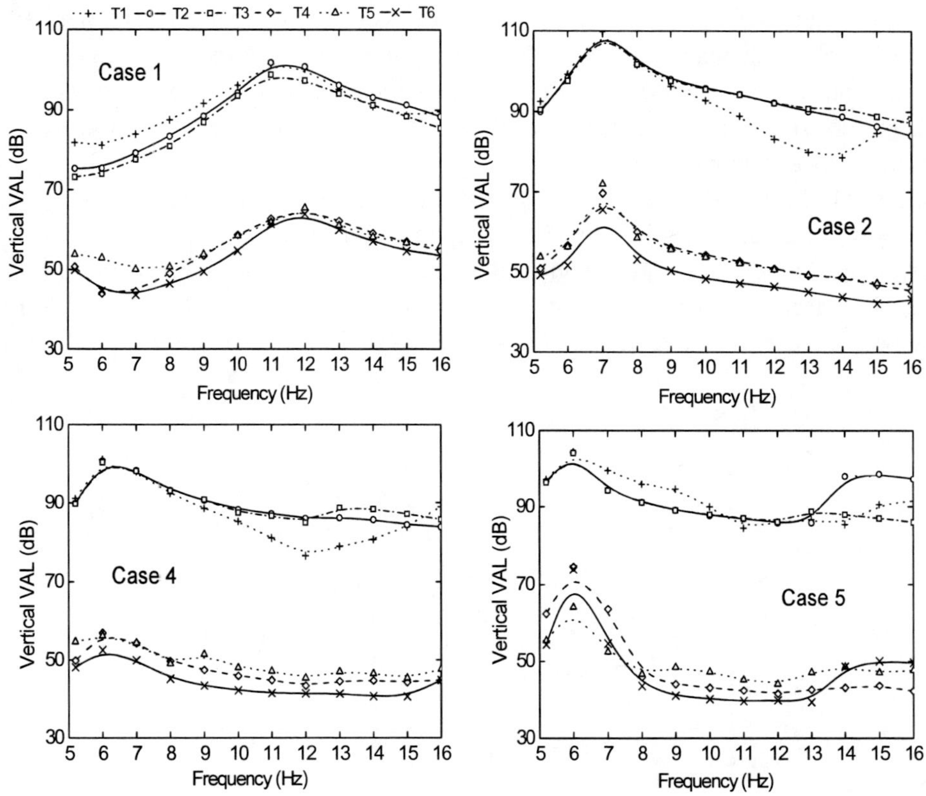

Figure 20. Vertical vibration acceleration levels at different positions in the tunnel in different cases.

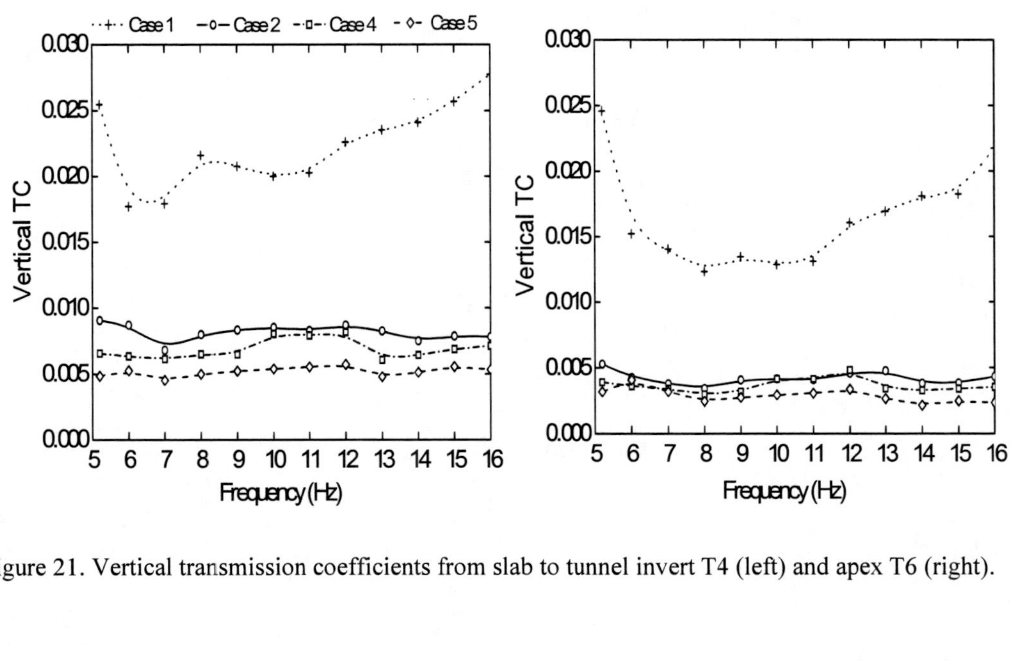

Figure 21. Vertical transmission coefficients from slab to tunnel invert T4 (left) and apex T6 (right).

An important question is how the horizontal distance affects the variation in VALs. Figure 23 shows the vertical and horizontal VALs as a function of the horizontal distance from the track at different frequencies in different cases.

For Case 1, the vertical component changed moderately with distance at frequencies below 10 Hz, but reduced slightly with distance at frequencies beyond 10 Hz; the horizontal component showed comparatively large variation within 40 m, and there was an amplification zone at around 20 m. For Cases 2, 4, and 5, the vertical component attenuated with distance overall; the horizontal component undulated with distance, and in general, there was also an amplification zone at about 20 m.

Figure 24 illustrates how the soil layers attenuate low frequency vibration. With the vibration on tunnel invert as input and the responses on the ground as outputs, the vertical TCs from the tunnel invert to the ground surface were obtained, and then averaged. In general, TCs decreased with distance, and ranged between 0 and 0.5. The TCs reduced slightly with frequency, indicating that the soil layers could not easily dissipate low frequency vibration.

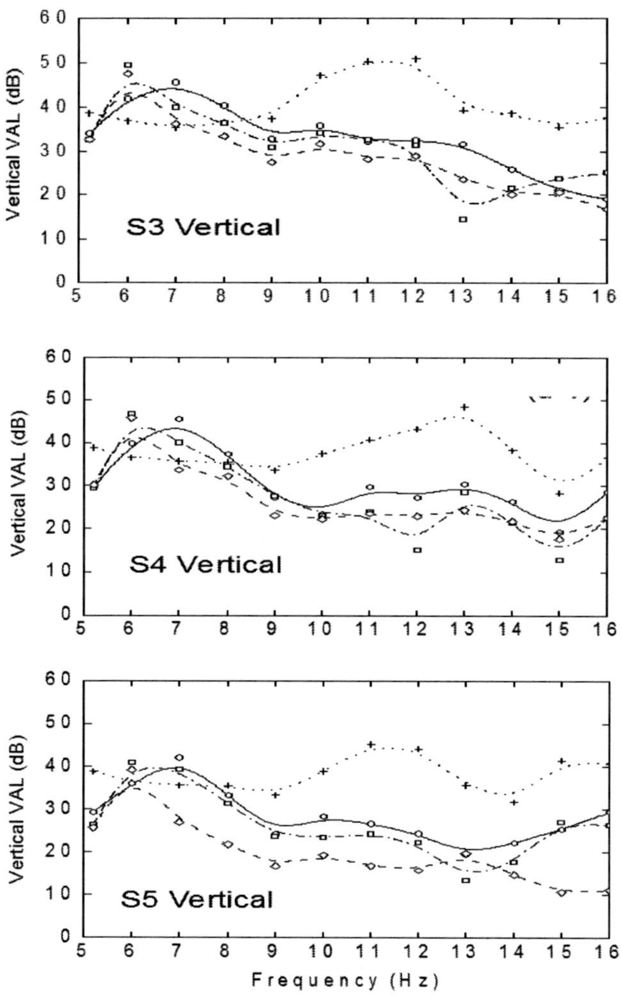

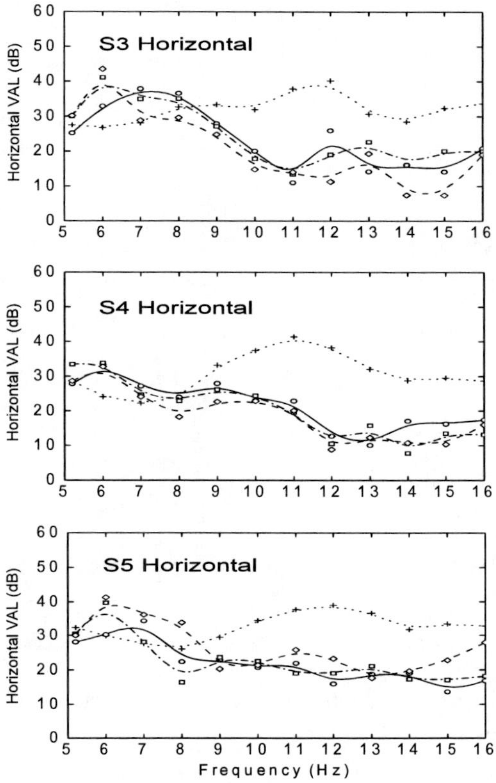

Figure 22. Vertical and horizontal acceleration levels for different cases at different positions.

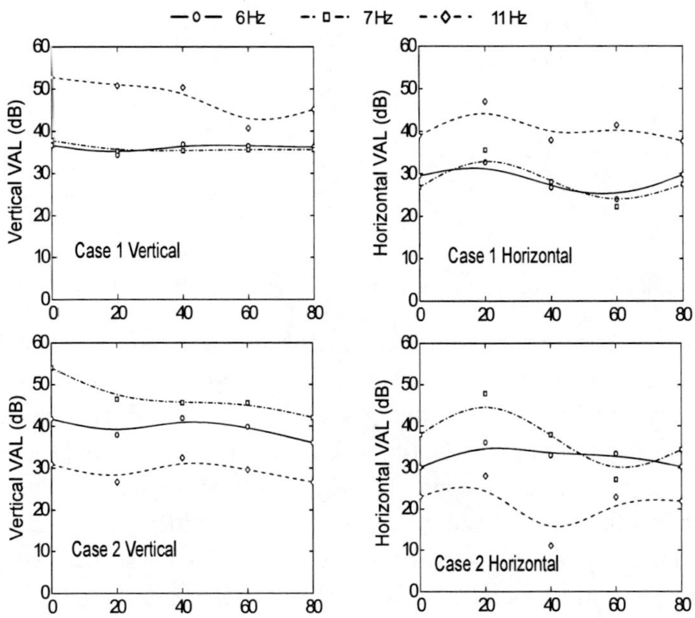

Figure 23. (Continued)

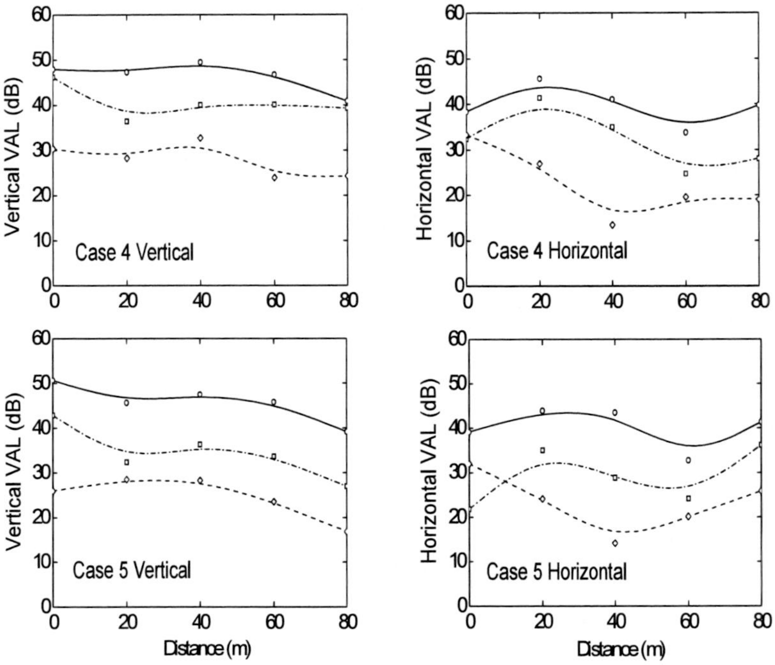

Figure 23. Vertical and horizontal acceleration levels versus the distance from the track at different frequencies in different cases.

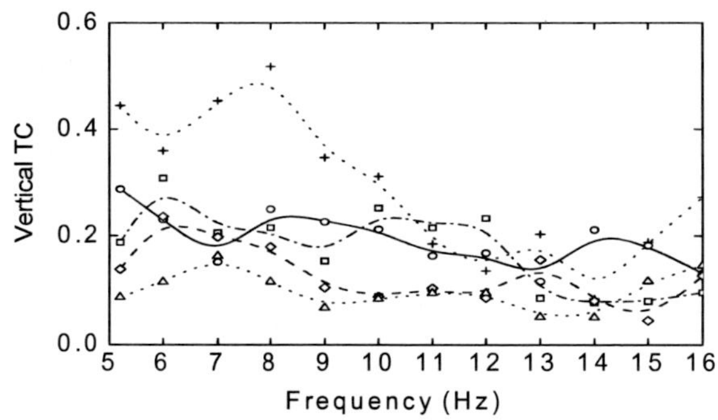

Figure 24. Vertical transmission coefficients from the tunnel invert to the ground surface.

The vibration responses on ground surface depend chiefly on the geologic characteristics of the soil. At the test site, there are seven layers from top to bottom: fill material, fine sand, sandy silt, silty clay, gravel, silty clay, and gravel (Figure 15), which is regarded as a typical stratum in Beijing. Therefore, the vibration characteristics on the ground surface at the LTVAC can be applied to describe approximately the vibration in other places of Beijing.

Figure 25 shows the vertical and horizontal VALs on different floors of the building in different cases. Generally, the horizontal response was greater than the vertical one, which is different from that on the ground surface. For vertical response, the VALs on all floors approximated each other, while the trends on different floors were the same for each case. For the horizontal response, the VALs changed markedly with the frequency: below 9 Hz, the VALs on the 4th floor were the highest, while between 10 and 15 Hz they were the lowest.

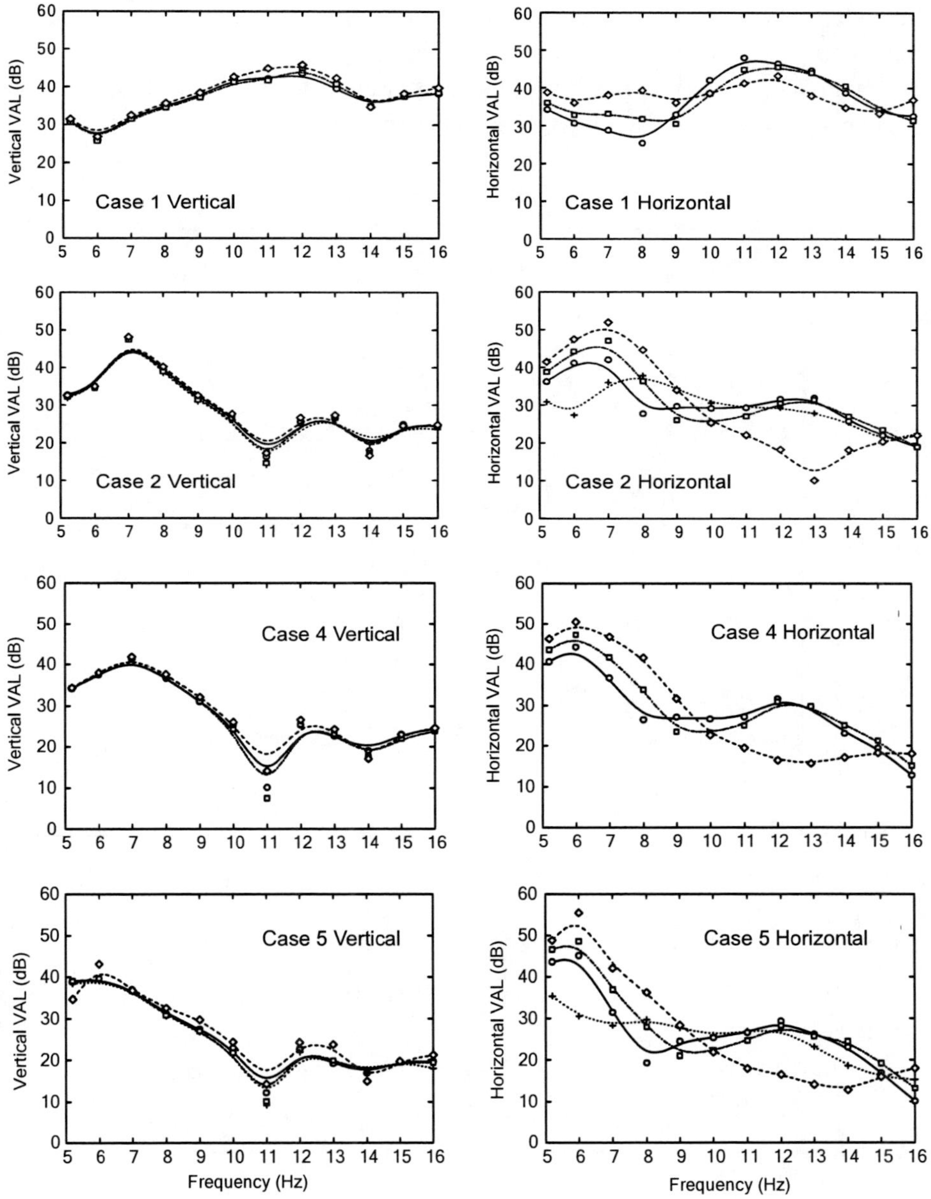

Figure 25. Vertical and horizontal acceleration levels on first (+), second (○), third (□) and fourth (◊) floors in different cases.

Figure 26 shows the vertical and horizontal VALs as a function of the floor number. In general, the vertical component varied moderately with the floor number and approximated on each floor. The horizontal component markedly increased with the floor number at frequencies below 10 Hz, and decreased at frequencies above 10 Hz.

The characteristics described above of low frequency vibration in the building are dependent mainly on the construction of the Tunnel Center Building. The experimental results in this study can be used as a reference for predicting the vibration of similar buildings. If a building had a different structure (foundation type, structure type, floor numbers, etc.), the vibration characteristics would change, therefore, it would be worth studying the low frequency vibration in different buildings.

The main conclusions drawn from the observations on the low frequency vibration measurements performed in the LTVAC can be summarized as follows.

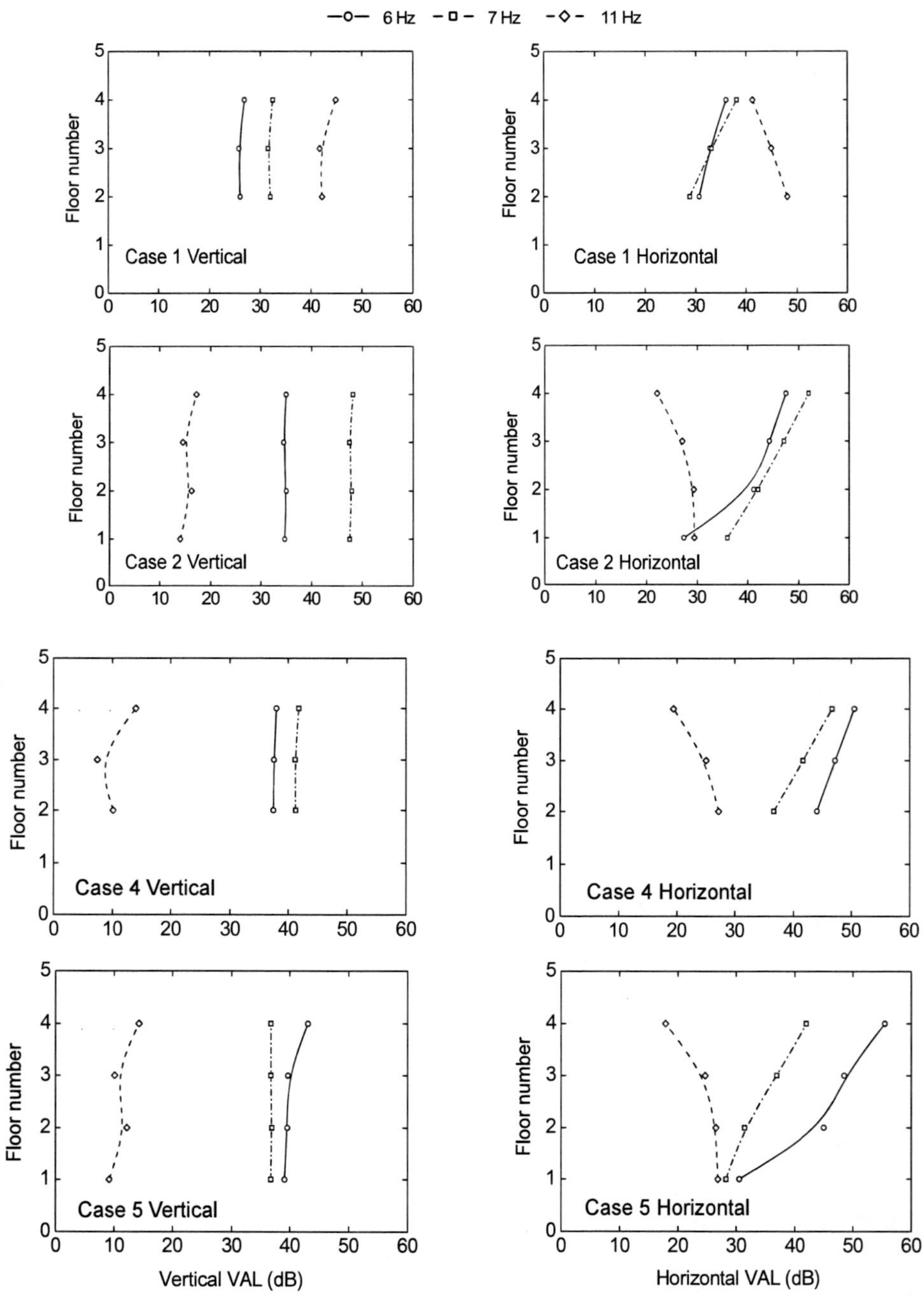

Figure 26. Vertical and horizontal acceleration levels on different floors at frequencies of 6 Hz (○), 7 Hz (□) and 11 Hz (◊) in different cases.

(1) The basic frequencies of the floating slab track (FST) used in the tests were in the range between 6 and 7 Hz. The larger the spacing and the smaller the stiffness of the springs, the lower the basic frequency.
(2) The transmission coefficients (TCs) could be greatly reduced by the FST. The TCs from the slab to the tunnel without isolation measures ranged between 0.012 and 0.028. However, with the FST these ranged between 0 and 0.01.
(3) The TCs of the soil at the test site ranged between 0 and 0.5, and are slightly reduced at low frequency, indicating that the soil could not easily dissipate the low frequency vibration.
(4) The vertical in low frequency vibration played a bigger role on the ground surface than the horizontal one. The vertical low frequency vibration attenuated with the distance to the track, while the horizontal one undulated with the distance and became amplified within 40 m to the track, showing an amplification zone at around 20 m.
(5) The horizontal low frequency vibration played a bigger role in the vicinity of the building than the vertical one. The vertical low frequency vibration varied moderately with the floor number and approximated on each floor, while the horizontal one increased markedly with the floor number at frequencies below 10 Hz, but decreased at above 10 Hz.

4. Prediction and Evaluation for Vibration Induced by Underground Trains

4.1. Prediction of Low-frequency Ground Vibration

Based on the low-frequency vibration tests on a direct-fixation-fastener track (DFT) and a steel spring floating slab track (SSFST) in the LTVAC, the coupled DFT-tunnel-soil and SSFST-tunnel-soil models are respectively established to further predict the low-frequency ground vibration, using the 3-D FEM. Using the regression analysis method, the prediction formulas for low-frequency ground acceleration levels are derived. The predicted acceleration levels are compared and verified with the test and simulated results.

4.1.1. Three-dimensional Numerical Simulation

In this section, the coupled 3-D track-tunnel-soil FE models are used (Figure 27) to further study the propagation laws of ground vibration under the conditions of different depths of the tracks and the discrete excitation frequencies in a range of 1–20 Hz.

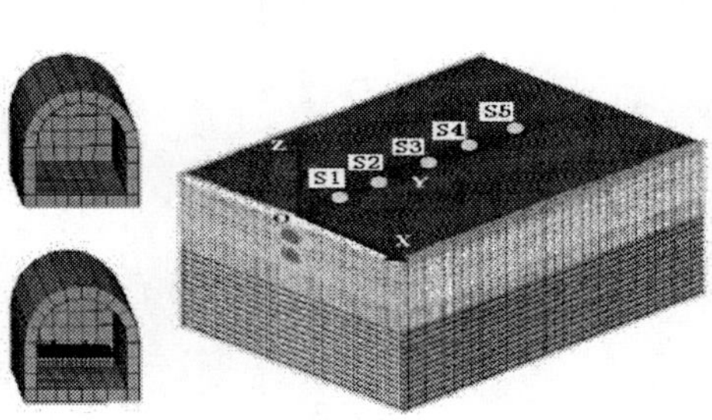

Figure 27. Three-dimensional coupled track-tunnel-soil model.

The geological section map in the area of the LTVAC indicates there are seven layers of soil from top to bottom: fill material, fine sand, sandy silt, silty clay, gravel, silty clay and gravel, as shown in Figure 15. According to the amplitude of shear wave velocity of each layer, the soil is simplified to 3 layers. The length (Y-direction), width (X-direction) and height (Z-direction) of the numerical models are respectively 110, 80 and 70 m, based on the following aspects: shear wave velocity of each soil layer, excitation frequency range, observation points on the ground surface and embedded depths of the track.

Four kinds of track cases including a DFT and three SSFSTs are considered in numerical models. As the embedded depth of tracks varies between 10 and 30 m in Beijing metro lines, three different depths of 10, 18 and 30 m are considered. Four unit harmonic loads are applied at four contact points of the rails and wheels of SBZ30. With the frequency changing from 1 to 20 Hz, the low frequency accelerations at observation points S1−S5 can be obtained.

The recorded and simulated ground vibration accelerations are processed and analyzed. In order to predict ground vibration accelerations under the conditions of the continuous excitation frequencies (1−20 Hz), the unbroken horizontal distances (0−100 m) to the track and the continuous embedded depths (10−30 m) of the track, a prediction formula for low frequency ground vibration is derived, using the regression analysis of the recorded and simulated ground vibration acceleration levels.

4.1.2. Mathematical Model

Assuming (a) L_{val} depends on the independent variables α, f, f_{tr}, D_t and D_s, where L_{val} is the VAL of ground observation point, α and f are the amplitude and frequency of the load, f_{tr} is the natural frequency of the track, D_t is the embedded depth of the track and D_s is the horizontal distance to the track; and (b) VAL $L_0(f, f_{tr})$ denotes the VAL of ground observation point at D_s=0 m, under a certain embedded depth D_{t0} of the track, the vertical low frequency ground VAL could be expressed as:

$$L_{val\_v} = 20\lg\alpha + L_{0\_v}(f, f_{tr})\left(\frac{D_t}{D_{t0}}\right)^a \times \left(1 - \frac{D_s}{D_{s0}}\right)^b \quad \text{(dB)} \tag{16}$$

where $L_{0\_h}(f, f_{tr})$ is the vertical reference VAL, D_{t0} = 18 m, D_{s0} = 100 m, a and b are respectively the influence coefficients for the depth and distance.

As the recorded and simulated horizontal accelerations are amplified at distance of around 20 m to the track, the correction value ΔL_{Ds} for the horizontal distance is supposed. Therefore, the horizontal VAL could be written as:

$$L_{val\_h} = 20\lg\alpha + L_{0\_h}(f, f_{tr})\left(\frac{D_t}{D_{t0}}\right)^c \times \left(1 - \frac{D_s}{D_{s0}}\right)^d + \Delta L_{Ds} \quad \text{(dB)} \tag{17}$$

where $L_{0\_h}(f, f_{tr})$ is the horizontal reference VAL, c and d are also the influence coefficients for the depth and distance.

4.1.3. Parameter Estimation

The natural frequencies of the tracks in Cases 1, 2, 3 and 4 are around 186, 7, 6 and 5 Hz, respectively. Analyzing the results by the least square method, these parameters $L_{0\_v}(f, f_{tr})$, $L_{0\_h}(f, f_{tr})$, a, b, c, d and ΔL_{Ds} could be determined.

By means of the regression analysis, Equations (16) and (17) can be expressed as:

$$L_{val\_v} = 20\lg\alpha + \left[\sum_{i=1}^{n-1} \Lambda_i (f - f_{tr})^{n+1-i}\right] \left(\frac{D_t}{D_{t0}}\right)^{-0.3023} \times \left(1 - \frac{D_s}{D_{s0}}\right)^{\sum_{j=1}^{m+1} \Lambda_j (f - f_{tr})^{m+1-j}} \quad \text{(dB)} \tag{18}$$

$$L_{val\_h} = 20\lg\alpha + \left[\sum_{i=1}^{n+1} \Theta_i (f - f_{tr})^{n+1-i}\right] \left(\frac{D_t}{D_{t0}}\right)^{-0.3023} \times \left(1 - \frac{D_s}{D_{s0}}\right)^{\sum_{j=1}^{m+1} \Theta_j (f - f_{tr})^{m+1-j}} + \Delta L_{D_s} \quad \text{(dB)}$$

$$\Delta L_{D_s} = \begin{cases} -0.0207 D_s^2 + 0.828 D_s & (0 \le D_s \le 40) \\ 0 & (40 < D_s < 100) \end{cases} \tag{19}$$

where α (unit: kN) and f ($1 \le f \le 20$ Hz) are the amplitude and frequency of the load, f_{tr} is the natural frequency of the track, D_t ($10 \le D_t \le 30$ m) is the embedded depth of the track, D_s ($0 \le D_s < 100$ m) is the horizontal distance to the track, D_{t0} = 18 m, D_{s0} = 100 m, n and m are the numbers of polynomial terms, Λ_i, Λ_j, Θ_i and Θ_j are the polynomial coefficients. Here, n, m, Λ_i, Λ_j, Θ_i and Θ_j could be determined by track types and dynamic characteristics of soil.

4.1.4. Verification for the Prediction Formula

The VALs calculated from the prediction formulas (18) and (19) are compared with the tested and simulated results. Figures 28−31 show, respectively, the predicted vertical and horizontal ground VALs in Cases 1 and 2 under the embedded depth D_t =18 m. It can be obviously seen that the trend of the predicted VALs agree well with the recorded and simulated results, and the difference is small, which verifies the prediction formulas.

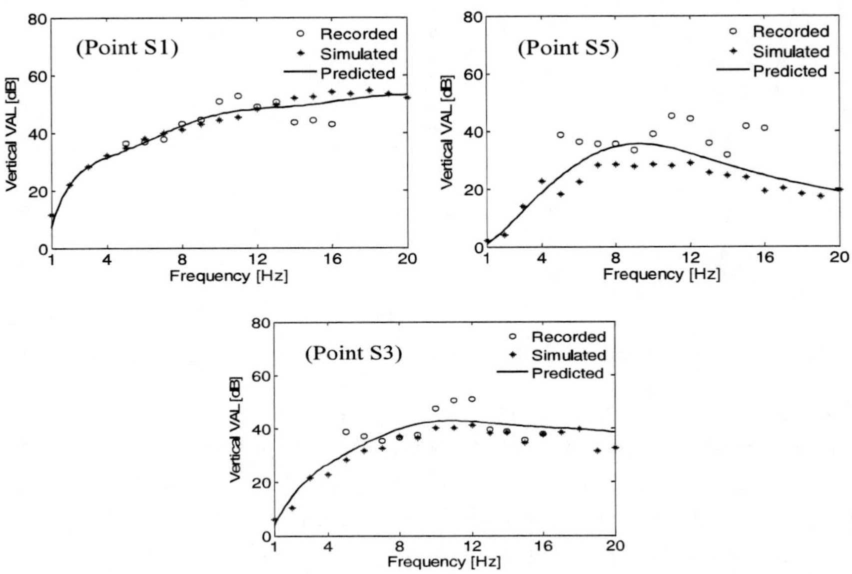

Figure 28. Comparison of predicted, simulated and recorded vertical acceleration levels in Case 1.

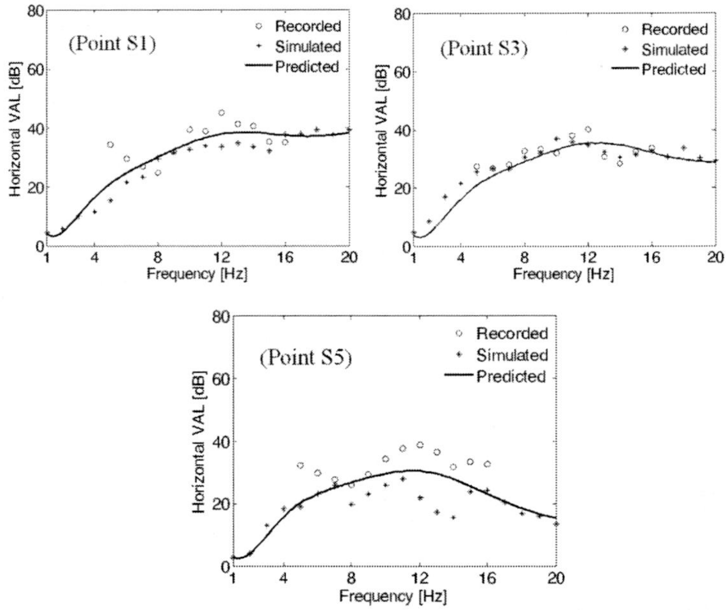

Figure 29. Comparison of predicted, simulated and recorded horizontal acceleration levels in Case 1.

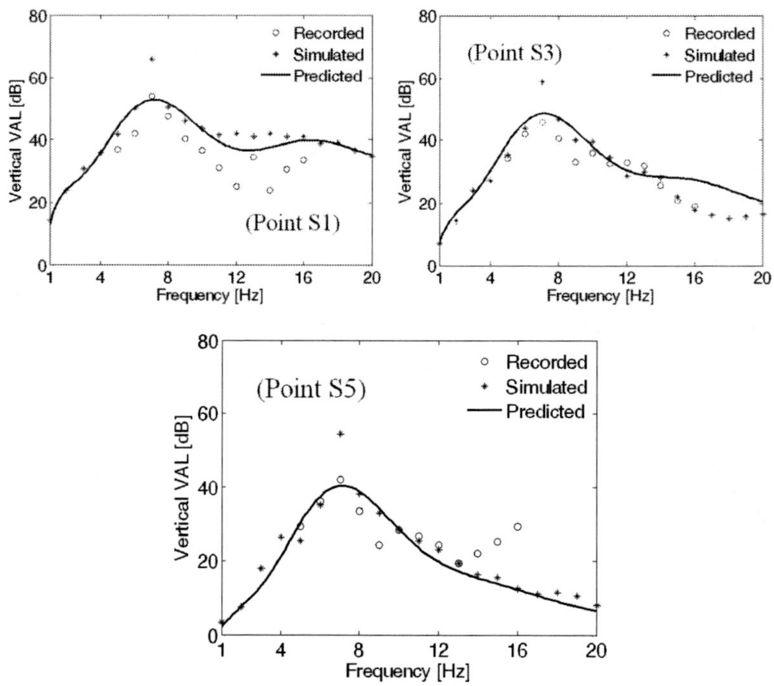

Figure 30. Comparison of predicted, simulated and recorded vertical acceleration levels in Case 2.

4.1.5. Vibration Acceleration Level Prediction

Shown in Figures 32 and 33 are, respectively, the predicted vertical and horizontal ground VALs in Case 1 under unit load $\alpha = 1$ kN and different embedded depths of the track. It can be observed that the vertical ground vibration goes down obviously with the embedded depth, and decreases in general with the distance to the track. The horizontal vibration also decreases with the embedded depth, while there is an amplification zone between 0 to 40 m.

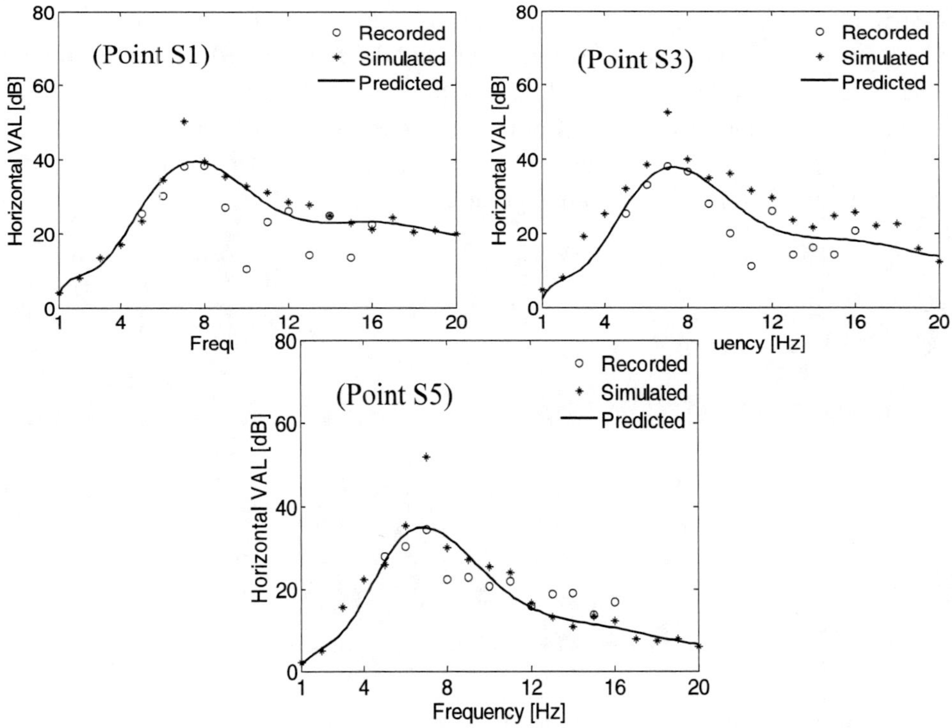

Figure 31. Comparison of predicted, simulated and recorded horizontal acceleration levels in Case 2.

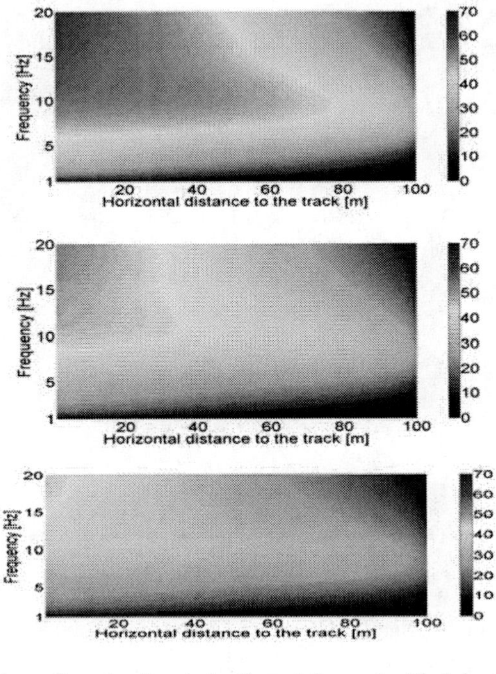

Figure 32. Predicted vertical acceleration levels in Case 1 for embedded depths of track: 10 m (top), 18 m (middle) and 30 m (bottom).

Shown in Figures 34 and 35 are, respectively, the predicted vertical and horizontal ground VALs in Case 4 under unit load $\alpha = 1$ kN and different embedded depths of the track. It can be seen that both the vertical and horizontal ground vibration decrease obviously with the embedded depth, and reach the maximums at frequency around 5 Hz. The vertical ground vibration decreases with the horizontal distance to the track, and there is also an amplification zone between 0 to 40 m for the horizontal ground vibration.

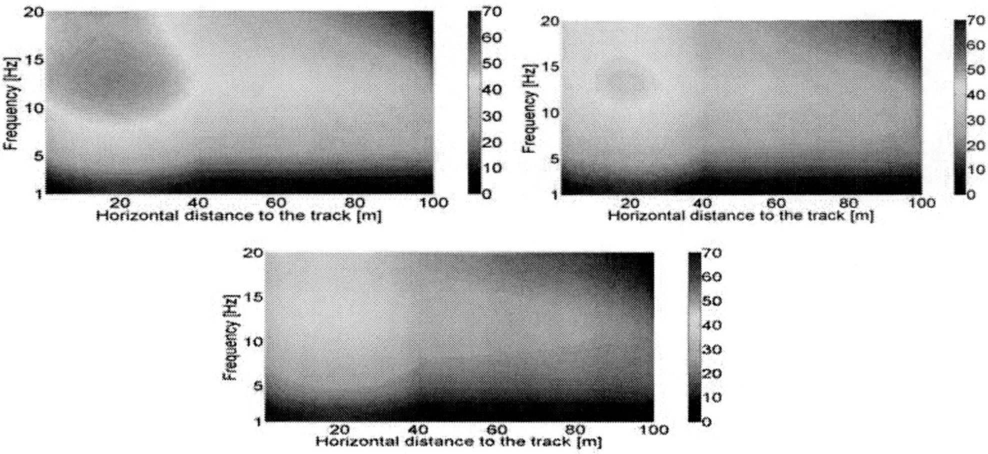

Figure 33. Predicted horizontal acceleration levels in Case 1 for embedded depths of track: 10 m (top), 18 m (middle) and 30 m (bottom).

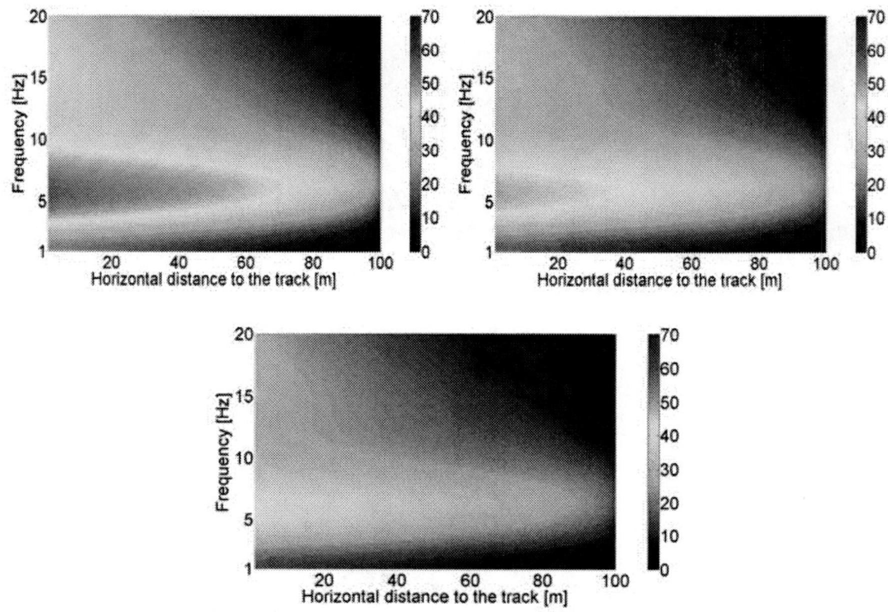

Figure 34. Predicted horizontal acceleration levels in Case 4 for embedded depths of track: 10 m (top), 18 m (middle) and 30 m (bottom).

Due to the restrictions of test condition, the excitation forces are non-moving harmonic loads, which are different from practical moving train loads. Therefore, a prediction formula considering moving loads should be further studied in future.

4.2. Prediction of Ground-borne Vibration

Prediction on the dynamic responses in the tunnel and free field north of the Chengfulu station on Beijing Metro Line, using a 3-D coupled periodic FE-BE model in the frequency-wavenumber domain. In this model, the periodicity of the tunnel and the soil in longitudinal direction is exploited with the Floquet transform to limit the discretization effort to a single bounded generic cell. The problem is solved with a subdomain formulation based on the FE method for the tunnel and the BE method for the soil.

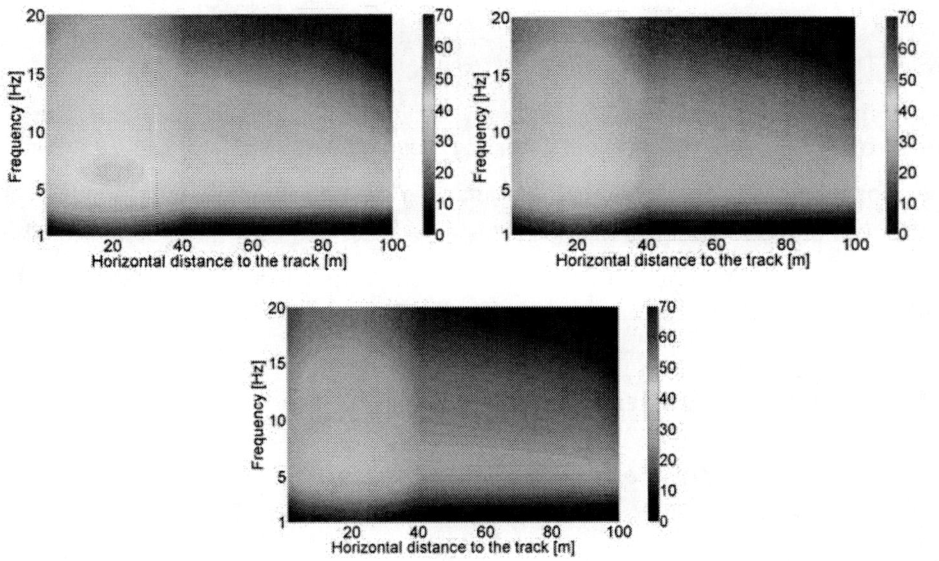

Figure 35. Predicted horizontal acceleration levels in Case 4 for embedded depths of track: 10 m (top), 18 m (middle) and 30 m (bottom).

4.2.1. Dynamic Track-tunnel-soil Interaction Model

The dynamic track-tunnel-soil interaction model is based on the following hypotheses:

(1) The tunnel is straight and periodic with a period L in the longitudinal direction e_y along the tunnel axis;
(2) The tunnel is embedded in a horizontally layered half space or a full space;
(3) All displacements remain sufficiently small so that the problem is linear and the superposition principle in the frequency domain is valid.

Thus, the problem can be restricted to periodic fields of the second kind defined on a generic cell $\widehat{\Omega}$ by application of Floquet transform (Figure 36). The generic cell is decomposed into two subdomains: the soil $\widehat{\Omega}_s$ and the tunnel $\widehat{\Omega}_t$. The boundary of it is decomposed into the free surface $\widehat{\Gamma}_{s\sigma}$ and the boundaries Σ_0 and Σ_L on which periodic conditions are imposed. The interface between tunnel and soil is denoted by $\widehat{\Sigma}_{ts}$. And a surface force $\widehat{f}_t$ is applied on $\widehat{\Gamma}_{t\sigma}$ in the tunnel.

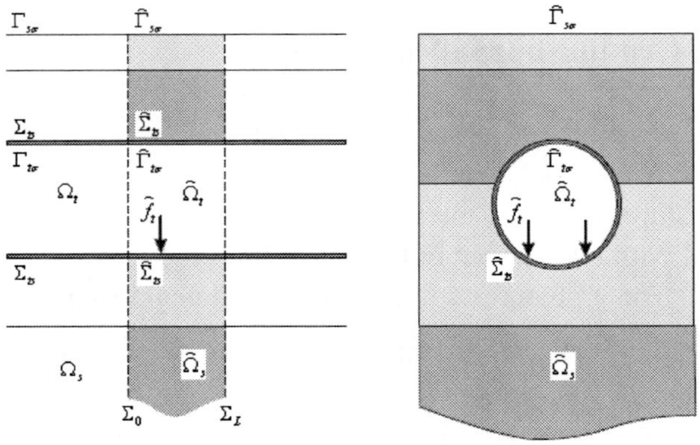

Figure 36. The periodic tunnel-soil interaction system.

An infinite periodic system can be analyzed using the Floquet theorem. The position vector $\boldsymbol{x}$ of any point in the problem domain Ω can be written as $\boldsymbol{x} = \widehat{\boldsymbol{x}} + nL\boldsymbol{e}_y$, where $\widehat{\boldsymbol{x}}$ is the position vector in the generic cell $\widehat{\Omega}$ and n is the cell number. The Floquet transform $\tilde{f}(\widehat{\boldsymbol{x}},\kappa)$ of a non-periodic function $f(x) = f(\widehat{x} + nLe_y)$ defined on a three-dimensional domain Ω, that is periodic with period of L in the direction e_y, transforms the distance nL between the n-th cell and the generic cell to the wavenumber κ and is defined as:

$$\tilde{f}(\widehat{\boldsymbol{x}},\kappa) = \sum_{n=-\infty}^{+\infty} f(\widehat{\boldsymbol{x}} + nL\boldsymbol{e}_y)\exp(+jnL\kappa) \tag{20}$$

where $j = \sqrt{-1}$, and the complex valued function $\tilde{f}(\widehat{\boldsymbol{x}},\kappa)$ is periodic of the second kind since the following condition holds for all $\widehat{\boldsymbol{x}}$:

$$\tilde{f}(\widehat{\boldsymbol{x}} + nL\boldsymbol{e}_y,\kappa) = \exp(-j\kappa L)\tilde{f}(\widehat{\boldsymbol{x}},\kappa) \tag{21}$$

The function $f(\widehat{x} + nLe_y)$ can be reconstructed for any $x = \widehat{x} + nLe_y$ using the inverse Floquet transform:

$$f(\widehat{\boldsymbol{x}} + nL\boldsymbol{e}_y) = \frac{L}{2\pi}\int_{-\pi/L}^{\pi/L} \tilde{f}(\widehat{\boldsymbol{x}},\kappa)\exp(-jnL\kappa)d\kappa \tag{22}$$

This definition can be used to synthesize the solution on the problem domain Ω if it is known on the generic cell $\widehat{\Omega}$.

The displacement and traction fields $\widehat{\boldsymbol{u}}(\widehat{x},\kappa,\omega)$ and $\widehat{\boldsymbol{t}}(\widehat{x},\kappa,\omega)$ on the subdomain $\widehat{\Omega}_s$ and $\widehat{\Omega}_t$ denote the dynamic perturbations of displacements and tractions in the soil and in the tunnel due to dynamic loads. By means of the Floquet transform, all displacement fields $\boldsymbol{u}(\boldsymbol{x},\omega)$ and traction fields $\boldsymbol{t}(\boldsymbol{x},\omega)$ defined on the periodic domain Ω are transformed to the fields $\hat{\boldsymbol{u}}(\hat{\boldsymbol{x}},\kappa,\omega)$ and $\hat{\boldsymbol{t}}(\hat{\boldsymbol{x}},\kappa,\omega)$ defined on the generic cell $\hat{\Omega}$. Subscripts s and t are used in the following to denote displacements, tractions and other field variables related to the soil and tunnel subdomain, respectively.

In the frequency-wavenumber domain, the Navier equations in the soil domain $\hat{\Omega}_s$ with the boundary conditions on $\hat{\Gamma}_{s\sigma}$ are written as follows for every frequency and wavenumber:

$$\operatorname{div}\hat{\sigma}_s(\hat{\boldsymbol{u}}_s) = -\rho^s\omega^2\hat{\boldsymbol{u}}_s \text{ in } \hat{\Omega}_s \tag{23}$$

$$\hat{\boldsymbol{t}}_s(\hat{\boldsymbol{u}}_s) = \boldsymbol{0} \text{ on } \hat{\Gamma}_{s\sigma} \tag{24}$$

$$\hat{\boldsymbol{u}}_s(\hat{\boldsymbol{x}}) = \exp(-j\kappa L)\hat{\boldsymbol{u}}_s(\hat{\boldsymbol{x}} - L\boldsymbol{e}_y) \text{ on } \Sigma_L \tag{25}$$

Likewise, the Navier equations in the tunnel domain $\hat{\Omega}_t$ with the boundary conditions on $\hat{\Gamma}_{t\sigma}$ are written as follows:

$$\operatorname{div}\hat{\sigma}_t(\hat{\boldsymbol{u}}_t) = -\rho^t\omega^2\hat{\boldsymbol{u}}_t \text{ in } \hat{\Omega}_t \tag{26}$$

$$\hat{\boldsymbol{t}}_t(\hat{\boldsymbol{u}}_t) = \hat{\boldsymbol{f}}_t \text{ on } \hat{\Gamma}_{t\sigma} \tag{27}$$

$$\hat{\boldsymbol{u}}_t(\hat{\boldsymbol{x}}) = \exp(-j\kappa L)\hat{\boldsymbol{u}}_t(\hat{\boldsymbol{x}} - L\boldsymbol{e}_y) \text{ on } \Sigma_L \tag{28}$$

where ρ^s and ρ^t are the density of soil and tunnel material. And continuity of displacements and equilibrium of stresses must hold on the tunnel-soil interface:

$$\hat{\boldsymbol{u}}_t = \hat{\boldsymbol{u}}_s \text{ on } \hat{\Sigma}_{ts} \tag{29}$$

$$\hat{\boldsymbol{t}}_t(\hat{\boldsymbol{u}}_t) + \hat{\boldsymbol{t}}_s(\hat{\boldsymbol{u}}_s) = \boldsymbol{0} \text{ on } \hat{\Sigma}_{ts} \tag{30}$$

The governing equation (26) and the boundary equation (27) are multiplied with the complex conjugate of any virtual field $\hat{\boldsymbol{v}}_t(\hat{\boldsymbol{x}},\kappa,\omega)$, and accounting for the stress equilibrium (30), the following weak variational form for the tunnel can be derived:

$$\int_{\hat{\Omega}_t} \overline{\hat{\varepsilon}(\hat{\boldsymbol{v}}_t)} : \hat{\sigma}(\hat{\boldsymbol{u}}_t)\mathrm{d}V - \omega^2 \int_{\hat{\Omega}_t} \rho^t \overline{\hat{\boldsymbol{v}}}_t \cdot \hat{\boldsymbol{u}}_t \mathrm{d}V + \int_{\hat{\Sigma}_{ts}} \overline{\hat{\boldsymbol{v}}}_t \cdot \hat{\boldsymbol{t}}_s[\hat{\boldsymbol{u}}_{sc}(\hat{\boldsymbol{u}}_t)]\mathrm{d}S = \int_{\hat{\Gamma}_{t\sigma}} \overline{\hat{\boldsymbol{v}}}_t \cdot \hat{\boldsymbol{f}}_t \mathrm{d}S \tag{31}$$

where $\hat{\boldsymbol{u}}_{sc}(\hat{\boldsymbol{u}}_t)$ denotes the wave field that is scattered by the tunnel into the soil.

Based on the mode superposition theorem, the displacement field of the tunnel can be decomposed as the following linear combination of functions $\hat{\psi}_m(\hat{\boldsymbol{x}},\kappa)$ that are periodic of the second kind:

$$\hat{\boldsymbol{u}}_{\mathrm{t}}(\hat{\boldsymbol{x}},\kappa,\omega)=\sum_{m=1}^{N}\hat{\psi}_m(\hat{\boldsymbol{x}},\kappa)\alpha_m(\kappa,\omega)=\hat{\boldsymbol{\Psi}}_{\mathrm{t}}\boldsymbol{\alpha}_{\mathrm{t}} \tag{32}$$

where the $\hat{\psi}_m(\hat{x},\kappa)$ is the m-th mode of the generic cell of tunnel, and $\alpha_m(\kappa,\omega)$ is the m-th modal coordinate. The soil displacements in the tunnel-soil interface can be written as the superposition of waves that are radiated by the tunnel into the soil:

$$\hat{\boldsymbol{u}}_{\mathrm{s}}(\hat{\boldsymbol{x}},\kappa,\omega)=\sum_{m=1}^{N}\hat{\boldsymbol{u}}_{\mathrm{sc}}(\hat{\psi}_m)(\hat{\boldsymbol{x}},\kappa,\omega)\alpha_m(\kappa,\omega) \tag{33}$$

The numerical solution of the dynamic tunnel-soil interaction problem is obtained using the classical domain decomposition approach based on the FE method for the tunnel and the BE method for the soil. The displacements in the tunnel structure are interpolated as:

$$\hat{\boldsymbol{u}}_{\mathrm{t}}\simeq N_{\mathrm{t}}\underline{\hat{\boldsymbol{u}}}_{\mathrm{t}}=N_{\mathrm{t}}\underline{\hat{\boldsymbol{\Psi}}}_{\mathrm{t}}\boldsymbol{\alpha}_{\mathrm{t}} \tag{34}$$

where N_{t} is a matrix of global finite element shape functions and $\underline{\hat{\boldsymbol{u}}}_{\mathrm{t}}$ is the N-D vector with nodal DOF. Employing the same interpolation for the virtual displacements $\hat{\boldsymbol{v}}_{\mathrm{t}}(\hat{\boldsymbol{x}},\kappa,\omega)$, the following system of equation can be obtained in the frequency-wavenumber domain:

$$\left[\boldsymbol{K}_{\mathrm{t}}(\kappa)-\omega^2\boldsymbol{M}_{\mathrm{t}}(\kappa)+\boldsymbol{K}_{\mathrm{s}}(\kappa,\omega)\right]\alpha(\kappa,\omega)=\boldsymbol{F}_{\mathrm{t}}(\kappa,\omega) \tag{35}$$

where, $\boldsymbol{K}_{\mathrm{t}}(\kappa)$, $\boldsymbol{M}_{\mathrm{t}}(\kappa)$, $\boldsymbol{K}_{\mathrm{s}}(\kappa,\omega)$ and $\boldsymbol{F}_{\mathrm{t}}(\kappa,\omega)$ can be written as:

$$\boldsymbol{K}_{\mathrm{t}}(\kappa)=\underline{\bar{\hat{\boldsymbol{\Psi}}}}_{\mathrm{t}}^{\mathrm{T}}\boldsymbol{K}_{\mathrm{t}}^{\mathrm{FE}}\underline{\hat{\boldsymbol{\Psi}}}_{\mathrm{t}}=\underline{\bar{\hat{\boldsymbol{\Psi}}}}_{\mathrm{t}}^{\mathrm{T}}\left[\int_{\hat{\Omega}_t}(\boldsymbol{L}\boldsymbol{N}_{\mathrm{t}})^{\mathrm{T}}\boldsymbol{D}(\boldsymbol{L}\boldsymbol{N}_{\mathrm{t}})\,\mathrm{d}V\right]\underline{\hat{\boldsymbol{\Psi}}}_{\mathrm{t}} \tag{36}$$

$$\boldsymbol{M}_t(\kappa)=\underline{\bar{\hat{\boldsymbol{\Psi}}}}_{\mathrm{t}}^{\mathrm{T}}\boldsymbol{M}_t^{\mathrm{FE}}\underline{\hat{\boldsymbol{\Psi}}}_{\mathrm{t}}=\underline{\bar{\hat{\boldsymbol{\Psi}}}}_{\mathrm{t}}^{\mathrm{T}}\left[\int_{\hat{\Omega}_t}\boldsymbol{N}_t^{\mathrm{T}}\rho'\boldsymbol{N}_t\,\mathrm{d}V\right]\underline{\hat{\boldsymbol{\Psi}}}_{\mathrm{t}} \tag{37}$$

$$\boldsymbol{K}_{\mathrm{s}}(\kappa,\omega)=\int_{\hat{\Sigma}_{\mathrm{ts}}}\underline{\bar{\hat{\boldsymbol{\Psi}}}}_{\mathrm{t}}^{\mathrm{T}}\boldsymbol{N}_{\mathrm{t}}^{\mathrm{T}}\hat{\boldsymbol{t}}_{\mathrm{s}}\left[\hat{\boldsymbol{u}}_{\mathrm{sc}}(\boldsymbol{N}_{\mathrm{t}}\underline{\hat{\boldsymbol{\Psi}}}_{\mathrm{t}})\right]\mathrm{d}S \tag{38}$$

$$\boldsymbol{F}_{\mathrm{t}}(\kappa,\omega)=\underline{\bar{\hat{\boldsymbol{\Psi}}}}_{\mathrm{t}}^{\mathrm{T}}\int_{\hat{\Gamma}_{\mathrm{t}\sigma}}\boldsymbol{N}_{\mathrm{t}}^{\mathrm{T}}\hat{\boldsymbol{f}}_{\mathrm{t}}\,\mathrm{d}S \tag{39}$$

where $\boldsymbol{K}_{\mathrm{t}}^{\mathrm{FE}}$ and $\boldsymbol{M}_{\mathrm{t}}^{\mathrm{FE}}$ are, respectively, the stiffness and mass matrices of the FE formulation of generic cell, and $\boldsymbol{L}$ is a matrix with differential operators that relates the strain vector to the displacement vector, while $\boldsymbol{D}$ is a matrix with constitutive coefficients that relates the stress vector to the strain vector.

In Equation (38) the stresses $\hat{\boldsymbol{t}}_{\mathrm{s}}[\hat{\boldsymbol{u}}_{\mathrm{sc}}(\boldsymbol{N}_{\mathrm{t}}\underline{\hat{\boldsymbol{\Psi}}}_{\mathrm{t}})]$ on the tunnel-soil interface are calculated by a periodic boundary element formulation with Green-Floquet functions [14].

The Craig-Bampton substructuring technique is used to analyze the responses of different track structures in the track-tunnel-soil interaction model. Thus the tunnel displacements $\widehat{\boldsymbol{u}}_{\text{tu}}$ and the track displacements $\widehat{\boldsymbol{u}}_{\text{tr}}$ can be written as follows:

$$\begin{Bmatrix} \widehat{\boldsymbol{u}}_{\text{tr}} \\ \widehat{\boldsymbol{u}}_{\text{tu}} \end{Bmatrix} = \begin{bmatrix} \widehat{\boldsymbol{\Psi}}_{\text{tr}} & \widehat{\boldsymbol{\Psi}}^{\text{s}}_{\text{tr}} \\ \mathbf{0} & \widehat{\boldsymbol{\Psi}}_{\text{tu}} \end{bmatrix} \begin{Bmatrix} \boldsymbol{\alpha}_{\text{tr}} \\ \boldsymbol{\alpha}_{\text{tu}} \end{Bmatrix} \tag{40}$$

where $\widehat{\boldsymbol{\Psi}}_{\text{tr}}$ and $\widehat{\boldsymbol{\Psi}}_{\text{tu}}$ are the modes of track on a rigid base and the modes of tunnel without track respectively, and $\widehat{\boldsymbol{\Psi}}^{s}_{\text{tr}}$ are the quasi-static transmission of tunnel modes into the track.

It can be seen that it is efficient to calculate the track displacements using the Craig-Bampton substructuring technique, because it can avoid the recomputation of soil impedance, which is only governed by the kinematics of the tunnel, when alternative track configurations are considered in the tunnel.

4.2.2. Vibration Isolation Efficiency of Floating Slab Track

The floating slab track (FST) is widely used as a measure of vibration isolation for the underground tunnels. Based on the coupled periodic FE-BE model and the Craig-Bampton substructuring technique, an FST with an isolation frequency of 7.87 Hz is studied inside the tunnel north of Chengfulu station on Beijing Metro Line 4.

The tunnels are two parallel bored tunnels with a segmented concrete lining at 13.5 m below the ground surface, having an internal radius of 2.7 m and a wall thickness 0.3 m. The concrete lining of tunnels has a Young's modulus 35000 MPa, a Poisson's ratio 0.25, a density 2500 kg/m^3 and a hysteretic material damping ratio 0.02. According to the geological map of the area, there are mainly three layers of soil: fill material, silty clay, gravel and pebble. The dynamic soil characteristics are listed in Table 5. In the numerical calculations, the tunnels are assumed to be embedded in a horizontally layered halfspace consisting of a shallow layer of fill material (thickness 2.0m) and a layer of silty clay (18.5m) on top of a homogeneous halfspace consisting of a stiffer layer with gravel and pebble.

Table 5. Dynamic soil characteristics

Layer	Soil type	D /m	C_s (m/s)	C_p /(m/s)	E /(MN/m^2)	Y	P/(kg/m^3)	B
1	Fill material	2.0	150	306	116.6	0.34	1900	0.05
2	Silty clay	18.5	230	443	289.0	0.31	2023	0.04
3	Gravel and pebble	∞	320	610	704.0	0.22	1963	0.03

The FST has the discontinuous concrete slabs with a length of 31.0 m, a width of 2.5 m and a height of 0.505 m, supported by two rows of 17 springs with spring stiffness of 6.9 MN/m each. The density of the concrete is 2450 kg/m^3 and a hysteretic material damping ratio 0.02 is considered.

4.2.3. Response to A Non-moving Harmonic Load

In the coupled periodic FE-BE formulation, the model is restricted to a single bounded generic cell due to the periodicity of tunnel and free field in the direction of tunnel axis. The tunnels can be assumed to be invariant in the axis direction, so that the period L is arbitrarily selected as 0.6 m. A FE model of generic cell of the tunnel is made here, and 8-node brick elements are adopted to discretize the tunnel structure, as is shown in Figure 37.

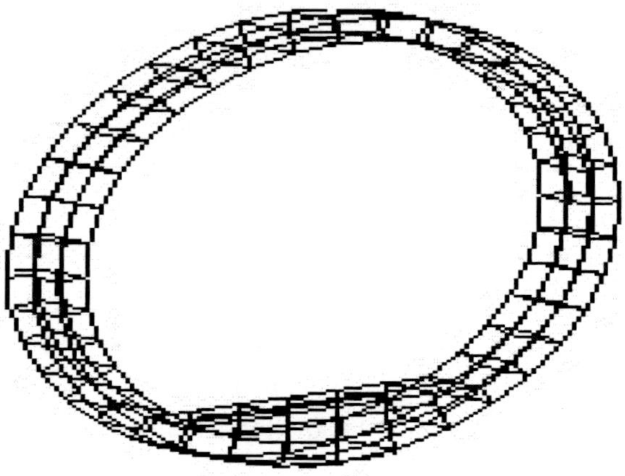

Figure 37. Finite element model of the generic cell of the tunnel.

In the following, the dynamic responses in the tunnel and free field due to a concentric harmonic vertical load on the floating slab and tunnel invert are computed in the frequency range up to 80 Hz, and the vibration isolation efficiency of the FST is analyzed.

Figure 38 shows the vertical displacements at the tunnel invert as a function of the frequency and the wavenumber for the cases without and with the FST, respectively. The maxima in the response, for the case with the FST, are found along the dispersion curve of the track on elastic foundation. In this case, the slab support is so soft that most of the energy of wave motion is transmitted along the slab, so it is why the floating slab track can reduce the vibration in the tunnel and soil.

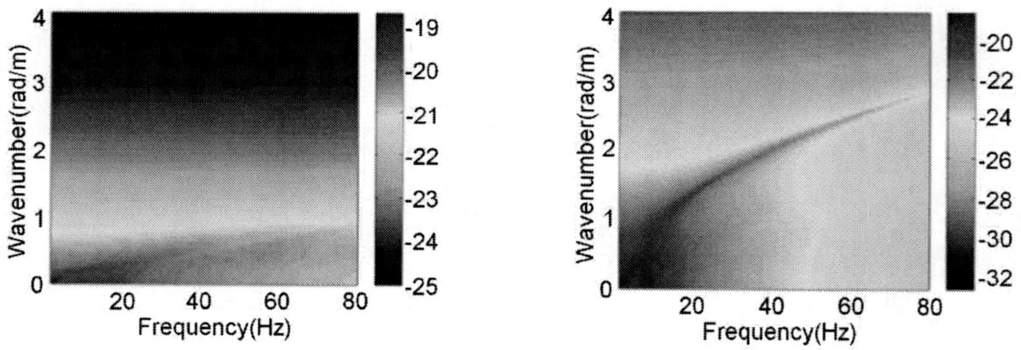

Figure 38. Vertical displacements [dB ref. 1m] at tunnel invert: without FST (left) abd with FST(right).

Compared in Figure 39 are the vertical displacements at 0 m and 20 m from the tunnel centerline on the ground surface, computed for the cases without and with the FSTs. The response in the case with the FST exhibits a peak at about 7.87 Hz, which corresponds to the uncoupling of the slab on the stiffness of the subgrade.

A useful way to evaluate the effectiveness of an FST is to take the ratio of the responses with and without the resilient bearings between the slab and the tunnel invert. Herein, the concept of insertion gain (IG) is used, which indicates the increase in vibration levels caused by the bearings. The IG is defined as the ratio of the response u^{iso} of the isolated system and the response u^{uniso} of the unisolated system, shown in Equation (41). The IG less than zero indicates the vibration isolation.

$$IG = 20\log\left(\frac{u^{iso}}{u^{uniso}}\right) \quad (\text{dB}) \tag{41}$$

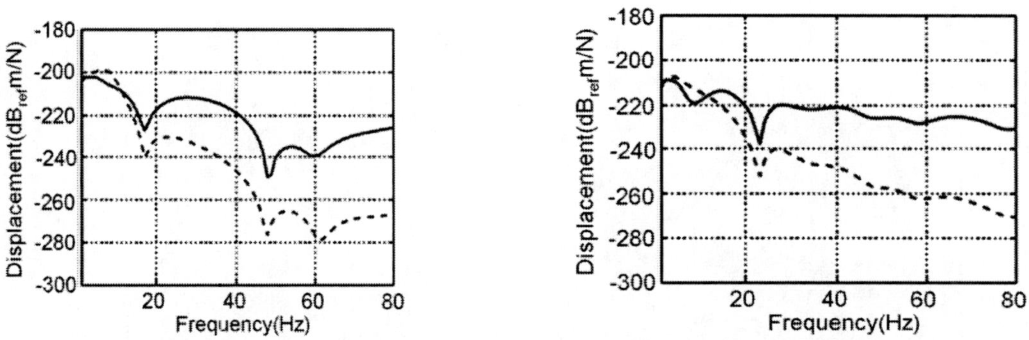

Figure 39. Vertical ground displacements vs. distances from tunnel center: 0m (left) and 20m (right).

Figure 40 shows the IGs in the frequency domain at 0 m and 20 m from the tunnel centerline. It can be seen that for a 7.87 Hz FST, the IG becomes negative at frequencies greater than 11 Hz. It means this FST can mitigate the vibration in the frequency range above 11 Hz, which is approximately equal to $\sqrt{2}$ times of 7.87 Hz.

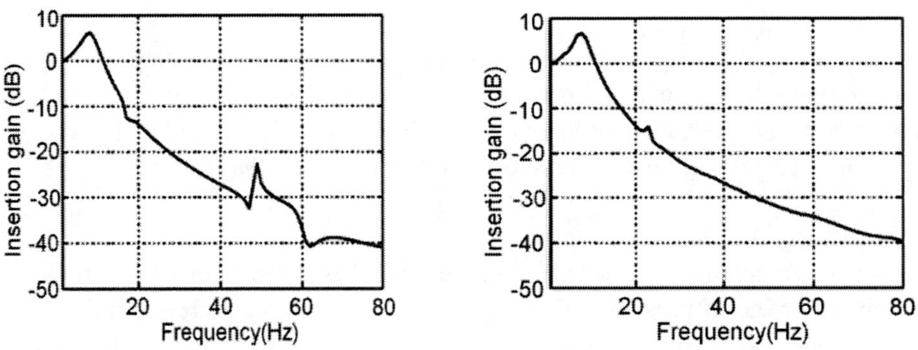

Figure 40. Insertion gains vs distances from tunnel centerline: 0m (left) and 20m (right).

Figure 41 shows the displacements of the tunnel and free field for a harmonic excitation at 30 Hz before and after the installation of a 7.87 Hz FST. The FST allows the wave energy to propagate along the slab before being transmitted to the tunnel and the soil, so there is a significant reduction in the displacements of the tunnel and the soil.

4.2.4. Conclusion

The coupled periodic FE-BE model offers full modeling flexibility on a generic cell of the tunnel and free field, which is an important advantage with respect to full 3-D or invariant tunnel models that are the two extremes of modeling complexity.

The results indicate that FST is an effective vibration mitigation measure, which can reduce the vibration in the tunnel and free field at frequencies above that of the FST itself. However, care is needed as amplification of vibration occurs around the isolation frequency of the FST.

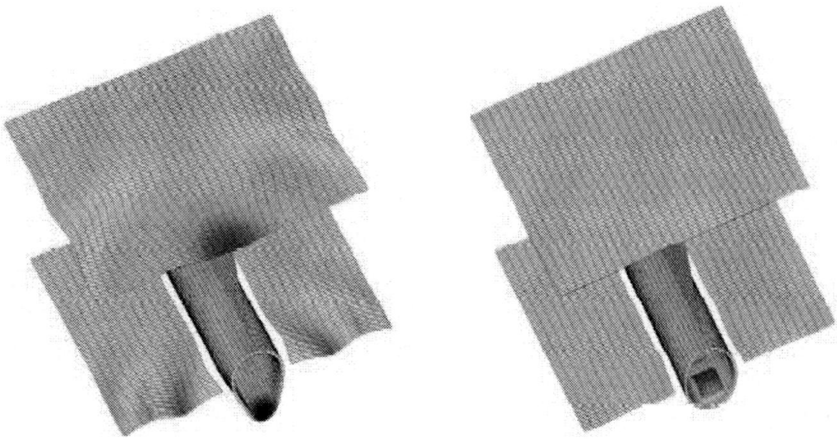

Figure 41. Displacements of tunnel and free field for a harmonic excitation at 30 Hz: without FST (left) and with FST (right).

4.3. FEM on Vibration Prediction

A case study on the vibration influence to a historic monument caused by Chengdu Metro Line 2 is presented, to predict the response in the soil and the structures with the dynamic finite element model. Both site measurements and FEM numerical calculations are performed to analysis the vibration influence caused by road traffic and subway trains.

4.3.1. Description of Historic Building

The historic monument is located in the People's Park, center of Chengdu City. It was built to remember the martyrs struggling for defending railway sovereign in the Autumn of 1911 AD. This brick-masonry structure with a height of about 30m was built in 1913, and was listed as a State Protected Historic Site in China. There are fine handwritings and relievos in the four faces. Due to its elaborate and unique art design, both structural and architectural damage should be avoided. In recent years, Chengdu Metro Line 2 is designed close to it. The distance between the monument and new metro station is about 50m.

4.3.2. Vibration Criteria for Historic Buildings

As Peak Particle Velocity (PPV) has a direct relation with the fatigue damage of buildings, it is often used to evaluate the vibration impacts on building structures. For historic buildings and memorials the criteria are from 3 to 10 mm/s PPV in different codes and guides

[15]. The Chinese National Code (GB/T 50452-2008) gives a criteria of 0.10−0.75 mm/s PPV, which is one of the strictest allowance in the world and could be used to evaluate architectural damage.

The allowable horizontal PPV at the top of the monument is 0.15−0.25 mm/s according to GB/T 50452-2008. Nevertheless, the vibration responses on the top of the monument are not easy to measure. To predict the comprehensive response under different traffics, the allowable root-mean-square (rms) velocity at the bottom of the building is more useful.

The relation between the horizontal PPV at the top and bottom can be expressed as:

$$V_{\max} = V_{\mathrm{r}} D = V_{\mathrm{r}} \sqrt{\sum_{j=1}^{n} (\gamma_j \beta_j)^2} \tag{42}$$

where $V_{\max}$ and V_{r} are PPVs at the top and foundation of the building, γ_j and β_j are participating factor and dynamic amplifying factor of the *j*th vibration mode, respectively; D is a dynamic factor of the ratio of $V_{\max}$ and V_{r}, which is about 4.5−5.5. Thus, the allowable PPV at the foundation could be calculated to be 27.27−55.56 μm/s.

The PPV is then changed into rms velocity with the crest factor. As the traffic-induced vibration could be regarded as a stationary narrow-band Gaussian process, whose mean value approaches 0 and variance is σ^2, the rms velocity of a group of sample data could be given as:

$$\mathrm{rms} = \sqrt{\frac{\sum_{i}^{n} x^2}{n}} = \sqrt{E\left(x_i^2\right)} = \sqrt{\sigma^2 + \mu^2} = \sigma \tag{43}$$

That is to say, rms is just the standard deviation σ of this signal. If the response obeys the Gaussian distribution absolutely, the crest factor larger than 3 could give a probability over 99%. Suppose metro operates 10 hours a day and the sampling frequency is 1280, the values of crest factor data larger than 4 and 5 are about 4600 and 41, respectively. For those important historic buildings, the former one is too large but the latter is acceptable. In order to define a conservative allowable rms velocity, the crest factor is defined as 5. Therefore, the maximum rms velocity in horizontal direction at the foot of the building (P1 in Figure 42) should be 5.45−11.11 μm/s, which is regarded as the vibration criteria in this case study.

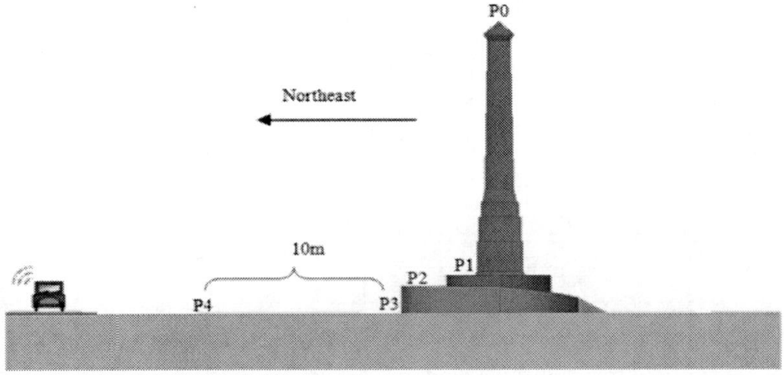

Figure 42. Schematic of measurement points.

4.3.3. Characteristics of Tunnel and Soil

According to the geologic survey on Chengdu Subway Line 2, the strata are simplified to four layers in this area. The soil characteristics are listed in Table 6. E_d and v_d are dynamic elastic modulus and Poisson's ratio, respectively. The station near the monument is 138 m long and the embedded depth of the track is about 13 m. The RC lining of the subway has a Young's modulus 35000 MPa, a Poisson' ratio 0.25, a density 2500 kg/m^3 and a hysteretic material damping ratio 0.02.

Table 6. Dynamic soil characteristics

Layer	Soil type	H /m	E_d /MPa	v_d	ρ /(kg/m^3)
1	Miscellaneous fill	5.5	150	0.38	1900
2	Loose gravel	5.7	425	0.28	1900
3	Dense gravel	16.4	893	0.27	2200
4	weathered mud stone	∞	1000	0.35	2350

4.3.4. Dynamic Train-track Models

To obtain dynamic train loads, numerical models are created in DARTS_NL developed at TU Delft, which is a specialised FE software for analysis of dynamic railway vehicle-track interaction. It has been successfully used for various railway applications [16].

The dynamic train-track model is shown in Figure 43. In this model, each vehicle is modeled by a mass-spring system that consists of a car-body, two bogies and four wheels, all of which are rigid bodies and connected to each other with the primary and secondary suspensions. The track is modeled of three hard and three soft layers. The hard layers are modeled as Timoshenko beams with both bending stiffness and shear stiffness, while the elastic layers are represented by spring-damper sets. The wheel/rail contact forces are modeled by the non-linear Hertzian contact spring [17] as:

$$K_{\mathrm{H}} = \sqrt[3]{\frac{6E^2P\sqrt{R_{\mathrm{w}}R_{\mathrm{r}}}}{4(1-\nu^2)^2}} \tag{44}$$

where P is the static wheel load, R_w and R_r are the radii of the wheel and rail profile, E is the Young's modulus of the wheel and rail material and v is the Poisson's ratio.

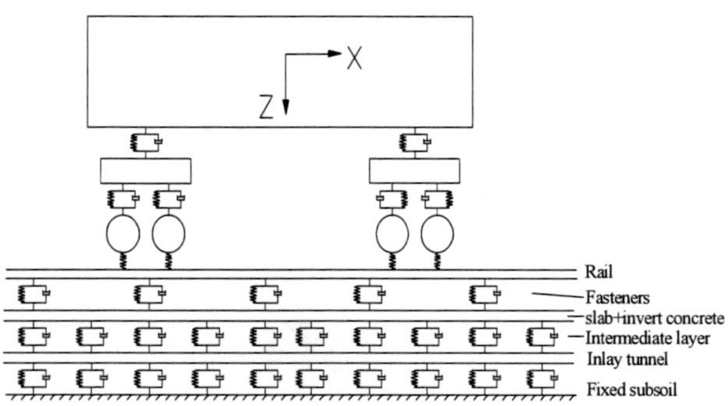

Figure 43. Schematic of train-track dynamic model.

4.3.5. Coupled Tunnel-soil-structure FE Model

To predict vibration response on the monument, the FE software MIDAS-GTS is used to build a 3-D FE model of tunnel-soil-structure, which is subjected to loads induced by trains with different kinds of fasteners and at different speeds. Soils are divided into 4 layers, with the material properties shown in Table 6. Special viscous-elastic artificial boundary is used to eliminate the reflection of propagation waves. The 3-D FE model is shown in Figure 44.

4.3.6. Numerical Results

The analysis is based on the case when two trains come across in the station at the same time, considering that may induce larger vibration than only one train.

Three kinds of fasteners are studied, which are widely used in metro track design: (1) $DTVI_2$ fastener, a kind of high resilience directly-fixed fastener; (2) shearing vibration absorption fastener which possesses double stiffness and good vibration reduction capacity; and (3) Pandrol Vanguard fastener, which was initially developed for British Rail. All the fasteners have a supporting interval of 0.6 m. The type of rail is T60. The stiffness of the rail pads of $DTVI_2$ fastener is 78 MN/m, the track damper fastening model III is chosen as a typical shearing vibration absorption fastener, whose equivalent stiffness is about 60 MN/m, and the stiffness of Vanguard fastener is 21 MN/m.

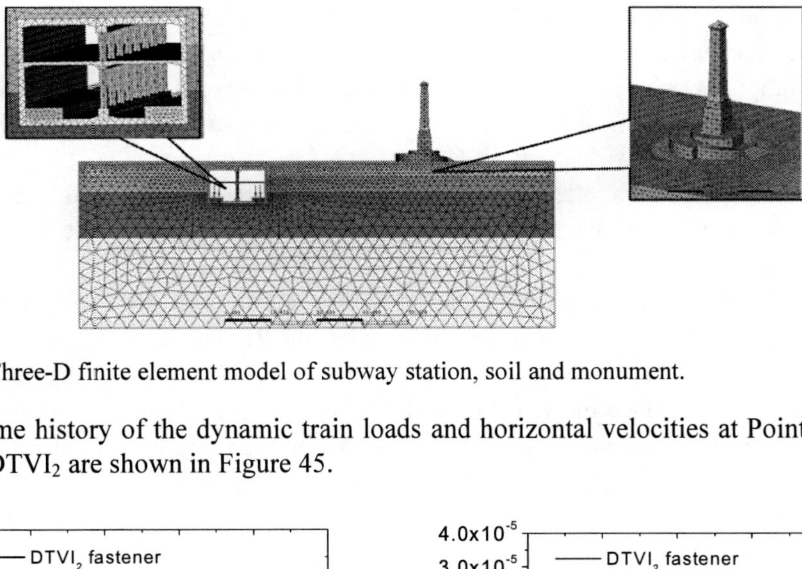

Figure 44. Three-D finite element model of subway station, soil and monument.

The time history of the dynamic train loads and horizontal velocities at Point 1 with the fastening $DTVI_2$ are shown in Figure 45.

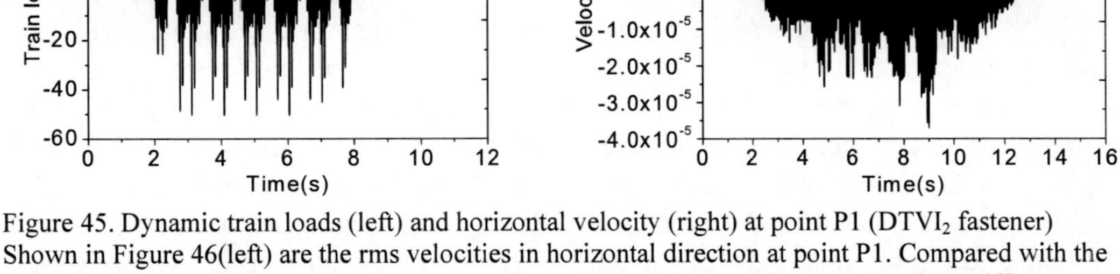

Figure 45. Dynamic train loads (left) and horizontal velocity (right) at point P1 ($DTVI_2$ fastener)

Shown in Figure 46(left) are the rms velocities in horizontal direction at point P1. Compared with the allowable rms velocity, vibration induced only by trains could be satisfactory when low-stiffness fasteners are used.

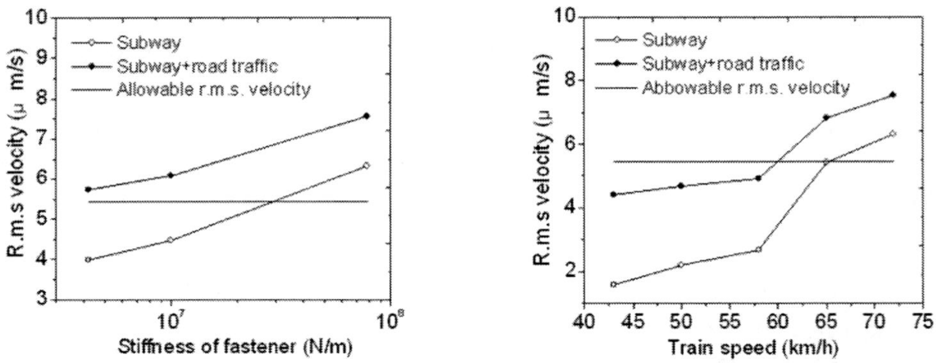

Figure 46. Horizontal rms velocity at point P1 vsfastener stiffness (left) and train speed (right).

As the square of rms velocity stands for the vibration energy, the following expression is used to calculate total environmental vibration by both road traffic and metro trains.

$$v_{\text{rms,t}} = \sqrt{v_{\text{rms,r}}^2 + v_{\text{rms,m}}^2} \tag{45}$$

where, v_{rms} is the rms velocity with the worst operation condition. Subscript t, r and m denote, respectively, total environmental vibration, road traffic-induced vibration and metro trains-induced vibration. By comparing the total response (subway+road traffic) and the allowable rms velocity, one can found for all kinds of fasteners the total vibration exceeds the limit, which is due to the road traffic.

Although the vibration attenuates a lot then track damper fastening model III and Vanguard fastener are used, the superposition rms velocities still exceed the low limit of the allowable value.

Another important factor affecting the vibration is the train speed. Figure 46 (right) shows the rms velocities in horizontal direction at point P1 with five different train speeds from 43 km/h to 72 km/h, in the case of $DTVI_2$ fastener. The results show that lowering train speed can effectively reduce the vibration: the rms velocities decrease sharply with the train speed from 65 km/h to 58 km/h, and when the train speed is lower than 58 km/h, the vibration is lower than allowable value even when the road traffic is considered.

CONCLUSION

In this chapter, the simulation of train loads, based on the analytical vehicle-track coupling system, is achieved. And also, the propagation of metro-induced ground vibration, especially in low frequency, is studied both by an experiment in the Laboratory of track vibration abatement and by a 3-D track-tunnel-soil FE model simulation. As an extension research, a prediction method of environmental vibration from metro trains is studied by the coupled periodic FE-BE model which offers full modeling flexibility on the tunnel and free field compared to traditional 3-D FE model. Based on the research, two case-studies of vibration effect prediction from metro trains, one for a laboratory equipped with high-precision transmission electron microscopy nearby a metro line in Beijing, another for a historic monument close to a metro line in Chengdu, are presented.

As a conlusion, in practice, although it is always difficult to evaluate and predict the environmental vibration induced by underground trains, a better prediction could be made by reasonable combination of theoretic analysis, numerical simulation, in-situ tests and empirical estimation.

REFERENCES

[1] Zhai W.M. (2002). *Vehicle-Track Coupling Dynamics(Second Edition)*, Beijing: China Railway Publishing House.(in Chinese)

[2] Wang L. (1988). *The Theory of Track Structure Random Vibration and Its Application in the Vibration Mitigation of Track Structure*. Ph.D Dissertion, Beijing: China Academy of Railway Sciences.

[3] Liu W.N. (1983). *Study on the Dynamic Responses of the Loess Tunnel under the Train Loads-In site Test and Numerical Analysis*. M.D Dissertion Lan Zhou Railway Collage.

[4] Zhai W.M. Cai C.B., Wang Kaiyun. (2005). Mechanism and Model High-speed Rail-track-bridge Dynamic Interaction. *China Civil Engineering Journal*, 38(11): 132−137.

[5] Jones CJC, Thompson DJ, Petyt M. (2002). A model for ground vibration from railway tunnels. *Transport, Proceedings of the Institution of Civil Engineers*, 153(2): 121−129.

[6] Lombaert G, Degrande G, Clouteau D. (2000). Numerical modelling of free field traffic induced vibrations. *Soil Dynamics and Earthquake Engineering*, 19(7): 473−488.

[7] Forrest J A, Hunt H E M. (2006). A three-dimensional tunnel model for calculation of train-induced ground vibration. *Journal of Sound and Vibration*, 294: 678−705.

[8] Pan C.H., Xie Z.G., (1990). Measurement and analysis of vibrations caused by passing trains in subway running tunnel. *China Civil Engineering Journal*, 23(2): 21−28. (in Chinese).

[9] Ruker W., (1977). Measurement and Evaluation of Random Vibrations. *Proceedings of DMSR 77*, Karlsruhe, Germany, p. 407−421.

[10] Degrande G., Schevenels, M., Chatterjee, P., Van de Velde, W., Hölscher, P., Hopman, V., Wang, A., Dadkah, N., (2006). Vibrations due to a test train at variable speeds in a deep bored tunnel embedded in London clay. *Journal of Sound and Vibration*, 293(3−5): 626−644.

[11] Liu W.N., Zhang H.R., Li, R.D., (2005). *Report on Environmental Vibration Measurements in Beijing Metro Line 1*. Report 2005-12, Beijing Jiaotong University, Beijing, China. (in Chinese).

[12] Degrande G., Liu W.N. (2002). *Joint research report of the bilateral project BIL 98/9*, Traffic induced vibrations in buildings.

[13] Ding D.Y., Liu W.N., Degrande G., Lombaert G., Gupta S. (2010). Prediction of vibrations induced by trains on line 8 of Beijing metro. *Journal of Zhejiang University-Science A*, 11 (4): 280−293.

[14] Clouteau D., Aubry D., Elhabre, M.L., et al. (1999). *Periodic and stochastic BEM for large structures embedded in an elastic half-space. Mathematical Aspects of Boundary Element Methods*. London: C.R.C Press.

[15] Ma M., Liu W.N., Ding D.Y. and et al. (2009). Vibration impacts on adjacent heritage buildings induced by metro trains. *4th International Symposium on Environmental Vibration*, Beijing, Science Press: p. 394–399.

[16] Markine V.L., Zwarthoed J.M. and Esveld C. (2001). Use Of Numerical Optimisation In Railway Slab Track Design. *Proceedings of the 3rd ASMO UK/ISSMO conference*, North Yorkshire, UK.

[17] Grassie SL. (1984). Dynamic modelling of railway track and wheelsets. *Proceedings of the 2nd International Conference on Recent Advances in Structural Dynamics*.

In: Traffic Induced Environmental Vibrations and Controls ISBN: 978-1-62618-240-0
Editors: H. Xia and R. Calçada

Chapter 8

NUMERICAL EVALUATION ON TRAIN-INDUCED GROUND VIBRATION AROUND HIGH-SPEED RAILWAY VIADUCTS AND ITS REDUCTION METHODS

***Xingwen He*[1*]*, Mitsuo Kawatani*[2] *and Toshiro Hayashikawa*[3]
[1,3]Faculty of Engineering, Hokkaido University, Sapporo, Japan
[2]Graduate School of Engineering, Kobe University, Kobe, Japan

ABSTRACT

In this chapter, taking advantage of three-dimensional dynamic analysis, a numerical approach to simulate the ground vibration around high-speed railway viaducts is established. In this approach, the dynamic interactions between the train and the bridge as well as between the foundation and the ground are considered. To simplify the problem, the entire interaction system is divided into two subsystems: train-bridge interaction and foundation-ground interaction. In the stage of the train-bridge interaction subsystem, the dynamic responses of viaducts are calculated using a developed computer program to obtain dynamic reaction forces at the pier bottoms. Then, applying those reaction forces as input excitation to the foundation-ground interaction subsystem, the ground vibration around the viaducts is simulated using a general-purpose program. The numerical results are compared with experimental values to confirm the validity of the established numerical approach. Then, based on the dynamic characteristics of the bridge vibration, countermeasures to reduce the excessive vibration of both the bridge and the ground are designed and numerically evaluated.

1. INTRODUCTION

Japanese high-speed railway system, so-called the Shinkansen, serves a vital role in the transportation network that connects its major cities. It has contributed importantly to the economic and social development of Japanese society since 1964. To this day, the

* E-mail address: xingwen_he@eng.hokudai.ac.jp.

transportation capacities of the high-speed railway system have been continually augmented through the use of new train models and improved equipment along with the increased number of trains.

The high-speed railway's main lines usually pass directly over densely populated urban areas, where the railway structure mainly comprises elevated bridges of reinforced concrete in the form of portal rigid frame. The bridge vibration caused by running trains is propagated to the ambient ground via footing and pile structures, thereby causing some environmental problems related to the ground vibration around the viaducts. Those vibrations can influence precision instruments in hospitals and laboratories, or people who are studying, working or resting in schools, offices, hospitals and residences, etc. Along with the further urbanization and development of more rapid transport facilities, there is rising public concern about the environmental problems in modern Japan [1]. Its vibration regulation law legislated in 1976 was the first one concerning the environmental vibration problems in the world. Almost concurrently, recommendations for countermeasures against vibration problems of the Shinkansen railway were also proposed, in which a vibration acceleration level limit was specified to allay the environmental impacts of train-induced vibration on important facilities surrounding the main lines.

Although the importance and urgency of environmental problems have been recognized, the development and propagation mechanism of ground vibration caused particularly by running vehicles on viaducts remains unclear because of its complicated nature. In Japan, environmental vibration problems caused by trains running along viaducts are mainly investigated by Railway Technical Research Institute (RTRI). Without a clear grasp of the ground vibration mechanism through theoretical studies, these problems are traditionally evaluated and predicted based on field test data, e.g. by Yoshioka and Ashiya [2]. The efficiency of such a process is limited to particular cases. For more general cases, essential information and reliable evaluation on ground vibrations are necessary to perform accurate predictions and develop effective countermeasures. For that purpose, a corresponding numerical approach to simulate the train-induced ground vibration problems is anticipated.

In recent years, effort has been devoted to studies of ground vibrations induced by trains moving on the ground surface. Fujikake [3] proposed a predictive method for vibration levels of the surrounding environment. Takemiya [4] conducted a simulation of track-ground vibrations caused by a high-speed train for predicting train-track and nearby ground-borne vibrations. Yang et al. [5] also examined train-induced wave propagation in layered soils using a 2.5-dimensional (2.5D) finite/infinite element approach. On the other hand, the ground vibration caused by trains moving over viaducts is more difficult to elucidate because of its complicated nature: vibrations are transmitted to the ground via piers, footings and piles. Recently, Xia et al. [6] evaluated the vibration-related effects of light-rail train-viaduct system on the surrounding environment using a two-dimensional (2D) interaction model of a train-bridge system for obtaining the dynamic loads of moving trains on bridge piers and a 2D dynamic model of pier-foundation-ground system for analyzing vibration responses of the ground. Wu et al. [7, 8] attempted to establish a semi-analytical approach to deal with ground vibration induced by trains moving over viaducts. In Japan, Hara et al. [9] attempted to clarify the ground vibration around Shinkansen viaducts by both experiments and numerical procedure. However in their analyses, the wheel load of the train is treated as simple equivalent moving force based on the measured results. Such approach not only cannot directly take consideration of the interaction between the bridge and train, but is incapable to set the wheel loads without experimental results. Based on the field test data of ground vibration around Shinkansen viaducts, Yoshida and Seki [10] indicated the importance of considering the coupled vibration of the train-bridge interaction system when evaluating the train-induced ground vibration.

In this study, taking advantage of three-dimensional (3D) dynamic analysis, an approach to simulate the ground vibration around Shinkansen viaducts is established. In this approach, the dynamic interactions between the train and the bridge as well as between the foundation and the ground are considered. Currently, it is still difficult to model the entire train-bridge-ground interaction system as a whole, because of not only the complexities of the problem but also the limit of the computational capacity. Therefore, to simplify the problem in this study, the entire interaction system is divided into two subsystems: the train-bridge interaction and the foundation-ground interaction. In the stage of the train-bridge interaction problem, the computer program to simulate the traffic-induced bridge vibration is developed, with which the dynamic responses of viaducts are calculated to obtain the dynamic reaction forces at the pier bottoms. Then, applying those reaction forces as input excitation forces in the foundation-ground interaction problem, the ground vibration around the viaducts is simulated and evaluated using a general-purpose program. The numerical results are compared with experimental values to confirm the validity of the developed numerical approach.

Based on the dynamic characteristics of the traffic-induced vibration of Shinkansen viaducts, the reason why excessive vibration occurs in specific parts is analyzed. Then, the countermeasures to reduce the excessive bridge vibration, which leads to ground vibration problems, are designed. The reduction effects of these proposed countermeasures are then simulated and evaluated, using the established numerical approach.

The numerical approach developed in this study can be used as a tool to perform the simulation, evaluation and prediction of the traffic-induced bridge and ground vibration. Furthermore, it also can be used to investigate in advance the effectiveness of designed countermeasures against bridge and ground vibration by reinforcing the bridge structure, improving the ground condition, or employing other means.

2. Theories for Bridge and Ground Vibration Analysis

2.1. Overview of the Proposed Numerical Approach

For the dynamic interaction problems between the train, bridge and ground, currently it is extremely complicated to handle the problem by treating all these factors as an integrated system considering their exact dynamic interaction. Therefore in this study, to simulate the bullet train-induced bridge vibration and related ground vibration problem of the bridge-train interaction system, endeavors are devoted to solve the problems by dividing the whole interaction system into two subsystems: the bridge-train interaction and the foundation-ground interaction. The two subsystems are connected with idealized ground springs obtained from the elastic effect of both the ground and foundation structure.

The bridge vibration caused by the running trains is propagated to the ambient ground via pier and foundation structures of the viaducts. To perform ground vibration analysis around the viaducts, the dynamic reaction forces at the pier bottoms are needed as input external excitations to the foundation-ground interaction system. Therefore, it is first necessary to simulate the dynamic responses of the viaducts to obtain those dynamic reaction forces. Undoubtedly, the more accurate simulation of the dynamic bridge responses will lead to the more reliable evaluation of the ground vibration.

In the first stage, the numerical approach to simulate the train-induced bridge vibration is established considering the train-bridge interaction, and using Fortran Language the computer program is developed [11]. The finite element method (FEM) is applied to model the bridge structures. The viaducts including the track structure are modeled as 3D beam elements. For the linear response analysis, simultaneous differential equations of the bridge are simplified using modal analysis technique. The elastic effect of the ground at the pier bottoms and the connection effect of the sleepers and ballast between the track and the deck slab are modeled with double nodes connected by spring elements. A 3D bullet train modeled as a dynamic system that can appropriately express its lateral, vertical and rotational motions is developed for the analyses. Newmark's β step-by-step integration method is applied to solve the differential equations of the coupled system. Then, the dynamic reaction forces at the pier bottoms are calculated based on the bridge dynamic responses using the influence value matrix. This section introduces the detailed theories used in this research to solve the bridge-train interaction problem.

The dynamic reaction forces at the pier bottoms obtained in train-bridge interaction analysis are applied as input excitation forces in the foundation-ground interaction problem. Modeling the foundation structure including footings and piles as finite elements and the ground as thin layer elements, the ground vibration around the viaducts is simulated and evaluated [12], using a general-purpose program named SASSI2000 [13, 14]. A brief introduction to the analytical theories of foundation-ground interaction problem employed in SASSI2000 software is described at the end of this section.

2.2. FEM for Dynamic Response Analysis

2.2.1. FEM Used to Establish the Bridge Model

The FEM used in this research is the typical theory that can be found in various references such as by Clough and Penzien [15], Washidu et al. [16, 17], Bathe [18], and Weaver and Johnston [19]. The whole bridge structure is modeled as 3D beam elements. The theories to obtain the stiffness and mass matrices of the beam elements, the transformation between local and global coordinates, and matrix condensation, etc., can be found in any FEM literatures , thus are not explained in detail here.

2.2.2. Calculation of Reaction Force at Pier Bottoms

In this study, the reaction forces at pier bottoms are calculated using the influence value matrix of the reaction force. The theoretical description is briefly represented as follows.

Assuming $\boldsymbol{f}_w$ is the external force vector acting on the free DOFs $\boldsymbol{w}$ of the bridge model, and $\boldsymbol{f}_R$ is the reaction force vector acting on the constrained ones, then the static equilibrium equation of the structure can be expressed as follows:

$$\begin{bmatrix} \mathbf{f}_w \\ \mathbf{f}_R \end{bmatrix} = \begin{bmatrix} \mathbf{K}_{11} & \mathbf{K}_{12} \\ \mathbf{K}_{21} & \mathbf{K}_{22} \end{bmatrix} \begin{bmatrix} \mathbf{w} \\ \mathbf{0} \end{bmatrix} \tag{1}$$

From the above equation, the following relationship between $\boldsymbol{f}_w$ and $\boldsymbol{f}_R$ can be derived.

$$\mathbf{f}_R = \mathbf{K}_R \mathbf{f}_w \tag{2}$$

where, $\mathbf{K}_R = \mathbf{K}_{21}\mathbf{K}_{11}^{-1}$ is called the influence value matrix of the reaction force.

For the dynamic problem, the dynamic external force can be treated as pseudo-static one in each integral step. The reaction force vector $\boldsymbol{f}_R$ can be further described as follows.

$$\mathbf{f}_R(t) = \mathbf{K}_R(\mathbf{P}_{vst} + \mathbf{P}_{vdy}) + \mathbf{K}_R\mathbf{P}_{sdy} \tag{3}$$

where, vectors $\mathbf{P}_{vst}$, $\mathbf{P}_{vdy}$ and $\mathbf{P}_{sdy}$ respectively denote the static and dynamic components of the wheel loads and the inertia forces of the bridge nodes.

2.3. Eigenvalue Analysis

In this study, the eigenvalue analysis is performed using QR method [20, 21]. This method was first published by J. G. F. Francis [20] and it has since been the subject of intense investigation. The QR method is quite complex in both its theory and application. The detailed description of the method can be found in the references. Also, there are many other new methods for eigenvalue analysis nowadays, which can be found in various literatures.

2.4. Formulation of Train-Bridge Interaction System

The train-bridge interaction system is formulated in this part, based on which the computer programs are developed. Cars of the train are idealized as 3D sprung-mass vibration systems, assuming that the car-body and the bogies are rigid bodies and that they are connected three-dimensionally by scalar spring and damper elements. The viaducts together with the rail structure are modeled as 3D beam elements and then formulated by FEM. Then the differential equations of train-bridge interaction system are derived as follows.

2.4.1. Formulation of the Train Motion

Figure 1 shows a train car modeled as 3D sprung-mass system. The variants employed in the car model are shown in Table 1. The notations of the train properties are indicated in Table 2. The definition of dimensions of the car is shown in Table 3. The formulation of the train motion can be obtained based on D'Alembert's Principle as follows.

a. Differential Equations of the Car-Body

y_{j1}: Lateral translation of the car-body

$$m_1\ddot{y}_{j1} - \sum_{l=1}^{2}\sum_{m=1}^{2}(-1)^m v_{jylm}(t) = 0 \tag{4}$$

z_{j1}: Bouncing of the car-body

$$m_1\ddot{z}_{j1} + \sum_{l=1}^{2}\sum_{m=1}^{2} v_{jzlm}(t) = 0 \tag{5}$$

θ_{jx1}: Rolling of the car-body

$$I_{x1}\ddot{\theta}_{jx1} - \sum_{l=1}^{2}\sum_{m=1}^{2}(-1)^m \lambda_{y3} v_{jzlm}(t) - \sum_{l=1}^{2}\sum_{m=1}^{2}(-1)^m \lambda_{z1} v_{jylm}(t) = 0 \tag{6}$$

θ_{jy1}: Pitching of the car-body

$$I_{y1}\ddot{\theta}_{jy1} + \sum_{l=1}^{2}\sum_{m=1}^{2}(-1)^l \lambda_{x1} v_{jzlm}(t) = 0 \tag{7}$$

θ_{jz1}: Yawing of the car-body

$$I_{z1}\ddot{\theta}_{jz1} + \sum_{l=1}^{2}\sum_{m=1}^{2}(-1)^{l+m} \lambda_{x1} v_{jylm}(t) + \sum_{l=1}^{2}\sum_{m=1}^{2}(-1)^m \lambda_{y4} v_{jxlm}(t) = 0 \tag{8}$$

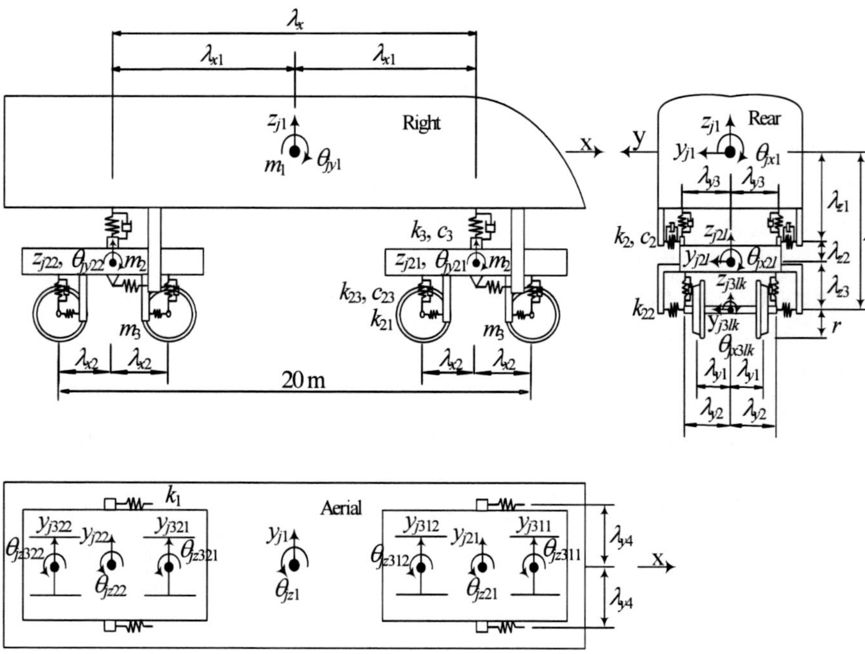

Figure 1. 3D train car model.

Table 1. Variants employed in the train model

Definition	*Notation*	*Definition*	*Notation*
Lateral translation of car-body	y_{jl}	*Axle tramp of bogie*	θ_{jx2l}
Sway of bogie	y_{j2l}	*Rolling of wheelset*	θ_{jx3lk}
Lateral displacement of wheelset	y_{j3lk}	*Pitching of car-body*	θ_{jyl}
Bouncing of car-body	z_{jl}	*Windup of bogie*	θ_{jy2l}
Parallel hop of bogie	z_{j2l}	*Yawing of car-body*	θ_{jzl}
Vertical displacement of wheelset	z_{j3lk}	*Yawing of bogie*	θ_{jz2l}
Rolling of car-body	θ_{jxl}	*Yawing of wheelset*	θ_{jz3lk}

Note: j: j^{th} car of the train; l = 1, 2: front and rear bogies; k = 1, 2: front and rear wheelsets.

Table 2. Definition of properties of the train model

Definition		*Notation*
Weight	*car-body*	W_1
	bogie	W_2
	wheelset	W_3
Mass moment of inertia	*car-body*	I_{x1}, I_{y1}, I_{z1}
	bogie	I_{x2}, I_{y2}, I_{z2}
	wheelset	I_{x3}, I_{z3}
Spring constant	*upper*	k_1, k_2, k_3
	lower	k_{21}, k_{22}, k_{23}
Damping coefficient	lateral upper	c_2
	vertical upper	c_3
	vertical lower	c_{23}

Table 3. Definition of dimensions of the train model

1/2 length of car-body in x-direction	λ_c
Distance of centers of bogies in x-direction	λ_x
1/2 distance of centers of bogies in x-direction	λ_{x1}
1/2 distance of axes in x-direction	λ_{x2}
1/2 width of track gauge	λ_{y1}
1/2 distance of vertical lower springs in y-direction	λ_{y2}
1/2 distance of vertical upper springs in y-direction	λ_{y3}
1/2 distance of longitudinal upper springs in y-direction	λ_{y4}
Distance from centroid of body to axis in z-direction	λ_z
Distance from centroid of body to lateral upper spring in z-direction	λ_{z1}
Distance from centroid of bogie to lateral upper spring in z-direction	λ_{z2}
Distance from centroid of bogie to lateral lower spring in z-direction	λ_{z3}
Radius of wheel	r

where,

$$v_{jxlm}(t) = k_1\left\{(-1)^m \lambda_{y4}(\theta_{jz1} - \theta_{jz2l})\right\} \tag{9}$$

$$\begin{aligned} v_{jylm}(t) = k_2\left\{-(-1)^m y_{j1} - (-1)^m \lambda_{z1}\theta_{jx1} + (-1)^{l+m}\lambda_{x1}\theta_{jz1} + (-1)^m y_{j2l} - (-1)^m \lambda_{z2}\theta_{jx2l}\right\} \\ + c_2\left\{-(-1)^m \dot{y}_{j1} - (-1)^m \lambda_{z1}\dot{\theta}_{jx1} + (-1)^{l+m}\lambda_{x1}\dot{\theta}_{jz1} + (-1)^m \dot{y}_{j2l} - (-1)^m \lambda_{z2}\dot{\theta}_{jx2l}\right\} \end{aligned} \tag{10}$$

$$\begin{aligned} v_{jzlm}(t) = k_3\left\{z_{j1} + (-1)^l \lambda_{x1}\theta_{jy1} - (-1)^m \lambda_{y3}\theta_{jx1} - z_{j2l} + (-1)^m \lambda_{y3}\theta_{jx2l}\right\} \\ + c_3\left\{\dot{z}_{j1} + (-1)^l \lambda_{x1}\dot{\theta}_{jy1} - (-1)^m \lambda_{y3}\dot{\theta}_{jx1} - \dot{z}_{j2l} + (-1)^m \lambda_{y3}\dot{\theta}_{jx2l}\right\} \end{aligned} \tag{11}$$

where, the subscripts relative to the motion of the car-body are described as: l=1, 2 respectively indicate the front and rear bogies; m=1, 2 respectively indicate the left and right sides of the train. j is the sequence number of the car. $v_{jxlm}(t)$, $v_{jylm}(t)$ and $v_{jzlm}(t)$ denote the forces caused by the extension of the upper springs in relative directions, respectively.

b. Differential Equations of the Bogies

y_{j2l}: Sway of the bogie

$$m_2\ddot{y}_{j2l} + \sum_{m=1}^{2}(-1)^m v_{jylm}(t) - \sum_{k=1}^{2}\sum_{m=1}^{2}(-1)^m v_{jylkm}(t) = 0 \tag{12}$$

z_{j2l} : Parallel hop of the bogie

$$m_2\ddot{z}_{j2l} - \sum_{m=1}^{2} v_{jzlm}(t) + \sum_{k=1}^{2}\sum_{m=1}^{2} v_{jzlkm}(t) = 0 \tag{13}$$

θ_{jx2l} : Axle tramp of the bogie

$$\begin{aligned} I_{x2}\ddot{\theta}_{jx2l} - \sum_{m=1}^{2}(-1)^m \lambda_{z2} v_{jylm}(t) + \sum_{m=1}^{2}(-1)^m \lambda_{y3} v_{jzlm}(t) \\ - \sum_{k=1}^{2}\sum_{m=1}^{2}(-1)^m \lambda_{z3} v_{jylkm}(t) - \sum_{k=1}^{2}\sum_{m=1}^{2}(-1)^m \lambda_{y2} v_{jzlkm}(t) = 0 \end{aligned} \tag{14}$$

θ_{jy2l} : Windup motion of the bogie

$$I_{y2}\ddot{\theta}_{jy2l} + \sum_{k=1}^{2}\sum_{m=1}^{2}(-1)^k \lambda_{x2} v_{jzlkm}(t) = 0 \tag{15}$$

θ_{jz2l} : Yawing of the bogie

$$I_{z2}\ddot{\theta}_{jz2l}-\sum_{m=1}^{2}(-1)^{m}\lambda_{y4}v_{jxlm}(t)+\sum_{k=1}^{2}\sum_{m=1}^{2}(-1)^{k+m}\lambda_{y2}v_{jxlkm}(t)+\sum_{k=1}^{2}\sum_{m=1}^{2}(-1)^{k+m}\lambda_{x2}v_{jylkm}(t)=0 \tag{16}$$

where,

$$v_{jxlkm}(t)=k_{21}\left\{(-1)^{k+m}\lambda_{y2}(\theta_{jz2l}-\theta_{jz3lk})\right\} \tag{17}$$

$$v_{jylkm}(t)=k_{22}\left\{-(-1)^{m}y_{j2l}-(-1)^{m}\lambda_{z3}\theta_{jx2l}+(-1)^{k+m}\lambda_{x2}\theta_{jz2l}+(-1)^{m}y_{j3lk}\right\} \tag{18}$$

$$\begin{aligned}v_{jzlkm}(t)=&k_{23}\left\{z_{j2l}-(-1)^{m}\lambda_{y2}\theta_{jx2l}+(-1)^{k}\lambda_{x2}\theta_{jy2l}-z_{j3lk}+(-1)^{m}\lambda_{y2}\theta_{jx3lk}\right\}\\&+c_{23}\left\{\dot{z}_{j2l}-(-1)^{m}\lambda_{y2}\dot{\theta}_{jx2l}+(-1)^{k}\lambda_{x2}\dot{\theta}_{jy2l}-\dot{z}_{j3lk}+(-1)^{m}\lambda_{y2}\dot{\theta}_{jx3lk}\right\}\end{aligned} \tag{19}$$

where, the subscripts relative to the motion of the bogies are described as: k=1, 2 respectively indicate the front and rear axles of the rear bogie, m=1, 2 respectively indicate the left and right sides of the bogie. $v_{jxlkm}(t)$, $v_{jylkm}(t)$ and $v_{jzlkm}(t)$ denote the forces caused by the extension of the lower springs in relative directions, respectively.

c) Differential Equations of the Wheelsets

y_{j3lk}: Lateral displacement of the wheelset

$$m_3\ddot{y}_{j3lk}+\sum_{m=1}^{2}(-1)^{m}v_{jylkm}(t)=-\sum_{m=1}^{2}P_{jylkm}(t) \tag{20}$$

z_{j3lk} : Vertical displacement of the wheelset

$$m_3\ddot{z}_{j3lk}-\sum_{m=1}^{2}v_{jzlkm}(t)=-\sum_{m=1}^{2}P_{jzlkm}(t) \tag{21}$$

θ_{jx3lk} : Rolling of the wheelset

$$I_{x3}\ddot{\theta}_{jx3lk}+\sum_{m=1}^{2}(-1)^{m}\lambda_{y2}v_{jzlkm}(t)=-r\sum_{m=1}^{2}P_{jylkm}(t)+(-1)^{m}\lambda_{y1}\sum_{m=1}^{2}P_{jzlkm}(t) \tag{22}$$

θ_{jz3lk} : Yawing of the wheelset

$$I_{z3}\ddot{\theta}_{jz3lk}-\sum_{m=1}^{2}(-1)^{k+m}\lambda_{y2}v_{jxlkm}(t)=-(-1)^{m}\sum_{m=1}^{2}P_{jxlkm}(t) \tag{23}$$

where, $P_{jxlkm}(t)$, $P_{jylkm}(t)$ and $P_{jzlkm}(t)$ respectively represent the dynamic wheel loads acting on the structure in horizontal, vertical and longitudinal directions. The depiction of the reaction forces from the rail treads acting on a wheelset is shown in Figure 2.

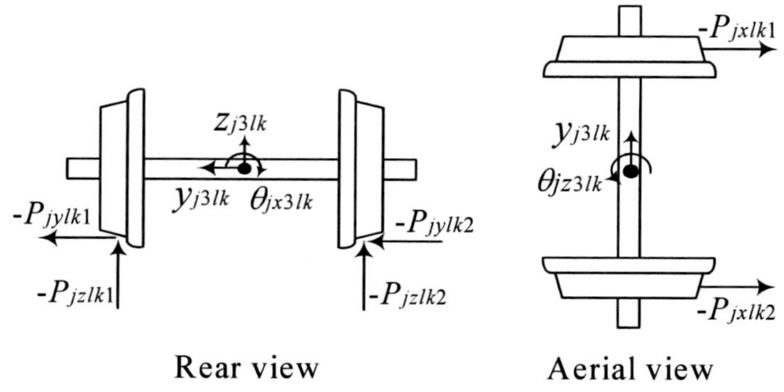

Figure 2. Contact forces acting on the wheelset.

Actually, $P_{jxlkm}(t)$ is the creeping force in longitudinal direction between the wheel and rail tread; $P_{jzlkm}(t)$ is the normal contact force; $P_{jylkm}(t)$ is the combination of the horizontal creeping force between the wheel and rail tread and the lateral force due to the contact of wheel flange and track, respectively. They can be calculated following the wheel-rail contact theory, which leads to a complicated numerical process [22].

In the present idealization of the train-bridge interaction system, to avoid its complication, instead of calculating the contact forces between the wheel and track structure, the motions of the wheelset are assumed to be compatible with the displacements of the rail structure, while its movement due to the rail irregularities or the wheelset hunting motion is directly inputted. If the yawing motion of the wheelset can be neglected, the motions of the wheelset are represented as follows.

$$y_{j3lk} = \frac{1}{2}\sum_{m=1}^{2} w_{jylkm} \tag{24}$$

$$z_{j3lk} = \frac{1}{2}\sum_{m=1}^{2} w_{jzlkm} \tag{25}$$

$$\theta_{jx3lk} = -(-1)^m \frac{1}{2\lambda_{y1}}\sum_{m=1}^{2} w_{jzlkm} \tag{26}$$

The variables w_{jylkm} and w_{jzlkm} denote the sum of the displacements and irregularities of the rail in y and z-direction, respectively.

$$w_{jylkm} = w_y(t, x_{jlkm}) + z_{0y}(x_{jlkm}) \tag{27}$$

$$w_{jzlkm} = w_z(t, x_{jlkm}) + z_{0z}(x_{jlkm}) \tag{28}$$

where, $w_y(t, x_{jlkm})$ and $w_z(t, x_{jlkm})$ are the displacements of the rail at the contact points of the wheel and the rail in y and z-direction, respectively; and $z_{0y}(x_{jlkm})$ and $z_{0z}(x_{jlkm})$ represent the rail irregularities in y and z-direction, respectively.

$$w_y(t, x_{jlkm}) = \mathbf{\Psi}^{\mathrm{T}}_{jylkm}(t)\mathbf{w}_{\mathrm{b}} \tag{29}$$

$$w_z(t, x_{jlkm}) = \mathbf{\Psi}^{\mathrm{T}}_{jzlkm}(t)\mathbf{w}_{\mathrm{b}} \tag{30}$$

where, $\mathbf{\Psi}_{jylkm}$ (t) and $\mathbf{\Psi}_{jzlkm}$ (t) respectively represent the distribution vectors of y and z-directions that distribute the wheel loads to the ends of the beam elements, and $\boldsymbol{w}_b$ denotes the nodal displacement vector of the bridge model.

$$\mathbf{\Psi}_{jylkm}(t) = \{0;\cdots;0;\psi_{p,jlkm};\psi_{p+1,jlkm};0;\cdots;0\}^{\mathrm{T}} \tag{31}$$

$$\mathbf{\Psi}_{jzlkm}(t) = \{0;\cdots;0;\psi_{q,jlkm};\psi_{q+1,jlkm};0;\cdots;0\}^{\mathrm{T}} \tag{32}$$

Expanding the equations described above, the differential equations of the train system can be expressed in matrix form as follows.

$$\mathbf{M}_{\mathrm{v}}\ddot{\mathbf{w}}_{\mathrm{v}} + \mathbf{C}_{\mathrm{v}}\dot{\mathbf{w}}_{\mathrm{v}} + \mathbf{K}_{\mathrm{v}}\mathbf{w}_{\mathrm{v}} = \mathbf{f}_{\mathrm{v}} \tag{33}$$

where, $\mathbf{M}_{\mathrm{v}}$, $\mathbf{C}_{\mathrm{v}}$, $\mathbf{K}_{\mathrm{v}}$ and $\mathbf{f}_{\mathrm{v}}$ respectively denote the mass, damping, stiffness matrices and the external force vector of the train system.

The wheel loads acting on the bridge, $P_{jylkm}(t)$ and $P_{jzlkm}(t)$, are simplified as follows:

$$P_{jylkm}(t) = -\frac{1}{2}m_3\ddot{w}_{jylkm} - (-1)^m v_{jylkm}(t) \tag{34}$$

$$P_{jzlkm}(t) = -\left(\frac{1}{8}m_1 g + \frac{1}{4}m_2 g + \frac{1}{2}m_3 g\right) - \frac{1}{2}m_3\ddot{w}_{jzlkm} + v_{jzlkm}(t) \tag{35}$$

2.4.2. Modal Analysis for the Bridge

The differential equation of the bridge is derived as follows, based on FEM and D'Alembert's Principle.

$$\mathbf{M}_{\mathrm{b}}\ddot{\mathbf{w}}_{\mathrm{b}} + \mathbf{C}_{\mathrm{b}}\dot{\mathbf{w}}_{\mathrm{b}} + \mathbf{K}_{\mathrm{b}}\mathbf{w}_{\mathrm{b}} = \mathbf{f}_{\mathrm{b}} \tag{36}$$

where, $\mathbf{M}_{\mathrm{b}}$, $\mathbf{C}_{\mathrm{b}}$ and $\mathbf{K}_{\mathrm{b}}$ denote mass, damping and stiffness matrices of the bridge system, respectively.

Herein, the damping matrix of bridge $\mathbf{C}_{\mathrm{b}}$ is assumed to be calculated by the linear relation between mass and stiffness matrices [23] as follows:

$$\mathbf{C}_\mathrm{b} = p_1\mathbf{M}_\mathrm{b} + p_2\mathbf{K}_\mathrm{b} \tag{37}$$

where, p_1 and p_2 are the ratio coefficients.

$$p_1 = \frac{2\omega_{\mathrm{b}1}\omega_{\mathrm{b}2}(h_{\mathrm{b}1}\omega_{\mathrm{b}2} - h_{\mathrm{b}2}\omega_{\mathrm{b}1})}{\omega_{\mathrm{b}2}^2 - \omega_{\mathrm{b}1}^2} \tag{38}$$

$$p_2 = \frac{2(h_{\mathrm{b}2}\omega_{\mathrm{b}2} - h_{\mathrm{b}1}\omega_{\mathrm{b}1})}{\omega_{\mathrm{b}2}^2 - \omega_{\mathrm{b}1}^2} \tag{39}$$

where, $\omega_{\mathrm{b}1}$ and $\omega_{\mathrm{b}2}$ respectively denote the first and second natural circular frequencies of the bridge model; $h_{\mathrm{b}1}$ and $h_{\mathrm{b}2}$ is the damping constants corresponding to $\omega_{\mathrm{b}1}$ and $\omega_{\mathrm{b}2,}$ respectively.

Assuming the total number of cars as h, the external force vector $\mathbf{f}_\mathrm{b}$ can be represented as follows:

$$\mathbf{f}_\mathrm{b} = \sum_{j=1}^{h}\sum_{l=1}^{2}\sum_{k=1}^{2}\sum_{m=1}^{2}\left\{\mathbf{\Psi}_{jylkm}(t)P_{jylkm}(t) + \mathbf{\Psi}_{jzlkm}(t)P_{jzlkm}(t)\right\} \tag{40}$$

where, $P_{jylkm}(t)$ and $P_{jzlkm}(t)$ are the wheel loads of the train and $\mathbf{\Psi}_{jylkm}(t)$ and $\mathbf{\Psi}_{jzlkm}(t)$ are the distribution vectors.

The vector of nodal displacement of the bridge, $\mathbf{w}_b$, is derived from modal analysis method and represented as follows.

$$\mathbf{w}_b = \sum_{i=1}^{n}\boldsymbol{\varphi}_i q_i = \mathbf{\Phi}\cdot\mathbf{q} \tag{41}$$

where, $\mathbf{q}$ is the generalized coordinate vector of the bridge and $\mathbf{\Phi}$ is the modal matrix composed of the natural modal vector of the bridge $\boldsymbol{\varphi}_i$.

$$\mathbf{q} = \{q_1 \quad q_2 \quad \cdots \quad q_n\}^\mathrm{T} \tag{42}$$

$$\mathbf{\Phi} = \{\boldsymbol{\varphi}_1 \quad \boldsymbol{\varphi}_2 \quad \cdots \quad \boldsymbol{\varphi}_n\} = \begin{bmatrix} \phi_{11} & \phi_{12} & \cdots & \phi_{1n} \\ \phi_{21} & \ddots & & \vdots \\ \vdots & & \ddots & \vdots \\ \phi_{m1} & \cdots & \cdots & \phi_{mn} \end{bmatrix} \tag{43}$$

where, the subscript m indicates the number of freedoms of the bridge finite element model after matrix condensation, and n denotes the highest mode number considered.

Substituting $\mathbf{w}_\mathrm{b}$ into Equation (36), the following equation can be derived,

$$\mathbf{M}_\mathrm{b}\mathbf{\Phi}\ddot{\mathbf{q}} + \mathbf{C}_\mathrm{b}\mathbf{\Phi}\dot{\mathbf{q}} + \mathbf{K}_\mathrm{b}\mathbf{\Phi}\mathbf{q} = \mathbf{f}_\mathrm{b} \tag{44}$$

Multiplying both sides by $\mathbf{\Phi}^{\mathrm{T}}$, the following equation is obtained,

$$\mathbf{\Phi}^T \mathbf{M}_{\mathrm{b}} \mathbf{\Phi} \ddot{\mathbf{q}} + \mathbf{\Phi}^T \mathbf{C}_{\mathrm{b}} \mathbf{\Phi} \dot{\mathbf{q}} + \mathbf{\Phi}^T \mathbf{K}_{\mathrm{b}} \mathbf{\Phi} \mathbf{q} = \mathbf{\Phi}^{\mathrm{T}} \mathbf{f}_{\mathrm{b}} \tag{45}$$

where,

$$\mathbf{\Phi}^{\mathrm{T}} = \begin{Bmatrix} \boldsymbol{\varphi}_1 \\ \boldsymbol{\varphi}_2 \\ \vdots \\ \boldsymbol{\varphi}_n \end{Bmatrix} = \begin{bmatrix} \phi_{11} & \phi_{21} & \cdots & \phi_{m1} \\ \phi_{12} & \ddots & & \vdots \\ \vdots & & \ddots & \vdots \\ \phi_{1n} & \cdots & \cdots & \phi_{mn} \end{bmatrix} \tag{46}$$

According to the orthogonality of the normal modal vectors, and that $\boldsymbol{C}_{\mathrm{b}}$ is linearly composed of $\mathbf{M}_{\mathrm{b}}$ and $\mathbf{K}_{\mathrm{b}}$,

while $i \neq j$, $\boldsymbol{\varphi}_i^{\mathrm{T}} \mathbf{M}_{\mathrm{b}} \boldsymbol{\varphi}_i = 0$, $\boldsymbol{\varphi}_i^{\mathrm{T}} \mathbf{K}_{\mathrm{b}} \boldsymbol{\varphi}_i = 0$, $\boldsymbol{\varphi}_i^{\mathrm{T}} \mathbf{C}_{\mathrm{b}} \boldsymbol{\varphi}_i = 0$;
while $i=j$, $\boldsymbol{\varphi}_i^{\mathrm{T}} \mathbf{M}_{\mathrm{b}} \boldsymbol{\varphi}_i = M_i$, $\boldsymbol{\varphi}_i^{\mathrm{T}} \mathbf{M}_{\mathrm{b}} \boldsymbol{\varphi}_i = K_i$, $\boldsymbol{\varphi}_i^{\mathrm{T}} \mathbf{M}_{\mathrm{b}} \boldsymbol{\varphi}_i = C_i$.

Assuming $\mathbf{\Phi}^{\mathrm{T}} \mathbf{f}_{\mathrm{b}} - \mathbf{\Phi}^{\mathrm{T}} \mathbf{M}_{\mathrm{b}} \ddot{\Delta}_{\mathrm{b}} = f_i$, the differential equation of the bridge with respect to generalized coordinate can be developed as follows:

$$M_i \ddot{q}_i + C_i \dot{q}_i + K_i q_i = f_i \tag{47}$$

2.4.3. Coupled Equation of Train-bridge System in Matrix Form

Based on the formulation developed above, the coupled equation of train-bridge interaction system can be expressed in matrix form as follows.

$$\begin{bmatrix} \mathbf{M}_{\mathrm{b}}^* & \mathbf{0} \\ \mathrm{Sym} & \mathbf{M}_{\mathrm{v}} \end{bmatrix} \begin{Bmatrix} \ddot{\mathbf{q}}_{\mathrm{b}} \\ \ddot{\mathbf{w}}_{\mathrm{v}} \end{Bmatrix} + \begin{bmatrix} \mathbf{C}_{\mathrm{b}}^* & \mathbf{C}_{\mathrm{bv}} \\ \mathrm{Sym} & \mathbf{C}_{\mathrm{v}} \end{bmatrix} \begin{Bmatrix} \dot{\mathbf{q}}_{\mathrm{b}} \\ \dot{\mathbf{w}}_{\mathrm{v}} \end{Bmatrix} + \begin{bmatrix} \mathbf{K}_{\mathrm{b}}^* & \mathbf{K}_{\mathrm{bv}} \\ \mathrm{Sym} & \mathbf{K}_{\mathrm{v}} \end{bmatrix} \begin{Bmatrix} \mathbf{q}_{\mathrm{b}} \\ \mathbf{w}_{\mathrm{v}} \end{Bmatrix} = \begin{Bmatrix} \mathbf{F}_{\mathrm{b}} \\ \mathbf{F}_{\mathrm{v}} \end{Bmatrix} \tag{48}$$

where, $\mathbf{M}_{\mathrm{b}}^*$, $\mathbf{C}_{\mathrm{b}}^*$ and $\mathbf{K}_{\mathrm{b}}^*$, respectively denote the mass, damping and stiffness components corresponding to the generalized coordinate of the bridge system; $\mathbf{M}_{\mathrm{v}}$, $\mathbf{C}_{\mathrm{v}}$ and $\mathbf{K}_{\mathrm{v}}$, respectively denote the mass, damping and stiffness components corresponding to the DOF of the train system; $\mathbf{C}_{\mathrm{bv}}$ and $\mathbf{K}_{\mathrm{bv}}$, respectively denote the coupled damping and stiffness components between the bridge and train systems. The detailed components of the matrix can accordingly be obtained by expanding the above formula, which are used to create the computer program.

2.5. Numerical Integration Methods

The simultaneous differential equations of the train-bridge interaction system are non-stationary dynamic problems because the coefficient matrices or the external vectors in the equations vary according to the train position. Such problems can be solved by step-by-step numerical integration methods, in which the changed components are updated at each time step.

Various step-by-step integration methods to solve the equations of structural dynamics have been developed. The central-difference method such as discussed by Dahlquist [24] was a widely used explicit two-step formula for solving the structural dynamics problems. However, the expressions of this type have a rather critical time step, above which the solution becomes numerically unstable and diverges [24]. For complicated finite element structural models, containing slender members exhibiting bending effects, such restriction is stringent and often entails using time steps that are much smaller than those needed for accuracy. For many cases only low-mode responses are of interest and a larger time step is desired. High-mode responses of complicated finite element models often cannot express the reasonable motions of the actual structures and should be eliminated. Therefore, it is often advantageous for an algorithm to possess some form of numerical dissipation to damp out any spurious participation of the higher modes.

For those reasons, unconditionally stable algorithms which achieve the optimal balance between effective numerical dissipation and loss of accuracy compared with trapezoidal rule are generally preferred. Although there is no universal consensus, it is generally agreed that for a method to be competitive, it should possess the following attributes [25]:

(1) Unconditional stability
(2) No more than one set of implicit equations
(3) Second-order accuracy
(4) Controllable algorithmic dissipation in the higher modes
(5) Self-starting

The Newmark family of algorithms enjoys wide use in structural dynamics. In his paper [26], Newmark generalized certain direct numerical integration methods that had been in use up to that time. For the damped equation of motion of a single degree-of-freedom (SDOF) system, he presented equations for approximating the velocity and displacement at time step t_{j+1} as the following equations.

$$\dot{x}^{j+1} = \dot{x}^{j} + \{(1-\gamma)\ddot{x}^{j} + \gamma\ddot{x}^{j+1}\}\Delta t \tag{49}$$

$$x^{j+1} = x^{j} + \dot{x}^{j}\Delta t + \left\{\left(\frac{1}{2}-\beta\right)\ddot{x}^{j} + \beta\ddot{x}^{j+1}\right\}(\Delta t)^2 \tag{50}$$

$$m\ddot{x}^{j+1} + c\dot{x}^{j+1} + kx^{j+1} = f^{j+1} \tag{51}$$

where, x denotes the displacement and superposed dots indicate time differentiation. m, c, k and f are respectively mass, damping coefficient, stiffness and external force of the SDOF system. j is the sequence number of time step and Δt represents the time interval. Actually the time interval can be set to vary at each step, whereas for most cases it is convenient to set the time interval constant.

The Newmark family of methods, so called Newmark's β method (or Newmark's generalized acceleration method), allows the amount of numerical (or algorithmic) damping to be continuously controlled by the parameter γ other than time step. If γ is taken to be less than 1/2, an artificial negative damping occurs. This will involve a self-excited vibration arising solely from the numerical procedure and should be avoided. On the contrary, if γ is greater than 1/2, such damping is positive. For example, set $\beta = (\gamma + 1/2)^2/4$ and $\gamma > 1/2$; then the amount of dissipation for a fixed time step is increased by increasing γ. The positive damping will reduce the magnitude of the response even without real damping. To avoid numerical damping altogether, the value of γ must be equal to 1/2; and Equation (49) becomes the trapezoidal rule.

On the other hand, the parameter β controls the variation of acceleration within the time step. For example, while β = 0, 1/4 and 1/6, the formula respectively become the constant-acceleration method, average-acceleration method and linear-acceleration method. It is well-known that the linear-acceleration method is somewhat more accurate than the average-acceleration method [27]. However, it has been shown [26] that the former technique is only conditionally stable and requires a critical time step. The average-acceleration method is unconditionally stable, although less accurate.

It should be noted that for the Newmark family of methods, the attributes (3) and (4) enumerated previously cannot exist simultaneously (Second-order accuracy requires γ = 1/2 which precludes numerical damping) [28]. Furthermore, the dissipative properties of this family of algorithms are considered to be inferior to both the Houbolt [29] and the Wilson [30] methods, since the lower modes are affected too strongly. It seems all of these algorithms adequately damp the highest modes [31].

In 1968 Wilson developed the first method, so called Wilson-θ method [30], which essentially satisfied attributes (1)–(5) [25], by extending the linear-acceleration method in a manner that makes it numerically stable. The basic assumption of the Wilson-θ approach is that the acceleration $\ddot{x}$ varies linearly over an extended time step $\Delta t^{\theta} = \theta\Delta t$. During that time step the acceleration is $\ddot{x}^{j+\theta} = \ddot{x}^{j} + \theta(\ddot{x}^{j+1} - \ddot{x}^{j})$. Not to explain, while θ=1.0 the Whilson-θ method is identical with the linear-acceleration method. The Wilson-θ method for solving a SDOF system can be represented as follows:

$$\dot{x}^{j+\theta} = \dot{x}^{j} + \frac{1}{2}(\ddot{x}^{j} + \ddot{x}^{j+\theta})\Delta t^{\theta} \tag{52}$$

$$x^{j+\theta} = x^{j} + \dot{x}^{j}\Delta t^{\theta} + \frac{1}{2}\ddot{x}^{j}(\Delta t^{\theta})^2 + \frac{1}{6}\{\ddot{x}^{j+\theta} - \ddot{x}^{j}\}(\Delta t^{\theta})^2 \tag{53}$$

$$m\ddot{x}^{j+\theta} + c\dot{x}^{j+\theta} + kx^{j+\theta} = f^{j+\theta} \tag{54}$$

where,

$$f^{j+\theta} = f^{j} + \theta(f^{j+1} - f^{j}) \tag{55}$$

Then the differential equation can be solved at time $t_{j+\theta}$ by the above formula just like the procedure of Newmark's β method. While $\ddot{x}^{j+\theta}$ is obtained, the acceleration at time t_{j+1} can be derived with Equation (56). Note here that the velocity and displacement responses at time t_{j+1} should be further calculated with Equations (52) and (53) by setting $\theta = 1.0$.

$$\ddot{x}^{j+1} = \ddot{x}^{j} + \frac{1}{\theta}(\ddot{x}^{j+\theta} - \ddot{x}^{j}) \tag{56}$$

In the Wilson method, θ must be selected greater than or equal to 1.37 to maintain unconditional stability. It is recommended [31] as $\theta = 1.4$ since further increasing θ reduces accuracy and further increases dissipation. However, even for $\theta = 1.4$ the Wilson method possess excessive low-mode dissipation (i.e. loss of accuracy), and requires a time step to be taken that is smaller than that needed for accuracy [28]. Another peculiar property of the Wilson method was shown numerically by Goudreau and Taylor [32] and Argyris et al. [33]. When large time steps are employed, the Wilson method has a tendency to overshoot significantly exact solutions to initial value problems in the early response, especially for applications involving impact or suddenly applied loads. Incidentally, the Houbolt's method [29] is even more highly dissipative than Wilson's method and does not permit parametric control over the amount of dissipation.

Considering the drawbacks possessed by the above described algorithms, Hilber et al. [28] developed an approach, so called Hilber-α method, to enhance the Newmark family of methods by employing a parameter α to improve control of numerical damping. The Hilber-α approach can be described as Equation (57) by introducing the parameter α into the equation of motion of a SDOF system. The rules for calculation of displacement and velocity responses are the same as those in Newark's method. In the Hilber method, it is recommended by Weaver and Johnston [19] that the optimum parameters is to let $\alpha = -0.1$, $\beta = 0.3025$, and $\gamma = 0.6$.

$$m\ddot{x}^{j+1} + c\dot{x}^{j+1} + (1+\alpha)kx^{j+1} - \alpha kx^{j} = f^{j+1} \tag{57}$$

The errors introduced by numerical integration methods mainly attribute to amplitude suppression and period elongation [19, 25, 28]. Generally the effect of amplitude suppression is considered more important. Here, to compare the significant differences of the approaches introduced so far, assuming T as the period of a SDOF system the diagram for the damping ratios h *vs* Δt/T of those methods are indicated in Figure 3 [28] (It is only an extraction from the references and the values indicated in the diagram are approximations.)

Desirable properties for an algorithmic damping ratio graph to possess are a zero tangent at the origin and subsequently a controlled turn upward. This ensures adequate dissipation in the higher modes and at the same time guarantees that the lower modes are not affected too strongly. In Figure 3, the dissipation ratio curve of the Newmark method with γ-damping has positive slope at the origin. This is why the Newmark family is felt to possess ineffective numerical dissipation. In the rest methods, the Houbolt method has the strongest dissipative property but affects lower modes too strongly. Compared with the Wilson method, the Hilber method seems to be the most accurate one in the lower modes. However, if the time step can be set sufficiently short, the Wilson method may be more effective to damp out the higher modes. Figure 3 also indicates the criterion to set the time interval if the highest mode to be considered is determined in a complicated structure.

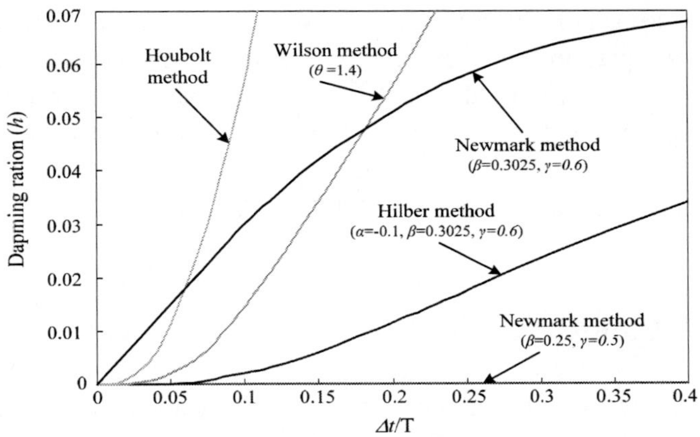

Figure 3. Damping ratio *vs* Δt/T for various numerical methods.

According to the discussion above, in the case of numerical damping is desired, the Hilber method or the Wilson method should be used. If large time step is preferred, the Hilber method seems to be the best choice. However, it is difficult to conclude that the Hilber method is superior to the Wilson method in all cases. On the other hand, though the Newmark method of the trapezoidal rule ($\gamma = 1/2$) offers no damping effect, it is considered the most accurate one if without the influence of higher modes. Therefore, for cases that the effect of higher modes can be eliminated, such as using modal analysis or viscous damping, the linear acceleration method or the average acceleration method may lead to more accurate results.

Incidentally, to ensure the stability and accuracy of numerical analysis in structural dynamics, some researchers continue to suggest that the viscous damping term, e.g. Rayleigh damping, be used to suppress higher modes [34]. The viscous damping is expected to have the same effect as the numerical damping. However, some researches indicated that the interaction between the viscous damping and typical implicit integrators results in virtually no dissipation in the higher modes [31, 35, 36].

As discussed above, the numerical integration method used to solve the differential equations of train-bridge interaction problem should be selected carefully to ensure the enginnering accuracy.

2.6. Analysis of Soil-Structure Interaction Using SASSI2000

For the foundation-ground interaction system in this study, employing the previously obtained reaction forces at the pier bottoms of the viaducts as input dynamic forces, ground vibrations around the viaducts are simulated using a general-purpose computer program named SASSI2000. The full theoretical information of this software can be referred to the references by Chin [37], Lysmer [38] and Lysmer et al. [13], thus is not introduced here in detail.

In this program, the soil-structure interaction (SSI) problem is analyzed conveniently using a substructuring approach by which the linear soil-structure interaction problem is subdivided into a series of simple sub-problems. Each sub-problem is solved separately and the results are combined in the final step of the analysis to provide a complete solution using the principle of superposition. In particular for ground response analysis, the 3D thin layer element method is adopted in this approach. This method is an extension from the study of the 2D problem that was originally developed by Lysmer and Drake [39], which can remove the limitation of half-space elastic theory of isotropic homogeneous media. The thin layer element method is expected to be widely applicable to different interaction problems and is studied by numerous researchers Shimizu et al. [40, 41] and Yang et al. [5].

3. Numerical Simulation and Evaluation

Focusing on a standard type of Shinkansen viaduct in Japan, the case studies of the train induced-bridge and ambient ground vibration problem are carried out using the developed numerical approach described previously. The reduction effect of designed countermeasures against the excessive bridge and ground vibration is also numerically simulated and evaluated.

3.1. Numerical Models

3.1.1 Bridge and Rail Model

A typical Shinkansen reinforced concrete viaduct in the form of portal rigid frame shown in Figure 4 is adopted for evaluation in this research. The viaducts are built with 24 m-length bridge blocks which are separated with each other and connected only by rail structure at adjacent ends. Each block consists of three 6 m-length center spans and two 3 m cantilever girders, so called hanging parts, at each end. Three blocks (72 m) of the bridge with 24 m length of each block are adopted as the numerical model. Considering the boundary condition of the bridge, only the dynamic response of the middle block will be examined.

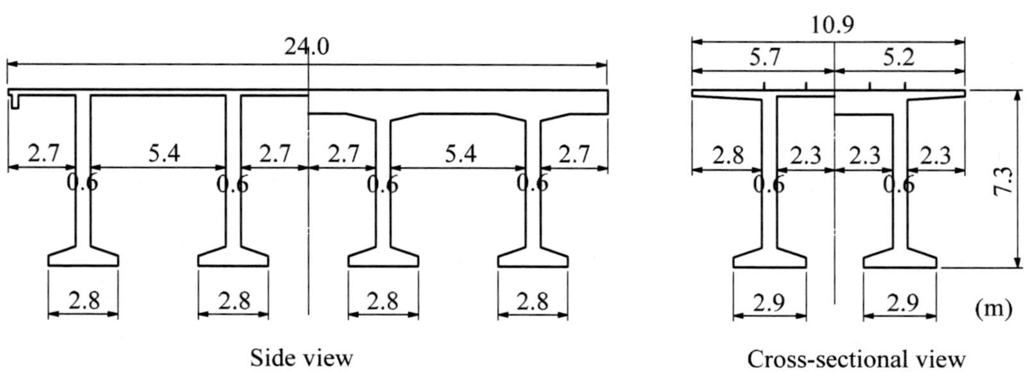

Figure 4. Dimensions of the bridge.

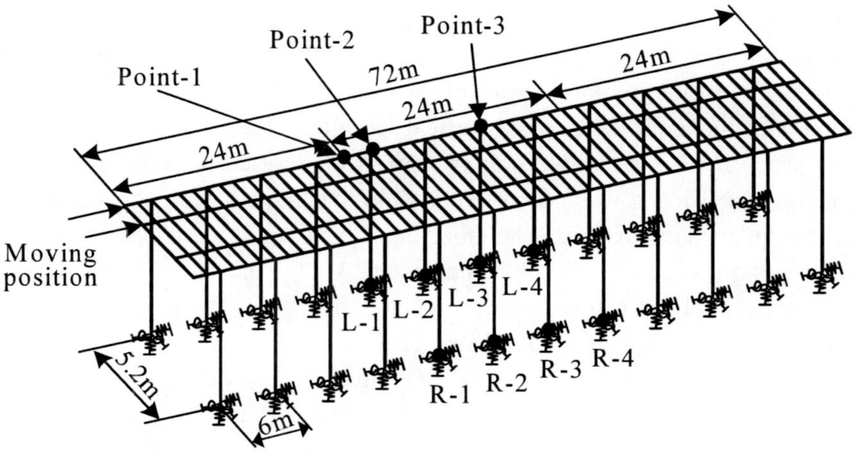

Figure 5. Finite element model of the bridge.

Figure 5 shows that the three-block bridge is modeled as 3D beam elements with six-DOF at each node. The lumped mass system is adopted for the beam elements. Mass of the ballast is also incorporated. Hanging parts also exist on both sides of the superstructure in the traverse direction as shown in Figure 4. The mass of these parts are attached to the nodes at the most outside line in the finite element model. Double nodes defined as two independent nodes sharing the same coordinate and connected by spring elements are adopted at the pier bottoms to simulate the effect of ground springs [42]. The ground springs are calculated considering the effects of the footing and pile structures as well as the elastic soil effect [11]. Rayleigh damping [23] is used for the structure. Based on the past field test results [43], the damping constant of 0.03 is assumed for the first and second natural modes of the structure.

The rail structure is also modeled as 3D beam elements with six-DOF at each node. Double nodes are also defined here to simulate the elastic effect of sleepers and ballast at the positions of sleepers. The vertical spring constant of the track is derived from the ratio of the wheel load to the rail's displacement in the vertical direction. The horizontal spring constant of the track is assumed to be 1/3 of the value in the vertical direction [43]. The measured values of vertical railway roughness are shown in Figure 6.

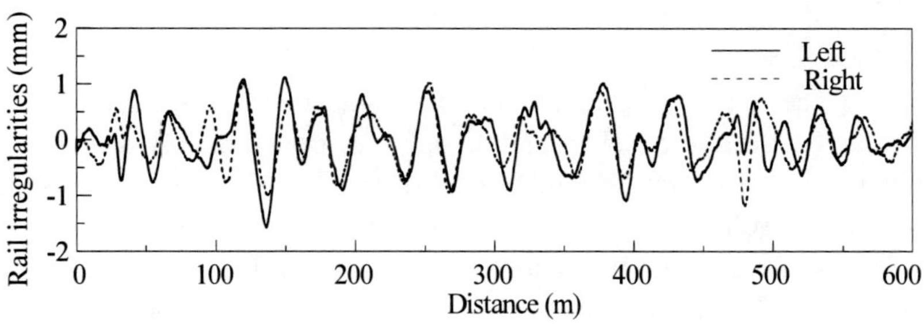

Figure 6. Measured value of rail surface roughness.

3.1.2. Train Model

In the field of vehicle engineering, vehicle model often comprises dozens of DOFs in order to accurately evaluate its dynamic responses. On the other hand, in many cases it is not necessary to use such detailed models in the field of civil engineering. The numerical models should be determined considering not only the accuracy but also the analytical efficiency and cost. In this study, considering the focuses of the analysis, the DOFs of the car model can be limited to the ones that contribute to the vertical vibration. The mass of the wheel-sets takes only a small proportion in the whole train system. It is reported by Shoji et al. [44] that the variation of the wheel loads was not notable while the train is running over a straight-line section under regular maintenances. Therefore in such cases the variation of the wheel loads caused by the motions of wheelsets can be neglected, and the enormous work of modeling the wheel-track contact problem can be avoided. Thus in this study, the motions of the wheelsets are assumed to be compatible with the rail structure.

Based on the above discussion, a nine-DOF train car model as shown in Figure 7 is conveniently used for the evaluation at first, considering that the vertical ground vibration is the main concern on environmental vibration problems. In the nine-DOF train model, only the freedoms contributing to the vertical vibration of the car-body and bogies are taken into account. The formulation process of the simple nine-DOF model can be easily derived from the complicated model described in Figure 1. The train is composed of 16 cars according to the actual operational case. The train velocity is set to be 270 km/h, referring to the actual Shinkansen operational speed. In addition, the numerical analyses using a more complicated train model that can consider horizontal vibrations of the train as shown in Figure 1 and formulated in Section 2 are also carried out.

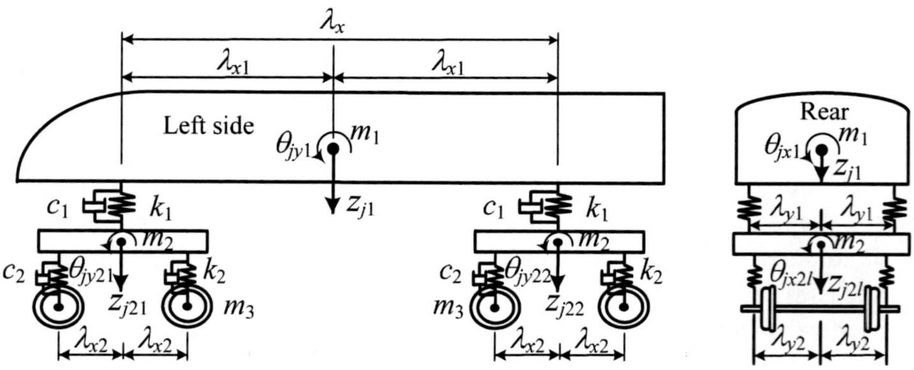

Figure 7. The nine-DOF train model.

3.1.3 Ground Model

Properties of the actual ground around the viaducts measured in the field test are employed to establish the numerical ground model. The ground mainly comprises three strata separated at depths of 6.8 m and 17.2 m. The velocity of the S-wave in the first stratum is 80 m/s, from which the soil can be considered as relatively soft. The damping constant is assumed as 5%, determined from experiential values. For analysis, the ground model is divided further into 21 thin layer elements, whose profiles are shown in Figure 8. The maximum thickness of each layer is determined in compliance with the criterion that it does not exceed 1/5 λ_s, where λ_s is the shortest S wavelength in that layer [13]. Layer elements are established down to the depth of 18.8 m, to which the structural model is embedded. The program then automatically adds some extra layer elements and the viscous boundary at the base to simulate the effect of half space.

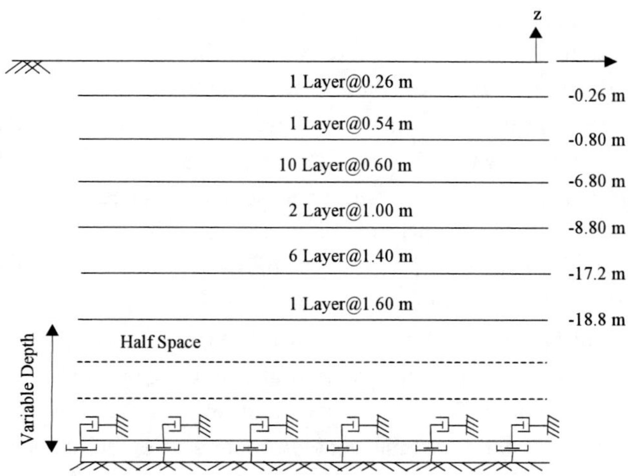

Figure 8. Profile of the layer element soil model.

3.1.4. Foundation Structure Model

The foundation structural model contains the footing and pile structures of the viaduct. One structural set consisting of one footing and seven piles is modeled as Figure 9. The actual footing structure is in the shape of rectangular parallelepiped at the base and a trapezoid at the top. To simplify the analyses in this analysis, the footing is approximated as a rectangular parallelepiped divided into solid elements according to the conversion of volume. The sizes of the solid elements also meet the criterion that they be less than 1/5 of the shortest S wavelength in the corresponding layer [13]. The upper footing surface is set to lie 0.26 m under the ground surface. The piles are divided into two types according to their length: Type 1 is 7 m long and Type 2 is 18 m. The ○ and × marks indicate the positions at which the piles are connected vertically to the footing. Herein, ○ represents 18-m-long piles and × represents 7-m-long piles. The piles are modeled as 3D beam elements. The ends of the beam elements are established at the soil layer interfaces.

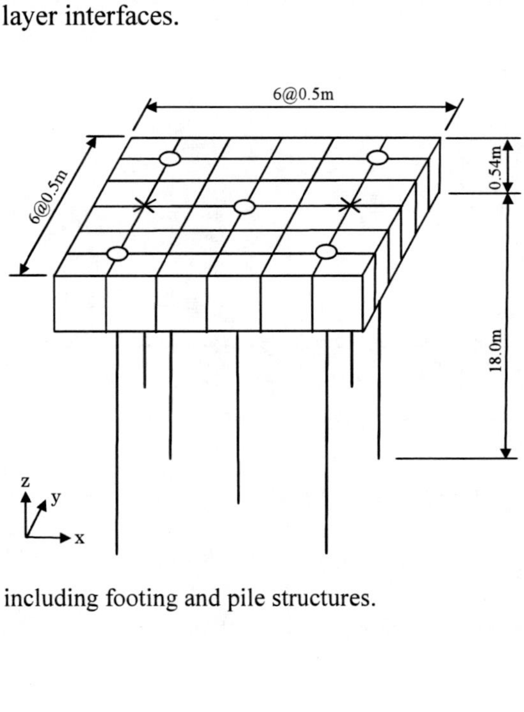

Figure 9. Foundation model including footing and pile structures.

3.1.5. Field Measurement of Bridge and Ground Vibration

The actual field tests were conducted on the Shinkansen line to measure the bridge vibration, whose information can be found in the references [10, 11]. Bullet trains composed of sixteen cars were running through the viaduct with its actual operational speed of 270 km/h. The bridge vibration was measured at several points of the viaduct during the bullet trains' passage using accelerometers. Then the acceleration responses of the viaduct were recorded on the data recorder from the accelerometers after being processed by amplifiers. The sampling rate of the data was 512 Hz. In this analysis, the bridge vibration recorded at point-1 through point-3 of the viaduct model indicated in Figure 5 will be examined.

The ground vibration around the viaduct was measured simultaneously in the same field test of the bridge vibration. The ground vibration was measured at several points around the viaduct using accelerometers. Then the acceleration responses of the ground were also recorded on the data recorder from the accelerometers after being processed by amplifiers. The sampling rate of the data was also 512 Hz. Before the field test to measure the train-induced ground vibration, the ground conditions [10] around the viaduct were investigated, which will be used to establish the ground model.

Figure 10 indicates the positions of the piers and the points of measurement. In all, 24 footings of the three blocks of bridges used in the previous bridge vibration analyses are adopted to be excited. Black rectangles in the figure indicate the footing positions. L and R denote the left and right sides of the bridge with respect to the moving direction of the train. The letters a–d and A–D and the numbers 1–4 respectively indicate the footing sequences in the three blocks of the bridge. The distances between the centers of neighboring footings on the same side are 6.0 m; those between the central lines of left and right footings are 5.2 m. Vicinity, 12.5 m and 25.0 m lying on the line passing through the centers of footings R-3 and L-3 denote the points at which the ground vibration is measured. They are respectively 3.5 m, 12.5 m and 25.0 m distant from the longitudinal central line of the bridge. In the analysis, the ground vibration response of a certain point is obtained from the superposition of those engendered by each footing. Numerical results of these points are compared with experimental ones.

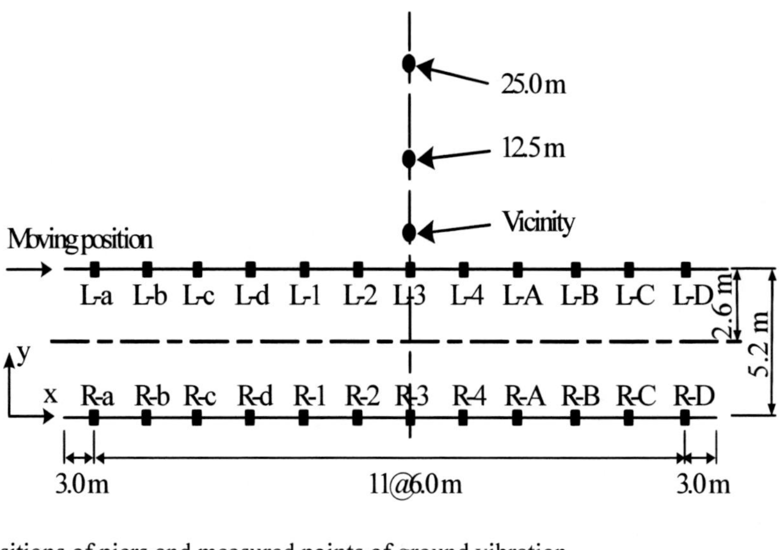

Figure 10. Positions of piers and measured points of ground vibration.

3.2. Numerical Evaluation on Bridge and Ground Vibration

3.2.1. Natural Modes and Frequencies of the Bridge

Eigenvalue analysis of the bridge model is performed and the mode shapes are shown in Figure 11. The first through the ninth modes indicate the vibration in horizontal direction of the bridge. The predominant frequency of the horizontal natural mode is observed to be 2.20 Hz, showing good agreement with the value of field test which is 2.19 Hz. Therefore, the validity of the bridge model can be confirmed. Predominant frequencies of the vertical bending natural mode and the torsional mode are about 11.9 Hz and 13.9 Hz, respectively. The highest frequency taken into account in this analysis is about 102 Hz corresponding to the 277th natural mode. The predominant frequencies of the rail mode are larger than 200 Hz, but it is considered that the influence of the rail vibration on the structures can be neglected.

3.2.2. Dynamic Responses of the Bridge

Based on the conditions of the field test described above and the actual properties of the viaduct and the bullet train, the analytical conditions were determined and the analysis was carried out with the developed analytical program. Since there are still many uncertainties in modeling either the bullet train or the viaduct, it is difficult to reproduce the completely accurate responses of the bridge vibration by numerical approaches. However, for actual discussion in civil engineering field, in most cases it is sufficient if the amplitudes and main vibrational components can be expressed.

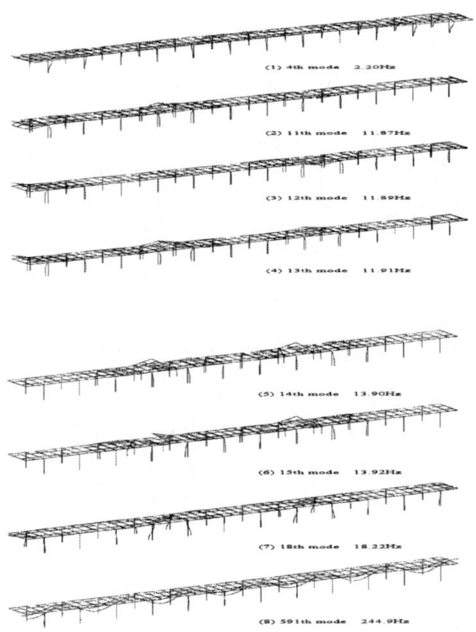

Figure 11. Natural mode shapes and Natural frequencies of bridge.

The numerical acceleration responses and the experimental ones, of point-1 through point-3 of the FE bridge model, are shown respectively in Figure 12 with Fourier spectra under trains with 270 km/h velocity. Herein, point-1, point-2, and point-3 respectively indicate the point of hanging part, the top of the first pier and the top of the third pier of the viaduct, with respect to the direction that the train runs towards. As shown in Figure 12, numerical results using the nine-DOF train model indicate good agreement with experimental results, thereby validating this numerical procedure. The hanging parts of the elevated bridge are connected with neighboring ones by rails and ballast in the actual structure, but only the connecting effect of the rails can be incorporated into analysis. Presumably for that reason, the vibrations are predominant at lower frequencies and numerical acceleration responses display larger amplitudes than do experimental ones at point-1 in Figure 12. More detailed description and discussion of the dynamic responses of the viaducts are conducted in Kawatani et al. [11].

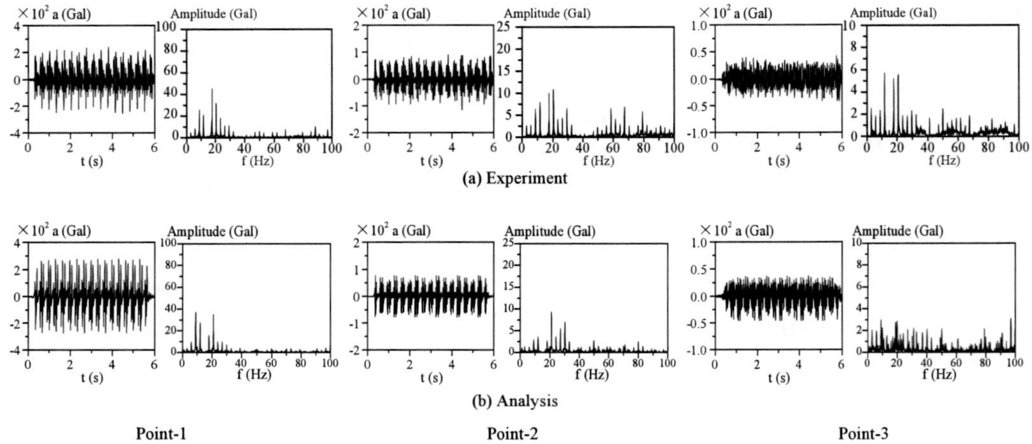

Figure 12. Bridge acceleration response under moving train.

3.2.3. Dynamic Reaction Forces at Pier Bottoms

Reaction forces at the pier bottoms, as shown in Figure 13, are calculated using the influence value matrix of the reaction force described previously. As shown in Figure 5, L-1 to L-4 and R-1 to R-4 respectively indicate the piers on the left and right sides of the middle block of the bridge, with respect to the train's direction. The dynamic reaction forces of the piers on the left side are much stronger than those on the right side because the trains are assumed to run along the left sides of the viaducts. In particular, as shown in Figure 13, the amplitude at L-1 is somewhat larger than that of L-2. Figure 12 shows the probable reason: the maximum acceleration response that engenders a larger inertia force appears at the hanging part of the bridge. Dynamic reaction forces at the bottoms of the piers obtained here are used as input external excitations in the consequent ground vibration analysis.

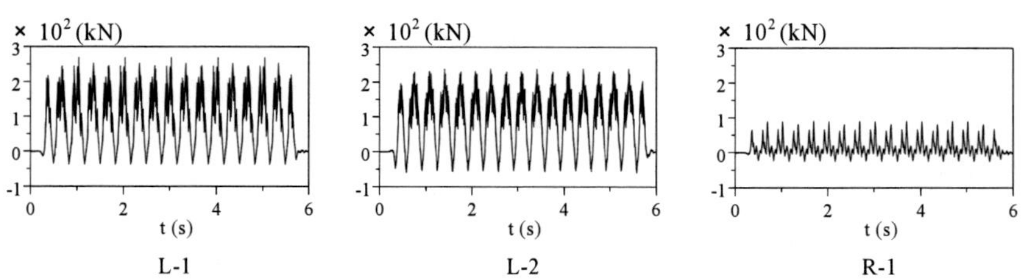

Figure 13. Reaction forces at pier bottoms of viaducts.

3.2.4. Ground Vibration Evaluation

Considering the predominant frequency components of the external forces that are confirmed within 15 Hz, the soil damping effect, and the efficiency of the analysis, the highest frequency considered in the analysis is determined as 25 Hz. By applying the dynamic reaction forces obtained in the bridge vibration analysis at 24 total footings, the ground response analyses using the numerical models described previously is carried out with the SASSI2000 computer program.

Numerical results of the points of Vicinity, 12.5 m and 25.0 m indicated in Figure 10 with the experimental values are shown in Figure 14 and are designated as (a) on the line. In spite of the complicated nature of the whole train-bridge-ground interaction system and the approximations or assumptions that have to be made in modeling the system, the numerical acceleration responses indicate comparatively good agreement with the experimental ones. The amplitudes of the numerical results are considerably coincident with the experimental ones. The components of the Fourier spectra of the numerical results to a certain extent re-create the experimental ones. Thereby the validity of the numerical procedure can be confirmed.

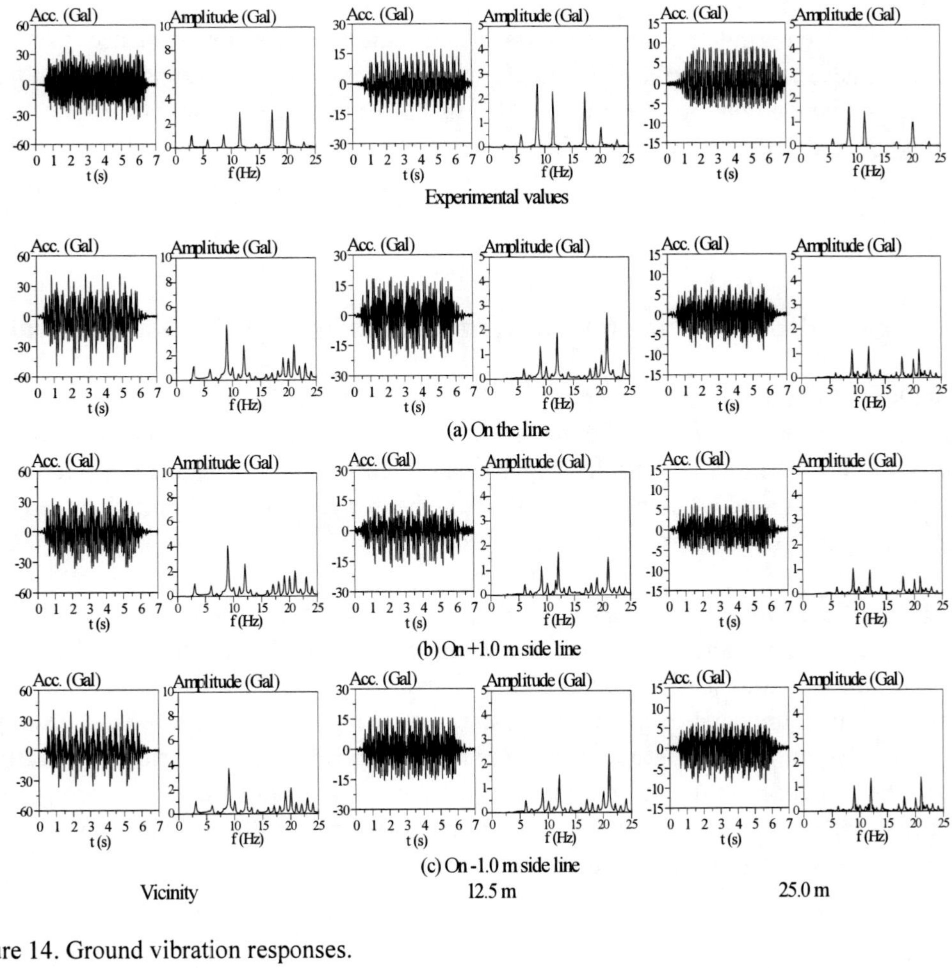

Figure 14. Ground vibration responses.

On the other hand, for points on the line, the predominant frequency components in the numerical results differ somewhat from those of the experimental ones, especially at around 20 Hz. In the actual field test, the surveyed points might differ slightly from those that are prearranged according to local conditions. In turn, that difference might affect the predominant frequency component of the vibration because of the phase difference of the waves. To investigate such influences, the responses of the points that are 1.0 m shifted from the points shown in Figure 10 in the x direction on right and left sides, respectively designated as (b) On +1.0 m side line and (c) On -1.0 m side line, are simulated. The results are also shown in Figure 14. As shown in the figure, the responses of the shifted points are similar to those of the points on the line, but they indicate some differences for amplitude and frequency components. Especially for the point of 12.5 m, the results of (b) indicate better agreement with experimental results. On the other hand, the predominant frequency components at about 10 Hz for the point of Vicinity and at about 20 Hz for the point of 12.5 m indicate some disagreement with that of the experimental ones. It can be considered that the discrepancies of the results arise from the difference between the actual ground properties and the idealized model. However, for frequency components of 3–12 Hz, which are mostly related to environmental problems, the numerical results are considered useful to evaluate environmental vibration problems.

It can be seen from the numerical results, the ground vibration response around the viaducts can have rather complicated natures. Through evaluating the ground vibration results of slightly shifted points on the line parallel to the bridge, the possibility exists that the vibration intensity may vary considerably even for near points. Such phenomenon should be carefully considered while evaluating the running train-induced environmental problems around the railway viaducts.

3.3. Design and Evaluation on Countermeasures against Excessive Vibration

Employing the numerical procedure established in this study, further investigation and evaluation as well as expectation of the bridge and related ground vibration can be conducted. Furthermore, it is also possible to investigate the effectiveness of designed countermeasures against bridge and ground vibration by reinforcing the bridge structure, improving the ground conditions, or employing other means.

3.3.1. Effect Evaluation on Simple Vibration Reduction Countermeasures

Dynamic characteristics of the high-speed railway elevated bridge due to bullet train are clarified in the previous part. Countermeasures against predominant vibration of the bridge are proposed and simulated in this part, considering such dynamic characteristics. Since the predominant vibrations occur at the hanging parts of the viaduct, it is considered effective to decrease the vibration by reinforcing the hanging parts. Two simple measures are proposed here at first. One is to connect the neighboring hanging parts rigidly and the other intends to raise the stiffness of the hanging parts by supporting them with H steel struts.

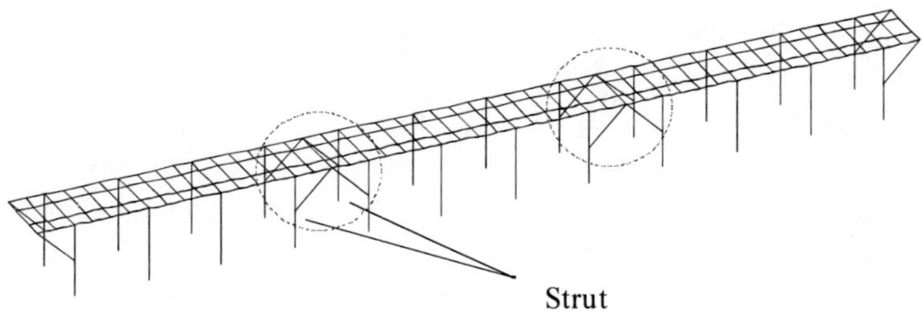

Figure 15. Numerical model of bridge with struts.

Proposal of reinforcing the hanging parts by H steels is depicted in Figure 15. The reinforcing material i.e. H steel is assumed to have 1/2 stiffness of the piers and is arranged to connect to the hanging part at one end and to the middle of the pier at the other end. The other analytical conditions, including the train model, are assumed the same with the previous analysis under the speed of 270 km/h.

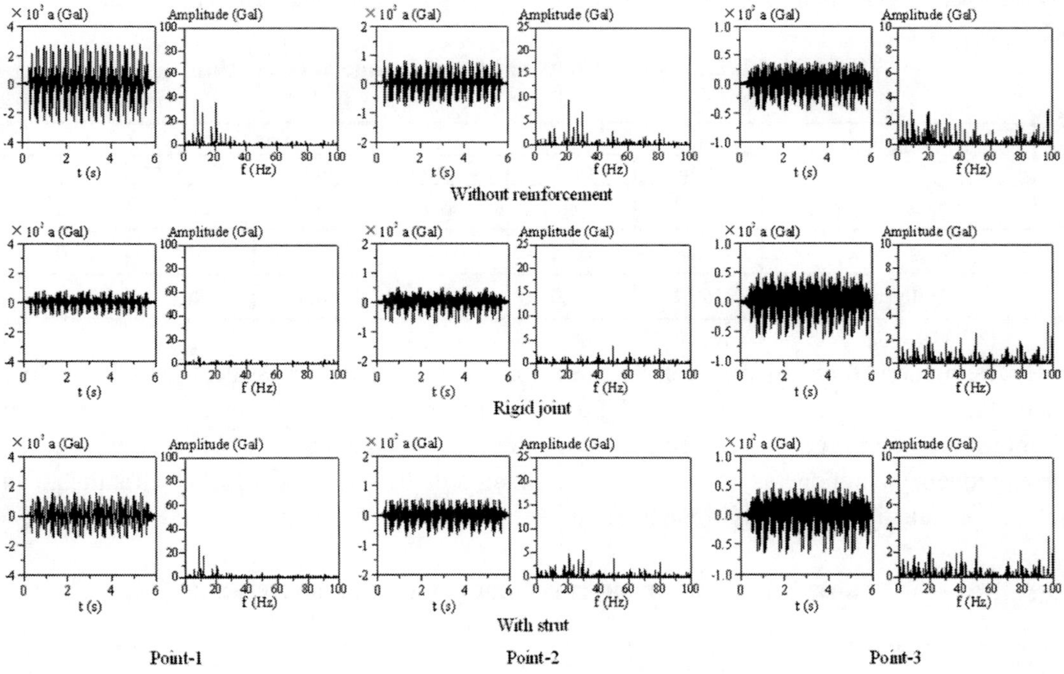

Figure 16. Acceleration of the bridge before and after reinforcement.

Eigenvalue analyses of both proposals are performed and the natural modal shapes simultaneously with the natural frequencies are confirmed. In the case of reinforcing with H steel, the predominant frequency of bending of the hanging part is observed to be 12.87 Hz that is about 1 Hz higher compared with that before reinforcement, and the predominant frequency of torsion is 14.09 Hz. In the case of reinforcement by connecting the hanging parts rigidly, the predominant frequency of bending mode of the hanging part is observed to be 17.86 Hz that is even higher than the case of reinforcing with H steel struts.

The numerical results of bridge acceleration response in vertical direction of the cases of H steel reinforcement and rigid joint simultaneously with those before reinforcement are shown in Figure 16. The maximum and rms values are shown in Table 4. As known from the comparison, at hanging parts of the bridge the acceleration responses are remarkably decreased after reinforcement for either reinforcement by H steel or rigid connection. From the Fourier spectra of acceleration responses, the low frequency components from 10 Hz to 30 Hz are observed extremely reduced, which is predicted to influence the ground vibration caused by the bridge vibration. Influenced by the vibration of hanging part, the response of point-2 is also observed to some extent reduced. On the other hand, the response of point-3 which is far from the hanging part indicates no notable change after reinforcement. According to the discussion, the effect on reducing bridge vibration of both reinforcement proposals can be expected. In addition, the effect of rigid connection is obviously conspicuous than that of H steel reinforcement, though it may bring some additional structural problems.

Table 4. Maximum and rms values of bridge acceleration

	Maximum (Gal)			rms value (Gal)		
	Point-1	Point-2	Point-3	Point-1	Point-2	Point-3
Without reinforcement	279.4	78.3	44.9	106.5	26.8	14.8
Rigid joint	93.1	65.8	62.4	24.5	14.5	17.0
With strut	157.2	72.0	65.4	55.4	20.1	16.6

The numerical results of acceleration response of the ground in vertical direction before and after reinforcement are shown in Figure 17. The maximum and rms values are shown in Table 5. The same reduction effect as that of the bridge can be confirmed. In particular, the low frequency components of the ground vibration are obviously reduced, which indicate an effective reduction of the environmental impact.

Table 5. Maximum and rms values of ground acceleration

	Maximum (Gal)			rms value (Gal)		
	Point-1	Point-2	Point-3	Point-1	Point-2	Point-3
Without reinforcement	48.8	21.7	9.0	14.0	7.3	3.0
With strut	38.2	18.1	6.25	10.6	6.2	2.3
Rigid joint	25.8	13.7	4.8	8.5	3.9	1.6

3.3.2. Effect Evaluation on an Actual Countermeasure

An actual constructed countermeasure to reduce the bridge vibration by Hara et al. [9] is examined using the developed numerical procedure. The outline of the reinforcement is shown in Figure 18. Here, the bridge model is established according to the information extracted from [9]. The properties of the bridge are almost the same as those of the previous model, except that the height of the bridge is 6.2 m and the ground condition is different.

To confirm the validity of the analysis, the numerical results and the experimental values (only Point-2 available for bridge vibration) are compared in Figure 19. The numerical results indicated relatively good agreement with the experimental ones. The difference between the experimental and numerical results is considered reasonable, because the complicated structure of the reinforcement cannot be fully reflected by the current numerical model.

Then, the numerical ground vibration results of the points of Vicinity, 12.5 m and 25 m before and after the reinforcement are shown in Figure 20. From the results, the effect of the countermeasure can be fully confirmed.

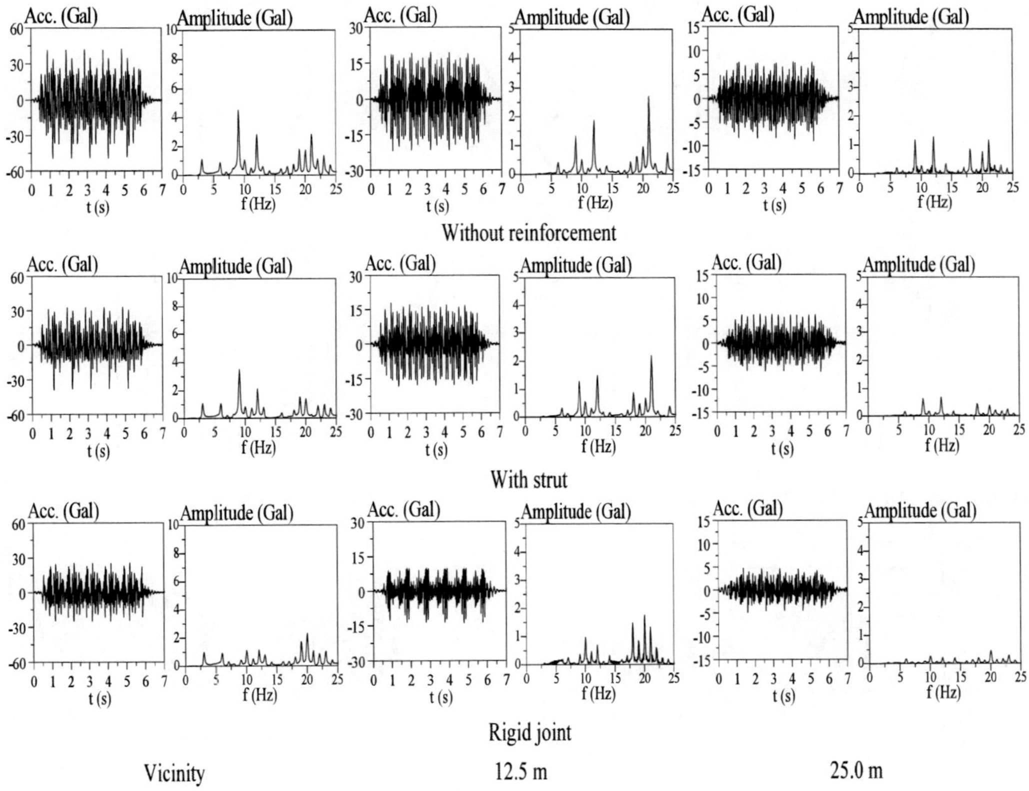

Figure 17. Acceleration of the ground before and after reinforcing the bridge.

3.4. Numerical Simulation Considering Horizontal Vibration

In the case of simulating the ground vibration around viaducts considering both vertical and horizontal components, it is necessary to use the horizontal dynamic reaction forces of the bridge simultaneously with the vertical ones. The horizontal response of the bridge may affect the horizontal ground vibration especially in the ambient area around the viaducts. Therefore in this part, the more complicated train model that can consider horizontal vibrations of the train as shown in Figure 1 and formulated in Section 2 is used to carry out the simulation of both vertical and horizontal bridge and ground vibration. By applying both the horizontal and vertical dynamic reaction forces at the pier bottoms obtained by train-bridge interaction analysis, the ground vibration of both directions can be simulated.

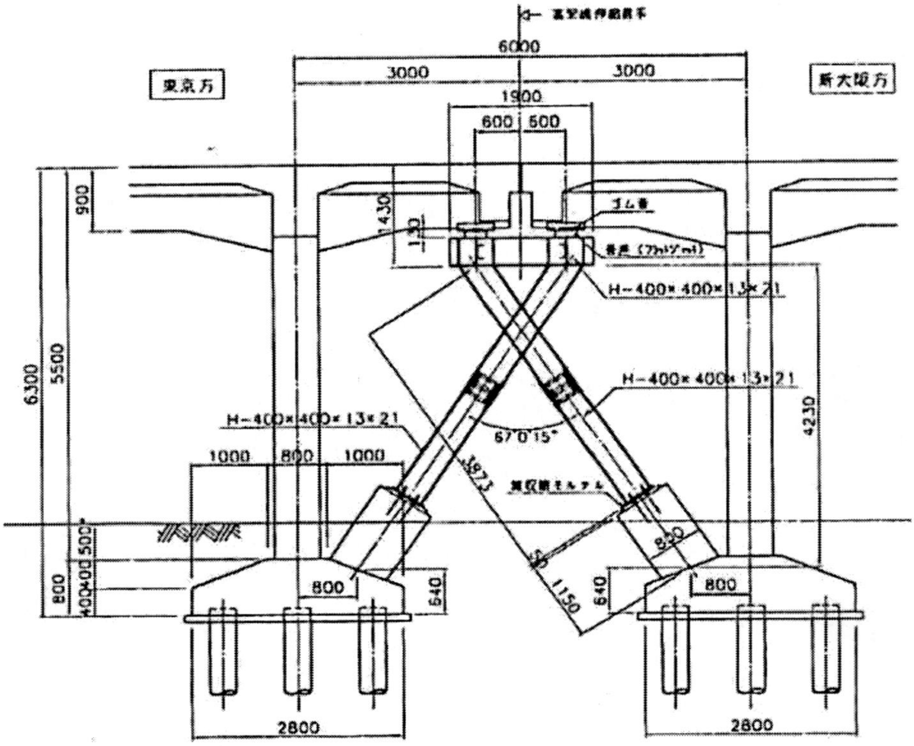

Figure 18. Outline of X-shape strut.

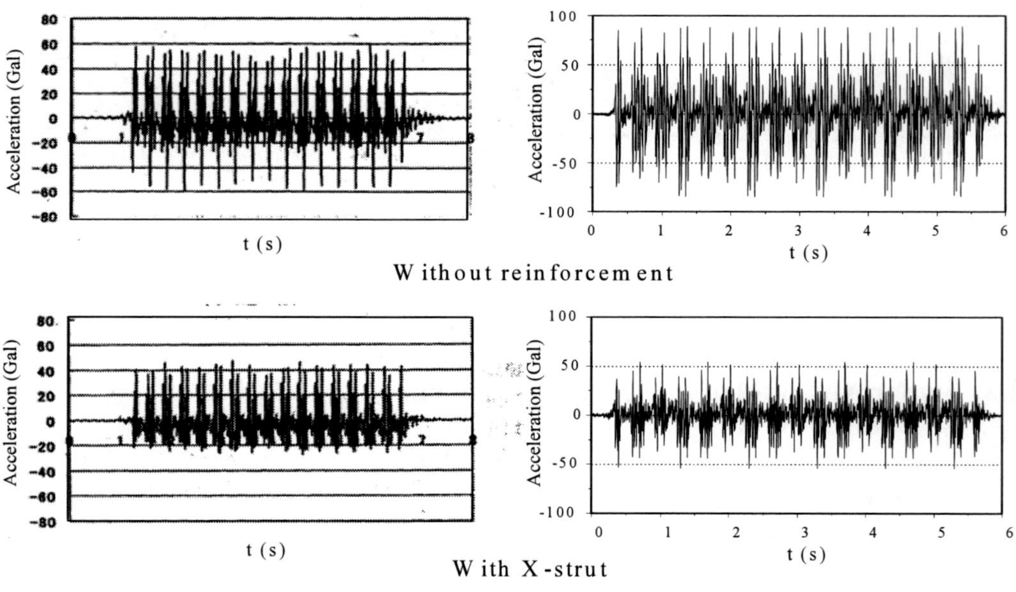

Figure 19. Bridge responses before and after actual reinforcement (Point-2).

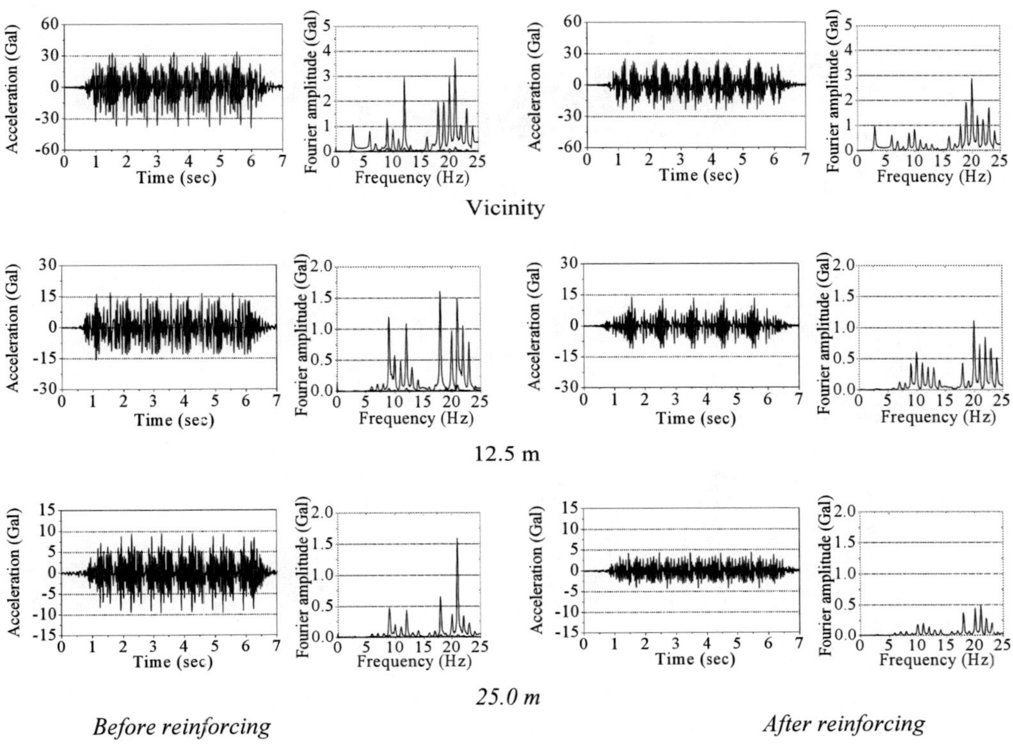

Figure 20. Comparison of ground responses before and after actual reinforcement.

3.4.1. Dynamic Responses of the Bridge

The numerical acceleration responses and the experimental ones in vertical direction, of point-1 through point-3 of the bridge, are shown respectively in Figure 21, and those in horizontal direction of point-3 are shown in Figure 22. As shown in these figures, numerical results using the complicated train model indicate good agreement with experimental results, thereby validating this numerical procedure. The relatively larger difference in the horizontal direction is due to the simplified contact relation between the wheel and rail, which can be improved by using more elaborated numerical models.

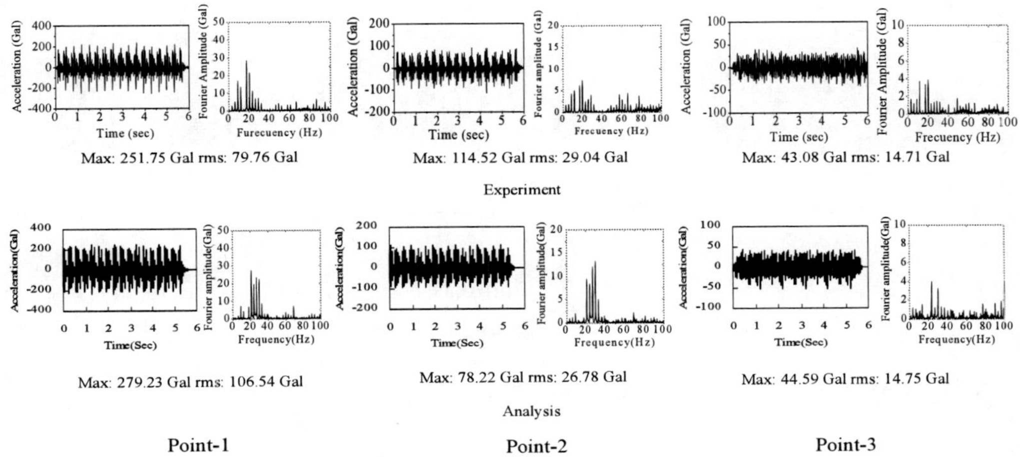

Figure 21. Acceleration of the bridge (vertical).

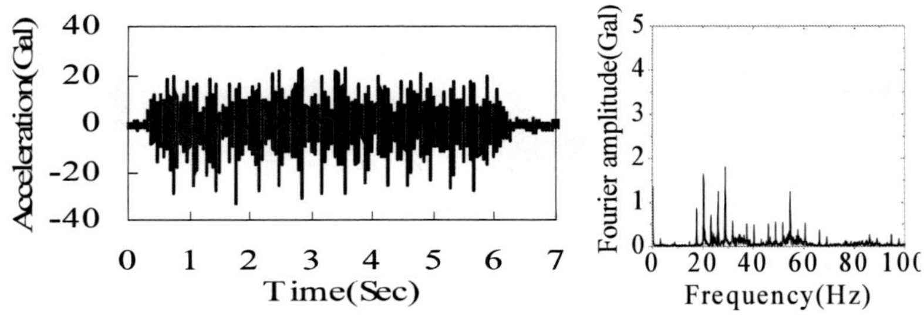

Max: 33.31 Gal rms: 7.41 Gal

(a) Experiment

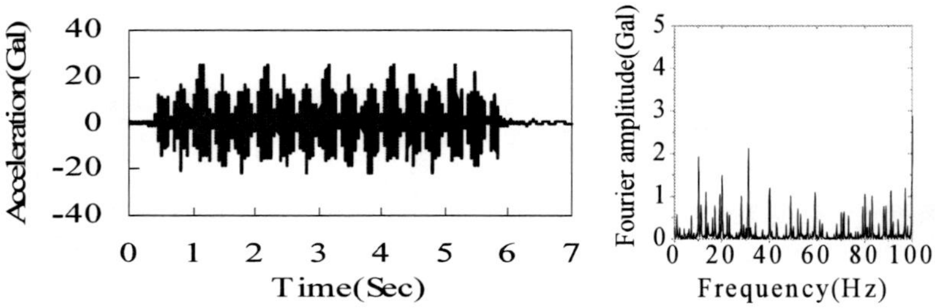

Max: 25.47 Gal rms: 6.81 Gal

(b) Analysis

Figure 22. Acceleration of bridge (Point-3, horizontal).

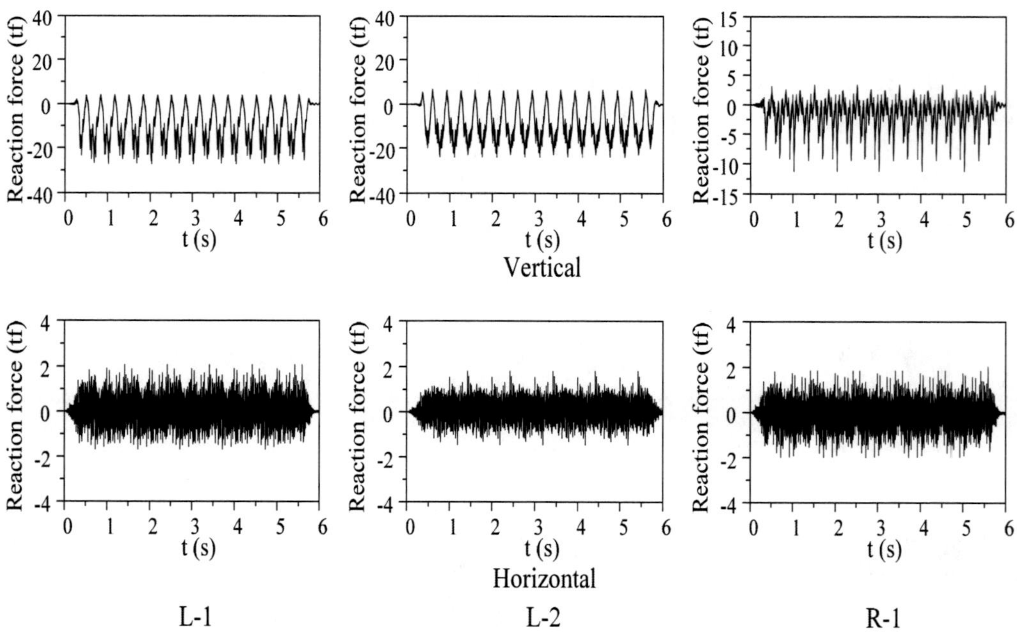

Figure 23. Reaction forces at pier bottoms of the bridge.

Reaction forces at the bottoms of the piers in vertical and horizontal directions, as respectively shown in Figure 23, are calculated using the influence value matrix of the reaction forces. The vertical reaction forces of the piers on the left side are much stronger than those on the right side because the trains are assumed to run along the left sides of the bridges. On the other hand, the reaction forces on the left and right sides in horizontal direction display similar amplitudes. In particular, for both directions the amplitude at L-1 is larger than that of L-2. The reason is the same as described previously that the maximum acceleration response engendering a larger inertia force appears at the hanging part of the bridge. Dynamic reaction forces obtained here are used as input external excitations in ground vibration problems.

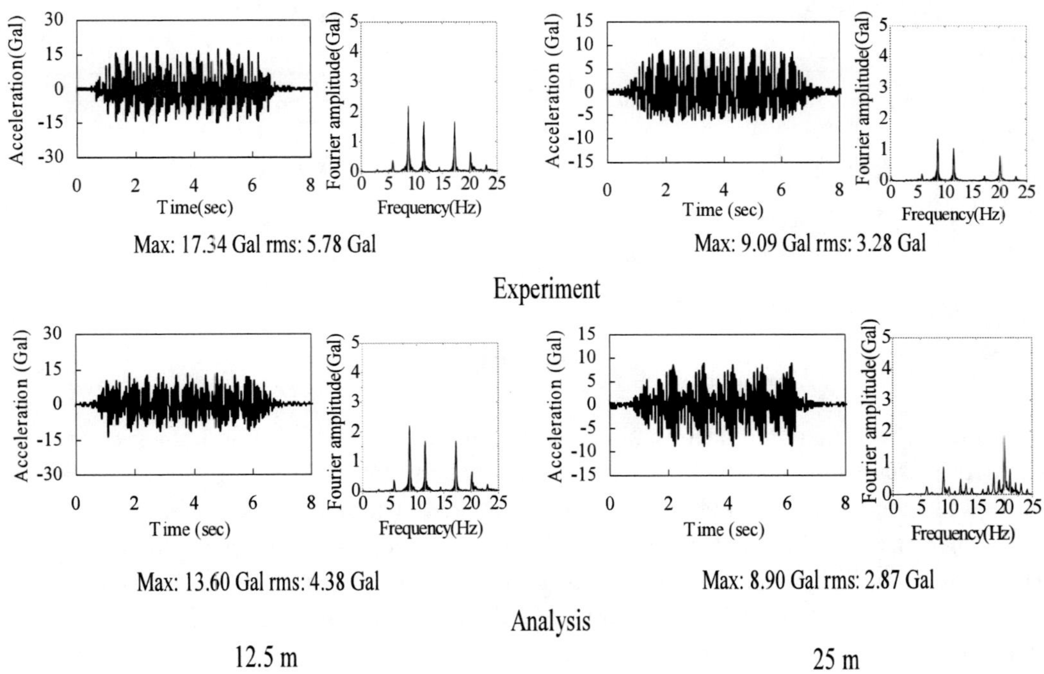

Figure 24. Acceleration of ground (vertical).

3.4.2. Numerical Results of the Ground Vibration

Numerical results and experimental values in vertical and horizontal directions simultaneously with the maximum and rms values of the points of 12.5 m and 25 m are shown in Figures 24 and 25. For both vertical and horizontal directions, the amplitudes of numerical results show relative agreement with the experimental ones. On the other hand, the predominant frequency components, particularly in the horizontal direction, indicate somewhat disagreement with those of the experimental ones. The discrepancies of the results may arise from the difference between the actual and idealized ground properties as well as the simplification of numerical models. However, considering the complicated nature of the whole train-bridge-ground system, the numerical results obtained here are considered useful to evaluate the ground vibration around viaducts.

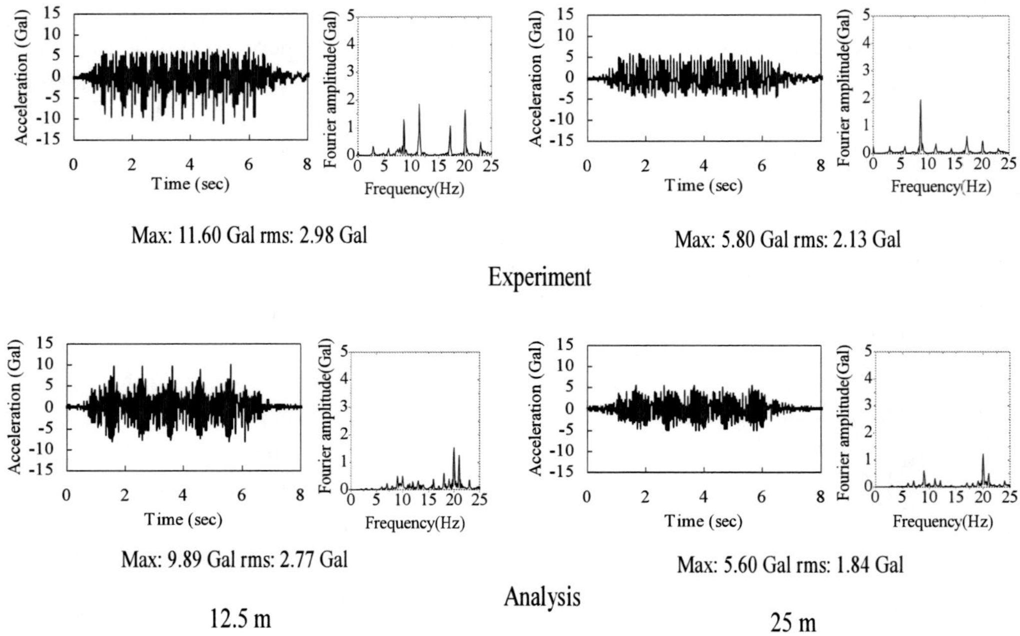

Figure 25. Acceleration of ground (horizontal).

CONCLUSION

In this chapter, an approach to simulate the ground vibration around high-speed railway viaducts is established, in which the dynamic interactions between the train and bridge as well as between the foundation and ground are considered. The numerical results are compared with experimental values and the validity of the developed numerical approach is confirmed. Then, based on the dynamic characteristics of the bridge vibration, countermeasures to reduce the excessive vibration of both the bridge and the ground are designed and numerically evaluated. This numerical approach can be used as a tool to perform the simulation, evaluation and prediction of the traffic-induced bridge and ground vibration. Furthermore, it is also possible using it to investigate in advance the effectiveness of designed countermeasures against bridge and ground vibration by reinforcing the bridge structure, improving the ground conditions, or employing other means.

In the current approach, to simplify the problem, the entire interaction system is divided into two subsystems: the train-bridge interaction and the foundation-ground interaction, under the assumption that the two subsystems are connected with idealized ground springs. This assumption can introduce some difference between the numerical model and the actual structure, which means that the numerical results cannot reproduce the real ground responses with complete accuracy. To obtain more accurate numerical results in the future study, it is recommended to model the whole train-bridge-ground interaction system as an integrated system, though theoretical and computational difficulties are expected.

REFERENCES

[1] Seki M., Inoue Y., Naganuma Y. (1997). Reduction of subgrade vibration and track maintenance for Tokaido Shinkansen. *WCRR'97*, E.

[2] Yoshioka O., Ashiya K. (1988). Dependence of Shinkansen-induced ground vibration upon their influence factors. *QR of RTRI*, 29(4): 176–183.

[3] Fujikake T.A. (1986). A prediction method for the propagation of ground vibration from railway trains. *Journal of Sound and Vibration*, 111(2): 289–297.

[4] Takemiya H. (2003). Simulation of track-ground vibrations due to a high-speed train: the case of X-2000 at ledsgard. *Journal of Sound and Vibration*, 261(3): 503–526.

[5] Yang Y.B., Hung H.H., Chang D.W. (2003). Train-induced wave propagation in layered soils using finite/infinite element simulation. *Soil Dynamics and Earthquake Engineering*, 23(4): 263–278.

[6] Xia H., Cao Y.M., Zhang N., Qu J.J. (2002). Vibration effects of light-rail train-viaduct system on surrounding environment. *International Journal of Structural Stability and Dynamics*, 2(2): 227–240.

[7] Wu, Y.S., Hsu, L.C., Yang, Y.B. (2002). Ground vibrations induced by trains moving over series of elevated bridge. *Proc. of the 10th Sound and Vibration Conference*, p.1–7.

[8] Wu Y.S., Yang Y.B. (2004). A semi-analytical approach for analyzing ground vibrations caused by trains moving over elevated bridges. *International Journal of Soil Dynamics and Earthquake Engineering*, 24(12): 949–962.

[9] Hara T., Yoshioka O., Kanda H., Funabashi H., Negishi H., Fujino Y., Yoshida K. (20-04). Development of a new method to reduce Shinkansen-induced wayside vibrations applicable to rigid frame bridges: bridge-end reinforcing method. *Journal of Structural and Earthquake Engineering*, JSCE, 68(766): 325–338. (in Japanese)

[10] Yoshida K., Seiki M. (2004). Influence of improved rigidity in railway viaducts on the environmental ground vibration. *Journal of Structural Engineering*, JSCE, 50A: 403–412. (in Japanese)

[11] Kawatani M., He X., Shiraga R., Seki M., Nishiyama S., Yoshida K. (2006). Dynamic response analysis of elevated railway bridges due to Shinkansen trains. *Journal of Struc-tural and Earthquake Engineering (Doboku Gakkai Ron-bunshuu A)*, JSCE, 62(3): 509–519. (in Japanese)

[12] He X., Kawatani M., Nishiyama S. (2010). An analytical approach to train-induced site vibration around Shinkansen viaducts. *Structure and Infrastructure Engineering*, 6(6): 689–701.

[13] Lysmer J., Ostadan F., Chin, C.C. (1999). *SASSI2000 theoretical manual – A system for analysis of soil-structure interaction*. Academic Version, University of California, Berkeley.

[14] Lysmer J., Ostadan F., Chin C.C. (1999). *SASSI2000 user's manual – A system for analysis of soil-structure interaction*. Academic Version, University of California, Berkeley.

[15] Clough R.W., Penzien J. (1975). *Dynamics of Structures*. McGraw-Hill, Inc.

[16] Washidu K., Miyamoto H., Yamada Y., Yamamoto Y., Kawai T. (1981). *A handbook of finite element method · I Theory*. BAIFUKAN. (in Japanese)

[17] Washidu K., Miyamoto H., Yamada Y., Yamamoto Y., Kawai T. (1981). *A handbook of finite element method · II Application*. BAIFUKAN. (in Japanese)

[18] Bathe K.J. (1982). *Finite element procedures in engineering analysis*. Prentice-Hall, Englewood Cliffs, N. J.

[19] Weaver W. Jr., Johnston P.R. (1987). *Structural Dynamics by Finite Elements*. Prentice-Hall, Inc.

[20] Francis J.G.F. (1961). The QR Transformation, I. *The Computer Journal*, 4(3): 265–271.

[21] Atkinson Kendall E. (1989). *An introduction to numerical analysis*. Second Edition, John Wiley & Sons, New York.

[22] Wakui H., Matsumoto N., Matsuura A., Tanabe M. (1995). Dynamic interaction analysis for railway vehicles and structures. *Journal of Structural and Earthquake Engineering*, JSCE, I-31(513): 129–138. (in Japanese)

[23] Agabein M.E. (1971). The effect of various damping assumptions on the dynamic response of structure. *Bulletin of International Institute of Seismology and Earthquake Eng.*, 8: 217–236.

[24] Dahlquist G.G. (1963). A special stability problem for linear multistep methods. Nord. Tidskr. Inf. behandling, 3: 27–43.

[25] Hilber H.M., Hughes T.J.R. (1978). Collocation, Dissipation, and 'Overshoot' for Time Integration Schemes in Structural Dynamics. *Earthquake Engineering & Structural Dynamics*, 6(1): 99–117.

[26] Newmark N.M. (1959). A Method of Computation for Structural Dynamics. *Journal of the Engineering Mechanics Division, ASCE*, 85(EM3): 67–94.

[27] Timoshenko S.P., Young D.H., Weaver W. Jr. (1974). *Vibration problems in engineering*. 4th ed., Wiley, New York.

[28] Hilber H.M., Hughes T.J.R., Taylor, R.L. (1977). Improved Numerical dissipation for Time Integration Algorithms in Structural Mechanics. *Earthquake Engineering & Structural Dynamics*, 5(3): 283–292.

[29] Houbolt J.C. (1950). A recurrence matrix solution for the dynamic response of elastic aircraft. *Journal of Aeronautical Sciences*, 17: 540–550.

[30] Wilson E.L., Farhoomand I., Bathe K.J. (1973). Nonlinear dynamic analysis of complex structures. *Earthquake Engineering & Structural Dynamics*, 1(3): 241–252.

[31] Bathe K.J., Wilson E.L. (1973). Stability and accuracy analysis of direct integration methods. *Earthquake Engineering & Structural Dynamics*, 1(3): 283–291.

[32] Goudreau G.L., Taylor R.L. (1972). Evaluation of numerical integration methods in elastodynamics. *Computer Methods in Applied Mechanics and Engineering*, 2: 69–97.

[33] Argyris J.H., Dunne P.C., Angelopoulos T. (1973). Non-linear oscillations using the finite element technique. *Computer Methods in Applied Mechanics and Engineering*, 2: 203–250.

[34] JSCE (2002). *Evaluation on checking approaches to earthquake performance of concrete structures based on damage investigation in Hanshin Awaji Great Earthquake*. Concrete Engineering Series 49. (in Japanese)

[35] Hilber H.M. (1976). *Analysis and design of numerical integration methods in structural dynamics*. Ph.D. Dissertation, Division of Structural Engineering and Structural Mechanics, Univ. of California, Berkeley.

[36] Hughes T.J.R. (1976). Stability, convergence and growth and decay of energy of the average acceleration method in nonlinear structural dynamics. *Computers and Structures*, 6: 313−324.

[37] Chin C.C. (1998). *Substructure subtraction method and dynamic analysis of pile foundations*. Ph.D. Dissertation, University of California, Berkeley.

[38] Lysmer J. (1978). *Analytical procedures in soil dynamics*. Report No. EERC 78/29, Earthquake Engineering Research Center, University of California, Berkeley.

[39] Lysmer J., Drake J.A., (1972). A finite element method for seismology. *Method in Computational Physics*, 198, Academic Press Inc.

[40] Shimizu N., Yamamoto S., Koori Y., Minowa N. (1977).Three-dimensional dynamic analysis of soil-structure system by thin layer element method (Part 1. Outline of analysis method and formulation of layered zone). *Trans. of A. I. J.*, 253.

[41] Shimizu N., Yamamoto S., Koori Y., Minowa, N. (1977). Three-dimensional dynamic analysis of soil-structure system by thin layer element method (Part 2. Formulation of irregular zone and energy transmitting boundary). *Trans. of A. I. J.*, 254.

[42] Kobori T., Kubo M., (1985). A practical dynamic analysis method of continuous girder bridge with spring connections on spring supports. *Proceedings of JSCE/Japan Society of Civil Engineers*, 356(I−3): 395–403. (in Japanese)

[43] Nishimura A. (1990). A study on integrity assessment of railway rigid frame bridge. *RTRI Report*, 4(9). (in Japanese)

[44] Shoji T., Yito Y., Seki M. (2004). Research on axle load distribution of high-speed train and stress distribution of steel girder. *Proceedings of the 59th Annual Conference of the Japan Society of Civil Engineers*, I-078. (in Japanese)

In: Traffic Induced Environmental Vibrations and Controls ISBN: 978-1-62618-240-0
Editors: H. Xia and R. Calçada

Chapter 9

VIBRATION ISOLATION BY GROUND BARRIERS

Guang-Yun Gao [1, 2*], Gongqi Chen [1, 2] and Chang Qiu [3]

[1]Department of Geotechnical Engineering, Tongji University, Shanghai, China
[2]Key Laboratory of Geotechnical and Underground Engineering of Ministry of Education, Tongji University, Shanghai, China
[3]Shanghai Pudong Construction Project Management Co. Ltd. Shanghai, China

ABSTRACT

Ground vibrations can be caused by train/road traffics, machine operations, and pile driving and blasting. Such ground vibrations can cause distress to adjacent structures and annoyance to residents. Generally, the adverse effects of ground vibrations can be eliminated or prevented with the installation of an adequate wave barrier between the vibration source and the concerned structure. In this chapter, the propagation properties for vibration waves and corresponding isolation measures are investigated in analytical, numerical and field measurement. Integral equations governing Rayleigh wave scattering are derived based on the Green's solution of Lamb problem. The integral equations, which are used to evaluate the complicated Rayleigh wave field generated by irregular scatter embedded in an elastic half-space solid, can be solved accurately and efficiently with an iteration technique. The passive isolation effectiveness of ground vibration by the piles and in-filled trench for screening Rayleigh wave is further studied in detail. In addition, the behavior of WIB isolation is conducted both by TLM and in-situ measurement. As case studies, the vibration isolation performance of these barriers are analyzed and explained.

INTRODUCTION

The systems of ground vibration isolation can be classified into two categories: active isolation system and passive isolation system. The active isolation barrier system is often used to reduce the ground vibrations at the source. It is normally installed either around the

* E-mail address: gaoguangyun@263.net.

Vibration source or at a close distance to the source. The passive isolation barrier system, on the other hand, is usually built away from the source, surrounding the protected structure.

Detailed descriptions on the barrier systems may be found in Barken [1] and Richarts et al. [2], where some practical, empirical and experimental methods are discussed to achieve effective wave screening by trenches or sheet piles. Woods [3] and Haupt [4,5] conducted a series of field tests, analytical studies and laboratory model experiments in order to examine the screening performance of open trenches and concrete walls.

For passive isolation barriers, an integral equation for Rayleigh wave scattering was deduced based on the Green's solution of Lamb problem. For the problem under consideration, only the Rayleigh surface wave contribution (the Rayleigh part of the Green's function) was considered [6, 7], since Rayleigh wave transferred most of the vibration energy to remote location. An iteration technique was employed to solve the integral equation. The integral equation were used to examine the complicated Rayleigh wave field due to the presence of a large number of irregular scatters embedded in an elastic half-space medium. The passive isolation effectiveness of ground vibration by the in-filled trench and piles for screening Rayleigh wave was further studied in detail by Gao et al. [7, 8]. Effects of relevant parameters on the effectiveness of vibration isolation were investigated.

For active isolation barriers, the investigation of the isolation barrier in multilayered half-space was accomplished with the aid of the semi-analytical boundary element method (BEM) in cooperation with the basic solution of the thin-layered method (TLM). Consequently, a 3D semi-analytical BEM model was proposed by Gao et al. [9−12], where the TLM basic solution was employed. Considering the dynamical interaction of soil-structure, a 3D axisymmetric model was developed to analyze WIB as active vibration barrier. Effects of the model parameters on WIB isolation effectiveness were discussed. In addition, an investigation on the actual active vibration isolation effect of WIB in multilayered ground under vertical excitation was accomplished, by means of a series of field vibration isolation experiments. Based on the experiments conducted under 6 different working conditions, the influences of WIB plane size, shear modulus and embedded depth on the wave mitigation effect were examined.

2. Three-Dimensional Analysis of Row-Piles as Passive Barriers for Ground Vibration Isolation

2.1. The Integral Equation Method for R-wave Scattering

2.1.1. Integral Equation Governing Elastic Waves Scattering

The partial differential equation governing motion in isotropic and homogeneous linear elastic solids can be expressed in terms of the displacements and in the frequency domain as follows:

$$C_{ijkl}u_{k,lj} + \rho\omega^2 u_i + \rho f_i = 0 \tag{1}$$

where $u_i = u_i(\mathbf{x}, \omega)$ is the amplitude of the displacement vector; $\mathbf{x}$ is a point of the elastic solid with volume V; ω is the circular frequency of vibration; ρ is the mass density of the elastic solid; f_i is a body force; C_{ijkl} is the elastic modulus tensor; and summation is assumed over repeated indices and $i, j, k, l = 1,2,3$, respectively.

It can be shown that the Green's function $G_{im}(\mathbf{x}, \mathbf{x}')$ of Lamb problem satisfies the following equation:

$$C_{ijkl} G_{km,lj} + \rho\omega^2 G_{im} + \delta_{im}\delta(\mathbf{x}-\mathbf{x}') = 0 \tag{2}$$

where, $G_{im}(\mathbf{x}, \mathbf{x}')$ is the im-th element of the Green's tensor and represents the displacement vector at the position $\mathbf{x}$ along direction i when a force is applied to the position $\mathbf{x}'$ along direction m; $\delta(\mathbf{x} - \mathbf{x}')$ is the three-dimensional (3D) Dirac delta function.

Multiplying Equation (1) with G_{im} and multiplying Equation (2) with u_i, then subtracting the two resulting equations, the following equation can be obtained

$$u_m \delta(\mathbf{x}-\mathbf{x}') = C_{ijkl}(G_{im} u_{k,lj} - u_i G_{km,lj}) + \rho f_i G_{im} \tag{3}$$

Integrating Equation (3) over the solid domain integral, and using the Gauss's integration theorem, the following equation can be derived

$$u_m(\mathbf{x}') = \oiint_{S} C_{ijkl}(G_{im} u_{k,l} - u_i G_{km,l}) n_j \mathrm{d}s + \iiint_{\Omega} \rho f_i G_{im} \mathrm{d}v \tag{4a}$$

The integration of the first term on the right-hand side is over the surface S and only depends on the displacement u_i and traction σ_{ij} of the surface. The integration is noted as follows:

$$u_m^0(\mathbf{x}') = \oiint_{S} C_{ijkl}(G_{im} u_{k,l} - u_i G_{km,l}) n_j \mathrm{d}s \tag{4b}$$

Further changing $\mathbf{x}'$ with $\mathbf{x}$ and using Equation (4b), Equation (4a) can be re-expressed as:

$$u_m^0(\mathbf{x}) = u_m(\mathbf{x}) - \iiint_{\Omega} \rho \cdot f_i(\mathbf{x}') G_{im}(\mathbf{x}'; \mathbf{x}) \mathrm{d}v(\mathbf{x}') \tag{5a}$$

Besides, we have the following equations to govern u_m^0:

$$\begin{aligned} C_{ijkl} u_{k,lj}^0 + \rho\omega^2 u_i^0 &= C_{ijkl} u_{k,jl} + \rho\omega^2 u_i - \iiint_{\Omega} f_p (C_{ijkl} G_{pi,jl} + \rho\omega^2 G_{pi}) \mathrm{d}v \\ &= C_{ijkl} u_{k,jl} + \rho\omega^2 u_i + \rho f_i = 0 \end{aligned} \tag{5b}$$

Equation (5b) shows that u_m^0 satisfy the motion equation with zero body force. In other words, it meets the transmitting equation of incident wave.

Generally, for each point on the boundary S, where either the displacement or the traction is known, so u_m^0 can be considered as the incident wave u_m^i for the problem with scattering. Thus, Equation (5a) can be re-expressed as

$$u_m(\mathbf{x}') = u_m^{(i)}(\mathbf{x}') + \iiint\limits_{\Omega} \rho f_i G_{im}(\mathbf{x};\mathbf{x}')\mathrm{d}v(\mathbf{x}) \tag{6a}$$

$$u_m^{(i)}(\mathbf{x}') = C_{ijkl} \iint\limits_{S} (G_{im}(\mathbf{x};\mathbf{x}')u_{k,l} - u_i G_{km,l}(\mathbf{x};\mathbf{x}')n_j \mathrm{d}s \tag{6b}$$

Equations (6a) and (6b) define the incident wave, which are the integral expression for the displacement field in an elastic isotropic medium with the body force. This expression shows the resultant displacement can be expressed as the summation of the incident wave field and a scattering field (a volume integration generated by the body force).

We further assume that the elastic solid contains N sub-domains Ω_1, Ω_2, … and Ω_N. The sub-domains represent obstacles, which are bounded by the surfaces S_1, S_2, … and S_N, with the elastic modulus matrices C_{hjkl}^1, C_{hjkl}^2, …, C_{hjkl}^N, and mass densities ρ_1, ρ_2, …, ρ_N, respectively.

Furthermore, the following 3D Heaviside function, H_1, H_2, …, H_N are introduced:

$$H_\alpha(\mathbf{x}) = \begin{cases} 1 & \mathbf{x} \in \Omega_\alpha \\ 0 & \mathbf{x} \notin \Omega_\alpha \end{cases} \quad \alpha = 1, 2, \ldots, N. \tag{7a}$$

The elastic constants over the entire region Ω are written as

$$\begin{cases} C_{ijkl}^t(\mathbf{x}) = C_{ijkl} + \sum\limits_{\alpha=1}^{N} \Delta C_{ijkl}^\alpha H_\alpha(\mathbf{x}) \\ \rho^t(\mathbf{x}) = \rho + \sum\limits_{\alpha=1}^{N} \Delta\rho_\alpha H_\alpha(\mathbf{x}) \end{cases} \quad \mathbf{x} \in \Omega \tag{7b}$$

where $\Delta C_{ijkl}^\alpha = C_{ijkl}^\alpha - C_{ijkl}$, $\Delta\rho_\alpha = \rho_\alpha - \rho$, and α=1, 2, …, N.

We consider the case without body force and with an infinite elastic solid region Ω. For this special case, the motion equation can be expressed as:

$$[C_{ijkl}^t(\mathbf{x})u_{k,l}(\mathbf{x})]_{,j} + \rho^t(\mathbf{x})\omega^2 u_i(\mathbf{x}) = 0\,. \tag{8}$$

Substituting Equation (7) into Equation (8) yields

$$C_{ijkl}u_{k,lj} + \rho\omega^2 u_i = -\sum_{\alpha=1}^{N}(\Delta C_{ijkl}^{\alpha} H_{\alpha} u_{k,lj} + \Delta C_{ijkl}^{\alpha} H_{\alpha,j} u_{k,l} + \Delta\rho_{\alpha} H_{\alpha}\omega^2 u_i). \quad (9)$$

All the terms on the right-hand side of Equation (9) can be treated as a body force, which can be substituted into Equation (6). We can further use the mathematical characteristics of H_{α} and the Green's theorem to make further derivations, resulting

$$u_m(\mathbf{x}') = u_m^{(i)}(\mathbf{x}') + \sum_{\alpha=1}^{N}\Delta\rho_{\alpha}\omega^2 \iiint_{\Omega_{\alpha}} G_{im}(\mathbf{x};\mathbf{x}')u_i \mathrm{d}v - \sum_{\alpha=1}^{N}\Delta C_{ijkl}^{\alpha}\iiint_{\Omega_{\alpha}} G_{im,j}(\mathbf{x};\mathbf{x}')u_{k,l}\mathrm{d}v. \quad (10)$$

Using the reciprocity theorem of Green's function, we obtain the integral equation as

$$u_i(\mathbf{x}) = u_i^{(i)}(\mathbf{x}) + \sum_{\alpha=1}^{N}\Delta\rho_{\alpha}\omega^2 \iiint_{\Omega_{\alpha}} G_{im}(\mathbf{x};\mathbf{x}')u_m(\mathbf{x}')\mathrm{d}v + \sum_{\alpha=1}^{N}\Delta C_{mjkl}^{\alpha}\iiint_{\Omega_{\alpha}} G_{mi,j}(\mathbf{x};\mathbf{x}')u_{k,l'}(\mathbf{x}')\mathrm{d}v. \quad (11)$$

Equations (10) and (11) are the integral equations for scattered displacement fields by many elastic obstacles embedded in an isotropic elastic solid. These equations can be solved by an iterative numerical method from the integral representation in Equation (11).

Evidently, this iterative process is very complex in computation. When the obstacles have sizes less than the incident wavelength, the Born approximation can be used to solve the scattered amplitudes for the obstacles. However, the Born approximation cannot be used to solve the strong scattering problem where the obstacles have great sizes. For example, piles installed in the ground are used as the obstacles for isolating barrier.

2.1.2. Integral Equation of Rayleigh Wave Scattering

Equations (10) and (11) are still ineffective for the present purpose. Further modifications are needed and presented in this section.

For the scattering problem, the resultant displacement can be expressed as a sum of the incident displacement and the scattered displacement, and the scattered displacement can be given as a series expansion as follows:

$$u_i(x) = u_i^{(i)}(x) + \sum_{p=1}^{M} u_i^{(p)}(x) \quad (12)$$

where $u_i^{(p)}(\mathbf{x})$ is the P-th scattered displacement.

Substituting Equation (12) into Equation (10) and considering the first iteration (i.e. Born approximation), we can have the following result:

$$u_i^{(1)}(\mathbf{x}') = \sum_{\alpha=1}^{N}\Delta\rho_{\alpha}\omega^2 \iiint_{\Omega_{\alpha}} G_{mi}(\mathbf{x};\mathbf{x}')u_m^{(i)}\mathrm{d}v - \sum_{\alpha=1}^{N}\Delta C_{mjkl}^{\alpha}\iiint_{\Omega_{\alpha}} G_{mi,j}(\mathbf{x};\mathbf{x}')u_{k,l}^{(i)}\mathrm{d}v. \quad (13)$$

Similarly, consider the (p-1)-th scattered fields as the incident wave of the p-th scattered fields and substitute the result into Equation (10), we can obtain the following equation:

$$u_i^{(p)}(\mathbf{x}')=\sum_{\alpha=1}^{N}\Delta\rho_\alpha\omega^2\iiint_{\Omega_\alpha}G_{mi}(\mathbf{x};\mathbf{x}')u_m^{(p-1)}\mathrm{d}v-\sum_{\alpha=1}^{N}\Delta C_{mjkl}^{\alpha}\iiint_{\Omega_\alpha}G_{mi,j}(\mathbf{x};\mathbf{x}')u_{k,l}^{(p-1)}\mathrm{d}v\,. \tag{14}$$

Details of the Green's function for an elastic half-space can be found in Aki and Richards [6]. In this section, only the Rayleigh surface-wave terms are considered. For isotropic or transversely isotropic elastic solid, the surface-wave terms of the Green's function in a half-space are expressed as:

$$G_{qi}(\mathbf{x};\mathbf{x}')=\sum_m Z_q^m(z)g_i^m(x,y;\mathbf{x}') \tag{15}$$

where $Z_q^m(z)$ is an operator notation with the form

$$Z_q^m(z)=\begin{pmatrix} \frac{1}{k_m}\left(-V^m(z)\frac{\partial}{\partial x}+W^m(z)\frac{\partial}{\partial y}\right)\\ \frac{1}{k_m}\left(-W^m(z)\frac{\partial}{\partial x}-V^m(z)\frac{\partial}{\partial y}\right)\\ U^m(z)\end{pmatrix} \tag{16}$$

and V^{m}(z), U^{m}(z) are the horizontal and vertical displacement induced by the m-th Rayleigh mode; W_m(z) is the horizontal displacement generated by the m-th Love mode; k_m is wave number of the m-th surface wave mode; X is the observation point and X' is the source point.

We note that $g_i^m(x,y;\mathbf{x}')$ is the displacement potential,

$$g_i^m(x,y;\mathbf{x}')=\frac{-i}{8c_m c_m^g J_1^m}Z_i^m(z')H_0^{(2)}(k_m R) \tag{17}$$

where c_m and c_m^g are the phase velocity and group velocity of the m-th surface wave mode, respectively; J_1^m is the normalized integral value given by Equation (18); $H_n^{(2)}$ is the Hankel function of second kind of order zero; R is the distance between the observation point and the source.

$$J_1^m=\frac{1}{2}\int_0^\infty \mathrm{d}z\rho[(U^m)^2+(V^m)^2+(W^m)^2] \tag{18}$$

where U^m, V^m, W^m, c_m and c_m^g are determined in detail in Aki and Richards [6].

Equation (15) is a general formula for the surface-terms of Green's function, which governs all the Rayleigh and Love modes. For isotropic elastic half-space, W^m=0 and m=1 in

Equation (16) because there is no Love wave. For layered elastic soil, W^m=0 for the m-th Rayleigh mode and U^m=V^m=0 for the m-th Love mode.

Similarly, the displacement of the p-th scattered field can be expressed as the same form as Equation (15),

$$u_i^{(p)}(\mathbf{x}) = \sum_m Z_i^m(z)\phi_m^{(p)}(x,y) \tag{19}$$

where $\phi_m^{(p)}(x,y)$ is the displacement potential of the m-th mode surface wave in the p-th scattered field.

Accordingly, the (p-1)-th scattered field can be written as the similar form if we change $\phi_m^{(p)}(x,y)$ as $\phi_m^{(p-1)}(x,y)$ in Equation (19).

So far, computation of displacement $\phi_i^{(p)}(x,y)$ is replaced by solving simpler equations with variance of potential $\phi_m^{(p)}(x,y)$. Substituting Equations (15) and (19) into (14) yields,

$$\begin{aligned}\phi_m^{(p)}(x,y) = &\sum_{\alpha=1}^{N} \Delta\rho_\alpha\omega^2 \iiint_{\Omega_\alpha} g_k^m \sum_n Z_k^n(z')\phi_n^{(p-1)}(x',y')\mathrm{d}v(\mathbf{x}') \\ &- \sum_{\alpha=1}^{N} \Delta C_{ijkl}^{\alpha} \iiint_{\Omega_\alpha} g_{i,j'}^m [\sum_n Z_k^n(z')\phi_n^{(p-1)}(x',y')]_{,l'}\mathrm{d}v(\mathbf{x}')\end{aligned} \tag{20}$$

Simplifying Equation (20) and taking out the summation notation related to n, the following equation is obtained:

$$\begin{aligned}\phi_m^{(p)}(x,y) = \sum_n \Bigg\{ &\sum_{\alpha=1}^{N} \Delta\rho_\alpha\omega^2 \iiint_{\Omega_\alpha} g_k^m Z_k^n(z')\phi_n^{(p-1)}(x',y')\mathrm{d}v(\mathbf{x}') \\ &- \sum_{\alpha=1}^{N} \Delta C_{ijkl}^{\alpha} \iiint_{\Omega_\alpha} g_{i,j'}^m [Z_k^n(z')\phi_n^{(p-1)}(x',y')]_{,l'}\mathrm{d}v(\mathbf{x}') \Bigg\}.\end{aligned} \tag{21}$$

Introducing a new potential function and writing Equation (21) as

$$\begin{aligned}\phi_{mn}^{(p)}(x,y) = &\sum_{\alpha=1}^{N} \Delta\rho_\alpha\omega^2 \iiint_{\Omega_\alpha} g_k^m Z_k^n(z')\phi_n^{(p-1)}(x',y')\mathrm{d}v(\mathbf{x}') \\ &- \sum_{\alpha=1}^{N} \Delta C_{ijkl}^{\alpha} \iiint_{\Omega_\alpha} g_{i,j'}^m [Z_k^n(z')\phi_n^{(p-1)}(x',y')]_{,l'}\mathrm{d}v(\mathbf{x}')\end{aligned} \tag{22}$$

$$\phi_m^{(p)} = \sum_n \phi_{mn}^{(p)} \tag{23}$$

where $\phi_{mn}^{(p)}$ is the displacement potential of m-th mode surface wave in p-th scattered field induced by the potential of n-th mode surface wave in (p-1)-th scattered field. Z_q^m is an operator including the partial differentiations with respect to x and y.

Assuming a harmonic Rayleigh wave travels along the x-axis direction, the variations of potential function in the y-axis direction can be neglected with respect to the one in the x-axis direction. Therefore, Equation (22) can be further simplified as

$$\begin{aligned}\phi_{mn}^{(p)}(x,y) &= \sum_{\alpha=1}^{N}\iiint_{\Omega_\alpha}(-\Delta C_{ijk3}^{\alpha} g_{i,j'}^{m} S_{k,3'}^{n} + ik_m \Delta C_{ijk1}^{\alpha} g_{i,j'}^{m} S_k^{n})\phi_n^{(p-1)}\mathrm{d}v(\mathbf{x}') \\ &+ \sum_{\alpha=1}^{N}\iiint_{\Omega_\alpha}\Delta\rho_\alpha \omega^2 g_i^m S_i^n \phi_n^{(p-1)}\mathrm{d}v(\mathbf{x}')\end{aligned} \quad . \tag{24}$$

where S_j^n has the form

$$S_j^n = U^n(z')\delta_{j3} + i\delta_{j1}V^n(z') + i\delta_{j2}W^n(z') \; . \tag{25}$$

Introducing a variable

$$K_{mn} = \sum_{\alpha=1}^{N}\int(-\Delta C_{ijk3}^{\alpha} g_{i,j'}^{m} S_{k,3'}^{n} + ik_m \Delta C_{ijk1}^{\alpha} g_{i,j'}^{m} S_k^{m})\mathrm{d}z' + \sum_{\alpha=1}^{N}\int\Delta\rho_\alpha \omega^2 g_i^m S_i^n \mathrm{d}z' \; , \tag{26}$$

Equation (25) can be re-written as:

$$\phi_{mn}^{(p)} = \iint K_{mn}(x,y;x',y')\phi_n^{(p-1)}(x',y')\mathrm{d}x'\mathrm{d}y' \; . \tag{27}$$

Hereinabove, we have simplified the 3D scattering problem of Rayleigh wave as a plane integral equation problem. The scattered field of Rayleigh wave can be obtained only by changing the integral kernel $K_{mn}(x,y;\ x',y')$ according to different obstacles with arbitrary shape and dimension.

It is noted that when Equation (26) is used, the obstacles must be identical in depth. Equation (27) shows when the incident waves contact the obstacles, every point in the obstacles is equivalent to a secondary wave source and induces scattered field $K_{mn}\phi_n^{(p-1)}$. The scattered displacement field can be obtained by integrating Equation (27) over the whole domains of all the obstacles. Comparing with Born approximation, Equation (27) is a process of continuous iteration instead of one times. Thus, it can be used to solve the scattering problem when the barriers are relative large in terms of the incident wavelength.

For an isotropic elastic half-space including obstacles, ΔC_{ijkl} can be expressed as:

$$\Delta C_{ijkl} = \Delta\lambda\delta_{ik}\delta_{jl} + \Delta\mu(\delta_{ij}\delta_{kl} + \delta_{il}\delta_{kj}) \tag{28}$$

where $\Delta\lambda$ and Δu are the variable values of Lame constants of the obstacles.

For having only one kind of Rayleigh mode in the isotropic elastic half-space, as shown in Figure 1, we can substitute Equation (28) into (26) and obtain the following equations:

$$K_{RR}(x,y;x',y') = -\frac{i}{8J_1^1cc^g}[H_0^{(2)}(kR)\times(-\Delta\lambda R_{\lambda 0}+\Delta\rho R_{\rho 0}-\Delta\mu R_{\mu 0}) \\ -iH_1^{(2)}(kR)\cos\varphi(\Delta\rho R_{\rho 1}-\Delta\mu R_{\mu 1})-H_2^{(2)}(kR)\cos 2\varphi(-\Delta\mu R_{\mu 2})] \quad (29)$$

where the angle φ is indicated in Figure 1; J_1^1 is the normalized integral value of Rayleigh wave in the elastic half-space expressed by Equation (18); c and c^g are the phase velocity and group velocity for the elastic half-space, respectively; k is Rayleigh wave number; $H_1^{(2)}$ and $H_2^{(2)}$ are Hankel functions of second kind of order one and two, respectively.

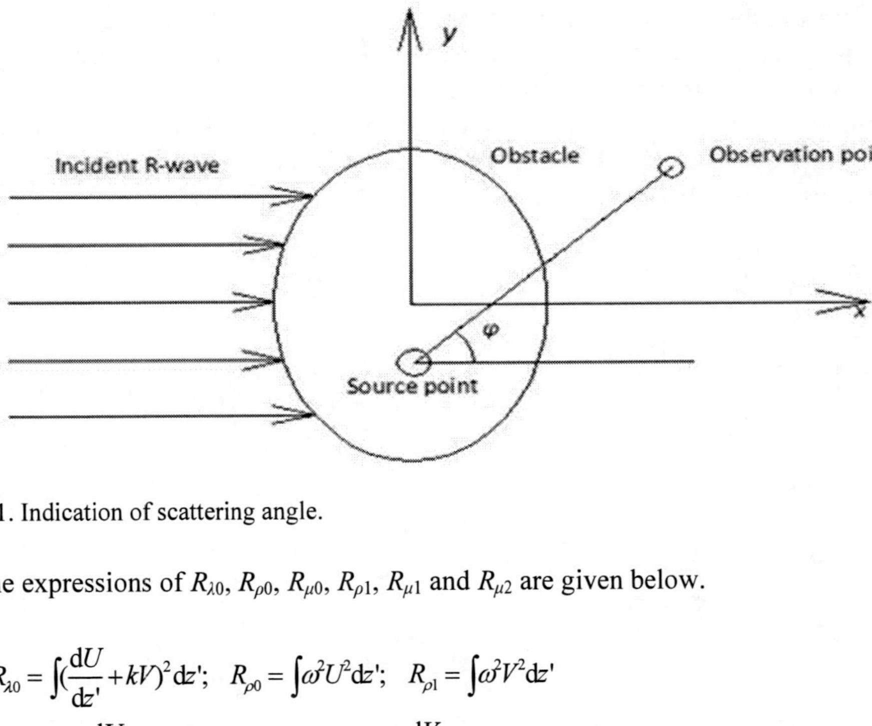

Figure 1. Indication of scattering angle.

The expressions of $R_{\lambda 0}$, $R_{\rho 0}$, $R_{\mu 0}$, $R_{\rho 1}$, $R_{\mu 1}$ and $R_{\mu 2}$ are given below.

$$\begin{cases} R_{\lambda 0} = \int(\frac{dU}{dz'}+kV)^2dz'; \quad R_{\rho 0} = \int\omega^2U^2dz'; \quad R_{\rho 1} = \int\omega^2V^2dz' \\ R_{\mu 0} = \int[2(\frac{dU}{dz'})^2+k^2V^2]dz'; \quad R_{\mu 1} = \int(kU-\frac{dV}{dz'})^2dz'; \quad R_{\mu 2} = \int k^2V^2dz' \end{cases} \quad (30)$$

Consequently, the problem of Rayleigh wave scattering in an elastic half-space can be governed by the following equation:

$$\phi_R^{(p)} = \iint K_{RR}(x,y;x',y')\phi_R^{(p-1)}dx'dy'. \quad (31)$$

Equation (31) can be solved by an iteration technique. A computer code in Matlab has been developed for the solution of the 3D scattering problem in the frequency domain by following the aforementioned integral equation procedure. The main process of this code is stated below:

- Compute the integral kernel $K_{mn}(x,y;x',y')$ according the situation and the dimension of the obstacles, e.g. piles in rows.
- Compute the incident wave field according the boundary conditions in the free field.
- Suppose the initial wave field is equal to the incident wave.
- Compute the next scattering field according Equation (31) by iteration until the scattered wave field is small enough.
- The resultant field is the summation of the incident field and the scattered field.

2.2. Numerical Verification

The key step for the iteration is the verification of the convergence. In this section, the code developed in the previous section for the complex 3D scattering problem is verified by checking its accuracy and iteration convergence. A simple problem is used, which has accurate numerical results available in Its [13].

The problem is shown in Figure 2, where a Rayleigh wave is incident normally to a narrow vertical layer inserted to between the two quarter-spaces. The parameters of the layer and the half-space are given in Table 1.

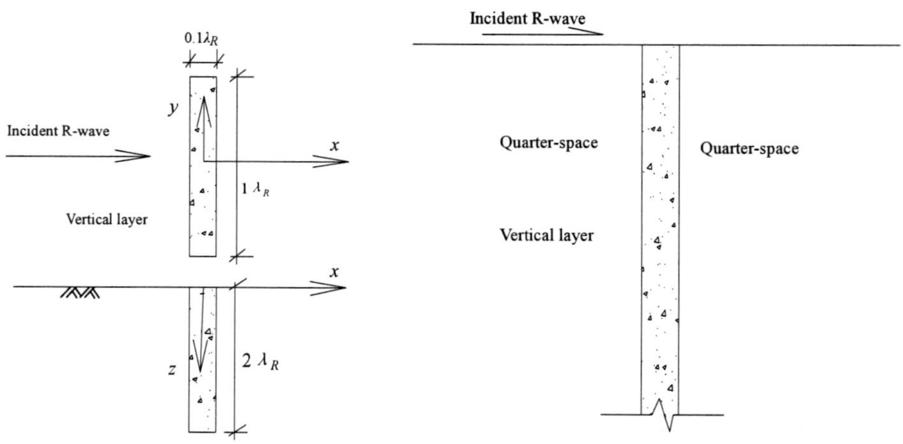

Figure 2. Dimensions of the layer and the model used by Its [13].

Table 1. Parameters of the quarter-space and the narrow vertical layer

		Young's modulus E (MPa)	Poisson ratio ν	Density γ (kg/m^3)
elastic quarter-space		70	0.35	2×10^3
layer	case 1	100	0.35	4×10^3
	case 2	140	0.35	4×10^3

Table 2 gives the maximum scattered amplitude along the x-axis for iterations calculated by Equation (31). It shows that accurate results can be obtained in the both cases with a maximum scattered amplitude yields to zero after 3 to 4 iterations. This result is also true in the study about the multi-rows of piles as passive barrier below.

Depicted in Figure 3 are the amplitude reduction factor w_b versus the non-dimensional distance $x^*=x/\lambda_R$ from the layer, which are solved by the present method and given by Its [13]. The factor w_b is defined as the ratio of the amplitude of resultant and incident wave field.

Table 2. Maximum scattered amplitudes along x-axis for times of iterations

The p-th scattered field		1	2	3	4	5	6
A^*	case 1	0.2903	0.0818	0.0164	0.0035	0.7734e-3	0.1347e-3
	case 2	0.3652	0.1151	0.0362	0.0109	0.0036	0.0011

A^* is defined as the ratio of the maximum scattering amplitude over the incident amplitude.

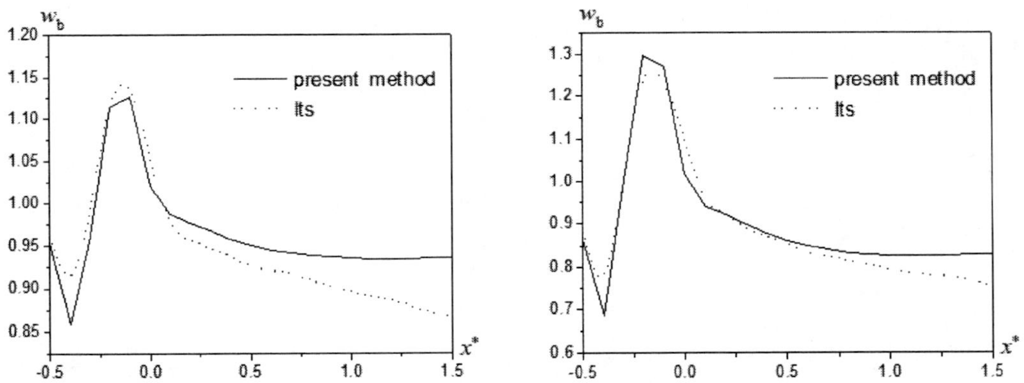

Figure 3. Rayleigh wave of the layer in Case 1 (left) and Case 2 (right).

It can be observed from the figure that the present results are in excellent agreement with those in Its [13] near the layer. The two results have some difference away from the layer, which may be due to the two assumptions, i.e., the plane strain and the infinite length of the barrier, used in Its [13], while the present method considers the full size of the layer and the scattered wave as 3D problem. The difference at the far distance indicates that the Rayleigh wave scattered by the obstacle is more pronounced in the far field. Thus, the model used in the present study can better simulate the actual isolation barrier.

2.3. Screening Effectiveness of Single Row of Piles

In this section, a row of eight concrete piles of square cross-section as passive barrier is examined and compared. The parameters of the piles and the soil in computation are the same as those of Kattis et al. [14], to be compatible with the analysis of them. on the same problem. As shown in Figure 4, every pile has a cross-section with normalized side b and length h, while the net spacing between two successive piles is s. The soil is assumed to be linear viscoelastic.

The material parameters of the soil are given as follows:

- shear modulus G_s=132 MPa,
- Poisson ratio v_s=0.25,
- mass density ρ_s=17.5 kg/m^3,
- damping coefficient β_s =5%,
- Rayleigh wavelength for the soil is computed to be λ_R=5 m.

The material parameters of the concrete piles are given as follows:

- shear modulus G_c=34.29G_s,
- Poisson ratio v_c=0.25,
- mass density ρ_c=1.37ρ_s,
- damping coefficient β_c=5.0β_s,
- The normalized parameters by Rayleigh wavelength λ_R are: $b^*=b/\lambda_R$ =0.2, $s^*=s/\lambda_R$=0.1, $h^*=h/\lambda_R$=1.0, $x^*=x/\lambda_R$, $y^*=y/\lambda_R$ and $l^*=l/\lambda_R$.

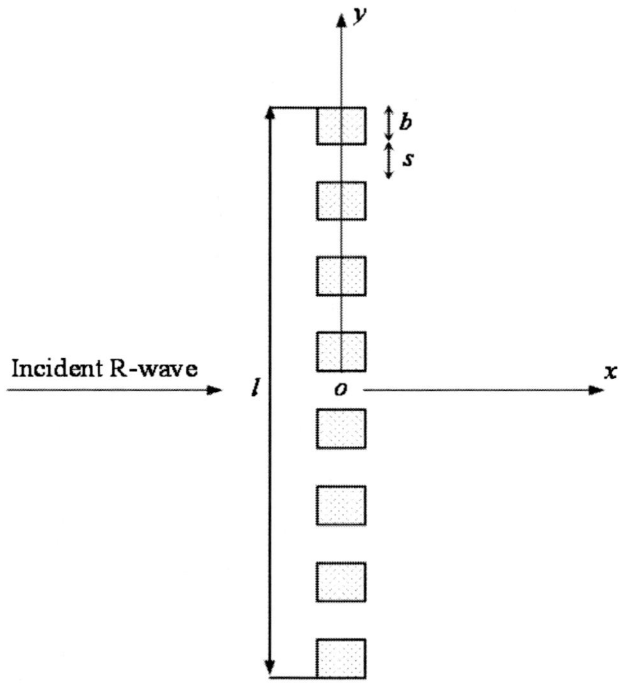

Figure 4. One row of pile barrier vibration system.

To represent the ratio of the spacing between two piles and the sectional length of the pile, a parameter r_{sb} is introduced as follows:

$$r_{sb} = \frac{s^* + b^*}{b^*}. \tag{32}$$

This is the amplitude attenuation ratio (ARF), defined as the ratio of the vertical displacement component of the soil surface in the presence of the piles over that displacement in the absence of the piles [15]. This ratio is frequently used as a measure of the screening effectiveness of an isolation barrier. A zero amplitude attenuation ratio indicates that the incoming wave is completely scattered and that the barrier is completely effective.

Figure 5 depicts the contour of the amplitude reduction factor for this case. The figure shows the maximum reductions occur near the center of the isolation system and at a distance of 1.0 λ_R behind the row.

It can also be observed that the screening effectiveness computed by the present method has some different with that of Kattis et al.[14]. This was expected in view of the fact that the codes and the assumptions of the two methods are different. The present method employs the half-space Green's function and only considers the incident Rayleigh wave, while Kattis et al. employed the full space to deal with all kinds of waves.

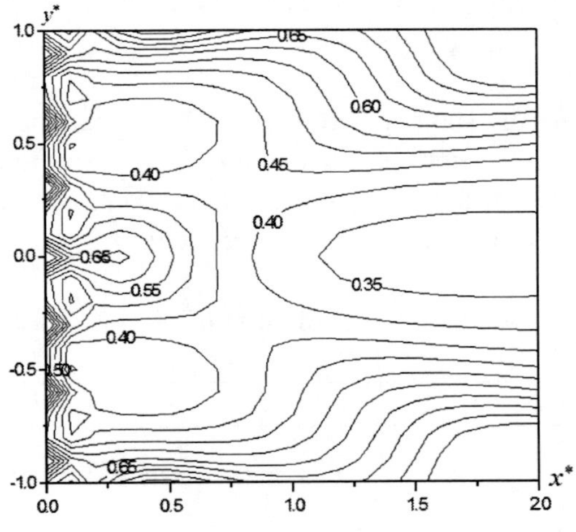

Figure 5. Contour of A_{RF} for the barrier vibration system.

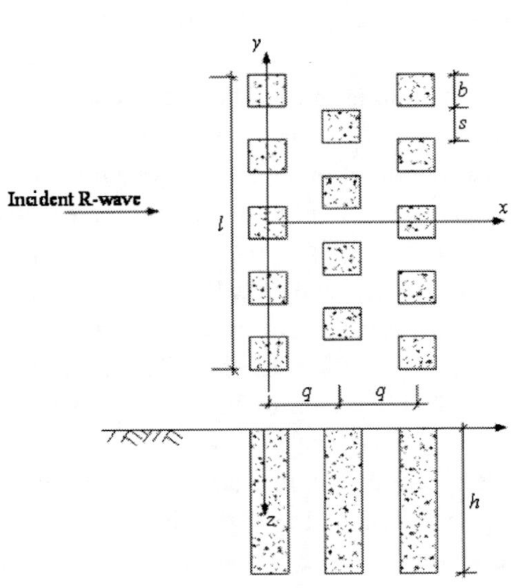

Figure 6. Model of rows of piles as barrier.

2.4. Screening Effectiveness of Multi-rows of Piles

In this section, the passive vibration isolation problem by multi-rows of stiff piles of square cross-section is examined in detail. As shown in Figure 6, we consider three rows of piles of square cross-section. Every pile has a sectional length b and depth h. The net spacing between two piles is s. The spacing between two rows is q.

The material parameters of the soil and piles are given as follows:

- the Poisson ratio of pile barrier v_b=0.25,
- the Poisson ratio of soil v_s=0.35,
- the velocity ratio of shear wave r_c=5, where r_c is the ratio of shear wave velocity of barrier over that of soil, which reflecting the stiffness of barrier. For r_c>1, the barrier is stiff than the surrounding soil and *vice versa.*
- The normalized parameters by Rayleigh wavelength λ_R are: $x^*=x/\lambda_R$, $y^*=y/\lambda_R$, $b=b^*/\lambda_R$, $s^*=s/\lambda_R$, $l^*=l/\lambda_R$, $h^*=h/\lambda_R$ and $q^*=q/\lambda_R$.

The geometric variables are summarized in Table 3. The parameters n_1, n_2 and n_3 in Table 3 are the numbers of piles in the first, second and third row, respectively. It is noted that when one parameter is examined, all the other parameters are kept constant. These three rows of piles are named as Row 1, Row 2 and Row 3, respectively.

Table 3. Parameters of rows of piles as barrier

Parameter	b^*	s^*	h^*	q^*	n_1	n_2	n_3
Value	0.1	0.1	1	0.2	5	4	5

2.4.1. Effect of the Rows

Depicted in Figure 7 (a)~(c) are the variations of the factor A_{RF} on the surface of the wave field for the above passive isolation. Figure 7(d) shows the variation of A_{RF} along x-axis for different rows, respectively.

These figures show that the maximum reductions occur near the center of the isolation system, representing the barrier behaves like a unit instead of a set of independent piles. Moreover, it can be observed that addition of the rows results in an increase of the obtained reduction, especially in the case shown in Figure 7(c) (for three rows).

The amplitude attenuation ratio A_{RF} is in the order of 33% at a distance of $1.0\lambda_R$ behind the rows (note: a trench barrier system with similar size in some cases cannot obtain such a ratio). The reductive effect can be further improved if the rows of piles increase. The results in these figures also show a double effect of the pile barrier as the isolation system: the amplification generated at the incident side of barrier and reduction of the field behind the barrier.

In addition, the isograms of A_{RF} in Figure 7 are similar to those of the continuous barriers such as opened trench or in-filled trench [16−18]. This finding again indicates that the multi-rows of piles isolate ground vibrations as an integrated system.

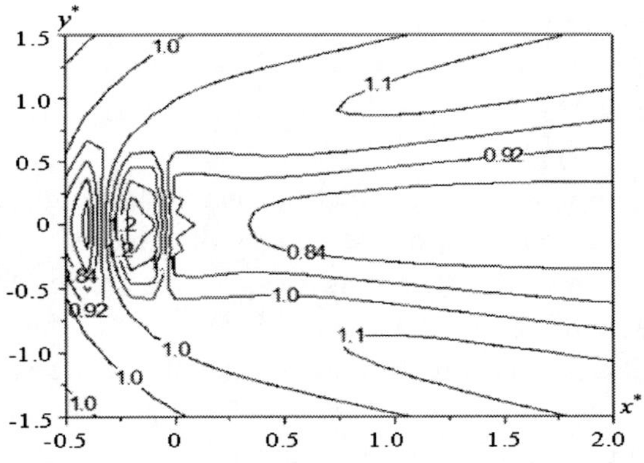

(a)

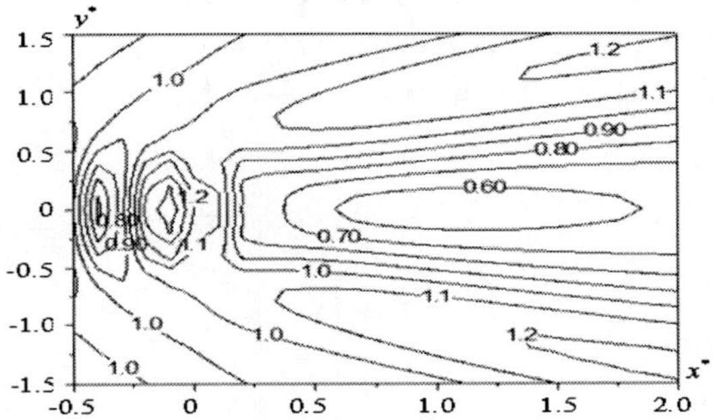

(b)

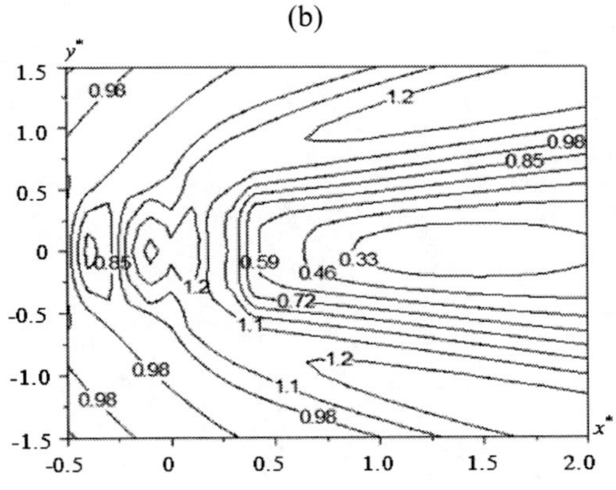

(c)

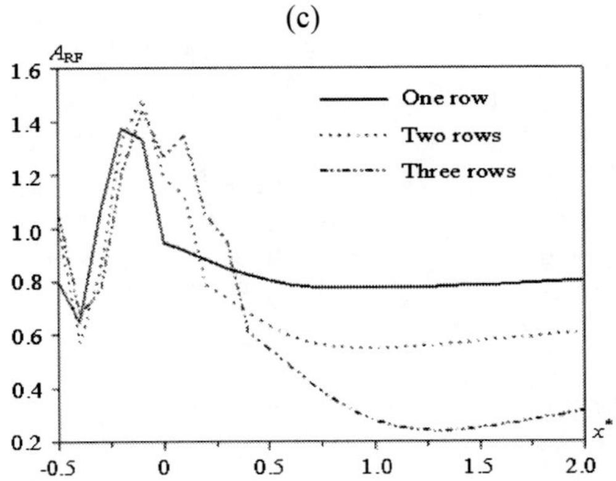

(d)

Figure 7. Contours of A_{RF} for (a) a single row, (b) two rows and (c) three rows, and graphs of A_{RF} for different rows along x-axis (d).

2.4.2. Effect of Normalized Spacing q^*

To examine the effect of spacing between rows, two rows of piles are considered, with the pile parameters in Table 3. The value of q^* varies as 0.15, 0.20, 0.30 and 0.40, respectively.

Shown in Figure 8 (a)~(c) are the amplitude attenuation ratios A_{RF} for different spacing of rows. One can observe that considerable reductions of screened zone behind barriers and decreasing screening efficiency are obtained as the spacing increases.

However, the variation of A_{RF} along the x-axis in Figure 8(d) shows the reduction can get slightly higher as q^* decreases. This result may indicate that decreasing the pile spacing has small influence on the reduction effects. Therefore, it is advisable to perform a feasibility study for each particular case to achieve the optimum space between pile rows. As a result, a maximum reduction can be achieved with the same construction cost.

The figures also show that the optimum screened zone is at a small distance behind the barrier. This finding is consistent with the result in Woods et al. [19].

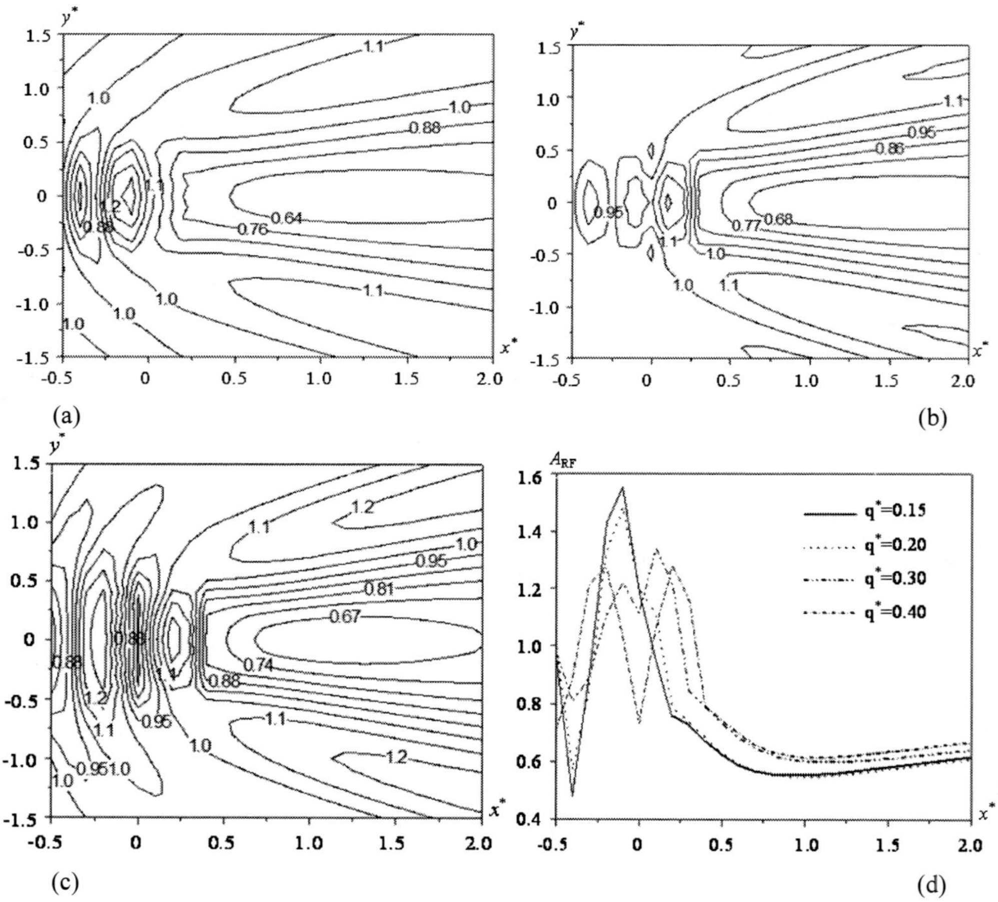

Figure 8. Contours of A_{RF} for (a) $q^*=0.15$, (b) $q^*=0.30$ and (c) $q^*=0.40$, and graphs of A_{RF} for different q^* along x-axis (d).

2.4.3. Effect of Normalized Net Spacing s^*

The net spacing is an important variable of practical concern for the vibration isolation by piles in rows. Two rows of piles are considered to study the influences of s^*, using the parameters in Table 3. The value of s^* varies as 0.1, 0.2 and 0.3, respectively.

Shown in Figure 9 are the isograms of A_{RF} for s^*=0.20 and 0.40. The isograms for s^*=0.10 are shown in Figure 7(b). Figure 10 is the variation graphs of A_{RF} with different s^* along x-axis.

These figures clearly reveal the decreasing magnification at the incident side of barriers and reducing screened zone behind barriers as the net spacing s^* increases. In Figure 10, the greater reductions does not appear in the center of the barrier (x^*=0). This result shows that the obstacle behaves as a set of independent piles instead of a unit system as the net space increases. The barrier isolating surface wave is no longer as an integrated system when the net spacing between piles is large, especially in the case s^*=0.4 (i.e. r_{sb}=5, the net spacing is large enough). The reductions are generally more efficient for smaller separation between piles. It is suggested that the separation satisfies $r_{sb} \leq 2.5$ in practical engineering to ensure a good screening effect.

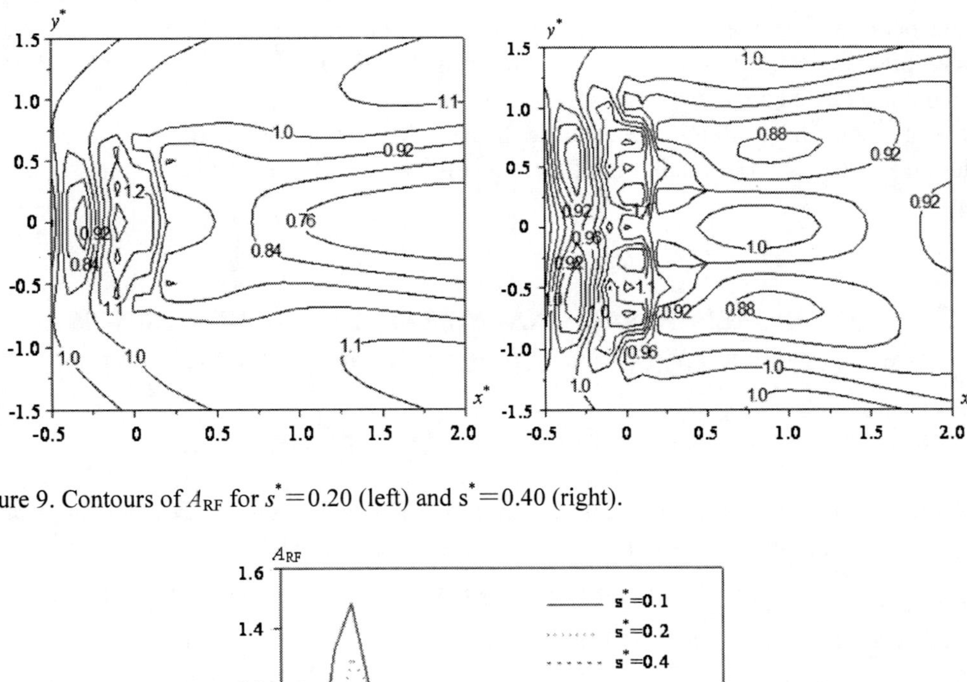

Figure 9. Contours of A_{RF} for $s^*=0.20$ (left) and $s^*=0.40$ (right).

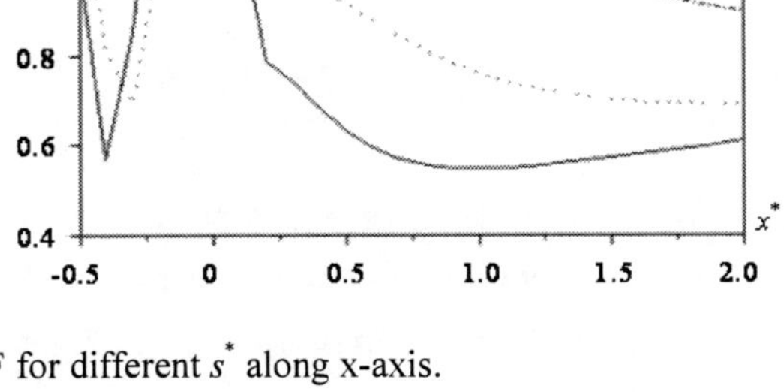

Figure 10. Graphs of ARF for different s^* along x-axis.

It is also observed that increasing pile length may improve the reduction effect of the surface wave field. However, due to the transmission of Rayleigh wave is confined near the surface, this effect will be less important when the length of the pile is beyond a Rayleigh wavelength. This observation is consistent with the finding in Woods et al. [19].

2.5. Conclusion

Row-piles installed in the ground as a barrier of vibration isolation in the far field can be modeled and predicted by the scattering problem of Rayleigh waves induced by the rows of obstacles. Based on the integral equation method of Rayleigh wave scattering, the screening effectiveness of single row and multi-rows of piles was studied as barrier in the far field.

The method can be used to solve the 3D problem of Rayleigh wave scattering from irregular multi-obstacles by iteration, and accurate results can be obtained after 3 to 4 numerical iterations.

Row-piles can work as an integrated system to isolate ground vibration. The screening effectiveness by rows of small cross-section piles can be as good as that of the trenches. The spacing between the pile rows has less influence on the reduction effect. The net spacing between piles in each row is an important factor for the vibration isolation. In order to obtain an effective screened zone, the criterion $r_{sb}\leq 2.5$ can be used.

The method and findings presented in this section are useful in the design of ground barrier and can be used to predict the functions of rows of piles as barrier system for ground vibration.

3. Three-Dimensional Analysis of In-Filled Trench as Passive Barriers for Ground Vibration Isolation

3.1. The Integral Equation Method for R-wave Scattering

Assume that the elastic solid contains N sub-domains Ω_1, Ω_2,... Ω_N, which represent obstacles. They are bounded by the surfaces S_1, S_2, ..., S_N, which have the elastic modulus matrices C^1_{hjkl}, C^2_{hjkl},..., C^N_{hjkl}, and mass densities ρ_1, ρ_2, ..., ρ_N, respectively. Considering the case without body force and using the reciprocity theorem of Green's function, the integral equations for scattered displacement fields by many elastic obstacles embedded in an isotropic elastic solid can be obtained [7, 8]:

$$u_i(\mathbf{x})=u_i^{(i)}(\mathbf{x})+\sum_{\alpha=1}^{N}\Delta\rho_\alpha\omega^2\iiint_{\Omega_\alpha}G_{im}(\mathbf{x};\mathbf{x}')u_m(\mathbf{x}')\mathrm{d}v+\sum_{\alpha=1}^{N}\Delta C^\alpha_{mjkl}\iiint_{\Omega_\alpha}G_{mi,j}(\mathbf{x};\mathbf{x}')u_{k,l'}(\mathbf{x}')\mathrm{d}v \quad (33)$$

where u_i is the amplitude of displacement vector; $\mathbf{x}$ is a point of the elastic solid with volume V; ω is the circular frequency of vibration; $G_{im}(\mathbf{x},\mathbf{x}')$ is the im-th element of the Green's tensor, representing the displacement vector at the position $\mathbf{x}$ along direction i when a force is applied to the position **x'** along direction *m*; and $\Delta C^{\alpha}_{ijkl} = C^{\alpha}_{ijkl} - C_{ijkl}$, $\Delta\rho_{\alpha} = \rho_{\alpha} - \rho$, α=1, 2, …, *N*, C_{ijkl} and ρ represent the elastic modulus matrix and density of the elastic solid, respectively.

For the scattering problem, the resultant displacement can be expressed as a sum of the incident displacement and the scattered displacement, and the latter is given as a series expansion:

$$u_i(\mathbf{x}) = u_i^{(i)}(\mathbf{x}) + \sum_{p=1}^{M} u_i^{(p)}(\mathbf{x}) \tag{34}$$

where $u_i^{(p)}(\mathbf{x})$ is the *p*-th scattered displacement.

Substituting Equation (34) into Equation (33) and considering the (*p*-1)-th scattered fields as the incident wave of the *p*-th scattered fields, we can obtain following equation:

$$u_i^{(p)}(\mathbf{x}) = \sum_{\alpha=1}^{N} \Delta\rho_{\alpha}\omega^2 \iiint_{\Omega_\alpha} G_{mi}(\mathbf{x};\mathbf{x}')u_m^{(p-1)}(\mathbf{x}')\,\mathrm{d}v - \sum_{\alpha=1}^{N} \Delta C^{\alpha}_{mjkl} \iiint_{\Omega_\alpha} G_{mi,j'}(\mathbf{x};\mathbf{x}')u_{k,l'}^{(p-1)}(\mathbf{x}')\,\mathrm{d}v \tag{35}$$

In this secession, only the Rayleigh surface-wave terms are considered as follows:

$$G_{qi}(\mathbf{x};\mathbf{x}') = \sum_{m} Z_q^m(z) g_i^m(x, y; \mathbf{x}'), \quad (36)$$

where $Z_q^m(z)$ is an operator notation of the form

$$Z_q^m(z) = \begin{pmatrix} \frac{1}{k_m}\left(-V^m(z)\frac{\partial}{\partial x} + W^m(z)\frac{\partial}{\partial y}\right) \\ \frac{1}{k_m}\left(-W^m(z)\frac{\partial}{\partial x} - V^m(z)\frac{\partial}{\partial y}\right) \\ U^m(z) \end{pmatrix} \tag{37}$$

and V^m(z), U^m(z) are the horizontal and vertical displacements induced by the *m*-th Rayleigh mode; W^m(z) is the horizontal displacement generated by the *m*-th Love mode; k_m is wave number of the *m*-th surface wave mode; **x** is the observation point and **x'** is the source point; $g_i^m(x, y; \mathbf{x}')$ is the displacement potential. Details of Green's function for an elastic half-space can be found in Aki and Richards [6].

Similarly, the displacement of the *p*-th scattered field can be expressed as the same form as Equation (36),

$$u_i^{(p)}(\mathbf{x}) = \sum_{m} Z_i^m(z)\phi_m^{(p)}(x, y) \tag{38}$$

where $\phi_m^{(p)}(x, y)$ is the displacement potential of the *m*-th mode surface wave in the *p*-th scattered field.

Then substituting Equations (36) and (38) into (35), and introducing a new potential function, yields,

$$\phi_m^{(p)} = \sum_n \phi_{mn}^{(p)} \tag{39}$$

$$\begin{aligned}\phi_{mn}^{(p)}(x,y) = &\sum_{\alpha=1}^{N} \Delta\rho_\alpha \omega^2 \iiint_{\Omega_\alpha} g_k^m Z_k^n(z')\phi_n^{(p-1)}(x',y')\mathrm{d}v(\mathbf{x}') \\ &-\sum_{\alpha=1}^{N} \Delta C_{ijkl}^{\alpha} \iiint_{\Omega_\alpha} g_{i,j'}^{m}[Z_k^n(z')\phi_n^{(p-1)}(x',y')]_{,l'}\mathrm{d}v(\mathbf{x}')\end{aligned} \tag{40}$$

where $\phi_m^{(p)}$ is the displacement potential of m-th mode surface wave in p-th scattered field induced by the potential of n-th mode surface wave in (p-1)-th scattered field.

Assuming a harmonic Rayleigh wave travels along the x-axis direction, the variation of potential function in the y-axis direction can be neglected with respect to the one in the x-axis direction. Therefore, Equation (40) can be further simplified as:

$$\phi_{mn}^{(p)} = \iint K_{mn}(x,y;x',y')\phi_n^{(p-1)}(x',y')\mathrm{d}x'\mathrm{d}y' \tag{41}$$

in which,

$$K_{mn} = \sum_{\alpha=1}^{N} \int (-\Delta C_{ijk3}^{\alpha} g_{i,j'}^{m} S_{k,3'}^{n} + ik_m \Delta C_{ijk1}^{\alpha} g_{i,j'}^{m} S_k^m)\mathrm{d}z' + \sum_{\alpha=1}^{N} \int \Delta\rho_\alpha \omega^2 g_i^m S_i^n \mathrm{d}z' \tag{42}$$

where S_j^n is in the form $S_j^n = U^n(z')\delta_{j3} + i\delta_{j1}V^n(z') + i\delta_{j2}W^n(z')$.

Equation (41) shows when the incident waves contact the obstacles, every point in the obstacles is equivalent to a secondary wave source and induces scattered displacement field $K_{mn}\phi_n^{(p-1)}$, which can be obtained by integrating Equation (41) over the whole domains of all the obstacles. Comparing with the Born approximation, Equation (41) is a process of continuous iteration instead of one times. Thus, it can be used to solve the scattering problem when the barriers are relative large in terms of the incident wavelength.

For an isotropic elastic half-space including obstacles, as shown in Figure 1, only one kind of Rayleigh mode is considered. Then Equations (41) and (42) can be simplified as

$$\phi_R^{(p)} = \iint K_{RR}(x,y;x',y')\phi_R^{(p-1)}\mathrm{d}x'\mathrm{d}y'. \tag{43}$$

$$\begin{aligned}K_{RR}(x,y;x',y') = -\frac{i}{8J_1^1 cc^g}&[H_0^{(2)}(kR)\times(-\Delta\lambda R_{\lambda 0} + \Delta\rho R_{\rho 0} - \Delta\mu R_{\mu 0}) \\ &-iH_1^{(2)}(kR)\cos\varphi(\Delta\rho R_{\rho 1} - \Delta\mu R_{\mu 1}) - H_2^{(2)}(kR)\cos 2\varphi(-\Delta\mu R_{\mu 2})]\end{aligned} \tag{44}$$

where the angle φ is indicated in Figure 1; J_1^1 is the normalized integral value of Rayleigh wave in the elastic half-space; c and c^g are the phase velocity and group velocity for the elastic half-space, respectively; k is Rayleigh wave number; and $\mathrm{H}_1^{(2)}$ and $\mathrm{H}_2^{(2)}$ are Hankel functions of second kind of order one and two, respectively. The $R_{\lambda 0}$, $R_{\rho 0}$, $R_{\mu 0}$, $R_{\rho 1}$, $R_{\mu 1}$ and $R_{\mu 2}$ are expressed as below.

$$\begin{cases} R_{\lambda 0} = \int (\frac{dU}{dz'} + kV)^2 dz'; \quad R_{\mu 1} = \int (kU - \frac{dV}{dz'})^2 dz'; \quad R_{\rho 0} = \int \omega^2 U^2 dz' \\ R_{\mu 0} = \int [2(\frac{dU}{dz'})^2 + k^2 V^2] dz'; \quad R_{\rho 1} = \int \omega^2 V^2 dz'; \quad R_{\mu 2} = \int k^2 V^2 dz' \end{cases} \tag{45}$$

Equation (43) can be solved by an iteration technique. Unlike the classic Born approximation [7], the proposed iteration technique can iterate enough times to obtain good solution. A computer code in Matlab has been developed for the solution of the 3D scattering problem in the frequency domain. The method is verified with a numerical solution available in the literature for a simplified Rayleigh wave scattering problem [7, 13].

3.2. Comparison with Published Experiments

Haupt [4] performed a series of laboratory tests on model scales for passive vibration isolation using in-filled trenches in a sand bin with 9.5 m by 9.5 m in plane area and 3 m in depth. The soil was dense medium sand with a moisture content of about 3%. The depths of the in-filled trenches were 0.06m, 0.12m and 0.30m, and the thickness were 0.30m, 0.12m, and 0.06m. The trenches were located at a distance of 0.165m, 0.175m from the vibrator located in the center of the sand bin, generating a concentric harmonic Rayleigh wave field. The ranges of nondimensional parameters for Haupt's passive isolation tests were: h^*=0.20~1.28, t^*=0.20~1.24. These model tests were numerically simulated utilizing the method of this section, and a comparison of the calculating results with Haupt's testing ones is presented herein.

The material parameters of soil and trenches needed for simulation were mass density ρ, shear wave velocity c and Poisson's ratio v, which utilize the published data as follows:

- soil: ρ_s=1720 kg/m^3, c_s=106 m/s, v_s=0.25;
- trench: ρ_b=2320 kg/m^3, c_b=530 m/s, v_b=0.25.

Figure 11 presents a comparison of the calculated results with Haupt's data for test No.M2/30 (t^*=1.0. l^*=0.4, f=320 Hz). The amplitude attenuation ratio (A_{RF}) is defined as the ratio of the vertical displacement component of the soil surface in the case with in-filled trench over that without the trench [15]. This ratio is often used as a measure of the screening effectiveness of an isolation barrier. A zero amplitude attenuation ratio indicates that the incoming wave is completely scattered and the barrier is completely effective.

It can be observed from Figure 11 that in spite of considerable variations, the calculated curve is in good agreement with the experimental one, if the mean value is considered.

However, the calculated value of A_{RF} in front of the trench is larger than that of the experiment, while the two curves coincide well concerning the location of the peaks. This may be accounted that the damping material is assumed to be zero in calculation.

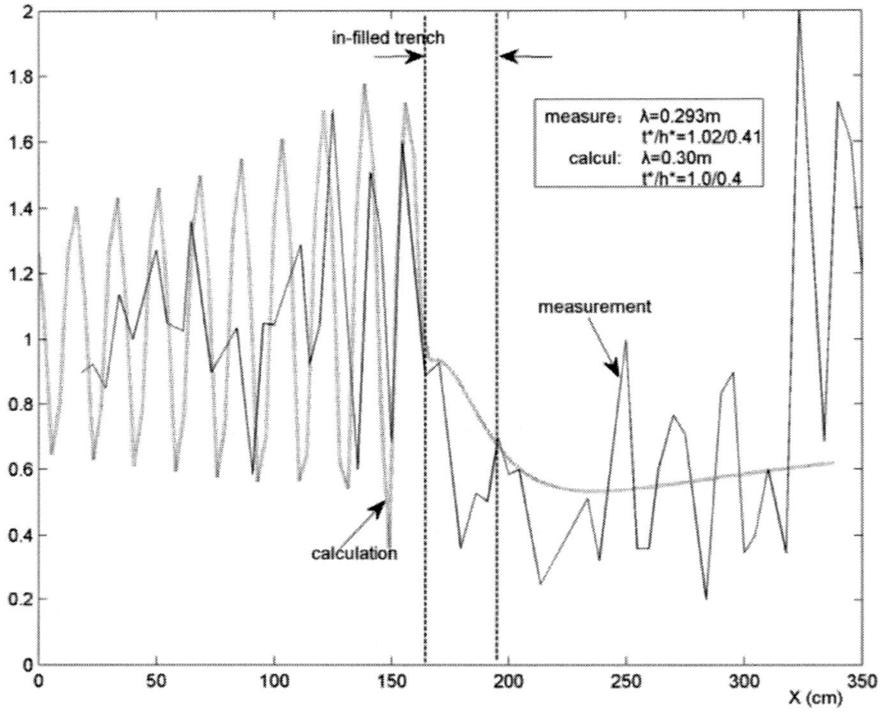

Figure 11. Comparative study for passive isolation by in-filled trench (t^*=1.0. l^*=0.4, f=320 Hz).

3.3. Screening Effectiveness of In-filled Trench

In this section, in-filled trenches with different backfill materials as passive barrier are examined and compared. The efficiency of vibration isolation by an in-filled trench mainly depends on the factors of the backfill material properties and the geometrical parameters of the trench. An in-filled trench with rectangular section is considered, as shown in Figure 12.

In the model, the soil has a shear modulus G_s, a Poisson ratio v_s, a mass density ρ_s, a shear wave speed c_{ss} and a Rayleigh wavelength λ_R. The trench has a shear modulus G_b, a Poisson ratio v_b, a mass density ρ_b and a shear wave speed c_{sb}. The normalized geometrical data of the trench read the depth $h^*=h/\lambda_R$, the thickness $t^*=t/\lambda_R$ and the length $l^*=l/\lambda_R$.

3.3.1. Effects of Backfill Material Properties

In what follows, the effects of the backfill material properties on vibration isolation are examined in detail, focusing on shear modulus G_b, Poisson ratio v_b and mass density ρ_b. A velocity ratio of shear wave (r_c) is defined as:

$$r_c = c_{sb} / c_{ss} \tag{46}$$

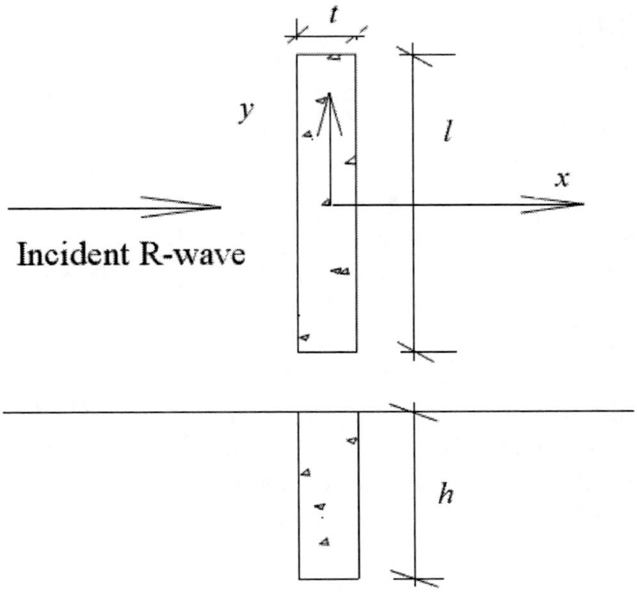

Figure 12. In-filled trench barrier as passive vibration isolation.

Herein, r_c reflects the difference between the backfill material and the soil, which can be used to determine whether the trench barrier is soft or stiff. When $r_c<1$, the trench is defined as soft barrier, meaning that it is softer than the surrounding soil, e.g. the barrier filled by sawdust or flyash. While $r_c>1$, the barrier is called as stiff one, e.g. concrete-filled trench.

Throughout the following studies, to examine the effect of the backfill material properties, the geometrical parameters of the trench are assumed to be constant, taking the values $t^*=0.10$, $l^*=1.0$ and $h^*=1.0$.

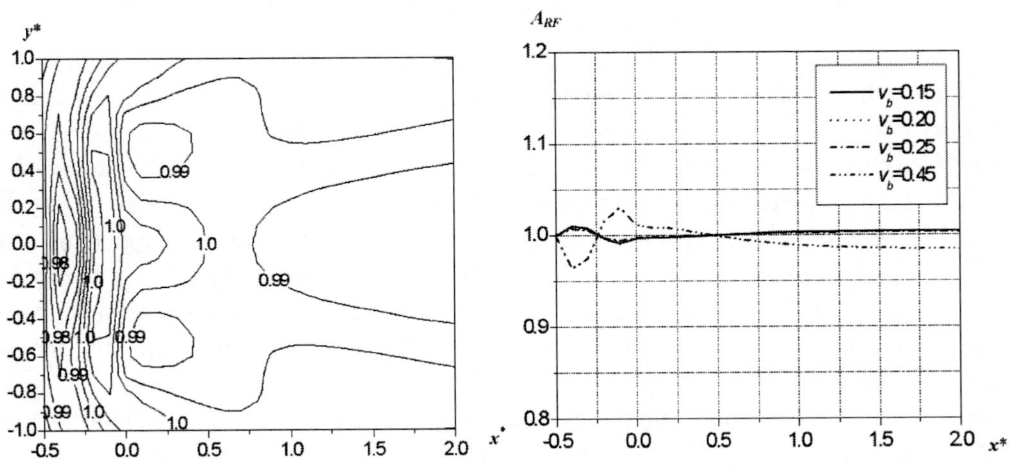

Figure 13. Contour of A_{RF} for the trench as passive barrier (v_b =0.45) (left) and graphs of A_{RF} for different v_b along x-axis (right).

3.3.1.1. Effect of Poisson's Ratio

The analysis is performed for the case of the soil with v_s=0.35. The Poisson's ratio of the backfill material varies as 0.15, 0.20, 0.30 and 0.45, respectively, while the velocity ratio of shear wave r_c is kept constant, taking the value of r_c=1.

Figure 13(left) depicts the variation of the factor A_{RF} on the surface of the wave field for above passive isolation at υ_b =0.45. It can be observed that the values of A_{RF} vary around 1.0 in most area, and there is not an effective isolated zone. Figure 13 (right) shows the variation of A_{RF} along x-axis for different Poisson' ratios. These figures show that the Poisson's ratio has only a slight influence on the screening effectiveness of the in-filled trench.

3.3.1.2. Effect of the Velocity Ratio of Shear Wave (r_c)

In what follows, the analysis is performed for the case of an in-filled trench with the same geometrical parameters as section 3.3.1.1. The Poisson' ratio of the backfill material is taken as 0.25 for a stiff barrier and 0.45 for a soft barrier. Consequently, the value of the velocity ratio of shear wave r_c is not a constant and equals to 0.15, 0.5, 0.75, 2, 5 and 7.5, respectively.

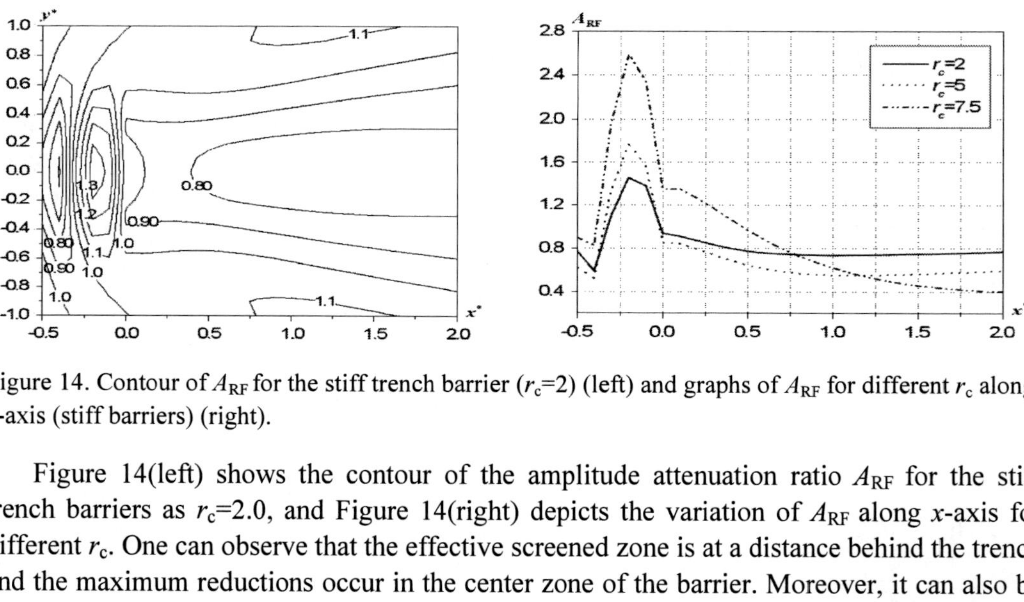

Figure 14. Contour of A_{RF} for the stiff trench barrier (r_c=2) (left) and graphs of A_{RF} for different r_c along x-axis (stiff barriers) (right).

Figure 14(left) shows the contour of the amplitude attenuation ratio A_{RF} for the stiff trench barriers as r_c=2.0, and Figure 14(right) depicts the variation of A_{RF} along x-axis for different r_c. One can observe that the effective screened zone is at a distance behind the trench and the maximum reductions occur in the center zone of the barrier. Moreover, it can also be found that increase of r_c results in an increase of the obtained reduction. The reduction effect can be further improved if the r_c increases. As a result, it is suggested to choose the backfill material with the value of $r_c \geq 5$. In addition, the results in these figures also show a dual effect, i.e., the amplification generated at the incident side of barrier and reduction of the field behind the barrier.

Figure 15(left) shows the amplitude attenuation ratio A_{RF} for the soft trench barriers with r_c=0.75, and Figure 15 (right) depicts the variation of A_{RF} along x-axis for different r_c. One can observe that the effective screened zone is small and it abuts upon the trench. Moreover, these figures clearly reveal that the increasing magnification at the incident barrier side and the slight increasing screening efficiency are generated as r_c decreases. Generally, to obtain efficiency in soft barrier case, r_c should be less than 0.25.

Comparing the results shown in Figures 14 and 15, it can be observed that the reduction values obtained using a soft barrier are greatly less than those using a stiff one. Moreover, the screened zone in the case of soft barrier is also greatly less than that of stiff one. Furthermore, the amplitude attenuation ratio A_{RF} obtained in the case of soft barrier tends to approach 1.0 at a small distance behind the barrier, meaning that the screened efficiency of the soft barrier is not steady. As a result, it can be concluded that velocity ratio of shear wave, contrary to the Poisson' ratio, has a greatly influence on the screening effectiveness of the in-filled trench, and using stiffer backfill material can increase the effectiveness.

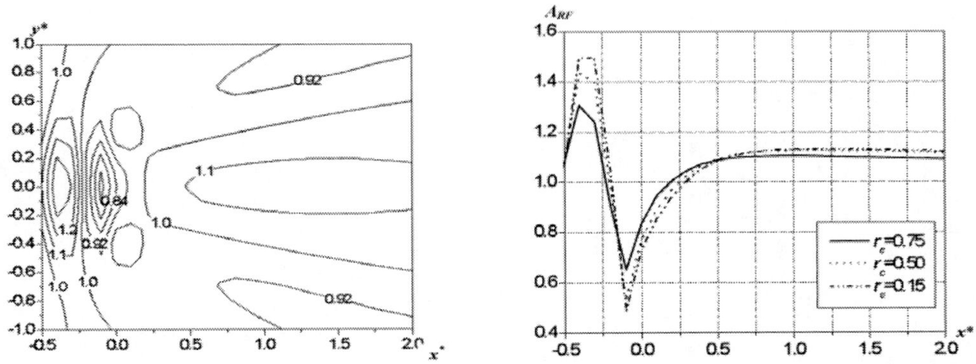

Figure 15. Contour of A_{RF} for the stiff trench barrier (r_c=0.75) (left) and graphs of A_{RF} for different r_c along x-axis (soft barriers) (right).

3.3.2. Effect of Geometrical Parameters

In this section, the passive vibration isolation problem by an in-filled trench with different geometrical parameters is examined in detail. Herein, only the stiff trench barrier (r_c>1) is considered. The Poisson's ratios of the trench and soil are ν_b=0.25 and ν_s=0.35, respectively, and the velocity ratio of shear wave is r_c =5. The geometric variables of the trench are summarized in Table 4. Note that when a certain parameter is investigated, the other two parameters are kept constant. Therefore, unless otherwise specified, the normalized depth h^* of the trench and its width t^* and length l^* are taken as 1.0, 0.1, 1.0, respectively.

Table 4. Geometric parameters of the trench barrier

Parameter	Calculating values
Depth of the in-filled trench (h^*)	0.2, 0.5, 1.0, 1.5
Width of the in-filled trench (t^*)	0.05, 0.10, 0.15, 0.20
Length of the in-filled trench (l^*)	0.5, 1.0, 1.5

3.3.2.1. Effect of the Trench Depth (h^*)

The trench depth is an important variable of practical concern for the vibration isolation by an in-filled trench. For the purpose of this investigation, it is assumed that the value of h^* is not a constant and equals to 0.2, 0.5, 1.0 and 1.5, respectively.

Figure 16(left) depicts the variation of the factor A_{RF} on the surface of the wave field for the above passive isolation, and Figure 16 (right) is the variation graphs of A_{RF} with different depth h^* along x-axis. One can observe an increase of screening effect and an increasing screened zone with the trench depth. However, it also can be found that when h^*>1.0, the increases are slight with the trench depth, due to the transmission of Rayleigh wave is confined to near the surface. Therefore, it is suggested that the depth of the trench should be equal to a Rayleigh wavelength to ensure a good screening effect. An excessive depth is not requisite since it may induce additional costs and the difficulties of construction.

3.3.2.2. Effect of the Trench Width (t^*)

In what follows, the width of the trench t^* is considered to be variable, taking the values given in Table 4.

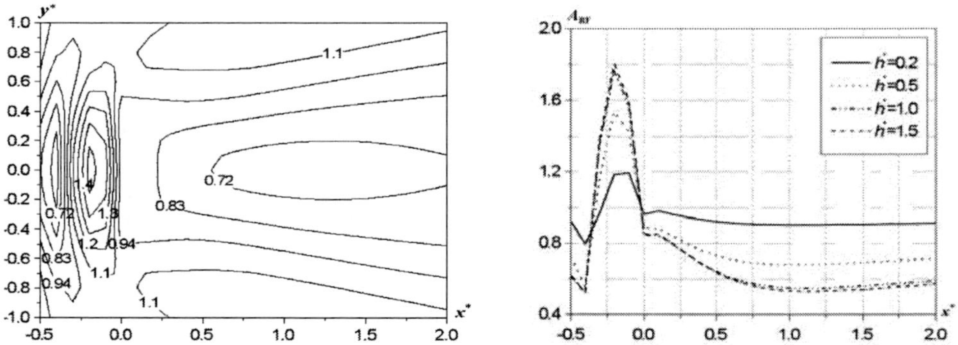

Figure 16. Contour of A_{RF} for the stiff trench barrier (h^*=0.5) (left) and graphs of A_{RF} for different h^* along x-axis (right).

The isogram of A_{RF} for t^*=0.10 is shown in Figure 17(left), and the variation graphs of A_{RF} with different t^* along x-axis in Figure 17(right). These figures indicate that the screened efficiency and screened zone behind barrier increase with the trench width. Moreover, the magnification at the incident side of barrier increases rapidly with the width t^*, while the A_{RF} behind the barrier decreases rapidly. Therefore, increasing the width of the trench can result in a better screening effect. When taking the costs and the difficulties of construction into account, it is suggested that trench depth should meet $t^*\geq 0.1$ to ensure a good screening effect.

3.3.2.3. Effect of the Trench Length (l^*)

In what follows, the length of the in-filled trench l^* varies, taking the values in Table 4.

Figure 18 shows the isograms of A_{RF} for l^*=0.5, 1.0 and 1.5, respectively. It can be observed that the optimum screened zone is at a small distance behind the barriers and the distance increases with l^*. The screening effect and the area of screened zone also increase with l^*.

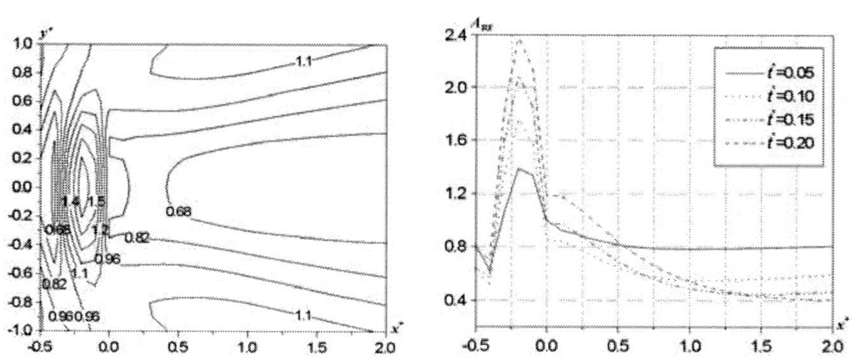

Figure 17. Contour of A_{RF} for the stiff trench barrier (t^*=0.10) (left) and graphs of A_{RF} for different t^* along x-axis (right).

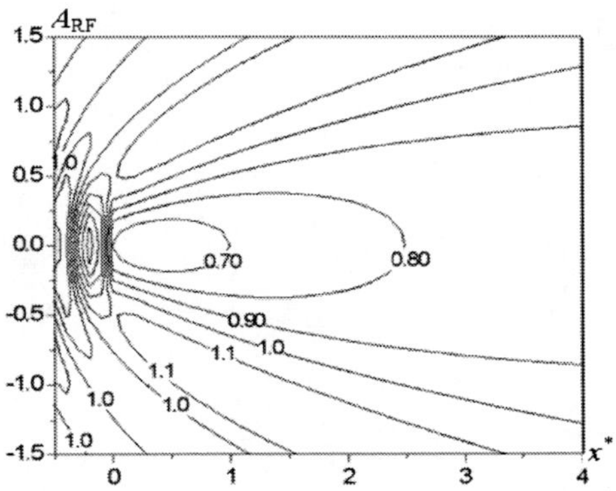

(a)

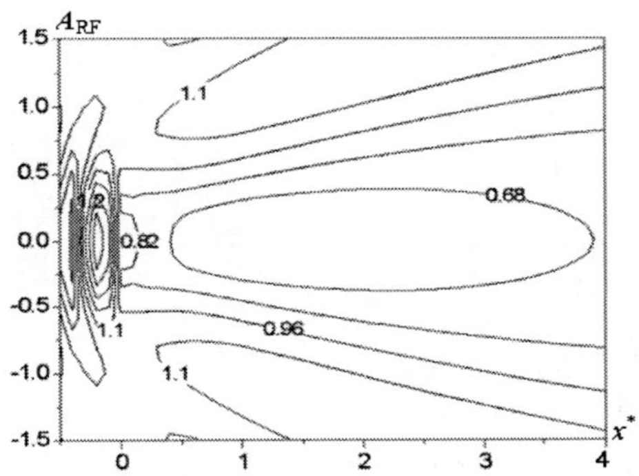

(b)

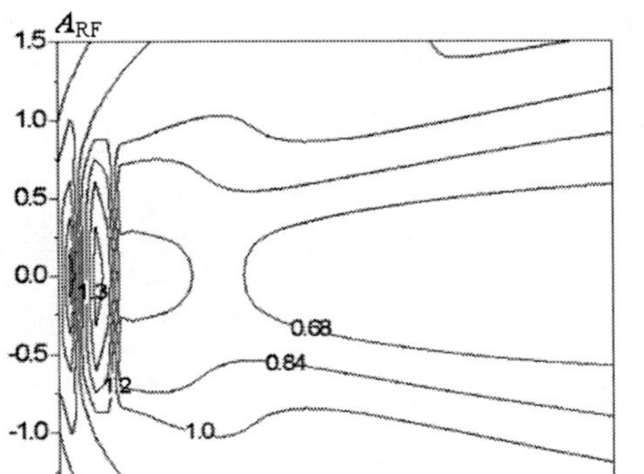

(c)

Figure 18. Contour of A_{RF} for the stiff trench barrier for: (a) l^*=0.5, (b) l^*=1.0 and (c) l^*=1.5.

Figure 19 depicts the contour of A_{RF} for the different value of y^* in the direction of x-axis. The figures show that the value of A_{RF} varies at different position from the center to the border of the trench barrier. The value of A_{RF} is a small amount and varies slowly near the center of the barrier, while it increases dramatically near the border of the barrier. This result indicates that the screening effect near the border is worse than the center. Therefore, it is suggested that the value of l^* should be large enough to avoid the protected object lying in the border region of the barrier.

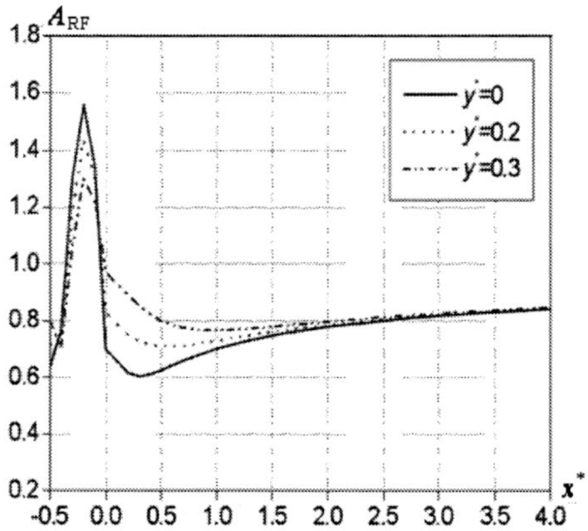

(a)

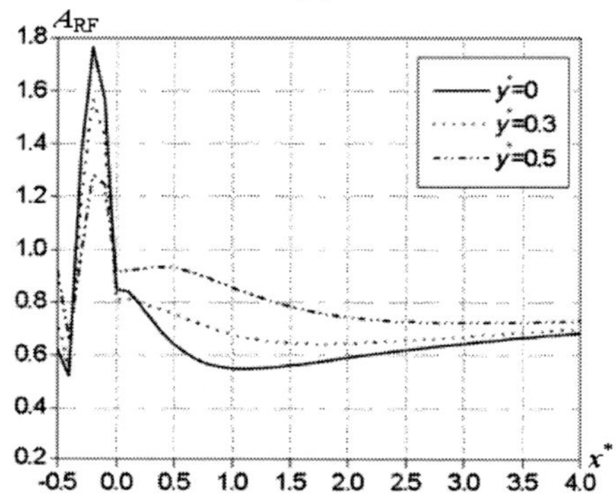

(b)

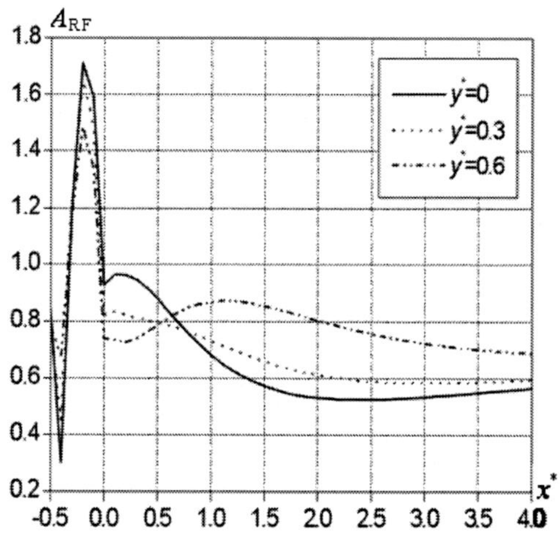

(c)

Figure 19. Graphs of A_{RF} for different y^* in x-axis direction for: (a) l^*=0.5, (b) l^*=1.0 and (c) l^*=1.5.

3.4. Conclusion

In-filled trench as a passive barrier of vibration isolation in the far field can be modeled and predicted by the scattering problem of Rayleigh waves induced by the trench obstacle.

Based on the integral equation method of Rayleigh wave scattering, the screening effectiveness of the in-filled trench as passive isolation barrier was studied. The method presented can be used to solve the 3D problem of Rayleigh wave scattering from irregular obstacle by iteration.

Using in-filled trench barrier can isolate the ground vibrations in the screened zone. The maximum reductions occur in the center zone of the barrier behind the trench. The Poisson's ratio of backfill material has slight influence on the screening effect. Moreover, the trench in-filled by stiff backfill material can get a better isolation effect and a larger screened zone than soft one. Thus, it is suggested to choose the backfill material with the value of $r_c \geq 5$ in practice.

Increasing the depth and width of the in-filled trench can result in an increase of screening effect, so the trench should meet the conditions of $h^*=1.0$ and $t^* \geq 0.1$ to achieve a satisfactory effect of isolation. The screening effect and area of screened zone increase with the trench length also, so the value of l^* should be large enough to avoid the protected object lying in the border region of the barrier.

4. Three-dimensional Analysis and Experimental Study on Active Vibration Isolation Using WIB in Layered Ground under Vertical Excitation

4.1. Fundamental Solutions of Layered Ground by TLM

It is a critical problem in BEM to find an appropriate fundamental solution, which can improve the calculation efficiency dramatically. Full-space or homogenous half-space Green's functions are often applied in study of dynamic soil-structure interaction, but they are unsuitable for problems of stratified half-space, which requires the discretization at the layered interfaces and increases the computational complexity dramatically.

In order to improve the calculation efficiency, an approximate method, named the thin layered method (TLM) was developed by Kausel [20]. TLM is a semi-discrete technique to handle the problem of wave propagation in soils consisting of semi-infinite horizontal layers on bedrock or an elastic half-space, as depicted in Figure 20. The soil layers are discretized into n_s thin layers with the thickness h_i, $i=1,2,\ldots n_s$ in vertical direction, while TLM yields an analytical solution in other directions.

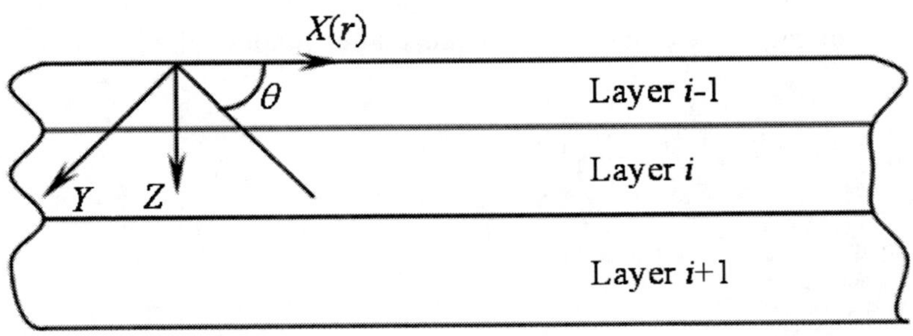

Figure 20. Sketch of layered ground.

The partial differential equation governing motion in each thin layer can be expressed in terms of the displacements in cylindrical coordinate system as follows:

$$\begin{cases} \mu\left[\nabla^2 u_r - \frac{1}{r}(2\frac{\partial u_\theta}{r\partial\theta} + \frac{u_r}{r})\right] + (\lambda+\mu)\frac{\partial\varepsilon_V}{\partial r} + f_r = \rho\frac{\partial^2 u_r}{\partial t^2} \\ \mu\left[\nabla^2 u_\theta - \frac{1}{r}(\frac{u_\theta}{r} - 2\frac{\partial u_r}{r\partial\theta})\right] + (\lambda+\mu)\frac{\partial\varepsilon_V}{r\partial\theta} + f_\theta = \rho\frac{\partial^2 u_\theta}{\partial t^2} \\ \mu\nabla^2 u_z + (\lambda+\mu)\frac{\partial\varepsilon_V}{\partial z} + f_z = \rho\frac{\partial^2 u_z}{\partial t^2} \end{cases} \tag{47}$$

in which u_r, u_θ, u_z represent the component of displacement in axial direction, tangential direction and vertical direction, respectively; f_r, f_θ, f_z are the body forces in the direction of axial, tangential, and vertical, respectively; ε_V, is the volume strain; and λ and μ are Lame constants; ∇^2 is an operator which is defined as:

$$\nabla^2 = \frac{\partial^2}{\partial r^2} + \frac{1}{r}\frac{\partial}{\partial r} + \frac{1}{r^2}\frac{\partial^2}{\partial\theta^2} + \frac{\partial^2}{\partial z^2}$$

First, assuming the thickness of each layer is small enough, the displacement in each layer can be approximated using the interpolation functions as follows:

$$\begin{cases} u_r = \mathbf{N}\mathbf{u}_{ri} \\ u_\theta = \mathbf{N}\mathbf{u}_{\theta i} \\ u_z = \mathbf{N}\mathbf{u}_{zi} \end{cases} \tag{48}$$

where $\mathbf{u}_{ri} = \{u_{ri}, u_{r(i+1)}\}^{\mathrm{T}}$, $\mathbf{u}_{\theta i} = \{u_{\theta i}, u_{\theta(i+1)}\}^{\mathrm{T}}$, $\mathbf{u}_{zi} = \{u_{zi}, u_{z(i+1)}\}^{\mathrm{T}}$ denote the node displacement vectors in the i-th layer; herein, linear interpolation scheme is adopted for simulation of displacements, and $\mathbf{N}=[1-\xi, \xi]$ is the interpolation function; ξ is the generalized coordinates in the i-th layer, where $\xi=(z-z_i)/(z_{i+1}-z_i)$ for $z_i \leq z \leq z_{i+1}$.

Then substituting Equation (48) into the differential Equation (47), and due to the work of weighted residuals formulation, in which the interpolation function **N** is chosen as tentative function, the following expressions can be obtained after simplification:

$$\sum_{i=1}^{n_s}\left\{\int_{V_i}\begin{bmatrix} \mu\left[\nabla^2 u_r - \frac{1}{r}\left(2\frac{\partial u_\theta}{r\partial\theta} + \frac{u_r}{r}\right)\right] + (\lambda+\mu)\frac{\partial\varepsilon_V}{\partial r} + f_r - \rho\frac{\partial^2 u_r}{\partial t^2} \\ \mu\left[\nabla^2 u_\theta - \frac{1}{r}\left(\frac{u_\theta}{r} - 2\frac{\partial u_r}{r\partial\theta}\right)\right] + (\lambda+\mu)\frac{\partial\varepsilon_V}{r\partial\theta} + f_\theta - \rho\frac{\partial^2 u_\theta}{\partial t^2} \\ \mu\nabla^2 u_z + (\lambda+\mu)\frac{\partial\varepsilon_V}{\partial z} + f_z - \rho\frac{\partial^2 u_z}{\partial t^2} \end{bmatrix} \mathrm{d}V\right.$$

$$+\int_{\Omega_i}\mathbf{N}^{\mathrm{T}}\mathbf{I}\mathbf{F}_{si}d\Omega+\int_{\Omega_i}\mathbf{N}^{\mathrm{T}}\mathbf{I}\begin{bmatrix}\mu\dfrac{\partial u_r}{\partial z}+\mu\dfrac{\partial u_z}{\partial r}\\ \mu\dfrac{\partial u_\theta}{\partial z}+\mu\dfrac{1}{r}\dfrac{\partial u_z}{\partial \theta}\\ \lambda\varepsilon_V+2\mu\dfrac{\partial u_z}{\partial z}\end{bmatrix}d\Omega\Bigg\}=0 \tag{49}$$

where V_i, Ω_i denote the domain and the boundary of the i-th soil layer, respectively; $\mathbf{F}_{\mathrm{si}}$ indicates the surface forces of the i-th soil layer; **I** is a 3×3 unit matrix.

Then Fourier series expansion is applied to Equation (49) with respect to variable θ, and only the m-th term is considered. Further, adding and subtracting the first row and the second row of the equations, respectively, a new set of equations are obtained. Thereafter, performing m-1 order, m order and m+1 order Hankel transform on each row of the new set of wave equations with respect to variable r, respectively, another new set of equations are obtained. Considering the harmonic vibration ($e^{-i\omega t}$), and substituting Equation (48) into the series of transformed ordinary differential equations mentioned above, yields the force-displacement relation as

$$(k^2\mathbf{A}-k\mathbf{B}+\mathbf{C})\bar{\mathbf{U}}_m=\bar{\mathbf{P}}_m \tag{50}$$

in which k is the wave number in the Hankel transformed domain, $\bar{\mathbf{U}}_m$ denotes the displacement vector in the transformed domain and $\bar{\mathbf{P}}_m$ the loading vector; **A**, **B** and **C** are the stiffness matrices of the layered ground. These matrices are assembled from the stiffness matrices for each layer in a finite-element sense, which are only dependent on the material properties and the thickness of the layer.

Equation (50) can be solved by modal superposition method. Once the loading vector $\bar{\mathbf{P}}_m$ is determined, the displacement vector $\bar{\mathbf{U}}_m$ can be obtained. However, larger errors may generate when the total discretized depth of the layered soils is insufficient. Though the solutions can approximate the analytical ones in half-space with increasing the discretized depth, the scale of the nodes and the computational time increases dramatically.

To this end, some measures can be adopted to achieve high computational accuracy. One is applying springs or dampers to the bottom boundary. This measure is easy to be coupled into Equation (50), however, its usage has some limitation and may fail in certain condition.

The Paraxial Approximation proposed by Kausel and Seal [21] is a relatively accurate method to overcome the problem of simulating half-space by TLM. In this method, the stiffness matrix of the homogeneous half-space in wave number domain is expanded by using second order Taylor's series, and yields the approximate solutions suited to be the bottom boundary of TLM. However, the paraxial solutions are available only in cylindrical coordinate system.

Based on the fore-mentioned deduction, the general approximate paraxial solutions to the *m-th* Fourier series, namely the variables in field vary with the $\sin(m\theta)$ or $\cos(m\theta)$ in tangential direction, have been derived in this section. The steady-state fundamental solutions (Green's function) are then evaluated by applying the inverse Hankel transform. Unlike the transfer matrices technique by Kausel and Seal [21], the components of Green's function in frequency domain can be obtained analytically, instead of being evaluated numerically, by the inverse Hankel transforms, which may improve the efficiency of calculations greatly. In addition, substituting the nodal displacement field into the constitutive equations, one can obtain the nodal stresses, and the stresses at any point in the ground can be gained by using the miscellaneous interpolation functions in each layer.

4.2. Three-dimensional Semi-analytical BEM

A boundary integral equation can be obtained by weighted residual formulation using the fundamental solutions u_{ij}^* and t_{ij}^* as weighting function. Hence, the integral representation of displacement in linear elastodynamics has the following form in the frequency domain, by neglecting the body force term:

$$C_{kj}(y_1)u_k(y_1)+\int_{\Omega} t_{kj}^*(y_2,y_1)u_k(y_2)\mathrm{d}\Omega=\int_{\Omega} t_k(y_2)u_{kj}^*(y_2,y_1)\mathrm{d}\Omega \tag{51}$$

in which $\boldsymbol{u}_k(\boldsymbol{y}_1)$ and $\boldsymbol{t}_k(\boldsymbol{y}_1)$ are the amplitude of the displacement and traction vectors, u_{ij}^* and t_{ij}^* are the fundamental displacement and traction tensors, respectively. $\boldsymbol{y}_1$ and $\boldsymbol{y}_2$ are the field and source points, respectively; the jump tensor or free term $c_{ij}(y_1)=\delta_{ij}/2$ for $\boldsymbol{y}_1\in\Omega$, being smooth, or $c_{ij}(y_1)$ is a function of the local geometry at $\boldsymbol{y}_1$ when the surface at that point is not smooth; summation is assumed over repeated indices and $i, j=1,2,3$.

Equation (51) can be solved numerically by the BEM in terms of the unknown vectors $\boldsymbol{u}_k(\boldsymbol{y}_1)$ and $\boldsymbol{t}_k(\boldsymbol{y}_1)$ on the boundary for given boundary data. Herein, the direct boundary element method is obtained by discretized the surface Ω into a finite number of boundary elements and through a collocation process, one reaches the matrix equation

$$\mathbf{U}_{\Omega}\mathbf{t}_{\Omega}=\mathbf{T}_{\Omega}\mathbf{u}_{\Omega} \tag{52}$$

where $\mathbf{U}_{\Omega}$ and $\mathbf{T}_{\Omega}$ are influence matrices with entries integrals over boundary elements with integrands involving the fundamental tensors u_{ij}^* and t_{ij}^*, respectively, while $\mathbf{t}_{\Omega}$ and $\mathbf{u}_{\Omega}$ denote the vectors of nodal boundary tractions and displacements, respectively.

Herein, a 3D semi-analytical boundary element method is put forward, where the fundamental solutions by TLM is employed, to investigate the isolation efficiency of WIB in multilayered half-space. Unlike other conventional BEM programs, this semi-analytical BEM does not require discretization at the ground surface and the soils interfaces, which decrease the computational complexity dramatically.

In order to validate the method, the study of Lamb's problem is carried out. Lamb [22] systemically investigated the problem of ground vibration due to a surface and internal vibration source in half-space. Later, Wang [23] achieved the closed solution of Lamb's problem, which avoids calculating the complex integration. The vertical surface displacements of the above- mentioned analytical solution are compared well to those obtained by the present method, which can be observed in Figure 21, where a very close agreement is found, except in the area immediately surrounding the point load.

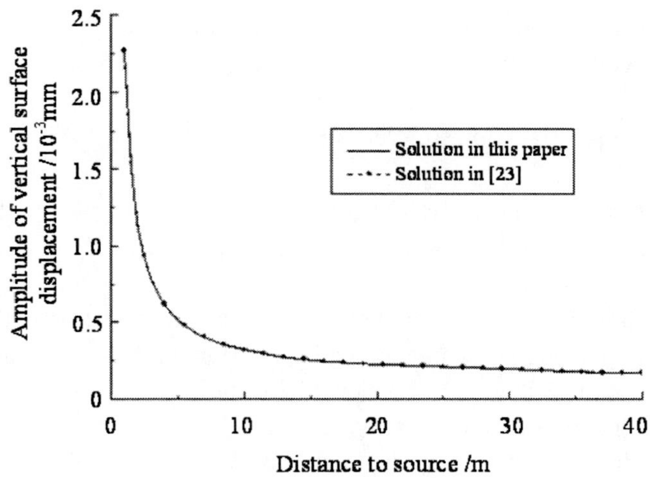

Figure 21. Contrast solutions of ground displacement due to SH loading.

4.3. Analysis of Active Isolation by WIB under Vertical Excitation

Based on the semi-analytical BEM, the vibration screening effect by WIB in 3D multilayered ground is analyzed in detail. Figure 22 shows a schematic diagram of vibration isolation system of WIB due to vertical excitation. The soil medium is regarded as layered half-space in the cases of softer upper layer on stiffer half-space and stiffer upper layer on softer half-space, respectively.

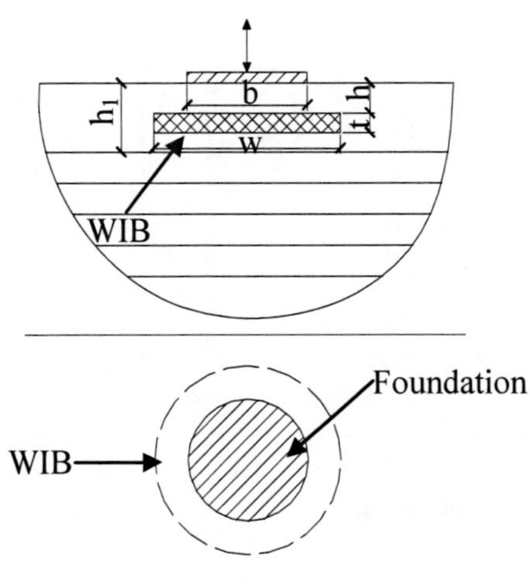

Figure 22. Sketch of vibration isolation by WIB.

Soils parameters for the BEM analysis are the shear modulus, mass density, Poisson's ratio and material damping coefficient. These parameters are given as follows:

Case A: Softer upper layer on stiffer half-space

- Top layer: μ_s=53 MPa, ρ_s=1800 kg/m^3, v_s=0.44, ξ=5%;
- Bottom layer: μ_s=120 MPa, ρ_s=2000 kg/m^3, v_s=0.33, ξ=5%.

Case B: Stiffer upper layer on softer half-space

- Top layer: μ_s=53 MPa, ρ_s=1800 kg/m^3, v_s=0.33, ξ=5%;
- Bottom layer: μ_s=20.5 MPa, ρ_s=1522 kg/m^3, v_s=0.44, ξ=5%.

The soil vibration is caused by a vertical, harmonic unit excitation with a frequency of 16 Hz. It acts on a circular surface foundation which is assumed to be rigid and massless, and has a diameter b of 2 m. The Rayleigh wavelength for the upper layer soil is L_R=10 m and the normalized (by L_R) top layer thickness read $H_1 = h_1/L_R$. The normalized geometrical data of the WIB read the diameter $W_1=w_1/L_R$, the thickness $T_1=t_1/L_R$, the embedded depth $H=h/L_R$. In addition, the WIB has a normalized shear modulus $M=(\mu_b/\mu_s)^{0.5}$, a density ρ_b=2400 kg/m^3, a Poisson's ratio v_b=0.2 and a material damping ξ=5%.

The amplitude attenuation ratio A_R is defined as the ratio of the displacement component of the soil surface in the presence of the WIB over that without WIB. This parameter is frequently used as a measure of the screening effectiveness of an isolation barrier. A zero amplitude attenuation ratio indicates that the incoming wave is completely scattered and that the barrier is completely effective [19].

In what follows, unless otherwise stated, the normalized geometrical parameters of the WIB are assumed to be constant with the values as: W = 1.0, H = 0.05, T = 0.1, M = 12 and H_1=1.0.

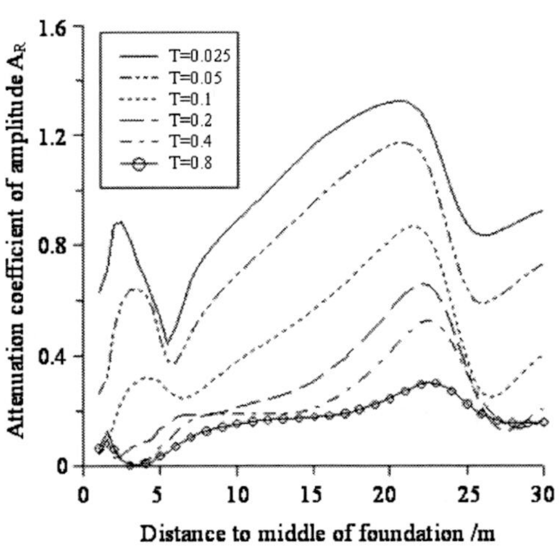

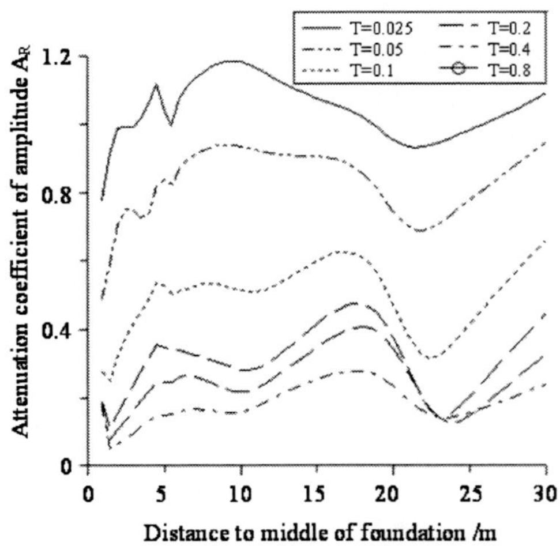

Figure 23. Graphs of radial (left) and vertical A_R for various T along distance (right).

4.3.1. Active Isolation by WIB for Case A: Softer Upper Layer over Stiffer Half-space Effect of the Thickness of WIB

The influence of the thickness of WIB on the screening effect is studied. The normalized thickness T varies as 0.025, 0.05, 0.1, 0.2, 0.4 and 0.8, respectively.

Figure 23 depict the variation of A_R of the radial and vertical displacement on the surface along r-axis for different thickness of WIB. One can observe that WIB can successfully isolate the vibration caused by a vertical vibration source, except for very small thickness $T \leq 0.025$. When $0.025 < T < 0.2$, increasing thickness of WIB is very efficient for reduction of the surface wave field. However only slightly higher reduction effect can be expected with increase of T when $T \geq 0.2$.

4.3.2. Active Isolation by WIB for Case B: Stiffer Upper Layer over Softer Half-space

Figure 24 shows the influence of normalized thickness of the top layer on the amplitude attenuation ratio A_R. Herein, the normalized thickness H_1 varies from 0.5 to 3.0.

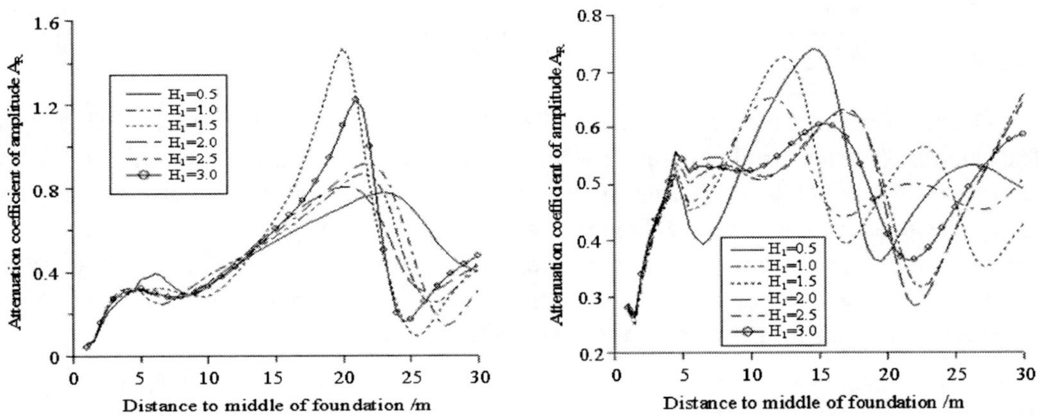

Figure 24. Graphs of radial (left) and vertical A_R for various H_1 along distance (right).

It is observed when the distance is less than 15 m, the variation of radial amplitude attenuation ratios A_R versus distance are almost identical for all top layer thicknesses. When the distance is beyond 15 m, the A_R curves show slight difference between each other. However, for vertical vibration, increasing top layer thickness has a great influence on reduction of surface wave field, only except for the case of near-field (<5 m), in which the variation of A_R is almost negligible.

4.4. Field Tests

4.4.1. Site Conditions

4.4.1.1. Geological Conditions

The geological conditions at the test site are shown in Table 5.

4.4.1.2. Background Vibration Conditions

In order to evaluate the real mitigation effects of WIB for the noise generated by the electric vibrator, the influence of background vibration should be taken into consideration before the field test. Three measuring points at the site were selected for the 24-hour ambient vibration test.

Table 5. Soil profile and properties

No.	Depth /m	Soil layer	Soil type	Thickness /m	Unit weight /kg/m^3	V_s /m/s
1	0.00~1.60	①	Fill	1.6	1830	85
2	1.60~2.50	②-1	Clayey silt	0.9	1939	90
3	2.50~9.00	②-3a	Sandy silt	6.5	1888	100
4	9.00~15.50	②-3b	Silt	6.5	1898	160
5	15.50~25.50	②-3c	Sandy silt	10.0	1837	130
6	25.50~30.00	⑤	Silty clay	4.5	1847	170

The measurements were conducted for the latitudinal, longitudinal and vertical directions. The one-third octave band response spectra for each point along the three directions are depicted in Figure 25 in the forms of monograph. As shown in the figure, the maximum velocity responses along the three directions were all less than 10^{-5} m/s. However, the velocity responses at different exciting frequencies induced by the electric vibrator were always to the amount of 10^{-4} m/s in Figure 29. They were not in the same order of magnitude, so the influence of the background vibration may be neglected.

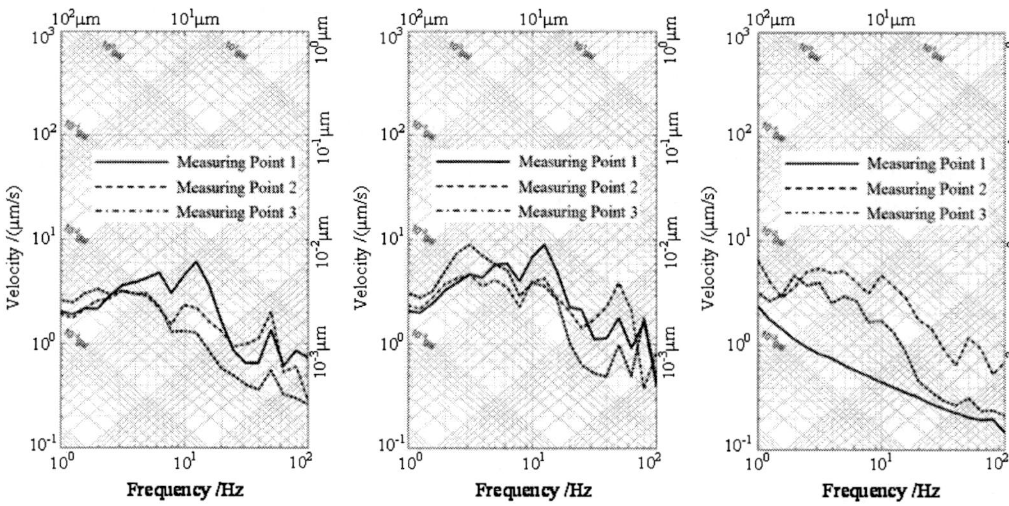

Figure 25. One-third octave band response spectra for 24-houred background vibration in latitudinal (left), longitudinal (middle) and vertical directions (right).

4.4.2. Vibration Source

A set of Continue Surface Wave System (CSWS) produced by the GDS Co. Ltd. was employed for the field tests. The CSWS comprises an electric vibrator, a surface wave system control unit, a surface wave system drive unit, 6 geophones and a generator as shown in Figure 26 (a), (b). With the aid of other devices of CSWS, the electric vibrator can generate harmonic waves from 10 Hz to 100 Hz with the step of 5 Hz by exerting periodic vertical loading, which was used to simulate the common industrial machine operations.

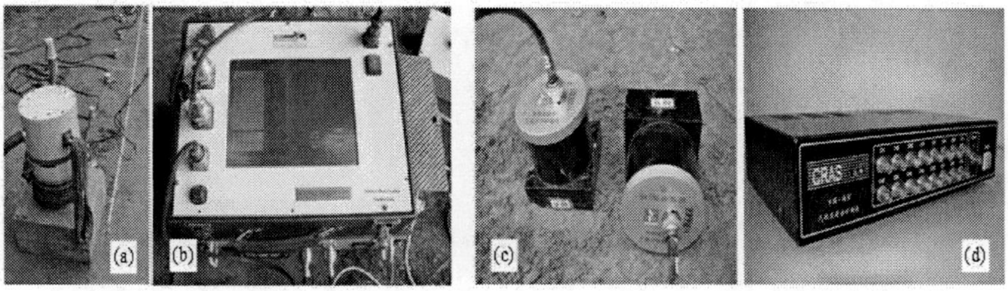

Figure 26. Continue Surface Wave System: (a) vibrator, (b) control-and-drive unit; and Date Acquisition System: (c) 941B velocity meter, (d) recording system.

4.4.3. Date Acquisition System

The sensors used to measure the ground vibration were the servo-type velocity meters 941B, and the recording device was QL-116 system, which are commonly used in China, as shown in Figure 26 (c) and (d). The data acquisition was performed for the ground response with the sampling frequency of 256 Hz. A total 128 data blocks could be recorded for each velocity sensor in every case with 1024 sample points in each data block.

4.4.4. Experimental Cases

Three WIB parameters (i.e., WIB plane size, shear modulus and embedded depth or in-filled thickness on WIB) were selected as the objective parameters for the experimental study. Six experimental cases with different types of WIB were installed for the field tests. The schematic diagram for the field test is depicted in Figure 27(left). The detailed information for each case is listed in Table 6, and the main dimensions of WIB are shown in Figure 27(right).

Table 6. Experimental cases for the field tests

Case	length a /mm	width b /mm	thickness t /mm	depth h /mm	Material	ρ /(kg/m^3)	G /MPa
Case 1	600	600	50	50	Concrete	2500	1.25×10^4
Case 2	1000	700	50	50	Concrete	2500	1.25×10^4
Case 3	710	710	25	50	Particleboard	713	654.76
Case 4	710	710	25	50	Steel	7850	7.86×10^4
Case 5	700	1000	50	200	Concrete	2500	1.25×10^4
Case 6	700	1000	50	400	Concrete	2500	1.25×10^4

Before the measurements, the ground response for free field, namely field without WIB, was measured first. By comparing the ground response of each case with that of free field, we knew the isolation effect of the designed case. On the other hand, the influences of WIB plane size, shear modulus and embedded depth on the mitigation effect were examined through comparisons between Cases 1 and Case 2, Case 3 and Case 4, Case 5 and Case 6, respectively.

As shown in Figure 27, a total of 8 measuring points with 0.5 m interval were located along the measuring line starting from the center of the concrete block (surface footing), which supported the electric vibrator. The WIB was buried at a proper depth under the block. The three geometric centers of the vibrator, surface footing and WIB were coincided along the vertical direction.

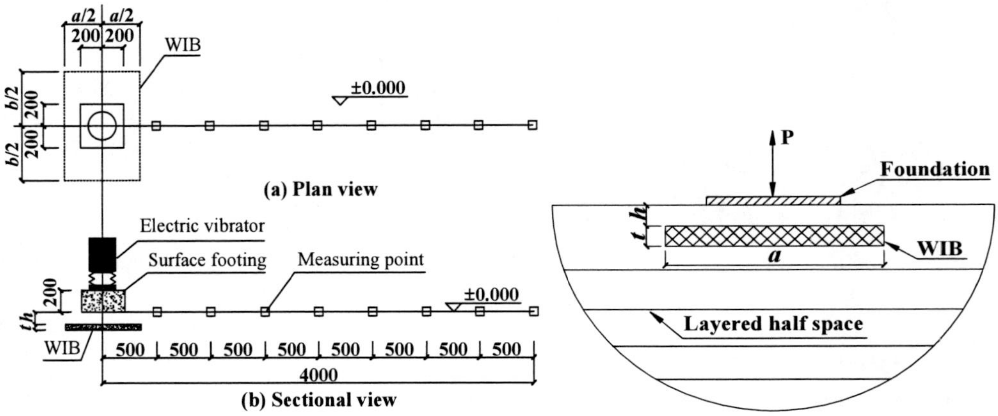

Figure 27. Schematic diagrams for the field test (left) and main dimensions of WIB (right).

4.4.5. Data Processing

The original data recorded by the devices are vibration velocities in time series, while the analysis needs the RMS velocity amplitude. Hence, the following steps are used to get RMS velocity amplitudes. Firstly, the velocity time history for each exciting frequency is subtracted from the original data with both low and high pass filters. Secondly, the DFT transformation is conducted for the selected velocity versus time history according to the flowing equations:

$$X(k)=\sum_{n=0}^{N-1} x(n)e^{-j\frac{2\pi}{N}nk} \tag{53}$$

$$A_{k\,\mathrm{rms}}=\frac{\sqrt{2}}{N}\left|X(k)\right| \tag{54}$$

where N is the number of sample points, $x(n)$ refers to the velocity value of the nth sample point and $A_{k\mathrm{rms}}$ means the RMS velocity amplitude in the frequency domain.

4.4.6. Evaluation on Wave Mitigation Effect

The mitigation effect on ground response can be evaluated by comparing the amplitude reduction ratio curves or the average amplitude reduction ratio within the length of measuring line for each of the experimental cases. Herein, the amplitude reduction ratio A_R proposed by Woods et al. [19] is defined as follows,

$$A_R=\frac{\text{Amplitude with WIB}}{\text{Amplitude without WIB}} \tag{55}$$

where the numerator and the denominator refer to, respectively, the RMS velocity responses before and after the WIB is installed.

The average amplitude reduction ratio $\bar{A}_R$ presented by Tsai et al. [24] can be obtained with the following formulation:

$$\bar{A}_R = \frac{1}{l}\int A_R \mathrm{d}l \tag{56}$$

where l is the length of measuring line.

4.5. Analysis of Test Results

4.5.1. Measured Response

The response data of the No. 3 measuring point for Case 2 is taken as the example. It locates at 1.5m away from the source. The electric vibrator works by means of frequency sweeping from 10 Hz to 100 Hz with the step 5 Hz. So the vibration is unsteady on the whole but harmonic for a certain frequency. The recorded velocity time history diagram is like a shuttle, as shown in Figure 28. Its peaks are reached at 65 Hz and 50 Hz along the radial and vertical directions, respectively. The data processing procedure is applied to the origin data, and the RMS velocity response spectra can be obtained for all the exciting frequencies. The results are depicted in Figure 29. Furthermore, A_R and $\bar{A}_R$ are calculated with Equations (55) and (56). The analysis of WIB parameters on the vibration screening effect can be performed in the next section.

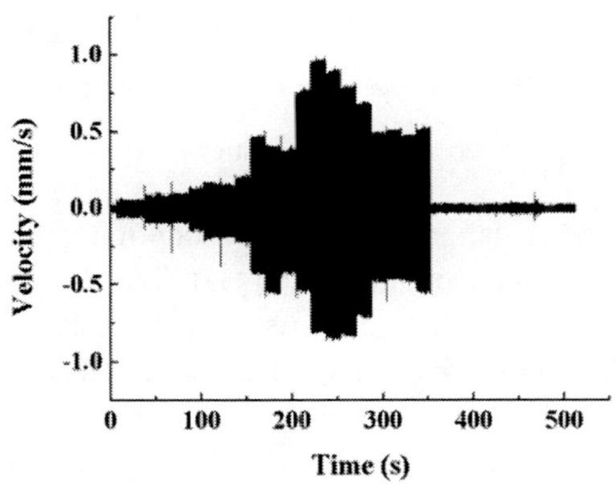

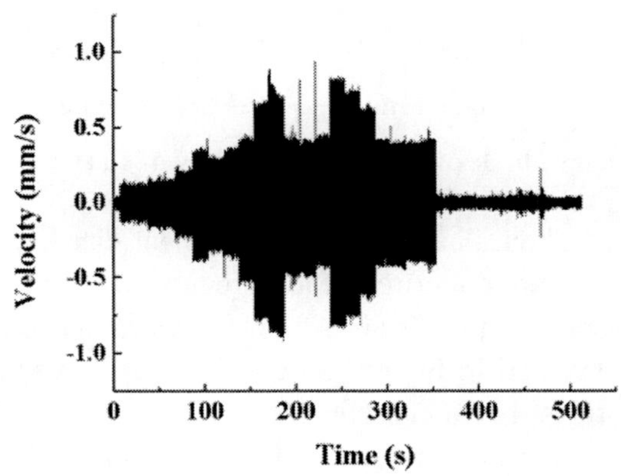

Figure 28. Recorded velocity time history: Radial (left) and Vertical (right).

4.5.2. Analysis of WIB Parameters

4.5.2.1. Effect of Plane Size

The WIB models are generally assumed to be circle and symmetric in Gao et al. [9−11]. However, the WIBs in the field experiment are designed to be rectangle. So an equivalent diameter D_e is calculated according to Equation (57) below to evaluate the effect of WIB plane size on wave impedance.

$$D_e = 2(a \cdot b / \pi)^{0.5} \tag{57}$$

where a and b are the length and width of the rectangle WIB, respectively. The equivalent diameter D_e is 677 mm for case 1 and 944 mm for case 2 in the test.

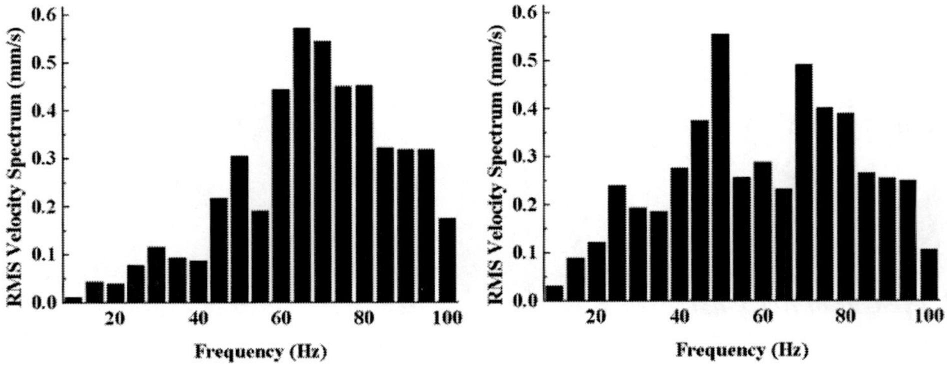

Figure 29. RMS velocity spectra: Radial (left) and Vertical (right).

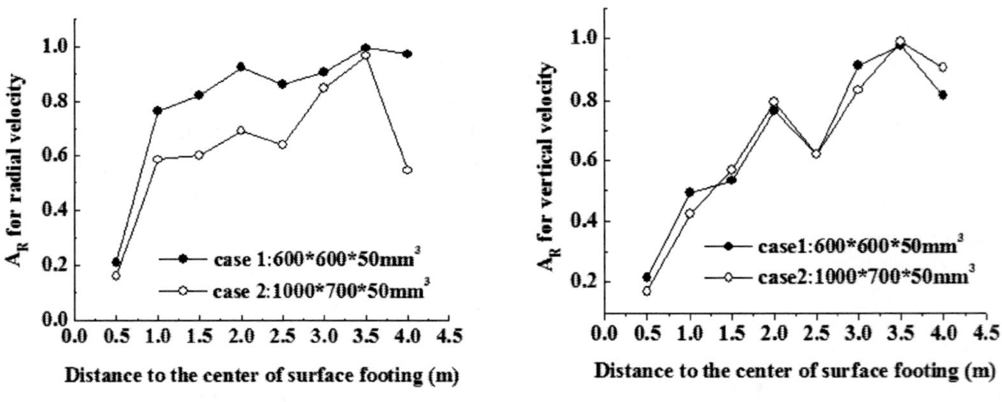

Figure 30. A_R of velocity at frequency of 10 Hz: radial (left) and vertical (left) velocities.

The influences of the equivalent diameter on the radial and vertical vibration mitigation effect are compared at low frequency (10 Hz) and medium frequency (60 Hz). The A_R with respect to the distance from the vibration source is used in the comparison. The results are illustrated in Figures 30 and 31, respectively. Meanwhile, the average amplitude reduction ratio is shown in Figure 32.

As shown in Figures 30 and 31, the A_R value fluctuates with the distance from the source, but it is always less than 1.0 for all the measuring points. Moreover, the A_R value is about 0.2 for the measuring point at 0.5 m from the source, which means about 80% vibrations are eliminated at this point. The result shows WIB has the advantage to reduce the low frequency vibrations over an open or in-filled trench and row-piles in near field, and better mitigation effects can be expected for medium frequency (60 Hz) vibration in Figure 31 than low frequency (10 Hz) vibrations in Figure 30. For example, the A_R curve for 60 Hz radial vibration (Figure 31 left) nearly swings around the line of A_R=0.2. The exact mean of the screening effectiveness reaches 0.24, and is much better than the corresponding one of only 0.63 at 10 Hz.

Also, from Figure 30 (left), it can be noticed that the A_R curves for the smaller WIB is almost always lower than those for the bigger one, except for the curves in Figure 30 (right). This may prove the theoretical finding in Gao et al. [9−11] that the WIB with bigger diameter can impede more vibrations while the other WIB parameters keep constant.

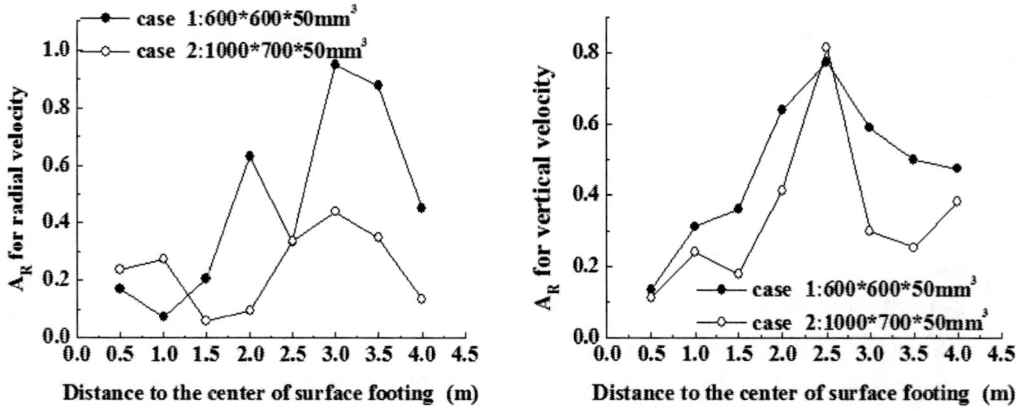

Figure 31. A_R of velocity at frequency of 60 Hz: radial (left) and vertical (left) velocities.

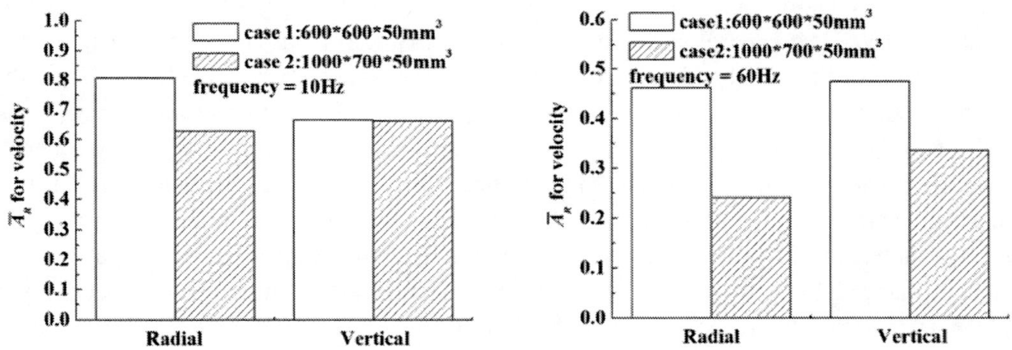

Figure 32. $\bar{A}_R$ of radial and vertical velocity at 10 Hz (left) and 60 Hz (right).

However, there is a little difference between the two A_R curves in Figure 49(b), which also can be seen from Figure 32 (left). This exceptional result may attribute to the reason that the difference in the WIB sizes cannot cause much difference in the A_R curves because of the longer Rayleigh wavelength (about 8.5m) at the low frequency of 10 Hz.

Besides, it seems that the wave mitigation effect in the radial direction is more sensitive to the WIB sizes than that in the vertical direction, which can be observed from a comparison between the results in Figures 30(left) and 30(right), or Figs 31(left) and 31(right). Therefore, if the radial vibration is taken as the control index, increasing the WIB size may be a good option.

4.5.2.2. Effect of Material

In Gao et al. [8−11], the WIB shear modulus was regarded as one of the most crucial factors for reducing wave transmission and relative vibration responses. Herein, the change of shear modulus is reflected by the two distinct kinds of materials (the particleboard in Case 3 and the steel in Case 4). Herein, the $\bar{A}_R$ curves of the two cases are plotted in Figure 33, and the average reduction ratio in the radial and vertical directions is depicted as histograms in Figure 34(left).

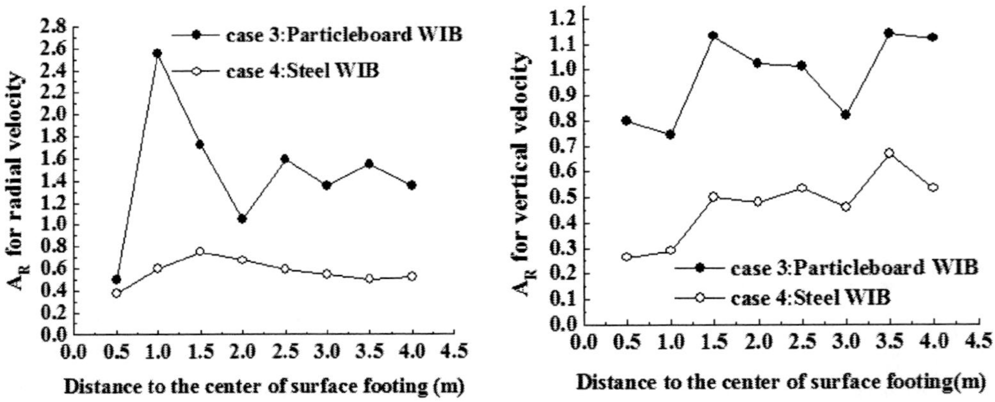

Figure 33. A_R of velocity at frequency of 30 Hz: radial (left) and vertical (left) velocities.

The figures show that steel WIB isolates more low frequency vibration than the particleboard WIB. It can be observed from 34(left) that the $\bar{A}_R$ values for the steel WIB are 0.57 and 0.47 for the radial and vertical directions, respectively, showing 43% and 53% of the average ground vibrations are reduced at the frequency of 30 Hz. In contrast, the $\bar{A}_R$ values for the particleboard WIB at 30 Hz are 1.46 in the radial direction and 0.97 in the vertical direction. Obviously, due to a much smaller shear modulus, the particleboard WIB screens almost no vibration in the vertical. It further amplifies the radial vibration, which is against the origin intension of vibration isolation. The results show that the wave mitigation effect can be enhanced by 61% and 52% for the radial and vertical directions respectively by replacing the particleboard WIB with the steel WIB.

A conclusion can be made that increasing the WIB shear modulus is an effective measure to improve the screening effect. When the shear modulus of WIB is decreased to some extent, the ground vibration is possible to be amplified rather than reduced.

4.5.2.3. Effect of Embedded Depth

According to the measurements of Cases 5 and 6, the curves for A_R of radial and vertical velocities versus the distance relationship at 35 Hz are plotted in Figure 35, and the corresponding $\bar{A}_R$ of the radial and vertical velocities for the two cases is shown in Figure 34(right).

It is noted when the WIB is deepened from 20 cm to 40 cm, the isolation effect gets worse, especially in the radial direction. The vibration is even amplified at some measuring points such as No. 1 to 3. As illustrated in Figure 34(right), the average amplitude reduction ratio rises from 0.53 to 1.05 in the radial direction and from 0.59 to 0.67 in the vertical direction. When the embedded depth is decreased from 40 cm to 20 cm, the screening effect is improved by 50% and 12% for the radial and vertical directions, respectively. This demonstrates the thinner the backfill on the WIB, the better the mitigation effect. When the WIB is embedded deep enough to some degree, the WIB would enlarge the vibration responses instead of screening them. This finding coincides with the theoretical findings in Gao et al. [9–11].

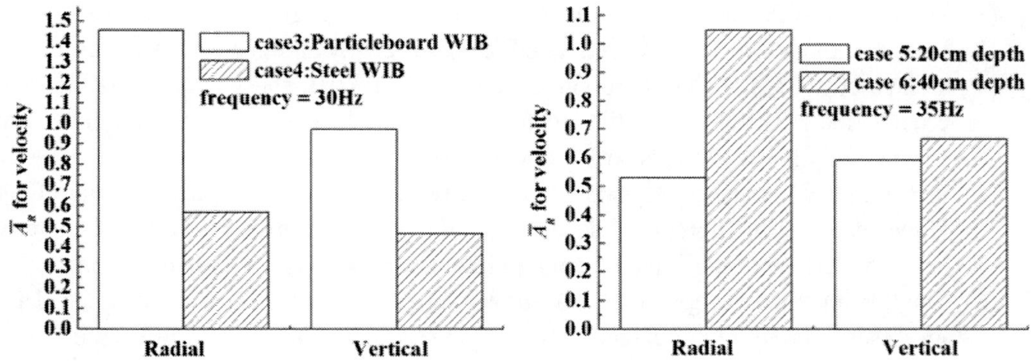

Figure 34. Measured $\bar{A}_R$ for 30 Hz (left) and 35 Hz (right).

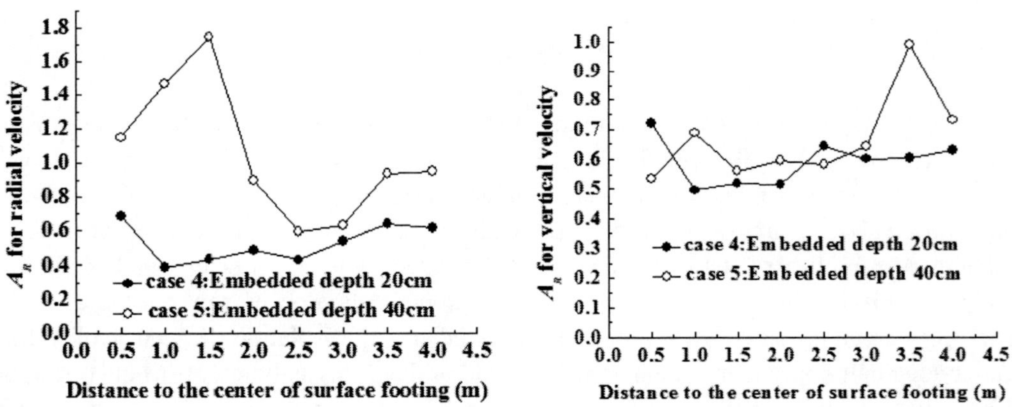

Figure 35. A_R of velocity at frequency of 35 Hz: radial (left) and vertical (left) velocities.

4.6. Conclusion

This section presents an analysis study on WIB in multi-layered half-space for active vibration isolation in a 3D context. A 3D axisymmetric model is developed to use WIB as active vibration isolation, taking into account the dynamical interaction of soil-structure, with the semi-analytical BEM in cooperation with the basic solution of the TLM. A series of field tests on active vibration isolation with WIB under vertical harmonic loading were conducted. Some design criteria for using an axisymmetric WIB as vibration isolation are developed and suggested.

Based on the analysis and test results, influences of the WIB parameters on the wave mitigation effect are investigated, and the following conclusions are drawn:

(1) The presented semi-analytical BEM can be used to solve the 3D problem of dynamic soil-structure interaction.

(2) WIB is an effective wave barrier for the medium to low frequency vibration generated by vertical loading, better than other measures such as open or in-filled trench and row-piles.
(3) Parameters of the layered soil affect the screening effect greatly. Particularly, the top layer thickness affects the screening effect significantly and in a complex way.
(4) Compared with plane size and embedded depth of WIB, shear modulus is the more crucial factor influencing the screening effect. Increasing shear modulus can substantially improve the isolating effectiveness.
(5) Expanding WIB plane size is also an effective way to increase the screening effect. However, its mitigation effect depends greatly on the vibration frequency. Generally, it is more effective to isolate medium to high frequency vibration than low one.
(6) Smaller embedded depth of WIB shows better screening effect. When the depth increases to an improper extent, WIB fails to screen and or even amplifies the ground vibration.

CONCLUSION

In this chapter, the waves scattering and the corresponding isolation measures are investigated by analytical method, numerical method and field measurement. The integral equations governing Rayleigh wave scattering are derived based on the Green's solution of Lamb problem, and are solved efficiently with an iteration technique. They are used to evaluate the complicated Rayleigh wave field generated by irregular scatter embedded in an elastic half-space solid. The passive isolation of ground vibration by row-piles and in-filled trench for screening Rayleigh wave is further studied. The isolation efficiency of WIB in multilayered half-space is investigated with the aid of the semi-analytical BEM in cooperation with the basic solution of the thin-layered method. Moreover, a series of field tests on active vibration isolation with WIB under vertical harmonic loading was conducted.

Considering the complexity of the problems of wave screening, there are still many problems require further research. And more in-situ tests and laboratory experiments should be conducted to reveal the law of ground vibration isolation.

ACKNOWLEDGMENTS

The research of this project is partly supported by the Natural Science Foundation of China (No. 51178342, No. 50538010, No. 50878155).

REFERENCES

[1] Barkan D.D. (1962). *Dynamics of bases and foundations*. Mc Graw-Hill: New York.
[2] Richard F.E., Hall J.R., Woods R.D. (1970). *Vibrations of soils and foundations*, Prentice-Hall, Inc. Englewood cliffs N J.

[3] Woods R.D. (1968). Screening of surface waves in soils. *Journal of Soil Mechanics and Foundation Engineering Division*, ASCE; 94 (4): 951–79.

[4] Haupt W.A. (1978). Surface waves in nonhomogeneous half-space. In: Prange B, editor. *Dynamical methods in soil and rock mechanics.* Rotterdam: Balkema: p.335–67.

[5] Haupt W.A. (1981). Model tests on screening of surface waves. *In: Proceedings of the 10th International Conference in Soil Mechanics and Foundation Engineering*, Stockholm, vol.3: p.215–22.

[6] Aki K, Richards P.G. (1980). *Quantitative seismology*. New York: Freeman W H.

[7] Gao G.Y., Li Z.Y., Qiu C., et al. (2006). Three-dimensional analysis of rows of piles as passive barriers for ground vibration isolation. *Soil Dynamics and Earthquake Engineering*, 26(11): 1015–1027.

[8] Gao G.Y., Shi G., Feng S.J., Qiu C. (2008). 3D analysis of in-filled trench as passive barriers for ground vibration isolation, *Science in China Series G: Physics, Mechanics & Astronomy*, 51(10): 1573–1585.

[9] Gao, G.Y., Feng S.J., Li W., Zheng J.G. (2007a). 2-D analysis of vibration isolation by wave impeding block in layered ground. *Journal of Vibration Engineering*, 20(2): 174–179. (in Chinese)

[10] Gao G.Y., Feng S.J., Li W., Zheng J.G. (2007b). 3D analysis of active vibration isolation with wave impeding block in layered ground under vertical loading. *Chinese Journal of Geotechnical Engineering*, 29(4): 471–476. (in Chinese)

[11] Gao, G.Y., Shi G., Xiong H., Yue, Z.Q. (2007c). 3-D analysis of active vibration isolation using WIB in multilayered ground under vertical excitation. *3rd International Symposium on Environmental Vibration: Prediction, Monitoring and Evaluation*: p.37–46.

[12] Gao G.Y., Li N, Yue Z.Q. (2009). Experimental study on active vibration isolation by WIB in layered ground under vertical loading, *4th International Symposium on Environmental Vibration: Prediction, Monitoring and Evaluation, Beijing, China*: p.245–253

[13] Its N.E. (1991). Surface wave propagation across a narrow vertical layer between horizontally homogeneous quarter-space. *Wave motion*; 14:11–23.

[14] Kattis S.E., Polyzos S., Beskos D.E. (1999). Vibration isolation by a row of piles using a 3-D frequency domain BEM. *International Journal of Numerical Methods in Engineering*. 46: 713–728.

[15] Liao S., Sangrey D.A. (1978). Use of piles as isolation barriers. *Journal of Geotechnical Engineering Division*, ASCE; 104(9): 1139–1152.

[16] Segol G., Lee C.Y., Abel J.F. (1978). Amplitude reduction of surface waves by trenches. *Journal of Geotechnical Engineering Division*, ASCE; 104: 621–641.

[17] Fuykui M., Matsumoto Y. (1980). Finite difference analysis of Rayleigh wave scattering at a trench. *Bullet of Seismological Society of America*; 70(6): 2051–2069.

[18] May T.W., Bolt B.A. (1982). The effectiveness of trenches in reducing seismic motion. *Earthquake Engineering and Structural Dynamics*; 10:195–210.

[19] Woods R.D., Barnet N.E., Sangesser R. Holography, (1974). A new tool for soil dynamics. *Journal of Geotechnical Engineering Division*, ASCE; 100(11): 1234–1247.

[20] Kausel, E. (1994). Thin-layer method, *International Journal for Numerical Methods in Engineering*, 37: 927–941.

[21] Kausel, E., Seal S.H. (1987). Static loads in layered half-spaces, ASME *Journal of Applied Mechanics*, 54(2): 403–408.

[22] Lamb H. (1904). On the propagation of tremors over the surface of an elastic solid. *Philosophical Transactions of the Royal Society of London Series A-Containing Papers of a Mathematical or Physical Character*, 203: 1–42.

[23] Wang Y.S. (1982). Some problems in the ground vibration analysis. *Journal of Building Structures*, 3(2): 56–67 (In Chinese).

[24] Tsai P.H., Fengb Z.Y., Jen T.L. (2008). Three-dimensional analysis of the screening effectiveness of hollow pile barriers for foundation-induced vertical vibration. *Computers and Geotechnics*, 35(3): 489–499.

In: Traffic Induced Environmental Vibrations and Controls ISBN: 978-1-62618-240-0
Editors: H. Xia and R. Calçada

Chapter 10

Railway Traffic Induced Ground Vibration and Its Prediction Approaches

Yanmei Cao**, Pengbo Wei, Hong Qiao and Jinbao Yao
School of Civil Engineering, Beijing Jiaotong University, Beijing, China

Abstract

In this chapter, the ground vibration and the building vibration caused by railway trains are studied by theoretical analysis, numerical simulation and field tests. Based on vehicle dynamics, track dynamics and the Green's functions of subsoil, the train-track-subsoil dynamic interaction model is established, in which the influence of track structure and wave motion in the three-dimensional layered-soil on ground vibrations are taken into account. The characteristics of ground vibrations are analyzed by case studies and experimental results. Combining with the train-track-subsoil model, the roadbed-subsoil-building dynamic interaction model is proposed by finite element method. Some laws of building vibrations are also summarized by results of field experiment. Based on the calculated results, the prediction formulae are proposed and validated by experimental results, which can be used to predict the ground vibration and building vibration.

1. Introduction

When the railway train runs along the track, the vibrations of train and track are caused and transferred to the supporting structure through the track, and then spread all around in the form of wave through subsoil layer, which may induce environmental vibrations.

Different analytical, semi-analytical or numerical methods for predicting train-induced vibrations have been developed. In a series of research, Krylov et al. [1–3] proposed an analytical prediction model for train induced ground vibrations, in which the quasi-static force transmitted by a sleeper is derived from the deflection curve of the track modeled as a beam on an elastic foundation. Jones and Block [4] divided the vibration observed at the track

* E-mail address: ymcao@bjtu.edu.cn.

induced by a train into two components: the quasi-static deformation due to successive passage of the train axles, and the dynamic response caused by the acceleration of the train masses over the combined track and wheels irregular profiles. Madshus and Kaynia [5] observed large amplitudes of displacement when the train ran on the track at high speed towards its critical speed of about 180 km/h at the measurement site. Along with their measurements, Kaynia et al. [6] reported the development of a FEM model composed of a moving load on a railway embankment, which was coupled to a layered ground model at a series of points via the Green's functions. Karlström and Boström [7] investigated ground vibrations from railways by an analytical approach, in which the subsoil was modeled as a stratified half-space with linearly viscoelastic layers and the solution based on the Fourier transforms in time along the track. Lombaert and Degrande [8] evaluated the contributions of quasi-static excitation related to the axle loads and the dynamic excitation of random track unevenness to the track and free-field responses. Sheng et al. [9−10] modeled the ground as a structure of three-dimensional viscoelastic layers overlying either a half-space or a rigid foundation, and the track as a rail beam, a rail pad and a sleeper mass supported by a viscoelastic ballast layer. He illustrated the existence of a wave in the combined track and ground system propagating along the track with a speed controlled by the lowest propagating wave in the layered ground.

Numerical procedures to analyze the railway-induced ground vibrations have also been presented by researchers. Yang [11, 12] established a 2.5D finite/infinite element approach to study the transmissibility of soils for vibrations induced by moving trains, by simulating the train with a sequence of moving axle loads. Auersch [13] presented a comprehensive model to analyze the ground vibration, using a combined FEM/BEM model for soil and track and a multi-DOF model for the vehicle. Galvín and Domínguez [14] presented a 3D model for analysis of soil motion and effects of high-speed train passage on nearby ground and underground structures, and studied the ground vibrations in zones with local discontinuities in the soil properties or the effects on structures such as underpasses, buildings and auxiliary structures around the track. Takemiya [15, 16] proposed a 2.5D FE model considering the irregular topographies of the track. Masafumi et al. [17] adopted a 3D viscoelastic staggered-grid finite difference method with fourth-order accuracy in space to study the mechanism of ground vibrations induced by a high-speed train, where the input to numerical simulations for a single wheel was a source function developed by the time-series data from the Shinkansen train.

2. TRAIN-TRACK-SUBSOIL DYNAMIC INTERACTION MODEL

2.1. Train Model

The train model is shown in Figure 1, and the motion equation of the i-th vehicle can be expressed by:

$$\mathbf{M}_i\ddot{\mathbf{Z}}_i + \mathbf{C}_i\dot{\mathbf{Z}}_i + \mathbf{K}_i\mathbf{Z}_i = \mathbf{P}_i(t) \qquad (i = 1, 2, \cdots, N_{\mathrm{v}}) \tag{1}$$

in which $\mathbf{M}_i$, $\mathbf{K}_i$ and $\mathbf{C}_i$ are, respectively, mass, stiffness and damping matrices for the i-th vehicle, whose detail expressions can be seen in Section 3.2 of Chapter 3; $\mathbf{Z}_i$ is the displacement vector of the i-th vehicle and can be written as:

$$\mathbf{Z}_i = \{z_{ci} \quad \phi_{ci} \quad z_{ti1} \quad \phi_{ti1} \quad z_{ti2} \quad \phi_{ti2} \quad z_{wi1} \quad z_{wi2} \quad \cdots \quad z_{wiN_{wi}}\}^{\mathrm{T}} \tag{2}$$

and $\mathbf{P}_i(t)$ is the excitation force vector of the i-th vehicle as follows:

$$\mathbf{P}_i(t)=\{0 \quad 0 \quad 0 \quad 0 \quad 0 \quad 0 \quad -P_{i1}^{\mathrm{w/r}}(t) \quad -P_{i2}^{\mathrm{w/r}}(t) \quad \cdots \quad -P_{iN_{\mathrm{w}}}^{\mathrm{w/r}}(t)\}^{\mathrm{T}} \tag{3}$$

where $P_{il}^{\mathrm{w/r}}(t)$ is the wheel-track interaction force of the il-th wheel-set (short for the l-th wheel-set of i-th vehicle).

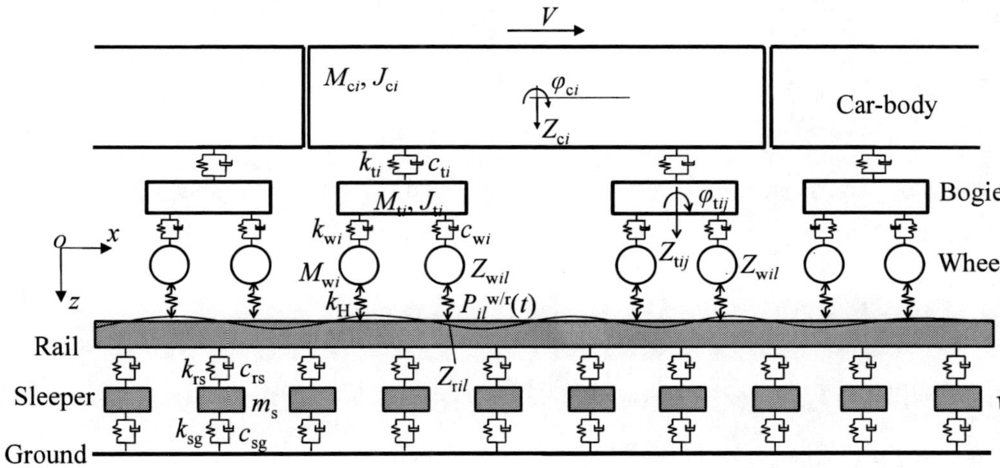

Figure 1. Illustration of train-track dynamic interaction model.

In order to study the frequency characteristics of environmental vibrations, the Fourier integral transform is carried out to Equation (1), and the problem can be transformed from the time domain to the frequency domain as:

$$(-\mathbf{M}_i\omega^2 + \mathrm{i}\mathbf{C}_i\omega + \mathbf{K}_i)\overline{\mathbf{Z}}_i(\omega) = \overline{\mathbf{P}}_i(\omega) \tag{4}$$

where $\overline{\mathbf{Z}}_i(\omega)$ is the displacement vector of vehicle in the frequency domain; $\overline{\mathbf{P}}_i(\omega)$ is the load vector in the frequency domain as follows:

$$\overline{\mathbf{P}}_i(\omega)=\{0 \quad 0 \quad 0 \quad 0 \quad 0 \quad 0 \quad -\overline{P}_{i1}^{\mathrm{w/r}}(\omega) \quad -\overline{P}_{i2}^{\mathrm{w/r}}(\omega) \quad \cdots \quad -\overline{P}_{iN_{\mathrm{w}}}^{\mathrm{w/r}}(\omega)\}^{\mathrm{T}} \tag{5}$$

On the basis of matrix analysis, Equation (5) can be written as $\overline{\mathbf{P}}_i(\omega) = -\mathbf{D}\overline{\mathbf{P}}_i^{\mathrm{w/r}}(\omega)$, in which $\overline{\mathbf{P}}_i^{\mathrm{w/r}}(\omega) = \{\overline{P}_{i1}^{\mathrm{w/r}}(\omega), \overline{P}_{i2}^{\mathrm{w/r}}(\omega), \cdots, \overline{P}_{iN_{\mathrm{w}i}}^{\mathrm{w/r}}(\omega)\}^{\mathrm{T}}$, and $\mathbf{D}$ is the factor matrix of vehicle load, which is $\mathbf{D} = \begin{bmatrix} \mathbf{0}_{6\times4} \\ \hline \mathbf{I}_{4\times4} \end{bmatrix}$ for a four-axle vehicle and $\mathbf{D} = \begin{bmatrix} \mathbf{0}_{6\times6} \\ \hline \mathbf{I}_{6\times6} \end{bmatrix}$ for a six-axle vehicle, with $\mathbf{I}$ being unit matrix. Therefore, Equation (4) can be rewritten as:

$$(\mathbf{K}_i - \mathbf{M}_i\omega^2 + i\mathbf{C}_i\omega)\overline{\mathbf{Z}}_i(\omega) = -\mathbf{D}\overline{\mathbf{P}}_i^{\mathrm{w/r}}(\omega) \tag{6}$$

Thus, the displacement of vehicle can be expressed as:

$$\overline{\mathbf{Z}}_i(\omega) = -(\mathbf{K}_i - \mathbf{M}_i\omega^2 + i\mathbf{C}_i\omega)^{-1}\mathbf{D}\overline{\mathbf{P}}_i^{\mathrm{w/r}}(\omega) \tag{7}$$

If the displacement vector of wheel-sets of the i-th vehicle are denoted by $\overline{\mathbf{Z}}_{\mathrm{w}i}(\omega)$, then

$$\overline{\mathbf{Z}}_{\mathrm{w}i}(\omega) = \{\overline{z}_{\mathrm{w}i1}(\omega) \quad \overline{z}_{\mathrm{w}i2}(\omega) \quad \cdots \quad \overline{z}_{\mathrm{w}iN_{\mathrm{w}i}}(\omega)\}^{\mathrm{T}}.$$

It can be seen from the expressions of $\overline{\mathbf{Z}}_{\mathrm{w}i}(\omega)$ and $\overline{\mathbf{Z}}_i(\omega)$ that they have the following relationship:

$$\overline{\mathbf{Z}}_{\mathrm{w}i}(\omega) = \mathbf{B}\overline{\mathbf{Z}}_i(\omega) \tag{8}$$

where $\mathbf{B} = [\mathbf{0}_{4\times6} \mid \mathbf{I}_{4\times4}]$ for a 4-axle vehicle or $\mathbf{B} = [\mathbf{0}_{6\times6} \mid \mathbf{I}_{6\times6}]$ for a 6-axle vehicle, and $\mathbf{B} = \mathbf{D}^{\mathrm{T}}$. So the displacement of wheel-sets can be written as:

$$\overline{\mathbf{Z}}_{\mathrm{w}i}(\omega) = -\mathbf{D}^{\mathrm{T}}(\mathbf{K}_i - \mathbf{M}_i\omega^2 + i\mathbf{C}_i\omega)^{-1}\mathbf{D}\overline{\mathbf{P}}_i^{\mathrm{w/r}}(\omega) \tag{9}$$

or

$$\overline{\mathbf{Z}}_{\mathrm{w}i}(\omega) = -\mathbf{A}_{\mathrm{w}i}\overline{\mathbf{P}}_i^{\mathrm{w/r}}(\omega) \tag{10}$$

in which the matrix $\mathbf{A}_{\mathrm{w}i} = \mathbf{D}^{\mathrm{T}}(\mathbf{K}_i - \mathbf{M}_i\omega^2 + i\mathbf{C}_i\omega)^{-1}\mathbf{D}$ is the complex matrix about ω with size of $N_{\mathrm{w}i} \times N_{\mathrm{w}i}$, and the minus means that the direction of wheel displacement is contrary to that of wheel-rail interaction force.

The element $a_{mn}^{\mathrm{w}i}$ $(m = 1, 2, \cdots, N_{\mathrm{w}i}; n = 1, 2, \cdots, N_{\mathrm{w}i})$ of matrix $\mathbf{A}_{\mathrm{w}i}$ represents the displacement of the n-th wheel-sets when the unit harmonic loads with frequency of ω acts at the position of the m-th wheel-sets, and $a_{mn}^{\mathrm{w}i} = a_{nm}^{\mathrm{w}i}$. Equation (10) can be expanded to the following vector form:

$$\begin{Bmatrix} \bar{z}_{wi1}(\omega) \\ \bar{z}_{wi2}(\omega) \\ \vdots \\ \bar{z}_{wiN_{wi}}(\omega) \end{Bmatrix} = -\begin{bmatrix} a_{11}^{wi} & a_{12}^{wi} & \cdots & a_{1N_{wi}}^{wi} \\ & a_{22}^{wi} & \cdots & a_{2N_{wi}}^{wi} \\ & & \ddots & \vdots \\ \text{sym} & & & a_{N_{wi}N_{wi}}^{wi} \end{bmatrix} \begin{Bmatrix} \bar{P}_{i1}^{w/r}(\omega) \\ \bar{P}_{i2}^{w/r}(\omega) \\ \vdots \\ \bar{P}_{iN_{wi}}^{w/r}(\omega) \end{Bmatrix} \tag{11}$$

from which the displacement of the il-th wheel-set can be expressed as:

$$\bar{z}_{wil}(\omega) = -\sum_{j=1}^{N_{wi}} a_{lj}^{wi} \cdot \bar{P}_{ij}^{w/r}(\omega) \tag{12}$$

in which the interaction between all the wheel-sets on the i-th vehicle is included, but the influence of other vehicles is not included due to the assumption of vibration independency between different vehicles.

Since each vehicle satisfies Equations (10) to (12), for the train composed by N_v number of vehicles, the displacements of all wheel-sets can be expressed as a large vector equation:

$$\begin{Bmatrix} \bar{\mathbf{Z}}_{w1}(\omega) \\ \bar{\mathbf{Z}}_{w2}(\omega) \\ \vdots \\ \bar{\mathbf{Z}}_{wN_v}(\omega) \end{Bmatrix} = -\begin{bmatrix} \mathbf{A}_{w1} & & \mathbf{0} & \\ & \mathbf{A}_{w2} & & \\ & & \ddots & \\ & \mathbf{0} & & \mathbf{A}_{wN_v} \end{bmatrix} \begin{Bmatrix} \bar{\mathbf{P}}_{1}^{w/r}(\omega) \\ \bar{\mathbf{P}}_{2}^{w/r}(\omega) \\ \vdots \\ \bar{\mathbf{P}}_{N_v}^{w/r}(\omega) \end{Bmatrix} \tag{13}$$

2.2. Track Model

Railway track is composed of rail, fasteners, sleepers, track-bed, and so on, which guides the running of trains, bears the pressure transmitted directly from wheel-sets and distributes the loads to the subsoil. In track model, the rail can be simulated with an Euler-beam or a Timoshenko-beam, depending whether shear stiffness and rotational moment of inertia are taken into account. Generally, the selection of rail model is determined by the actual problem: when the dynamic behaviors of track structure are more concerned, the Timoshenko beam is often adopted, while in studying the environmental vibrations induced by rail traffic, the Euler beam is sufficient.

In simulating the supports of track structure, the following two aspects should be considered:

(1) Layers of track support. In the single-layer track model, the rail and the sleeper are assumed to contact closely. In the two-layer track model, the rail-sleeper interaction is simulated with spring and dampers. In the three-layer track model, in addition to the rail-sleeper interaction, the ballast elasticity is also taken into account.
(2) Patterns of track support, i.e. continuous support or discrete support. The continuous supporting model neglects spacing of sleepers, and simplifies the rail as a rigid body or long beam with distributed mass and stiffness. The discrete support model considers the

role of sleeper spacing, which reflects the periodic property caused by distribution of discrete sleepers, and is often adopted. Knothe and Grassie [18] drew the conclusion that when the vibration frequency of rail is less than 500 Hz with vertical load or 400 Hz with lateral load, there is little difference between the ground vibration levels calculated with continuous support model and discrete support model.

Herein, the track model is simplified as a two-layer mass-spring-damper system, as shown in Figure 1. The rail is modeled as an infinite Euler-beam on elastic supports (rail-pads and fasteners), and the sleeper as a mass with only vertical movement. The spring-damper system between the rail and the sleepers are represented by k_{rs} and c_{rs}, and that between the sleepers and roadbed represented by k_{sg} and c_{sg}, respectively.

The sample series of track irregularity $z_{\mathrm{r}}(x)$ can be obtained by the trigonometric series superposition method as follows:

$$z_{\mathrm{r}}(x)=\sum_{k=1}^{N}\sqrt{2S(\Omega_k)\Delta\Omega}\cdot \mathrm{e}^{i(\Omega_k x+\theta_k)} \tag{14}$$

where $S(\Omega_k)$ is the power spectrum density function of track irregularity; the exponential term is expressed by triangle function in complex domain; Ω_k (k=1, 2,..., N) is the spatial angle frequency, and $\Delta\Omega$ is the band width of spatial frequency of track irregularity; θ_k is the random phase corresponding to the k-th spatial frequency, uniformly distributed in [0−2π].

Let $\tilde{z}_{\mathrm{r}}(\Omega_k)=\sqrt{2S(\Omega_k)\Delta\Omega}\cdot e^{i\theta_k}$, Equation (14) can be rewritten as:

$$z_{\mathrm{r}}(x)=\sum_{k=1}^{N}\tilde{z}_{\mathrm{r}}(\Omega_k)\mathrm{e}^{i\Omega_k x} \tag{15}$$

in which $\tilde{z}_{\mathrm{r}}(\Omega_k)$ is the amplitude of track irregularity with spatial frequency Ω_k.

According to the definition of spatial angle frequency, when the train runs at speed V, the wavelength of track irregularity is $\lambda=2\pi/\Omega$, and the circular frequency of irregularity is $\omega=2\pi f=2\pi V/\lambda$. So, the track irregularity at position x can be changed into that at position $x_{\mathrm{p}il}=x_{\mathrm{p}il}^{0}+Vt$ of wheel-set, in which $x_{\mathrm{p}il}^{0}$ is the initial coordinate of the il-th wheel-set, i.e.

$$z_{\mathrm{r}il}^{k}(t)=\bar{z}_{\mathrm{r}il}(\omega_k)\mathrm{e}^{i\omega_k t} \tag{16}$$

in which $\bar{z}_{\mathrm{r}il}(\omega_k)=\tilde{z}_{\mathrm{r}}(\Omega_k)\cdot\exp(i\frac{\omega_k}{V}x_{\mathrm{p}il}^{0})$ represents the amplitude of track irregularity with frequency ω_k at the position of the il-th wheel-set.

For the whole train, the track irregularity vector at the positions of all wheel-sets can be expressed as:

$$\begin{Bmatrix} \overline{\mathbf{z}}_{r1}(\omega_k) \\ \overline{\mathbf{z}}_{r2}(\omega_k) \\ \vdots \\ \overline{\mathbf{z}}_{rN_v}(\omega_k) \end{Bmatrix} = \tilde{z}_r(\Omega_k) \cdot \begin{Bmatrix} \mathbf{e}_1(\omega_k) \\ \mathbf{e}_2(\omega_k) \\ \vdots \\ \mathbf{e}_{Nv}(\omega_k) \end{Bmatrix} \tag{17}$$

in which $\overline{\mathbf{z}}_{ri}(\omega_k) = \{\overline{z}_{ri1}(\omega_k), \overline{z}_{ri2}(\omega_k), \cdots, \overline{z}_{riN_{wi}}(\omega_k)\}^{\mathrm{T}}$ represents the displacement vector of track irregularity at the positions of all wheel-sets for the i-th vehicle; the element of the last vector is $\mathbf{e}_i(\omega_k) = \{\exp(i\frac{\omega_k}{V}x^0_{\mathrm{p}i1}), \exp(i\frac{\omega_k}{V}x^0_{\mathrm{p}i2}), \cdots, \exp(i\frac{\omega_k}{V}x^0_{\mathrm{p}iN_{wi}})\}^{\mathrm{T}}$.

2.3. Vibration of Track Subjected to a Moving Vehicle

2.3.1. Coupling of Vehicle and Track

The coupling of vehicle and track is accomplished by the wheel-rail contact relationship assumed by the Hertz contact theory, as shown in Figure 2. Although the actual wheel-rail contact is nonlinear, it is often simplified to be linear, which has already been proved to be effective in dealing with environmental vibration [19].

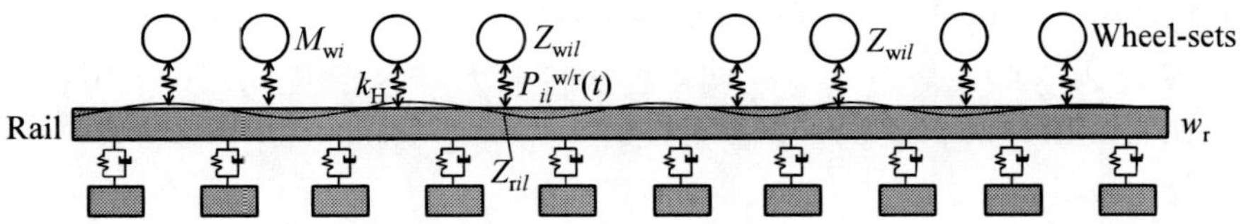

Figure 2. Wheel-rail interaction model.

Based on the wheel-rail contact condition, the relationship between wheel-rail interaction force and wheel-rail relative displacement can be expressed as:

$$P_{il}^{\mathrm{w/r}}(t) = k_{\mathrm{H}}[z_{\mathrm{w}il}(t) - w_{\mathrm{r}il}(t) - z_{\mathrm{r}il}(t)] \tag{18}$$

in which $P_{il}^{\mathrm{w/r}}(t)$ represents the wheel-rail interaction force of the il-th wheel-set; k_{H} is the stiffness of Hertz contact; $z_{\mathrm{w}il}(t)$ is the displacement of the il-th wheel-set; $w_{\mathrm{r}il}(t)$ is the displacement of rails induced by moving loads; $z_{\mathrm{r}il}$ is the track irregularity at the contact point of the il-th wheel-set and rail.

Since the contact point of wheel-sets and rail changes with train position, a moving coordinate system with $x_{\mathrm{p}il} = x^0_{\mathrm{p}il} + Vt$ can be established so that the positions of variables in Equation (18) are fixed. According to structural dynamics, when the wheel-rail interaction force with frequency ω_k is applied to the vehicle and the rail, they vibrate with a stationary response with frequency ω_k and a transient response with their natural frequency. Since the transient response attenuates rapidly due to damping, only the stationary one is considered herein, i.e. $z_{\mathrm{w}il}(t) = \overline{z}_{\mathrm{w}il}(\omega_k)e^{i\omega_k t}$, $w_{\mathrm{r}il}(t) = \overline{w}_{\mathrm{r}il}(\omega_k)e^{i\omega_k t}$. Equation (18) can be rewritten as:

$$\overline{P}_{il}^{\mathrm{w/r}}(\omega_k) = k_{\mathrm{H}}[\overline{z}_{\mathrm{w}il}(\omega_k) - \overline{w}_{\mathrm{r}il}(\omega_k) - \overline{z}_{\mathrm{r}il}(\omega_k)] \tag{19}$$

where $\overline{z}_{\mathrm{r}il}(\omega_k)$ is the track irregularity in frequency domain and can be calculated by Equation (16); $\overline{z}_{\mathrm{w}il}(\omega_k)$ is the displacement of the il-th wheel-set and can be obtained by Equation (12); $\overline{w}_{\mathrm{r}il}(\omega_k)$ is the rail displacement at the position of the il-th wheel-set, which can be calculated by the following track-supporting structure model.

Since each wheel-rail contact point satisfies Equation (19), all the wheel-rail interaction forces can be combined into the following equations in the form of vector:

$$\begin{Bmatrix} \overline{\mathbf{P}}_1^{\mathrm{w/r}}(\omega_k) \\ \overline{\mathbf{P}}_2^{\mathrm{w/r}}(\omega_k) \\ \vdots \\ \overline{\mathbf{P}}_{N_\mathrm{v}}^{\mathrm{w/r}}(\omega_k) \end{Bmatrix} = k_{\mathrm{H}} \cdot \left[\begin{Bmatrix} \overline{\mathbf{z}}_{\mathrm{w}1}(\omega_k) \\ \overline{\mathbf{z}}_{\mathrm{w}2}(\omega_k) \\ \vdots \\ \overline{\mathbf{z}}_{\mathrm{w}N_\mathrm{v}}(\omega_k) \end{Bmatrix} - \begin{Bmatrix} \overline{\mathbf{w}}_{\mathrm{r}1}(\omega_k) \\ \overline{\mathbf{w}}_{\mathrm{r}2}(\omega_k) \\ \vdots \\ \overline{\mathbf{w}}_{\mathrm{r}N_\mathrm{v}}(\omega_k) \end{Bmatrix} - \begin{Bmatrix} \overline{\mathbf{z}}_{\mathrm{r}1}(\omega_k) \\ \overline{\mathbf{z}}_{\mathrm{r}2}(\omega_k) \\ \vdots \\ \overline{\mathbf{z}}_{\mathrm{r}N_\mathrm{v}}(\omega_k) \end{Bmatrix} \right] \tag{20}$$

where $\overline{\mathbf{P}}_i^{\mathrm{w/r}}(\omega_k) = \{\overline{P}_{i1}^{\mathrm{w/r}}(\omega_k), \overline{P}_{i2}^{\mathrm{w/r}}(\omega_k), \cdots, \overline{P}_{iN_{\mathrm{w}i}}^{\mathrm{w/r}}(\omega_k)\}^{\mathrm{T}}$ is the column vector combined by all the wheel-rail interaction forces of the i-th vehicle; $\overline{\mathbf{w}}_{\mathrm{r}i}(\omega_k) = \{\overline{w}_{\mathrm{r}i1}(\omega_k), \overline{w}_{\mathrm{r}i2}(\omega_k), \cdots, \overline{w}_{\mathrm{r}iN_{\mathrm{w}i}}(\omega_k)\}^{\mathrm{T}}$ is the column vector combined by the rail displacements at the position of all wheel-sets of the i-th vehicle; $\overline{\mathbf{z}}_{\mathrm{w}i}(\omega_k)$ and $\overline{\mathbf{z}}_{\mathrm{r}i}(\omega_k)$ are the displacements of wheel-sets and track irregularity at the position of all wheel-sets of the i-th vehicle in frequency domain, respectively.

2.3.2. Vertical Displacement of Rail Subjected to a Single Wheel-set

When a single wheel-set load is applied to the rail, the force analysis and deformation of the rail is illustrated in Figure 3.

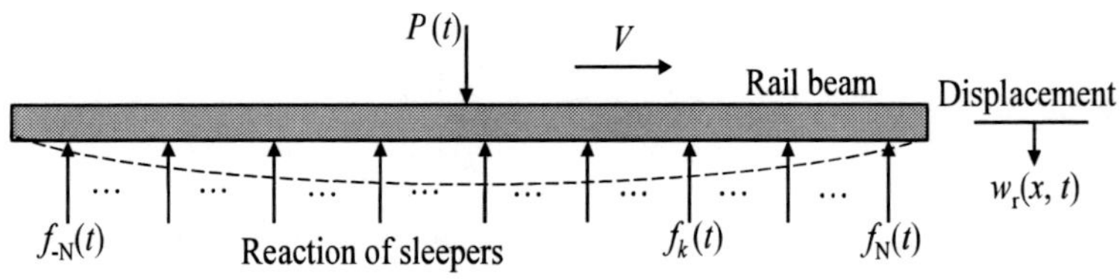

Figure 3. Force and deformation of rail under the action of a single wheel-set.

The forces from the sleepers to the rail can be represented by the discrete supporting actions $\sum_{k=-N}^{N} f_k(t)\delta(x - x_k)$, in which $2N$+1 sleepers are included in the calculation, x_k is the x-direction coordination of the k-th sleepers, and the Dirac function $\delta(x\text{-}x_k)$ indicates the acting point of $f_k(t)$ is located at the center of each sleeper.

When a single wheel-set load $P(t)$ moves along the rail at speed V, the motion equation of the rail can be expressed by:

$$E^{*}I\frac{\partial^{4}w_{\mathrm{r}}(x,t)}{\partial x^{4}}+\bar{m}_{\mathrm{r}}\frac{\partial^{2}w_{\mathrm{r}}(x,t)}{\partial t^{2}}=P(t)\delta(x-x_{\mathrm{p}}^{0}-Vt)-\sum_{m=-N}^{+N}f_{m}(t)\delta(x-x_{m}) \tag{21}$$

in which $w_{\mathrm{r}}(x,t)$ is the vertical displacement of rail; $E^{*}=E(1+2\xi_{\mathrm{r}}i)$ is the complex elastic modulus taking the internal damping of rail structure into account, where E is the real elastic modulus and ξ_{r} is the damping ratio of rail structure; I is the inertia moment of rail section; $\bar{m}_{\mathrm{r}}$ is the rail mass of unit length; x_{p}^{0} is the initial position of load $P(t)$; $f_m(t)$ is the supporting force from the m-th sleeper to rail induced by moving load $P(t)$ with invariable position.

In order to solve Equation (21), the rail displacement under unit impulse load is firstly obtained by solving the following equation of rail displacement:

$$E^{*}I\frac{\partial^{4}B_{\mathrm{r}}(x,t)}{\partial x^{4}}+\bar{m}_{\mathrm{r}}\frac{\partial^{2}B_{\mathrm{r}}(x,t)}{\partial x^{2}}=\delta(t) \tag{22}$$

in which $B_{\mathrm{r}}(x, t)$ represents the rail displacement under unit impulse load, called *the first kind of Green's function of rail displacement.*

Equation (22) can be changed from time domain to frequency domain by the Fourier integral transform $\bar{F}(\omega)=\int_{-\infty}^{+\infty}F(t)\mathrm{e}^{-i\omega t}\,\mathrm{d}t$ with respect to t as follows:

$$E^{*}I\frac{d^{4}\bar{B}_{\mathrm{r}}(x,\omega)}{\partial x^{4}}-\bar{m}_{\mathrm{r}}\omega^{2}\bar{B}_{\mathrm{r}}(x,\omega)=1 \tag{23}$$

Then the Fourier transform $\tilde{F}(k_x)=\int_{-\infty}^{\infty}F(x)\cdot\mathrm{e}^{ik_x x}\,\mathrm{d}x$ is carried to Equation (23) about x, and Equation (22) can be changed into wavenumber-frequency domain as follows:

$$E^{*}Ik_x^{4}\tilde{\bar{B}}_{\mathrm{r}}(k_x,\omega)-\bar{m}_{\mathrm{r}}\omega^{2}\tilde{\bar{B}}_{\mathrm{r}}(k_x,\omega)=1 \tag{24}$$

Then, we have

$$\tilde{\bar{B}}_{\mathrm{r}}(k_x,\omega)=\frac{1}{E^{*}Ik_x^{4}-\bar{m}_{\mathrm{r}}\omega^{2}} \tag{25}$$

The inversal Fourier transform with respect to k_x is carried to Equation (25), *the first kind of Green's function of rail displcement* in frequency domain can be obtained as:

$$\bar{B}_{\mathrm{r}}(x,\omega)=\frac{1}{2\pi E^{*}I}\int_{-\infty}^{+\infty}\frac{e^{-ik_x x}}{k_x^{4}-\beta}dk_x=\frac{1}{4E^{*}I\alpha^{3}}[-e^{-\alpha|x|}-ie^{-i\alpha|x|}] \tag{26}$$

in which $\beta = \overline{m}_{\mathrm{r}}\omega^2 / E^* I$; $E^* = E(1+2\xi_{\mathrm{r}} i)$; $\alpha = \sqrt[4]{\beta} = \sqrt[4]{\overline{m}\omega^2 / E^* I}$; the integral expression $\int_{-\infty}^{+\infty} e^{-ik_x x} / (k_x^4 - \beta)\, \mathrm{d} k_x$ is a kind of integral about the complex variable function, which can be integrated by using Cauchy residue theory [20].

Therefore, when the unit impulse load is applied to the origin of rail, the vertical displacement of the rail is as follows:

$$\overline{w}_{\mathrm{r}}(x,\omega) = \overline{B}_{\mathrm{r}}(x,\omega) - \sum_{m=-N}^{N} \overline{f}_m^{\delta}(\omega) \overline{B}_{\mathrm{r}}(x - x_m, \omega) \tag{27}$$

in which $\overline{f}_m^{\delta}(\omega)$ represents the supporting force from the m-th sleeper to rail induced by unit impulse load. Equation (27) is the *second kind of Green's function of rail displacement* in frequency domain, i.e.

$$\overline{G}_{\mathrm{rail}}(x,\omega) = \overline{B}_{\mathrm{r}}(x,\omega) - \sum_{m=-N}^{N} \overline{f}_m^{\delta}(\omega) \overline{B}_{\mathrm{r}}(x - x_m, \omega) \tag{28}$$

For the first item on the right side of Equation (21), the rail displacement under moving load can be obtained by using the generalized Duhamel integral as follows:

$$\overline{w}_{\mathrm{r1}}(x,\omega) = \frac{1}{2\pi V} \int_{-\infty}^{+\infty} \overline{P}(\overline{\omega}) \cdot \tilde{\overline{B}}_{\mathrm{r}}\left(\frac{\omega - \overline{\omega}}{V}, \omega\right) \cdot \exp\left[-i(x - x_{\mathrm{p}}^0)\frac{\omega - \overline{\omega}}{V}\right] \mathrm{d}\overline{\omega} \tag{29}$$

in which $\overline{\omega}$ is the loading frequency of external loads; $\tilde{\overline{B}}_{\mathrm{r}}(k_x, \omega)$ is the Green's function of rail displacement in the wavenumber-frequency domain.

When $P(t)$ is a quasi-static axle loading P_0, the rail displacement induced by moving constant loads can be obtained as:

$$\overline{w}_{\mathrm{r1}}^{\mathrm{axle}}(x,\omega) = \frac{P_0}{V} \tilde{\overline{B}}_{\mathrm{r}}\left(\frac{\omega}{V}, \omega\right) \exp\left[-i(x - x_{\mathrm{p}}^0)\frac{\omega}{V}\right] \tag{30}$$

When $P(t)$ is a wheel-rail interaction force $\overline{P}^{\mathrm{w/r}}(\omega_k) e^{i\omega_k t}$, the rail displacement induced by moving harmonic loading can be obtained as:

$$\overline{w}_{\mathrm{r1}}^{\mathrm{w/r}}(x,\omega) = \frac{\overline{P}^{\mathrm{w/r}}(\omega_k)}{V} \tilde{\overline{B}}_{\mathrm{r}}\left(\frac{\omega - \omega_k}{V}, \omega\right) \exp\left[-i(x - x_{\mathrm{p}}^0)\frac{\omega - \omega_k}{V}\right] \tag{31}$$

For the second item on the right-hand side of Equation (21), the rail displacement induced by supporting actions $\sum_{m=-N}^{N} f_m(t)\delta(x - x_m)$ of sleepers under single wheel-sets loading can be obtained by using the generalized Duhamel integral as:

$$\bar{w}_{r2}(x,\omega) = \sum_{m=-N}^{N} \bar{f}_m(\omega)\bar{B}_r(x - x_m,\omega) \tag{32}$$

2.3.3. Vertical Displacement of Rail Subjected to All Wheel-sets

Figure 4 shows the arrangement of all wheel-sets of a train.

According to Equations (30) and (31), the rail displacement induced by quasi-static axle loading and wheel-rail interaction force of running train can be respectively written as:

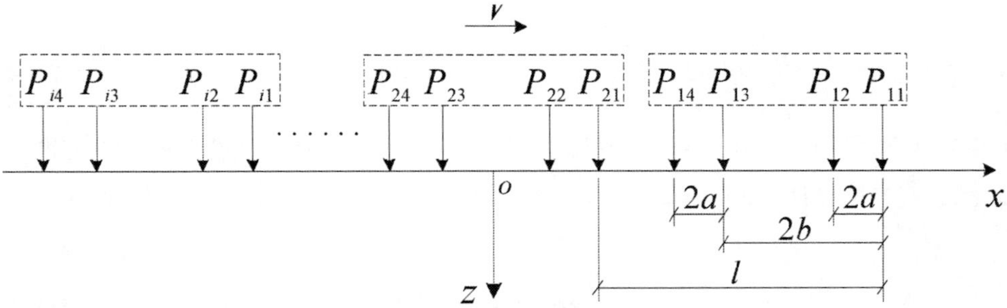

Figure 4. Arrangement of all wheel-sets of a train.

$$\bar{w}_r(x,\omega) = \frac{1}{V}\tilde{\bar{B}}_r\left(\frac{\omega}{V},\omega\right)e^{-ix\frac{\omega}{V}}\left[\sum_{i=1}^{N_v}\sum_{l=1}^{N_{wm}} P_{0il}e^{ix^0_{pil}\frac{\omega}{V}}\right] - \sum_{m=-N}^{+N}\bar{f}_m^{\text{final}}(\omega)\bar{B}_r(x-x_m,\omega) \tag{33}$$

$$\bar{w}_r(x,\omega) = \frac{1}{V}\tilde{\bar{B}}_r\left(\frac{\omega-\omega_k}{V},\omega\right)e^{-ix\frac{\omega-\omega_k}{V}}\left[\sum_{i=1}^{N_v}\sum_{l=1}^{N_{wm}} \bar{P}_{il}^{\text{w/r}}(\omega_k)e^{ix^0_{pil}\frac{\omega-\omega_k}{V}}\right] - \sum_{m=-N}^{+N}\bar{f}_m^{\text{final}}(\omega)\bar{B}_r(x-x_m,\omega) \tag{34}$$

in w425hich $\sum_{m=-N}^{N}\bar{f}_m^{\text{final}}(\omega)\bar{B}_r(x-x_m,\omega)$ represents the supporting force from sleepers to rail under the action of all wheel-sets loading for a running train, ω is the vibration frequency of rail structure, and ω_k represents the loading frequency excited by running train.

The contribution of all quasi-static axle loads and dynamic loads to rail displacement are respectively $\bar{C}_1(\omega) = \sum_{i=1}^{N_v}\sum_{l=1}^{N_{wi}} P_{0il}\exp(ix^0_{pil}\frac{\omega}{V})$ and $\bar{C}_2(\omega) = \sum_{i=1}^{N_v}\sum_{l=1}^{N_{wi}} \bar{P}_{il}^{\text{w/r}}(\omega_k)\exp(ix^0_{pil}\frac{\omega-\omega_k}{V})$, thus Equations (33) and (34) can be written by:

$$\bar{w}_r(x_q,\omega) = \frac{1}{V}\tilde{\bar{B}}_r\left(\frac{\omega}{V},\omega\right)e^{-ix_q\frac{\omega}{V}}\bar{C}_1(\omega) - \sum_{m=-N}^{N}\bar{f}_m^{\text{final}}(\omega)\bar{B}_r(x_q - x_m,\omega) \tag{35}$$

$$\bar{w}_r(x_q,\omega) = \frac{1}{V}\tilde{\bar{B}}_r\left(\frac{\omega-\omega_k}{V},\omega\right)e^{-ix_q\frac{\omega-\omega_k}{V}}\bar{C}_2(\omega) - \sum_{m=-N}^{N}\bar{f}_m^{\text{final}}(\omega)\bar{B}_r(x_q - x_m,\omega) \tag{36}$$

in which $x=x_q=qd$, and d is the spacing of adjacent sleepers; $\bar{w}_r(x_q,\omega)$ represents the rail displacement at position of the q-th sleeper induced by running train.

The rail displacement at the position of each sleeper satisfies Equations (27), (35) and (36), so for all (2N+1) sleepers the equations can be combined into the following vector forms:

$$\overline{\mathbf{W}}_{\mathrm{r}}^{\delta} = \overline{\mathbf{B}}_{\mathrm{rs}} - \overline{\mathbf{B}}_{\mathrm{r}} \cdot \overline{\mathbf{f}}^{\delta} \tag{37}$$

$$\overline{\mathbf{W}}_{\mathrm{r}} = A(k_x, \omega)\mathbf{E}_s - \overline{\mathbf{B}}_{\mathrm{r}} \cdot \overline{\mathbf{f}} \tag{38}$$

in which $A(k_x,\omega) = \frac{1}{V}\tilde{\overline{B}}_{\mathrm{r}}(k_x,\omega)\overline{C}(\omega)$; $k_x = \frac{\omega}{V}$ and $\overline{C}(\omega) = \overline{C}_1(\omega)$ when considering the axle loading, and $k_x = (\omega - \omega_k)/V$ and $\overline{C}(\omega) = \overline{C}_2(\omega)$ when considering the wheel-rail interaction force; and $\tilde{\overline{B}}_{\mathrm{r}}(k_x,\omega)$ is the Green's function of rail displacement in wavenumber-frequency domain. $\mathbf{E}_{\mathrm{s}} = \{e^{-ix_{-N}k_x},\ldots,e^{-ix_0k_x},\cdots,e^{-ix_Nk_x}\}^{\mathrm{T}}$, $\overline{\mathbf{W}}_{\mathrm{r}} = \{\overline{w}_{\mathrm{r}}(x_{-N},\omega),\ldots,\overline{w}_{\mathrm{r}}(x_0,\omega),\cdots,\overline{w}_{\mathrm{r}}(x_N,\omega)\}^{\mathrm{T}}$, and $\overline{\mathbf{f}} = \{\overline{f}_{-N}^{\mathrm{final}}(\omega),\ldots,\overline{f}_0^{\mathrm{final}}(\omega),\cdots,\overline{f}_N^{\mathrm{final}}(\omega)\}^{\mathrm{T}}$ are, respectively, column vectors composed of the position coordinates, vertical rail displacements and supporting forces at the positions of all 2N+1 sleepers induced by running train. $\overline{\mathbf{W}}_{\mathrm{r}}^{\delta}$ and $\overline{\mathbf{f}}^{\delta}$ mean those induced by unit impulse load acting on the origin of rail. $\overline{\mathbf{B}}_{\mathrm{rs}} = \{\overline{B}_{\mathrm{r}}(x_{-N},\omega),\ldots,\overline{B}_{\mathrm{r}}(x_0,\omega),\cdots,\overline{B}_{\mathrm{r}}(x_N,\omega)\}^{\mathrm{T}}$ consists of the first kind of Green's function of rail displacement in frequency domain. $\overline{\mathbf{B}}_{\mathrm{r}}$ has the size of (2N+1)×(2N+1) with the element $\overline{B}_{\mathrm{r}}^{qm} = \overline{B}_{\mathrm{r}}(x_q - x_m,\omega)$, expressed in detail as follows:

$$\overline{\mathbf{B}}_{\mathrm{r}} = \begin{bmatrix} \overline{B}_{\mathrm{r}}^{(-N)(-N)} & \cdots & \overline{B}_{\mathrm{r}}^{(-N)0} & \cdots & \overline{B}_{\mathrm{r}}^{(-N)N} \\ & \ddots & \vdots & \vdots & \vdots \\ & & \overline{B}_{\mathrm{r}}^{00} & \cdots & \overline{B}_{\mathrm{r}}^{0N} \\ \text{sym} & & & \ddots & \vdots \\ & & & & \overline{B}_{\mathrm{r}}^{NN} \end{bmatrix} \tag{39}$$

2.3.4. Coupling of Sleepers and Rail

The force analysis of single sleeper is illustrated in Figure 5 (left), in which $w_{\mathrm{s}}(x_q, t)$ is the vertical displacement of the q-th sleeper induced by unit impulse loading at the origin of rail surface; $f_q(t)$ is the pressure from the rail to the q-th sleeper; $F_q(t)$ is the reacting force from roadbed to the q-th sleeper.

So the dynamic equilibrium equation of a single sleeper can be written as:

$$f_q(t) - F_q(t) = -m_{\mathrm{s}}\ddot{w}_{\mathrm{s}}(x_q,t) \tag{40}$$

$$f_q(t) = [w_{\mathrm{s}}(x_q,t) - w_{\mathrm{r}}(x_q,t)]k_{\mathrm{rs}} + [\dot{w}_{\mathrm{s}}(x_q,t) - \dot{w}_{\mathrm{r}}(x_q,t)]c_{\mathrm{rs}} \tag{41}$$

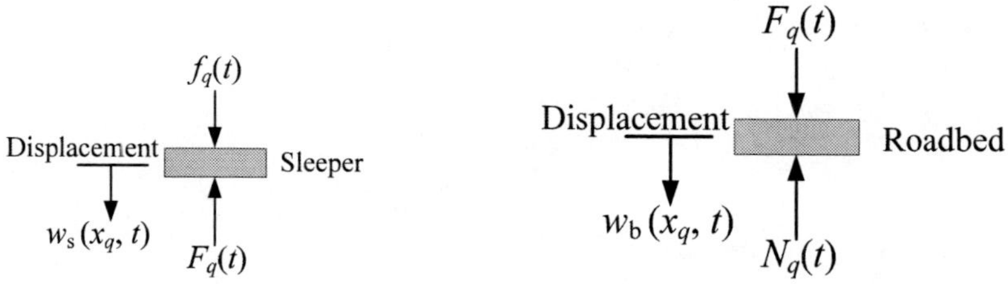

Figure 5. Force analysis of a single sleeper (left) and mass element of roadbed (right).

Equations (40) and (41) can be changed into the frequency domain by using the Fourier transform with respect to t as follows:

$$\bar{f}_q(\omega) - \bar{F}_q(\omega) = -m_s\omega^2\bar{w}_s(x_q,\omega) \tag{42}$$

$$\bar{f}_q(\omega) = (k_{rs} + i\omega c_{rs})[\bar{w}_s(x_q,\omega) - \bar{w}_r(x_q,\omega)] \tag{43}$$

Since each sleeper satisfies Equations (42) and (43), the following vector equation can be obtained by combing the displacement equations of the (2N+1) sleepers:

$$\bar{\mathbf{f}} - \bar{\mathbf{F}} = m_s\omega^2\bar{\mathbf{W}}_s \tag{44}$$

$$\bar{\mathbf{f}} = (k_{rs} + i\omega c_{rs})(\bar{\mathbf{W}}_s - \bar{\mathbf{W}}_r) \tag{45}$$

where $\bar{\mathbf{F}} = \{\bar{F}_{-N}(\omega),\cdots,\bar{F}_0(\omega),\cdots,\bar{F}_N(\omega)\}^T$ is the column vector consisting of interaction forces between 2N+1 sleepers and subsoil; $\bar{\mathbf{W}}_s = \{\bar{w}_s(x_{-N},\omega),\ldots,\bar{w}_s(x_0,\omega),\cdots,\bar{w}_s(x_N,\omega)\}^T$ is the column vector composed of 2N+1 sleeper displacements.

2.3.5. Coupling of Roadbed and Sleeper

The force analysis of a single element of roadbed is shown in Figure 5 (right), in which $w_b(x_q,t)$ is the vertical displacement of roadbed element under the q-th sleeper induced by unit impulse loading at the origin of rail surface; $F_q(t)$ is the action force from the q-th sleeper to roadbed; and $N_q(t)$ is the reaction of subsoil to roadbed.

The dynamic equilibrium equation of a single roadbed element can be written as:

$$F_q(t) - N_q(t) = -m_b\ddot{w}_b(x_q,t) \tag{46}$$

$$F_q(t) = [w_b(x_q,t) - w_s(x_q,t)]k_{sb} + [\dot{w}_b(x_q,t) - \dot{w}_s(x_q,t)]c_{sb} \tag{47}$$

$$N_q(t) = w_b(x_q,t)k_{bg} + \dot{w}_b(x_q,t)c_{bg} \tag{48}$$

Equations (46) to (48) can be changed into the frequency domain by using the Fourier transform with respect to t as follows:

$$\overline{F}_q(\omega) - \overline{N}_q(\omega) = -m_{\mathrm{b}}\omega^2\overline{w}_{\mathrm{b}}(x_q,\omega) \tag{49}$$

$$\overline{F}_q(\omega) = (k_{\mathrm{sb}} + i\omega c_{\mathrm{sb}})[\overline{w}_{\mathrm{b}}(x_q,\omega) - \overline{w}_{\mathrm{s}}(x_q,\omega)] \tag{50}$$

$$\overline{N}_q(\omega) = (k_{\mathrm{bg}} + i\omega c_{\mathrm{bg}})\overline{w}_{\mathrm{b}}(x_q,\omega) \tag{51}$$

Since each roadbed element satisfies Equations (49) to (51), the following vector equations can be obtained by combing the displacement equations of the $2N+1$ roadbed elements:

$$\overline{\mathbf{F}} - \overline{\mathbf{N}} = m_{\mathrm{b}}\omega^2\overline{\mathbf{W}}_{\mathrm{b}} \tag{52}$$

$$\overline{\mathbf{F}} = (k_{\mathrm{sb}} + i\omega c_{\mathrm{sb}})(\overline{\mathbf{W}}_{\mathrm{b}} - \overline{\mathbf{W}}_{\mathrm{s}}) \tag{53}$$

$$\overline{\mathbf{N}} = (k_{\mathrm{bg}} + i\omega c_{\mathrm{bg}})\overline{\mathbf{W}}_{\mathrm{b}} \tag{54}$$

in which $\overline{\mathbf{N}} = \{\overline{N}_{-N}(\omega),\cdots,\overline{N}_0(\omega),\cdots,\overline{N}_N(\omega)\}^{\mathrm{T}}$ represents interaction forces between $2N+1$ roadbed elements and subsoil; $\overline{\mathbf{W}}_{\mathrm{b}} = \{\overline{w}_{\mathrm{b}}(x_{-N},\omega),\cdots,\overline{w}_{\mathrm{b}}(x_0,\omega),\cdots,\overline{w}_{\mathrm{b}}(x_N,\omega)\}^{\mathrm{T}}$ is composed of the displacements of the $2N+1$ roadbed elements.

2.3.6. Interaction Forces between Adjacent Structures

By combining Equations (37)–(38), (44)–(45) and (52)–(54), the following relationship formulas can be obtained:

$$\overline{\mathbf{F}}^{\delta} = (\mathbf{T}_{\mathrm{r}} + \overline{\mathbf{B}}_{\mathrm{r}}\mathbf{T}_{\mathrm{f}})\mathbf{B}_{\mathrm{rs}} \tag{55}$$

$$\overline{\mathbf{F}} = A(k_x,\omega)(\mathbf{T}_{\mathrm{r}} + \overline{\mathbf{B}}_{\mathrm{r}}\mathbf{T}_{\mathrm{f}})^{-1}\overline{\mathbf{E}}_{\mathrm{s}} \tag{56}$$

$$\overline{\mathbf{W}}_{\mathrm{r}} = \mathbf{T}_{\mathrm{r}}\overline{\mathbf{F}} \tag{57}$$

$$\overline{\mathbf{W}}_{\mathrm{s}} = \mathbf{T}_{\mathrm{s}}\overline{\mathbf{F}} \tag{58}$$

$$\overline{\mathbf{W}}_{\mathrm{b}} = \mathbf{T}_{\mathrm{b}}\overline{\mathbf{F}} \tag{59}$$

$$\overline{\mathbf{f}} = \mathbf{T}_{\mathrm{f}}\overline{\mathbf{F}} \tag{60}$$

$$\overline{\mathbf{N}} = \mathbf{T}_{\mathrm{n}}\overline{\mathbf{F}} \tag{61}$$

in which $A(k_x,\omega)=\frac{1}{V}\tilde{\bar{B}}_{\mathrm{r}}(k_x,\omega)\bar{C}(\omega)$, $\mathbf{T}_{\mathrm{b}}=(k_{\mathrm{bg}}+i\omega c_{\mathrm{bg}}+m_{\mathrm{b}}\omega^2)^{-1}$,

$\mathbf{T}_{\mathrm{s}}=\mathbf{T}_{\mathrm{b}}-(k_{\mathrm{sb}}+i\omega c_{\mathrm{sb}})^{-1}$, $\mathbf{T}_{\mathrm{f}}=1+m_{\mathrm{s}}\omega^2\mathbf{T}_{\mathrm{s}}$, $\mathbf{T}_{\mathrm{r}}=\mathbf{T}_{\mathrm{s}}-(k_{\mathrm{rs}}+i\omega c_{\mathrm{rs}})^{-1}\mathbf{T}_{\mathrm{f}}$ and

$\mathbf{T}_{\mathrm{n}}=1-m_{\mathrm{b}}\omega^2\mathbf{T}_{\mathrm{b}}$.

Therefore, the interaction force $\bar{\mathbf{F}}$ between sleepers and roadbed, the rail displacement vector $\bar{\mathbf{W}}_{\mathrm{r}}$ at the position of sleepers, the sleeper displacement vector $\bar{\mathbf{W}}_{\mathrm{s}}$, the roadbed displacement vector $\bar{\mathbf{W}}_{\mathrm{b}}$, the supporting force vector $\bar{\mathbf{f}}$ of sleepers, the force vector $\bar{\mathbf{F}}$ from sleepers to roadbed surface and the reaction force $\bar{\mathbf{N}}$ of subsoil to roadbed can be solved.

2.3.7. Subsoil Modeling

Good prediction for the ground vibration alongside track requires a reasonable subsoil model. The generation of eigenwaves and their propagation are uniquely characterized by the constituting layer properties of the ground. Herein, the subsoil under the track is modeled as a three-dimensional layered viscoelastic medium with the bottom modeled as halfspace, as shown in Figure 6, by assumming each of the parallel layers isotropic and homogeneous. The wave propagation in the soil is characterized by the dynamic Green's function of ground displacement, which can be seen in Chapter 3.

The track model and the soil model are connected with roadbed, and the interaction forces between roadbed and subsoil are as follows:

$$\bar{\mathbf{N}}=(k_{\mathrm{sg}}+i\omega c_{\mathrm{sg}})(\bar{\mathbf{U}}_z-\bar{\mathbf{W}}_{\mathrm{s}}) \tag{62}$$

where $\bar{\mathbf{U}}_z=\{\bar{u}_z(x_{-N},\omega),\cdots,\bar{u}_z(x_0,\omega),\cdots,\bar{u}_z(x_N,\omega)\}^{\mathrm{T}}$ is a column vector consisting of $2N+1$ subsoil displacements at the position of sleepers.

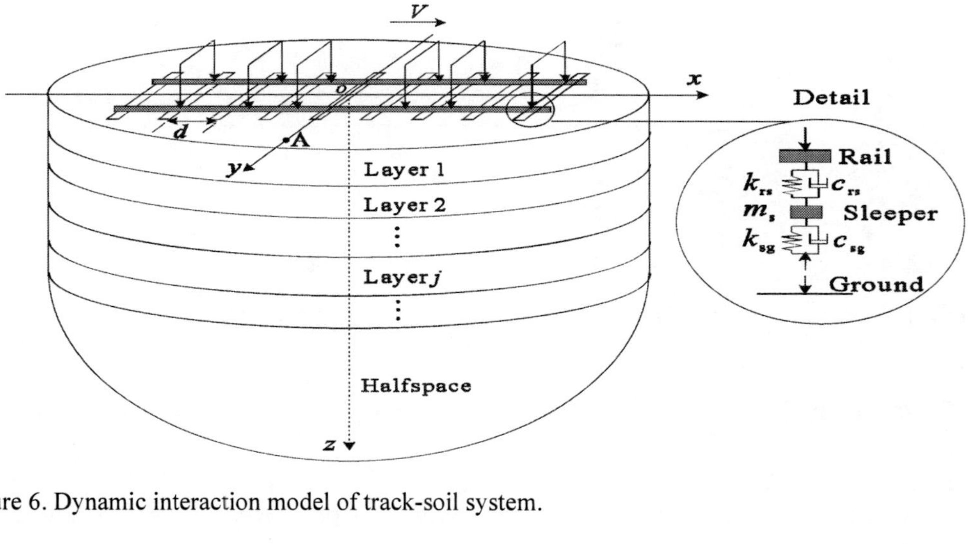

Figure 6. Dynamic interaction model of track-soil system.

On the other hand, for a layered-soil bottomed of a half-space, the $2N+1$ ground displacements at the position of sleepers can be expressed in a vector equation:

$$\bar{\mathbf{U}}_z = \bar{\mathbf{G}}_{zz}\bar{\mathbf{F}} \tag{63}$$

where $\bar{\mathbf{G}}_{zz}$ is the $(2N+1)\times(2N+1)$ order matrix that consists of the Green's functions of ground displacements, which can be found in Chapter 3.

2.3.8. Ground Displacement

When the reaction vector $\bar{\mathbf{N}}$ of soil under moving train-loads is obtained by solving the simultaneous Equations (20), (38) and (55) to (63), the total ground displacement at the receiver point A can be acquired by the superposition of all the displacement components induced by each of the subsoil reactions under the $2N+1$ sleepers. The solution expression is as follows:

$$\bar{u}_{zz}(y,0,\omega) = \bar{\mathbf{F}}\cdot\bar{\mathbf{G}}_{zz}^{*} \tag{64}$$

where $\bar{\mathbf{G}}_{zz}^{*}$ is the vector consisting of the Green's functions of ground displacements at the receiver point A induced by $2N+1$ subsoil reactions:

$$\bar{\mathbf{G}}_{zz}^{*} = \{\bar{G}_{zz}(r_{-N}, 0, \omega), \cdots, \bar{G}_{zz}(r_0, 0, \omega), \cdots, \bar{G}_{zz}(r_N, 0, \omega)\} \tag{65}$$

where $r_m = \sqrt{(md)^2 + y^2}$ is the distance between the m-th sleeper and the receiver point A, y is the distance of A from the track, and d is the sleeper spacing (Ref. Figure 7).

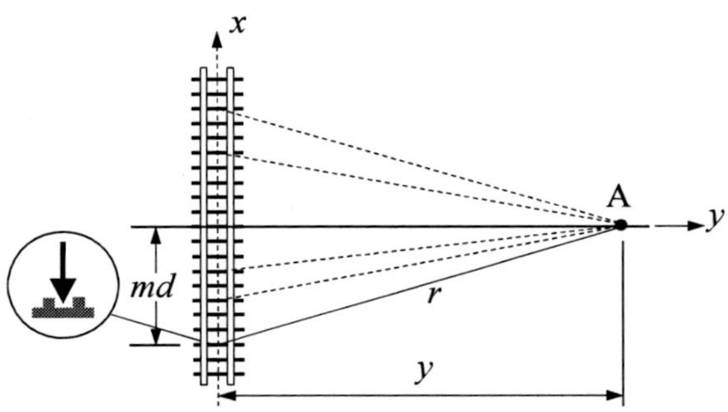

Figure 7. Influence of different subsoil reactions on ground vibration at receiver point A.

2.4. Case Study

2.4.1. Calculation Parameters

The China Star train composed of 1 locomotive + 4 passenger cars + 1 locomotive is taken as the train load, detailed parameters of the vehicles can be found in Cao [20]. The track parameters include: Rail: mass per unit length $\bar{m}_r$=60 kg/m, section stiffness EI=6.63 MN·m^2, damping ratio ξ_r=0.01; Sleeper: mass m_s=250 kg, spacing d=0.55m; Ballast bed: mass m_b=560 kg; Spring-damper coefficients: k_{rs}=78 MN/m, c_{rs}=50 kN·s/m, k_{sg}=65 MN/m and c_{sg}=90 kN·s/m. The track irregularity is simulated by a single wavelength harmonic profile with the irregularity amplitude $\tilde{z}_r(\Omega_k)$ equal to 0.2 mm in Equation (15).

The soil parameters are listed in Table 1. The first three cut-off frequencies of P-SV waves in the top-layer soil are, respectively, 24.57 Hz, 43.99 Hz and 81.78 Hz, and the Rayleigh wave speed is 151 m/s.

2.4.2. Vibration Analysis of Track Structure and Ground Surface

Plotted in Figure 8 are the vertical ground acceleration histories of rail, sleeper and receiver point A at 0m, 5m, 20m, 30m from the track, under the train speed of 200 km/h. The attenuation and propagation of vibration in the track-soil system can be observed from the figures as follows:

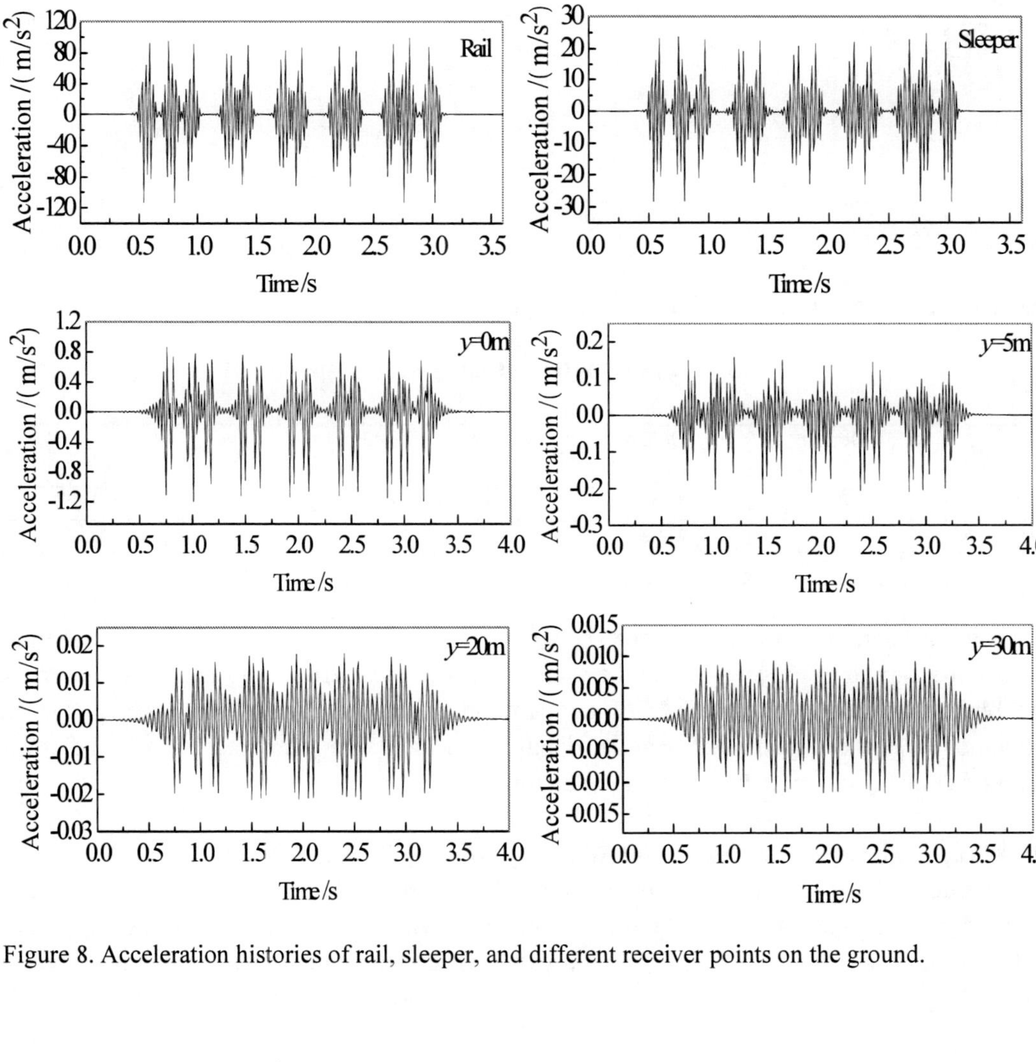

Figure 8. Acceleration histories of rail, sleeper, and different receiver points on the ground.

(1) The acceleration amplitudes of the track-soil system decrease successively from rail to sleeper and from sleeper to ground soil: the maximum acceleration of rail is 120 m/s^2, and it decreases to about 25 m/s^2 for sleeper and 1.2 m/s^2 for ground surface under sleeper.

(2) The ground acceleration attenuates with the distance to the track. With the increase of distance, the relevance between the ground accelerations and the train excitation becomes less obvious. This observation is in accordance with the experimental results in Xia [21].

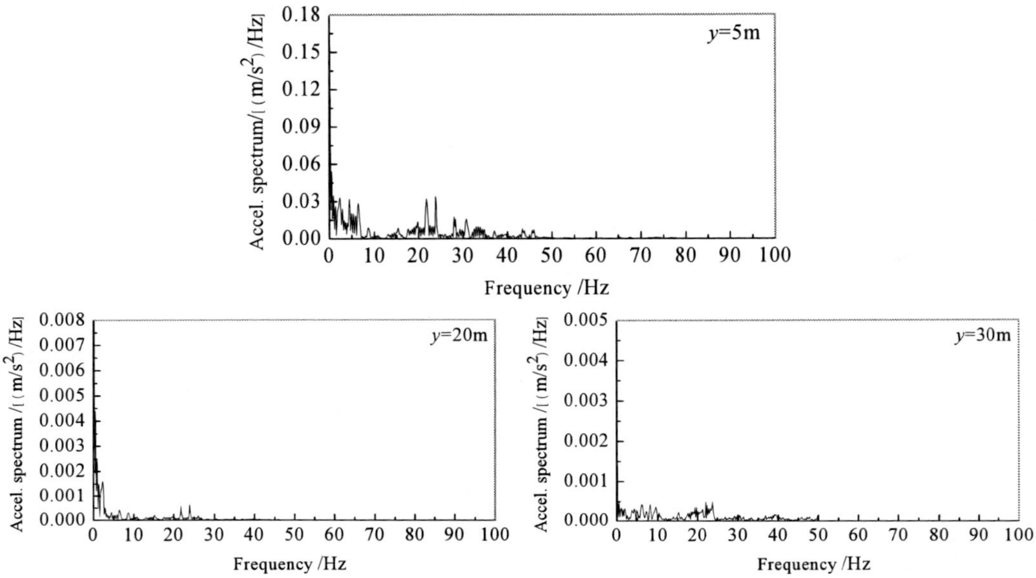

Figure 9. Acceleration spectra of different receiver points on the ground.

Table 1. Parameters of layered half-space soil

Layer	Depth /m	Density /(kg/m^3)	Elastic Modulus /MPa	Poison ratio	Damping ratio
1	3.0	1350	95.59	0.37	0.05
Half-space	—	1898	506.97	0.48	0.05

The acceleration spectra for receiver point A at y=0 m, 5m and 20 m on the ground surface are shown in Figure 9. It can be found that with the increase of distance to the track, the higher frequency components of acceleration spectra become lesser, which can be explained by the dissipation of material damping and radiation damping in soil medium to higher frequency components in vibration energy.

2.4.3. Influence of Train Speed and Soil Property on Ground Vibration

The forces applied to the track structure by a moving train mainly consist of two components: the quasi-static excitation induced by moving gravity axle-loads and the wheel-rail dynamic excitation induced by irregularity of wheel-rail contact interface. In this section, the moving unit-constant-load and the moving unit-harmonic-load are adopted to analyze the influence of train speed and soil property on ground vibrations, respectively, in the cases of quasi-static excitation and dynamic excitation.

2.4.3.1 Ground Vibration Induced by Moving Unit-constant-load

The vertical ground displacements induced by moving unit-constant-loads with the speed of 100 m/s, 151 m/s, 200 m/s, 250 m/s and 280 m/s are calculated by using the above dynamic interaction model, and the displacement spectra of two receiver points (y=0 m and y=8 m) are respectively plotted in Figure 10.

It can be observed from the figures that there exist amplification phenomena of ground vibration corresponding to different frequencies for different moving speeds, which may be explained hereafter as due to the soil resonance induced by loads moving at certain speeds. According to soil dynamics and structure dynamics, the resonance frequencies of layered-soil corresponding to the moving loads are the intersection frequencies of the moving loads speed-lines and the subsoil dispersion curves. Plotting the first four orders of P-SV dispersion curves of the layered-soil and the five speed-lines of moving constant-loads that are graphed on the basis of $k=2\pi f/V$ in the same coordinate system (Figure 11 (left)), one can find their intersection frequencies, i.e. resonance frequencies of the layered-soil. The detailed values of them are listed in Table 2.

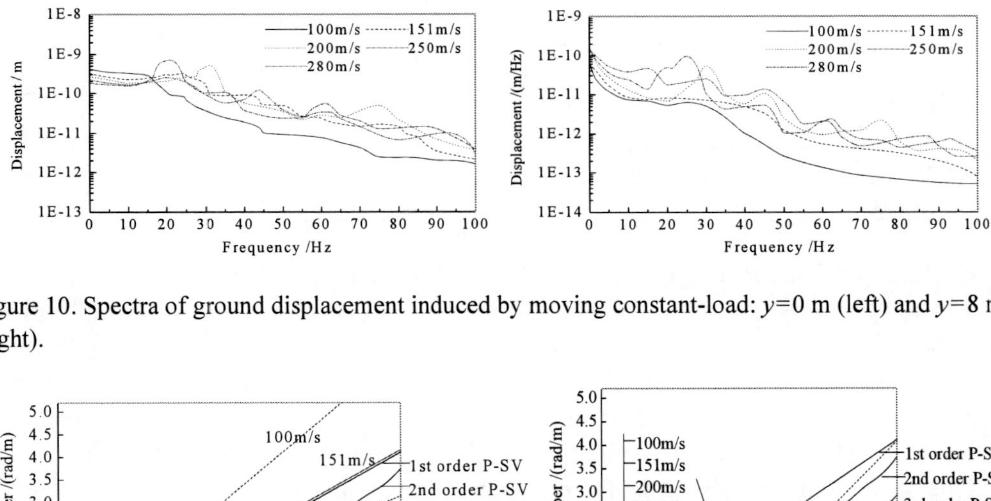

Figure 10. Spectra of ground displacement induced by moving constant-load: y=0 m (left) and y=8 m (right).

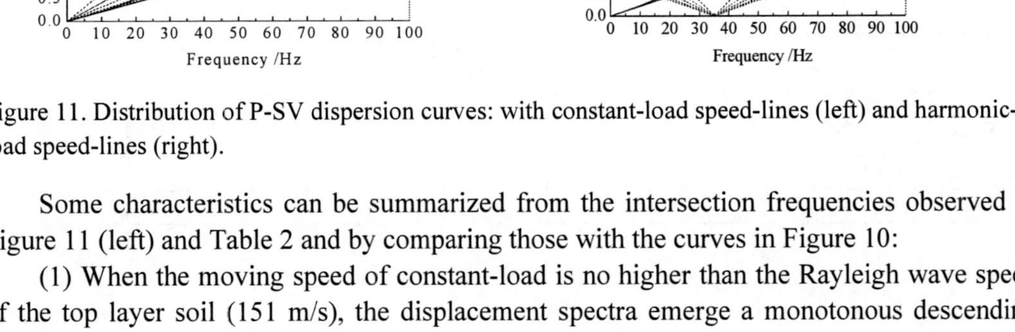

Figure 11. Distribution of P-SV dispersion curves: with constant-load speed-lines (left) and harmonic-load speed-lines (right).

Some characteristics can be summarized from the intersection frequencies observed in Figure 11 (left) and Table 2 and by comparing those with the curves in Figure 10:

(1) When the moving speed of constant-load is no higher than the Rayleigh wave speed of the top layer soil (151 m/s), the displacement spectra emerge a monotonous descending

tendency. In this case, there are no intersections of speed-lines and dispersion curves appears, which shows that the ground vibration frequencies excited by the moving loads cannot reach the resonance frequencies of the soil. This phenomenon indicates that when the speed of moving load is much smaller than the Rayleigh's wave speed, the soil vibration induced by the moving load is close to that induced by a static load.

(2) When the moving speed of constant-load is higher than the Rayleigh's wave speed of the top layer soil, one or more vibration peaks may appear. In this case, there exist several intersections for the speed-lines and the dispersion curves, the number of which is related to the speed of moving loads, and the frequencies corresponding to the vibration peaks are close to the resonance frequencies of the soil.

(3) When the resonance frequencies of the soil corresponding to the speed of moving loads are close to the natural frequencies (i.e. the cutoff frequencies) of the layered-soil, the ground vibration is amplified. This explains why the maximal displacement peaks shown in Figure 10 occur under the moving load speed of 250 m/s: the resonance frequency corresponding to this load-speed is 21 Hz and it is close to the first-order cutoff frequency 24.57 Hz of the soil.

Table 2. Intersection frequencies of constant-load speed-lines and P-SV dispersion /Hz

Modes of layer soil	Speeds of moving load V/(m/s)				
	100	151	200	250	280
First-order	—	—	31	21	—
Second-order	—	—	75	61	44
Third-order	—	—	—	87	62.3
Fourth-order	—	—	—	—	93

2.4.3.2. Ground Vibration Induced by Moving Unit-harmonic-load

Since the first-order and the second-order cutoff frequencies of the subsoil are respectively 24.57 Hz and 43.99 Hz, the vibration frequency 35 Hz in this range for the unit-harmonic-load is selected to well observe the intersection frequencies. The speed-lines of moving harmonic-loads and the P-SV dispersion curves of the subsoil model are plotted in Figure 11 (right), and the corresponding intersection frequencies are listed in Table 3. In this case, the speed-lines of moving harmonic-loads are determined by $k=2\pi|f-f_0|/V$, thus one can find two groups of speed-lines starting from the loading frequency f_0=35 Hz.

Table 3. Intersection frequencies of P-SV dispersion and harmonic-load speed-lines /Hz

Modes of layer soil	Speeds of moving load V/(m/s)				
	100	151	200	250	280
First-order	24.67; 101	21.72	19.59	17.89	16.91
Second-order	26; 57.4	—	—	—	—
Third-order	53.83	—	—	—	—
Fourth-order	—	—	—	—	—

The vertical ground displacements induced by the moving unit-harmonic-loads with the speed of 100 m/s, 151 m/s, 200 m/s and 250 m/s are calculated, and the displacement spectra of two receiver points (y=0 m and y=8 m) are illustrated in Figure 12.

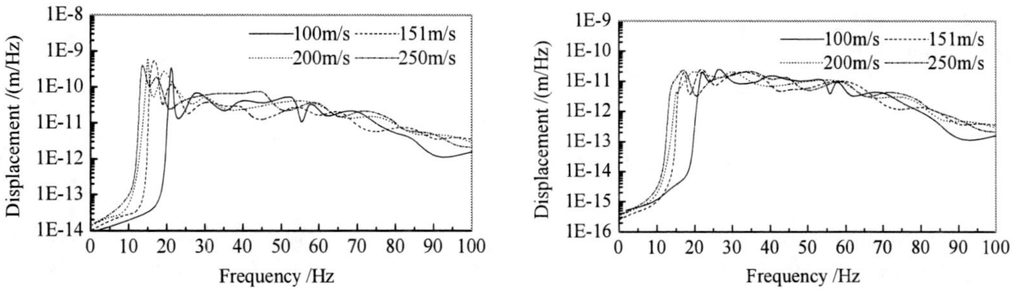

Figure 12. Spectra of ground displacement induced by moving harmonic-load: y=0 m (left) and y=8 m (right).

It can be found that the frequencies corresponding to displacement peaks in Figure 12 are close to the intersection frequencies listed in Table 3, which shows that in the case of moving harmonic-loads, the appearance of ground vibration amplification is also at the resonance frequency of the soil and the moving loads. Moreover, the displacement spectra of ground vibrations induced by moving harmonic-loads are mostly at the same order of magnitude, in which the higher frequency components are more abundant than those in the case of moving constant-loads, and the influence of their moving speed on ground vibration is smaller than that by constant-loads. The frequencies of ground vibration induced by harmonic-loads are within a certain range, and the higher the moving speed is, the wider the frequency range is. These observations are in accordance with the results of Lombaert [22] that the frequency of ground vibration induced by moving harmonic-loads mainly concentrates on the range of $[\omega_0/(1+V/c),\ \omega_0/(1+V/c)]$, in which ω_0 is the vibration frequency of harmonic-load (35Hz herein), V is the moving speed, and c is the Rayleigh's wave speed of top-layer soil model (151 m/s herein).

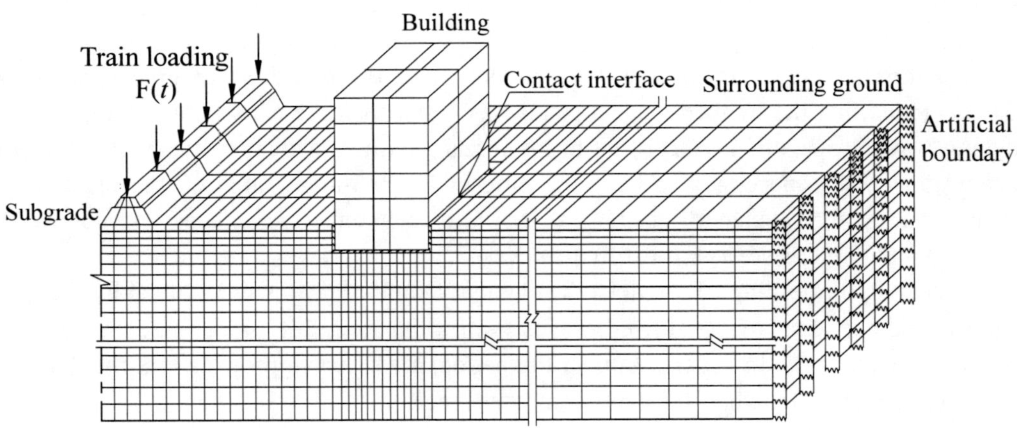

Figure 13. Three-dimensional roadbed-subsoil-building interaction model.

3. Roadbed-Subsoil-Building Interaction Model

The roadbed-subsoil-building dynamic model is established with 3D finite element method, as shown in Figure 13, in which the reaction force $F(t)$ applied to the roadbed surface can be obtained from the train-track model in Section 2. The dynamic responses of the surrounding soil layer and the building can be solved by numerical integral method.

3.1. Model Sizing and FE Meshing Strategies

The discretization method of a FE model should be able to reproduce the wave features in continuous media as much as possible, while research has shown that the propagation laws of elastic waves change when a discrete model is adopted to substitute the continuum. This is because the phase speed of plane wave propagating in discrete grid is not always equal to its physical wave velocity, but related to the direction and frequency of wave propagation, which causes additional dispersion and cut-off frequencies of wave in the discrete model. The wave energy with a frequency greater than the cut-off frequencies cannot propagate, but mixes with some low-frequency waves in the form of parasitic oscillation. For homogeneous continuous media with a wave speed of c, if the discrete size of FE grid is Δx, the minimum period of wave propagation in one-dimensional FE discrete model is $T_{max}=\pi\Delta x/c$, meaning this model cannot be used to study the propagation of waves with a period smaller than $\pi\Delta x/c$. If the wavelength of a harmonic wave with the period of T is $\lambda=cT$, the essential condition of wave propagating in discrete grid is $\Delta x \leq \lambda/\pi$.

By taking a unit harmonic-load acting on free surface of an infinite half-space as an example, Yang [11] performed an analysis on the meshing of finite element. He analyzed the convergence situations of horizontal and vertical displacements of free surface. For the distance R between loading point and boundary, the displacement converged to a good result when $R=1.5\lambda_{smax}$, in which λ_{smax} is the maximum shear wavelength of oscillatory waves concerned, and there was a slight error when $R=0.5\lambda_{smax}$. For the element length L, an enough accurate result was obtained when $L=\lambda_{smin}/12$, in which λ_{smin} is the minimum shear wavelength concerned, and good results was obtained except at the points $0.5\lambda_{smin}$ away from the wave source when $L=\lambda_{smin}/6$.

By comparing wave phenomena in FE models with different discrete dimensions, Guan [23] concluded that the lumped-mass method was better than the consistent-mass method in studying wave propagation. According to his research, when analyzing transient wave with discrete model, it is necessary to keep the ratio of lumped-mass finite element size to the effective minimum wavelength no more than 1/6, to guarantee the accuracy, to control the dispersion and to increase the cut-off frequency.

As a result, in meshing FE models, the element length of grid is determined by the high-frequency components of considered waves, while the range of model size by the low-frequency components. In order to reflect the periodicity of wave propagation, the discreted area should be larger than half the maximum wavelength.

3.2. Boundary Conditions of Subsoil

When using numerical methods, the effective artificial boundaries should be established, otherwise the stress wave will reflect on the truncated boundaries, which leads to distortion of calculation. Currently, the artificial boundaries applicable in engineering include viscous boundary, consistent boundary, stack boundary, paraxial boundary, transmitting boundary and dynamic mapping FE boundary. In this chapter, the 3D viscous-spring dynamic artificial boundary is adopted, which is proposed by Liu [24] on the basis of 3D spherical waves. The reliability and practical applicability of this boundary have been verified by the numerical simulation results of Lamb surface source and concentrated force source in homogeneous isotropic linear elastic infinite medium as well as other analytical solutions.

The 3D viscous-spring dynamic artificial boundary can be easily combined with FEM only by setting normal and tangential spring and damper elements on the artificial boundaries. Their elasticity and damping coefficient can be determined by the characteristic parameters of the elements. For the normal boundary,

$$k_{\mathrm{N}i} = \frac{2G_i}{r} A_i \text{ ; } c_{\mathrm{N}i} = \rho_i c_{\mathrm{p}i} A_i \tag{66}$$

And for the tangential boundary,

$$k_{\mathrm{T}i} = \frac{3G_i}{2r_i} A_i \text{ ; } c_{\mathrm{T}i} = \rho_i c_{\mathrm{s}i} A_i \tag{67}$$

where the subscripts "N" and "T" mean, respectively, the normal boundary and the tangential boundary; k_i and c_i are stiffness coefficient and damping coefficient; G_i is the shear modulus; ρ is the mass density of soil; c_{pi} and c_{si} are the velocity of P-wave and S-wave, respectively; A_i is the equivalent area occupied by boundary point i; r_i is the distance between source and boundary point i.

If the displacement shape functions of artificial boundary are the same as those used in FE model, the discrete artificial boundary is called equivalent viscoelastic artificial boundary. It can be obtained by extending a layer with same type of element along the normal direction on the boundary of the established FE model, fixing the outer boundary, and defining the material properties of equivalent elements so as to make it equal to the viscoelastic artificial boundary element. Based on the above concepts, the equivalent shear modulus and elasticity modulus of boundary element can be expressed as:

$$\tilde{G} = \alpha_{\mathrm{T}} h \frac{G}{R} \text{ ; } \tilde{E} = \alpha_{\mathrm{N}} h \frac{G}{R} \cdot \frac{(1+\tilde{\nu})(1-2\tilde{\nu})}{(1-\tilde{\nu})} \tag{68}$$

where $\tilde{G}$, $\tilde{E}$ and $\tilde{\nu}$ are equivalent shear modulus, elasticity modulus and Poisson ratio of the equivalent viscoelastic artificial boundary element, h is the thickness of the equivalent element, R is the distance between source and artificial boundary point, G is the shear elasticity of medium, and α_{T} and α_{N} are tangential and normal artificial boundary parameters, respectively.

Thus, the equivalent Poisson ratio can be expressed by:

$$\tilde{\nu} = \begin{cases} \dfrac{\alpha - 2}{2(\alpha - 1)} & , \ \alpha \geq 2 , \\ 0 & , \ \text{others} \end{cases} \tag{69}$$

where $\alpha = \alpha_N / \alpha_T$.

The damping matrix of equivalent viscoelastic boundary element can be obtained by the damping proportional to the stiffness, and the proportionality factor can be expressed as:

$$\tilde{\eta} = \frac{\rho R}{2G}\left(\frac{c_s}{\alpha_T} + \frac{c_p}{\alpha_N}\right) \tag{70}$$

where $\alpha_T = [0.35, 0.65], \alpha_N = [0.8, 1.2]$. For the 3D model, $\alpha = \alpha_N / \alpha_T = 2$ is suitable.

3.3. Consideration of Damping Characteristics

When the wave propagates in soil medium, the energy dissipates gradually due to two kinds of damping effects: radiation damping and material damping. For the material damping of soil, linear viscoelastic damping and linear hysteretic damping can be considered under micro strain condition. The linear viscoelastic damping is related to the vibration speed of particles and easy to realize in numerical integration, so it is often used in time domain analysis. While the linear hysteretic damping is related to the strain and strain rate of medium, which accords with the dissipation mechanism of soil to wave energy, and it is often used in frequency-domain analysis because of its relevency with frequency. As a result, in time-domain analysis, the performances of the two kinds of material damping are combined using the principle of equivalence, and the equivalent damping coefficient can be defined as:

$$C_{eq} = 2D\mathbf{K} / \omega \tag{71}$$

where D is hysteretic damping ratio of soil medium; $\mathbf{K}$ is stiffness matrix corresponding to soil; ω is dominant frequency of wave proagation. In calculation, the damping effect can be considered by inputting stiffness factors of different materials.

The Rayleigh damping is often used in structural analysis, which assumes that the damping matrix is proportional to the mass matrix and stiffness matrix as follows:

$$[C] = \alpha[M] + \beta[K] \tag{72}$$

where α is the factor of viscous damping and β the factor of hysteresis damping or stiffness damping, which satisfy the following relationship with the damping ratio:

$$\xi = \frac{\alpha}{2\omega} + \frac{\beta\omega}{2}; \quad 0 \leq \omega \leq \omega_{max} \tag{73}$$

Since the natural frequencies of two modes ω_i and ω_j and corresponding damping ratios ξ_i and ξ_j can be obtained by testing, α and β can be determined by:

$$\alpha = \frac{2(\xi_i\omega_j - \xi_j\omega_i)}{(\omega_j^2 - \omega_i^2)}\omega_i\omega_j \; ; \; \beta = \frac{2(\xi_j\omega_i - \xi_i\omega_j)}{(\omega_j^2 - \omega_i^2)} \tag{74}$$

3.4. Contact Interface between Subsoil and Building

In the roadbed-subsoil-building model, the contact interface between soil and structure may cause reflection and refraction, leading to an energy exchange phenomenon. Therefore, the interface problem should be concerned and a reasonable continuity condition of contact interface should be given.

In general, the contact interface can be simply treated using two extreme assumptions: the rough interface and the smooth interface. The rough interface assumes the contact interface is rough enough, meaning there is no relative sliding between the soil and the structures, so that the displacement on the interface is continuous and a common node can be used in FE modeling. The smooth interface assumes the contact interface is smooth enough, meaning there is no shear stress between the soil and the structures, so there exists relative sliding at the interface, which can be simulated by adding linear springs between the nodes of soil and structure, and setting zero modulus of elasticity for the tangential springs to simulate sliding.

Actually, the contact interface is neither absolutely rough nor absolutely smooth. At the soil-structure interface, the shear stress may prevent the relative sliding, which may occur only when the stress exceeds a certain value, while the tensile stress may depart the soil and the structure on the contact interface, and damage the continuous displacement between the two adjacent points at the beginning. Under dynamic loads, due to the repeated loading and unloading, there will be a varying state of closing, opening and again closing at the interface. The related research shows this complex state can be conducted in two ways: one is adopting a reasonable surface element to simulate the interfacial deformation; the other is establishing a model to reflect the constitutive relation of the interface medium, especially the relationship between the shear stress and the shearing deformation.

In this chapter, the thin layer contact element shown in Figure 14 (left) is adopted, by assuming a thin layer with a certain thickness near the interface. The thin layer element has different performance with the surrounding medium, which can be reflected by the specially defined constitutive relation.

The composition of the thin-layer element can be expressed in matrix form as follows:

$$[\mathbf{K}] = \begin{bmatrix} \mathbf{K}_{\mathrm{nn}} & \mathbf{K}_{\mathrm{ns}} \\ \mathbf{K}_{\mathrm{sn}} & \mathbf{K}_{\mathrm{ss}} \end{bmatrix} \tag{75}$$

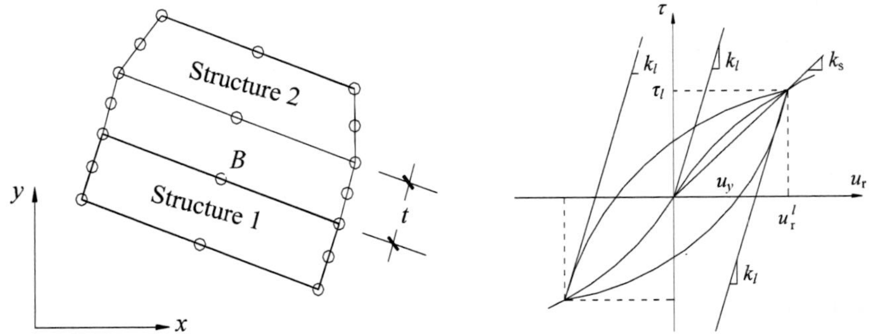

Figure 14. Thin layer element (left) and shear stress-displacement curve (right).

where $\mathbf{K}_{nn}$ and $\mathbf{K}_{ss}$ are, respectively, the normal part and the tangential part of interface; $\mathbf{K}_{ns}$ and $\mathbf{K}_{sn}$ represent the coupling of the two parts.

Even though the coupling of the normal and the tangential exists, in practical application it is always neglected. Denoted by $\mathbf{K}_{nn}=\mathbf{K}_n$ and $\mathbf{K}_{ss}=\mathbf{K}_s$, Equation (75) can be rewritten as:

$$[\mathbf{K}]=\begin{bmatrix}\mathbf{K}_n & \mathbf{0}\\ \mathbf{0} & \mathbf{K}_s\end{bmatrix} \tag{76}$$

where $\mathbf{K}_n$ and $\mathbf{K}_s$ are, respectively, normal and tangential stiffness of the interface.

By means of direct shear test or torsion test to the material on the interface, the shear force T and shear relative displacement u_r can be obtained and consequently the T–u_r curve can be determined, as shown in Figure 14 (right). For the normal stiffness of interface material, the following estimated value is adopted: when there is no sliding on the interface, the normal stress $\sigma_r>0$, and the elements of $\mathbf{K}_n$ take a large value ($\geq 10^8$); when there is uplifting or departing on the interface, they fall to a small value (≤ 100).

For the thin-layer element, thickness is the most important parameter. After numerical calculation and a shear test on the interface of sand-concrete, the suitable depth range of thin-layer element is as follows:

$$0.01 \leq \frac{t}{B} \leq 0.1 \tag{77}$$

where t is the thickness of thin-layer element, and B is the average width of element connecting soil and structures.

4. FIELD EXPERIMENT OF TRAIN-INDUCED ENVIRONMENTAL VIBRATIONS

An in-situ experiment was carried out in the suburb of Shijiazhuang City, near the Beijing–Guangzhou Railway. It is one of the busiest railways in China, where trains pass by at every 4–10 minutes, day and night. The residents living in the building nearby often complained of the vibration induced by the running trains. The purpose of this experiment was to study the vibration levels at the ground surface and different locations in a building induced by running trains with different speeds.

4.1. Layout of Experiment

At the experimental site, a six-story masonry structure near the railway and an open field in the north of the building were selected for the experiment, as shown in Figure 15. This building built in 2000 has a total height of 17.25 m, width of 12 m and length of 58 m, consisting of four units, Unit 1 to Unit 4, with a longitudinal axis perpendicular to the railway track as shown in Figures 16 and 17. The floor height of each story is 2.8 m, and the first floor elevation is 0.45 m higher than the outdoor ground surface. The floors are made of cast-in-place C25 concrete slabs of 120 mm thick. The parameters of different soil strata at the test site are shown in Table 4.

Figure 15. Experimental site near the existing railway.

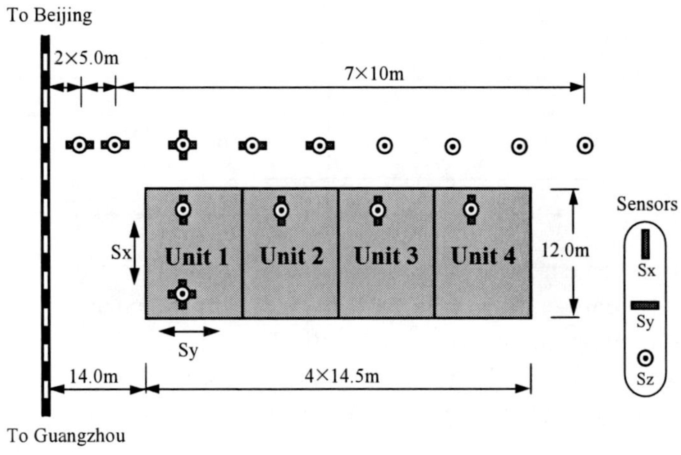

Figure 16. Layout of the experimental site.

Close to the building, there was a small apron made of 18 cm thick C25 plain concrete with 30 cm thick lime earth rammed base.

This experiment was carried out on Dec. 6–7, 2006. Along the ground surface there were 9 measuring points, and each was installed with one (in vertical direction), two or three velocity sensors. Meanwhile, measuring points were arranged at different locations of the building focusing on the different floors of the Unit 1 and second floor of all four units. As there were not enough sensors available, the measurements for the ground and for the building were conducted separately. A total of 69 velocity sensors were installed, among which 28 for the ground, 21 for the vertical vibration and 20 for the lateral vibration of the building.

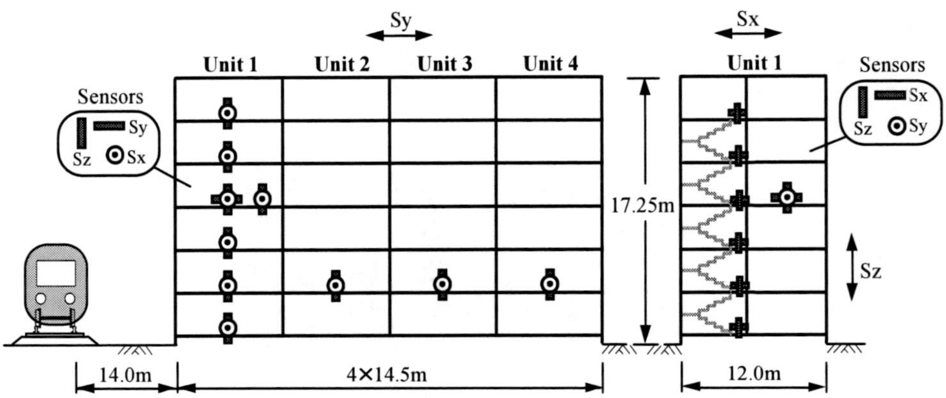

Figure 17. Arrangement of measuring sensors in the building.

Table 4. Soil Parameters in the experimental site

Soil layer	Thickness /m	Bearing capacity /kPa	Compression modulus /MPa	Moisture content /%	Density /(g/cm^3)
Silty clay	0.5~2.0	150	17.0	19.6	1.91
Silt	2.0~2.4	130	17.5	24.0	1.82
Silty clay	2.0~2.3	150	17.5	21.2	1.75
Silt	2.2~3.0	160	18.5	22.1	1.90
Silty fine sand	1.9~2.5	200	18.0	22.5	1.95
Medium-coarse sand	5.4~5.6	320	22.0	22.0	2.11
Silt	2.8~3.0	150	16.5	23.6	1.90

A detailed arrangement of the measuring sensors can be found in Figures 16 and 17, where S_x and S_y represent sensors parallel and perpendicular to the railway track, respectively, and S_z represents sensor for the vertical vibration.

The experimental loads are the normal trains running on the Beijing-Guangzhou Railway. In the experiment, the speed range was 40–115 km/h for passenger trains and 26–57 km/h for freight trains. From the train-loading conditions investigated at the Shijiazhuang Railway Administration, the vehicle axle-loads were 230 kN for each 6-axle locomotive, 200 kN for each 4-axle freight car, and 140–160 kN for each passenger car.

The instruments used in the test are 32 CH and 16 bit AD inverter INV303/306 Data Acquisition & Signal Processing System, and 891-IV velocity sensor and DLF-8 amplifier. Besides, an LD-98-II P radar speedometer was used to measure the running speed of trains. All instruments and sensors were systematically calibrated before the measurement.

4.2. Analysis of Experimental Results

4.2.1. Ground Vibrations

(1) History curves and 1/3 octave spectrum analysis of ground vibrations

Shown in Figures 18 to 20 are the measured vertical (direction S_z) and horizontal (directions S_x and S_y) acceleration history curves of the ground surface, in which the train speed is 67 km/h, and the distance between the measuring points and the track are D=10m and D=40m, respectively. It can be seen that the vertical vibrations of ground are greater than the horizontal vibrations, and in both directions, the ground vibrations at D=10m are greater than those at D=40m.

Plotted in Figures 21 to 23 are, respectively, the one-third octave spectra of ground accelerations in three directions, when a passenger train ran on the track at 104 km/h and a freight train at 34 km/h. As can be seen, the frequency of ground vibration focus on the range of 10−100 Hz, and the peak accelerations of main frequencies decrease with the distance of receiver point to the track, but the main frequencies at different points vary little.

(2) Influence of train speeds on main frequencies of ground vibrations

The ground accelerations at D=10m are measured to study the effect of train speeds on the frequencies of environmental vibration, as shown in Figures 24 to 26, for passenger trains and freight trains, respectively. When the passenger train passes through, the main frequencies of ground accelerations in S_x and S_y directions have a decreasing trend with train speeds in certain degree, while those in S_z-direction change very little. Although the freight train speeds are slower, the acceleration spectra related to the main frequencies are significantly greater than in the case of passenger trains, which is mainly due to the heavier axle-loads and larger impact forces of freight trains.

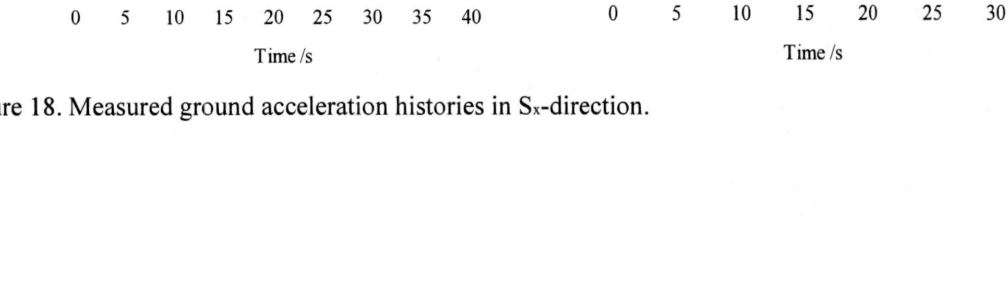

Figure 18. Measured ground acceleration histories in S_x-direction.

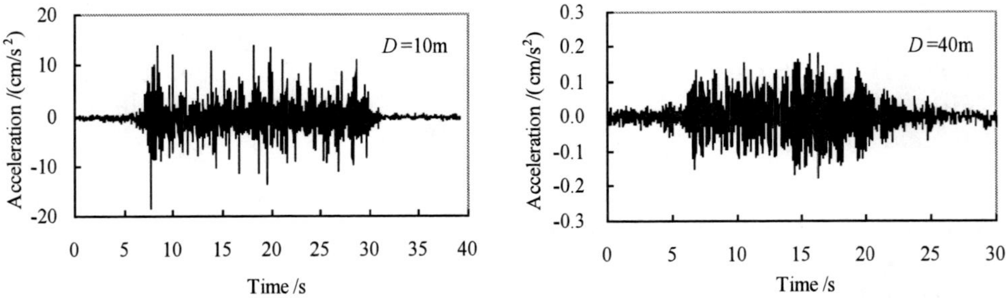

Figure 19. Measured ground acceleration histories in S_y-direction.

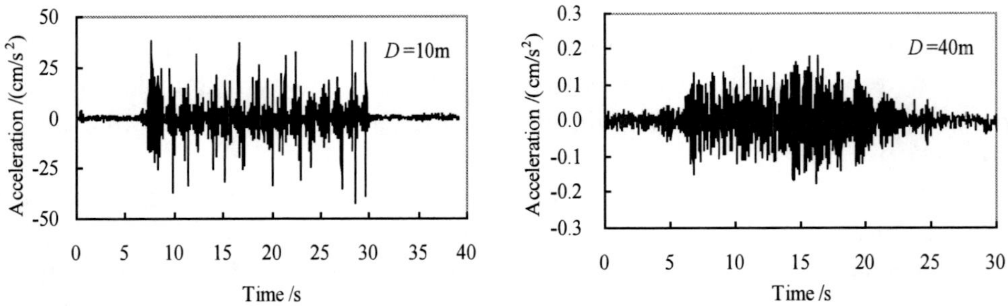

Figure 20. Measured ground acceleration histories in S_z-direction.

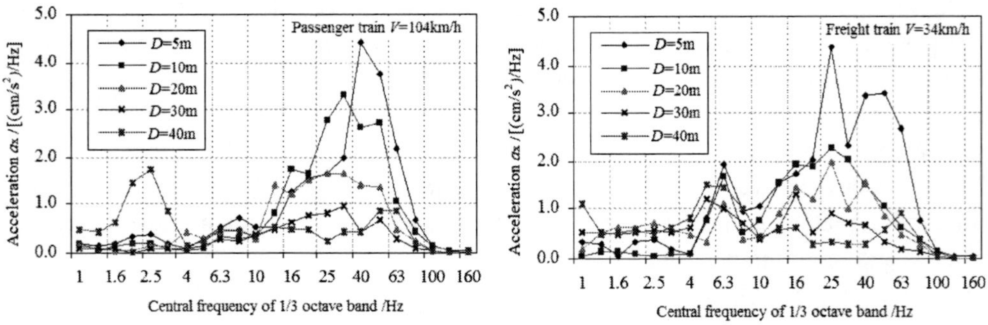

Figure 21. One-third octave spectrum of ground acceleration in S_x-direction.

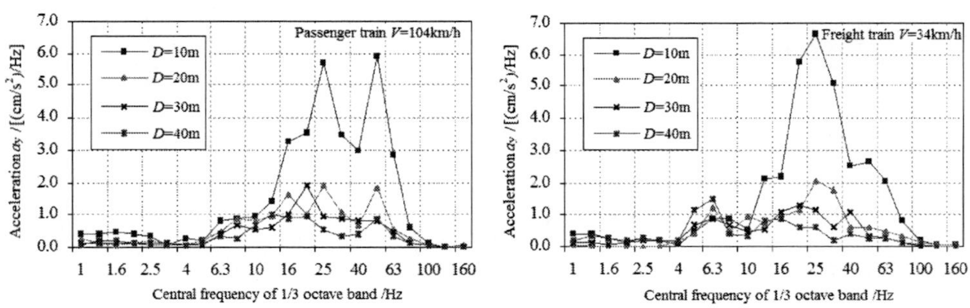

Figure 22. One-third octave spectrum of ground acceleration in S_y-direction.

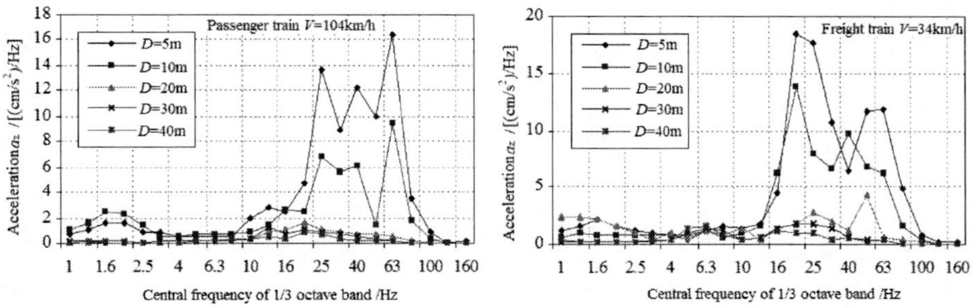

Figure 23. One-third octave spectrum of ground acceleration in S_z-direction.

(3) Variation of ground vibration levels with distance

Shown in Figures 27 to 29 are, respectively, the distributions of vibration levels of ground acceleration with distance of receiver point to the track in three directions. It can be seen that the vertical vibration levels of the ground range from 60 dB to 110 dB in the speed of 21−128 km/h. In addition, the vibration in S_x-direction parallel to the horizontal track ranges from 50 dB to 95 dB, and that in S_y-direction perpendicular to the track ranges from 55 dB to 80 dB. The vibration levels of ground in three directions have the attenuation trend with the distance, simultaneously showing some fluctuation.

4.2.2. Vibrations of Building Floors

Figures 30 to 31 give the typical velocity histories of building floors when the passenger train passes through. It is obvious that for the multi-story building, the velocity of top floor is greater than that of ground floor, in both vertical vibration and horizontal vibration.

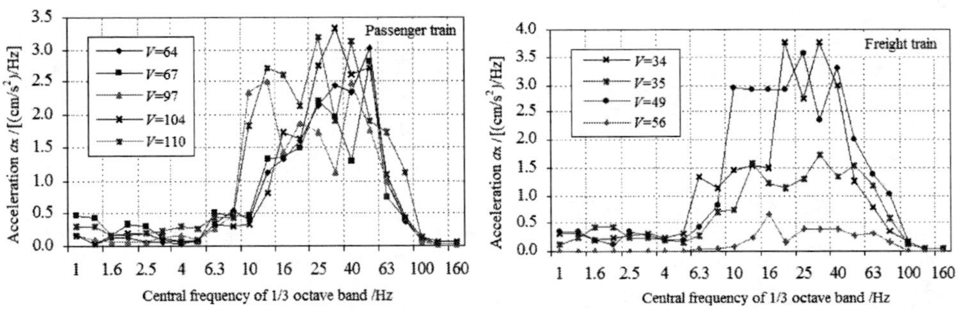

Figure 24. One-third octave spectrum of ground acceleration in S_x-direction (D=10m).

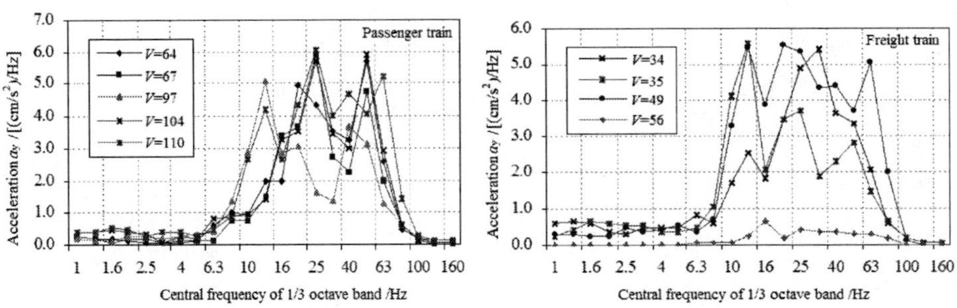

Figure 25. One-third octave spectrum of ground acceleration in S_y-direction (D=10m).

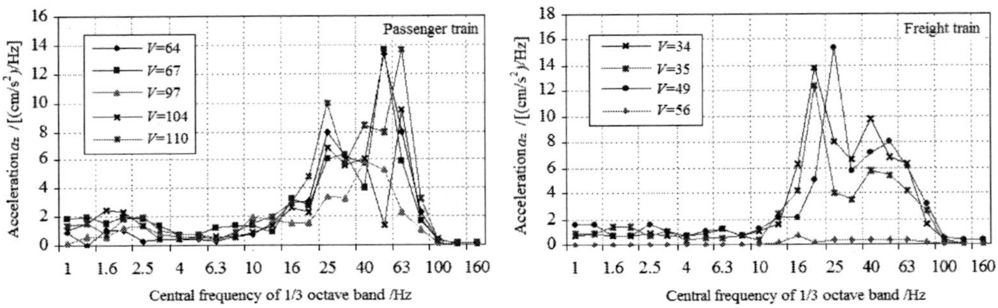

Figure 26. One-third octave spectrum of ground acceleration in S_z-direction (D=10m).

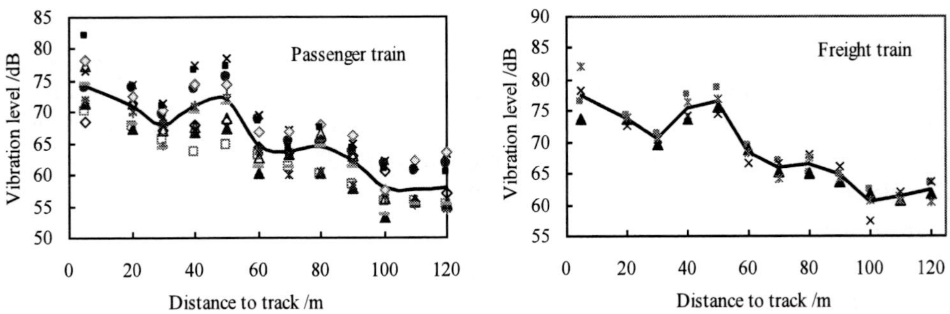

Figure 27. Variation of ground vibration level with distance (S_x-direction).

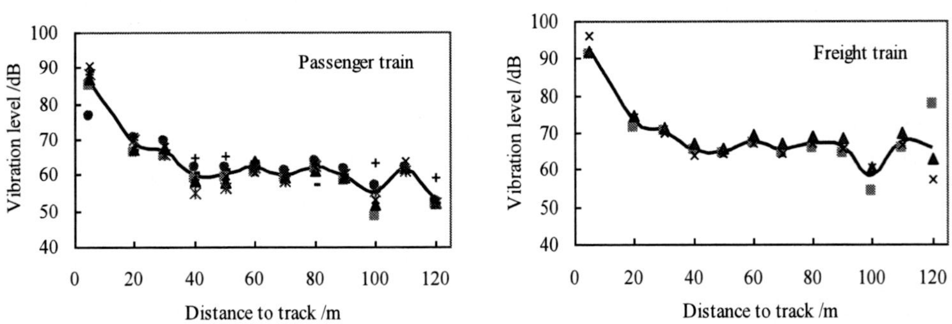

Figure 28. Variation of ground vibration level with distance (S_y-direction).

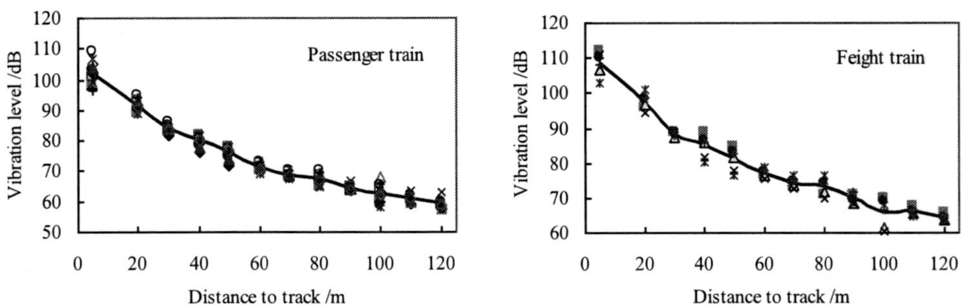

Figure 29. Variation of ground vibration level with distance (S_z-direction).

In order to observe the vibration characteristics of all floors, the vertical vibration levels of floor with number of floor are plotted in Figure 32, in which $L_v=20\times\lg(v/v_{ref})$ (dB) with $v_{ref}=2.54\times10^{-8}$ m/s. It can be seen that under the measured train speeds, the vertical vibration levels of floor velocity caused by passenger train range from 60 dB to 75 dB, and those caused by freight trains range from 70 dB to 81 dB. Moreover, for each train speed, the vertical vibration levels of floors change little with height of building, and the level of top floor is 2−5 dB larger than that of ground floor. In addition, the vibration levels of floors increase with the train speed, and increased by about 12 dB when the passenger train speed increases from 46 km/h to 99 km/h, and by about 10 dB when the freight train speed increases from 26 km/h to 57 km/h, respectively.

Different from the ground vibration, the horizontal vibration of building floor is not always smaller than the vertical vibration, as shown in Figure 33, in which the solid lines are fitted curves based on discrete experimental data. One can see that with the increase of train speed, the vibration levels in three directions increase. In addition, the horizontal vibration level of building floor in the direction (S_x) parallel to track is larger than that in the direction (S_y) perpendicular to line direction: for passenger trains, the maximum and average differences are respectively 6.9 dB and 4.5 dB, while for freight trains the values are 10.4 dB and 6.9 dB. For this phenomenon, it is mainly because of the stiffness difference of the building in the directions parallel to the track and perpendicular to the track.

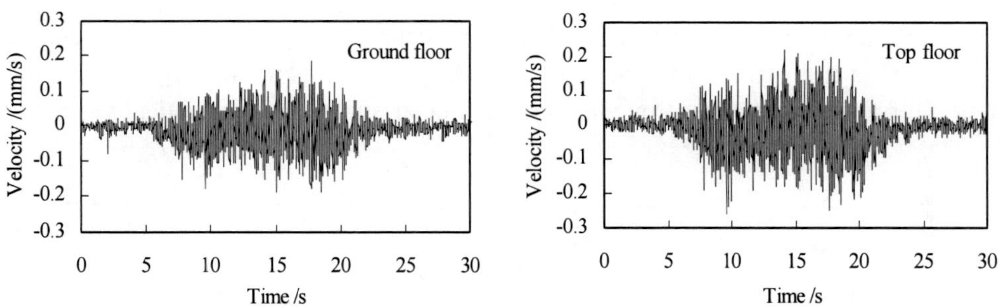

Figure 30. Histories of building floor vibrations in S_z-direction (passenger train, V = 91 km/h).

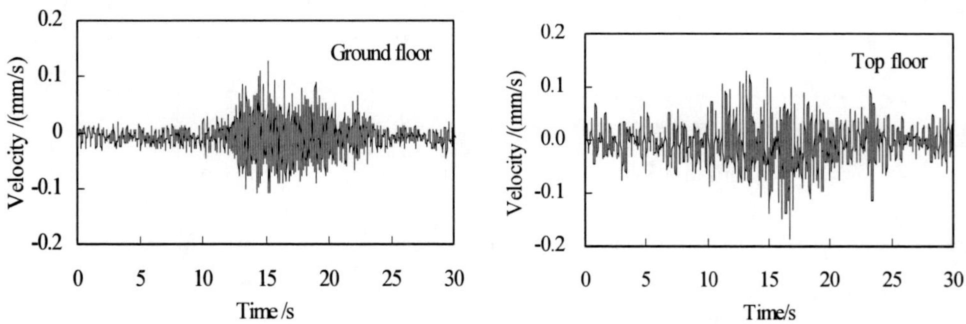

Figure 31. Histories of building floor vibrations in S_x-direction (passenger train, V = 115 km/h).

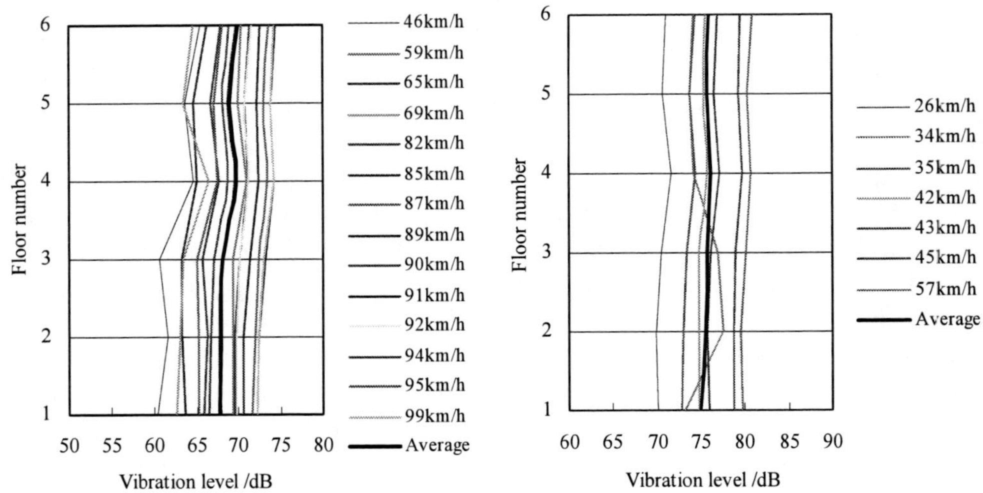

Figure 32. Vertical vibration levels of building floors under different train speeds: (left) passenger train and (right) freight train.

5. Prediction Formulas of Environmental Vibrations

The dynamic interaction models established in Section 2 and Section 3 of this chapter are used to calculate and predict the ground and building vibrations induced by railway, and the prediction formulas are proposed by means of parameter analysis and statistical regression.

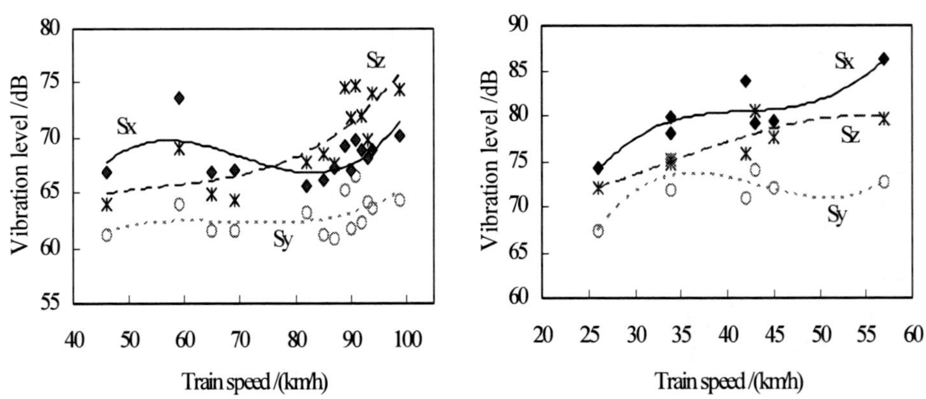

Figure 33. Comparison of the vibrations in three directions for the multi-story building.

5.1. Prediction Formula of Ground Vibration

Taking the train speed, the characteristics of track and soil and the axle-load of train into consideration, a prediction formula of free-field vibration due to railway trains running on ground, which can be easily applied in practical engineering, is derived as follows:

$$\mathrm{VL_g} = f(r) + X_{\mathrm{V}} + X_{\mathrm{A}} + X_{\Lambda} + X_{\mathrm{Aw}} + X_{\mathrm{T}} \tag{78}$$

where $\mathrm{VL_g}$ is the predicted level of ground vibration (dB); r is the distance between the receiver point and vibration source (m); $f(r)=A_1+A_2\times\ln(r)$ represents the base curve of attenuation of ground vibration with distance; $A_1=121.6+e^{0.5T}$ and $A_2=7.2+29.2T$ are two coefficients determined by properties of subsoil, which are functions of predominant period $T(s)$ of soil; $X_V=3.57+0.001r\times(V/70-1)$ is the adjusted value of vibration level considering train speed, and V is train speed (km/h); $X_A=6.15e^{-0.04r}\times(2.1+\ln A)$ is the adjusted value of vibration level considering amplitude A(mm) of track irregularity; $X_\Lambda=(35.14-r)\times(0.015\lambda-0.018)$ is the adjusted value of vibration level considering wavelength λ(m) of track irregularity; $X_{Aw}=0.09+0.01A_w$ is the adjusted value of vibration level considering passenger capacity A_w (number of passengers); X_T is the adjusted value of vibration level considering train composition: for a train composed of 6 cars, $X_T=0$, and for a train composed of 8 cars, $X_T=3$.

5.2. Experimental Verification of Prediction Formula on Ground Vibration

The prediction formula on ground vibration is herein verified by using the measured results induced by a passenger train and a freight train in the experiment of Section 4. By inputting the train parameters, corresponding speed and subsoil parameters into Equation (78), the vibration levels of vertical ground accelerations over a range of 150 m from the track are calculated. Both the predicted and the measured results are illustrated in Figure 34, showing the predicted values are in good agreement with the measured ones.

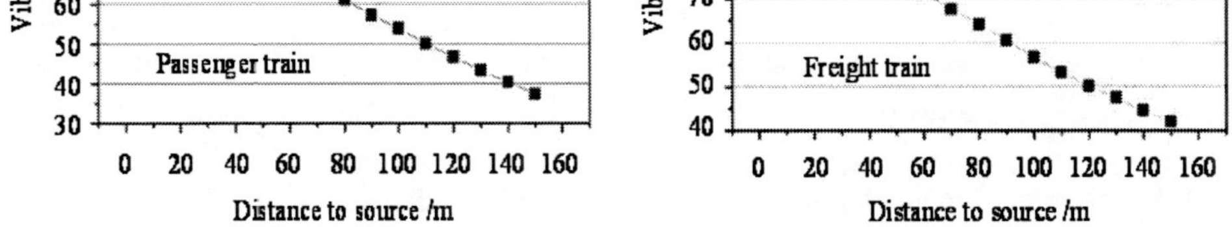

Figure 34. Verification of prediction model of train induced free-field vibration.

Figure 34 also shows the results calculated by other prediction formulas proposed by Bornitz [25] and Mao [26]. Compared with the two formulas, Equation (78) proposed in this chapter has a better precision not only in amplitudes but also in attenuation trend, for it considers more influencing factors.

5.3. Prediction Formula of Building Vibration

The vibration of buildings is very complex, so research on prediction formula of railway induced building vibrations is less than those of ground vibrations. Herein, a prediction formula of building vibration near railway track is proposed, based on the results of parameter analysis and field tests, which can be expressed as:

$$\mathrm{VL_B}(r) = \mathrm{VL_g}(r) + Y_{\mathrm{gb}} + Y_{\mathrm{bt}} + Y_{\mathrm{fh}} \tag{79}$$

where $\mathrm{VL_B}$ is the predicted vibration level of building (dB); $\mathrm{VL_g}(r)$ is the vibration level of free-field at the position of predicted building, which can be calculated by Equation (78); $Y_{\mathrm{gb}}=23.34-4.51\times\ln r$ (dB) is the coupled term of building vibration and free-field vibration; Y_{bt} is the adjusted value for structural type of building (dB): $Y_{\mathrm{bt}}=-4.5$ dB for a frame building with a good foundation (e.g. a high-rise building), $Y_{\mathrm{bt}}=0$ for a masonry-and-concrete building with an ordinary foundation (e.g. a multi-story building), and $Y_{\mathrm{bt}}=3.5$ dB for a low building with a poor foundation; $Y_{\mathrm{fh}}=0.04\alpha(n-1)$ is the adjusted value for vibration level for the n-th floor to the 1st floor; α is the coefficient considering structure of building: $\alpha=1$ for a frame building with a good foundation and $\alpha=2$ for a masonry-and-concrete building with an ordinary foundation.

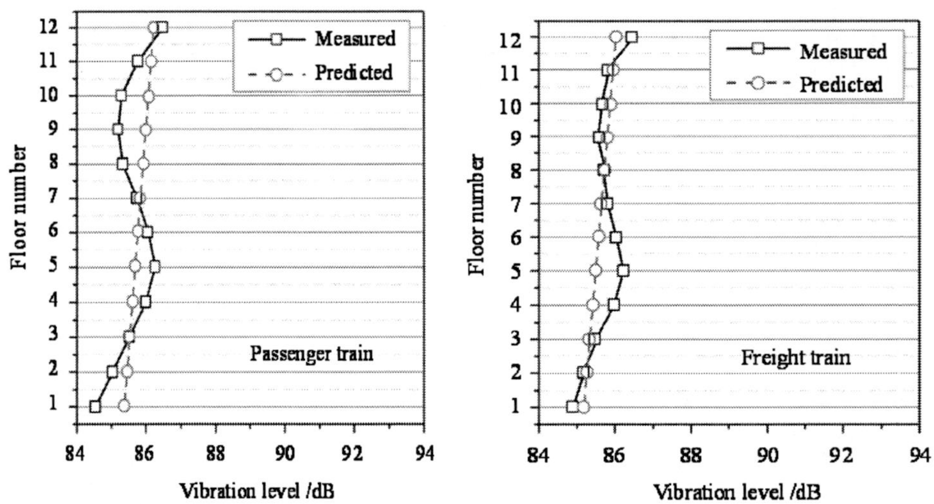

Figure 35. Verification of prediction formula of train induced building vibrations.

5.4. Experimental Verification of Prediction Formula on Building Vibration

The prediction formula on building vibrations due to ground trains is herein verified by the measured results of a 12-story RC frame structure [20]. The vibration levels of the building induced by a passenger train (1 locomotive + 17 cars, V=40 km/h) and a freight train (1 locomotive + 54 empty wagons, V=40 km/h) are predicted by Equation (79), using the actual parameters of trains, track, subsoil and building structures. The predicted vertical vibration levels of floor acceleration and the measured ones are shown in Figure 35. It can be seen that the predicted results are in good agreement with the measured data. Although the predicted vertical levels increase linearly with the floor height, slightly different from the measured ones, it does not affect the prediction of building vibration levels using Equation (79), which indicates that the proposed prediction formula can be used to predict the vibration of buildings in the primary planning phase of a railway.

CONCLUSION

In this chapter, the environmental vibrations induced by railway trains running on ground are studied by establishing the dynamic interaction model of train-track-subsoil and roadbed-subsoil-building, and by field experiments.

The train-track-subsoil model is based on vehicle dynamics, track dynamics and the Green's functions of subsoil, in which are taken into account the influences of track structure and wave motion in the 3D layered-soil on ground vibrations. The characteristics of ground vibrations are analyzed by case studies and experimental results. The results show that the ground vibrations have a close relationship with train speed and soil properties; the dynamic responses excited by track irregularity have big influence on the high-frequency components of ground vibration. With increase of distance to the track, the ground acceleration has the tendency of decrease, and the relevance of acceleration curves and train excitation becomes less obvious. The intersections of moving load speed-lines and subsoil dispersion curves are some resonance frequencies that cause the amplification of ground vibrations. There exists a critical speed for moving train, which is close to the minimum speed of the Rayleigh's wave in the subsoil. Furthermore, the roadbed-subsoil-building FE model is established to predict building vibration, in which the model meshing principles, boundary conditions and damping effect are discussed. Some characteristics of building vibrations are studied by field experiment: the horizontal vibrations of building floors are not always smaller than the vertical ones; the train speed, the type of building structure and foundation, and property of subsoil have great influence on train-induced building vibrations. Based on the numerical simulation results, the prediction formulae for free-field ground and building vibration levels are proposed and validated by experimental results, which has good applicability in practical engineering and can be used to predict the ground vibration and building vibration in the primary planning phase of a railway line.

ACKNOWLEDGMENTS

This study is sponsored by the National Natural Science Foundation of China (Grant No. 51108023), the Ph.D. Programs Foundation of Ministry of Education of China (20090009120021), and the Fundamental Research Fund of Beijing Jiaotong University (2011JBM088).

REFERENCES

[1] Krylov V.V., Ferguson C. (1994). Calculation of low-frequency ground vibrations from railway trains. *Applied Acoustics*, 42: 199–213.

[2] Krylov V.V. (1995). Generation of ground vibration by superfast trains. *Applied Acoustics*, 44: 149–164.

[3] Krylov V.V. (1998). Effects of track properties on ground vibrations generated by high-speed trains. *ACUSTICA· acta acustica*, 84: 78–90.

[4] Jones C. J. C., Block J. R. (1996). Prediction of ground vibration from freight trains. *Journal of Sound and Vibration*, 193(1): 205–213.

[5] Madshus C., Kaynia A.M. (2000). High-speed railway lines on soft ground: dynamic behavior at critical train speed. *Journal of Sound and Vibration*, 231(3): 689–701.

[6] Kaynia A.M., Madshus C., Harvik L., et al. (2000). Ground vibration from high-speed trains: prediction and countermeasure. *Journal of Geotechnical and Geoenvironmental Engineering, American Society of Civil Engineers*, 120 (6): 531–537.

[7] Karlström A., Boström A. (2006). An analytical model for train-induced ground vibrations from railways. *Journal of Sound and Vibration*, 292: 221–241.

[8] Lombaert G., Degrande G. (2009). Ground-borne vibration due to static and dynamic axle loads of InterCity and high-speed trains. *Journal of Sound and Vibration*, 319(3–5): 1036–1066.

[9] Sheng X., Jones C.J.C., Petyt M. (1999). Ground vibration generated by a load moving along a railway track. *Journal of Sound and Vibration*, 228(1): 129–156.

[10] Sheng X., Jones C.J.C., Thompson D.J. (2004). A theoretical model for ground vibration from trains generated by vertical track irregularities. *Journal of Sound and Vibration*, 272: 937–965.

[11] Yang Y.B. (1995). *Analysis of Soil Vibrations induced by High-speed Train*. Taipei: Taiwan University.

[12] Yang Y.B. (2003). Train-induced wave propagation in layered soils using finite/infinite element simulation. *Soil Dynamics and Earthquake Engineering*, 23: 263–278.

[13] Auersch L. (2005). The excitation of ground vibration by rail traffic: theory of vehicle–track–soil interaction and measurements of high-speed lines. *Journal of Sound and Vibration*, 284: 103–132.

[14] Galvín P., Domínguez J. (2007). Analysis of ground motion due to moving surface loads induced by high-speed trains. *Engineering Analysis with Boundary Elements*, 31: 931–941.

[15] Takemiya H. (2001). Ground vibrations alongside tracks induced by high-speed trans: prediction and mitigation, in Krylov V. V. (Ed.), *Noise and Vibration from High-speed Trains*, Thomas Telford, London, 347–393.

[16] Takemiya H. (2003). Simulation of track-ground vibrations due to a high-speed train: the case of X-2000 at Ledsgard. *Journal of Sound and Vibration*, 261: 503–526.

[17] Masafumi K., Toshifumi M., Osamu Y., et al. (2008). Numerical simulation study of ground vibrations using forces from wheels of a running high-speed train. *Journal of Sound and Vibration*, 318: 830–849.

[18] Knothe K.L., Grassie S.L. (1993). Modeling of railway track and vehicle/track interaction at high frequencies. *Vehicle System Dynamics*, 22: 209–262.

[19] Xia H., Zhang N. (2005). *Dynamic Interaction of Vehicle and Structure*. Beijing: Science Publishing House.

[20] Cao Y.M. (2006). *Theoretical and Experimental Study on Train-induced Vibrations of Free Field and Buildings*. Beijing, Beijing Jiaotong University.

[21] Xia H., Zhang N., Cao Y.M. (2005). Experimental study of train-induced vibrations of environments and buildings. *Journal of Sound and Vibration*, 280: 1017–1029.

[22] Lombaert G. (2001). *Development and experimental validation of a numerical model for the free vibrations induced by road traffic*. Belgium: Leuven University.

[23] Guan H.M., Liao Z.P. (1994). A comparison between the discrete local transmitting boundary and the superposition boundary in wave propagation. *ACTA Mechanica Sinica*, 26(3): 303–311.

[24] Liu J.B., Wang Z.Y., Du X.L., et al. (2005). Three-dimensional visco-elastic artificial boundary in time domain of wave problem. *Engineering Mechanics*, 22(6): 46–51.

[25] Bornitz G. (1931). *Uber die Ausbreitung der von Groszkolbenmashinen erzeugten Bodenschwing-ungen in die Tiefe.* Berlin : Springer.

[26] Mao Y.Q. (1987). Characteristics and attenuation of ground vibration induced by railway train. *Journal of Building Structures*, 1: 67–77.

In: Traffic Induced Environmental Vibrations and Controls ISBN: 978-1-62618-240-0
Editors: H. Xia and R. Calçada

Chapter 11

ENVIRONMENTAL VIBRATION INDUCED BY ELEVATED RAIL TRANSIT

Weiwei Guo**, Tian Zhang, Jianguo Chen and Huile Li
School of Civil Engineering, Beijing Jiaotong University, Beijing, China

ABSTRACT

To investigate the environmental vibration caused by the elevated rail transit, a 3D interaction model of coupled train-bridge-ground-building vibration system is established, which is composed of the train-track-bridge submodel and the pier-foundation-soil-building submodel. The two submodels interact with each other through the contact forces on the pier tops of the viaduct. A field experiment was carried out at the elevated trial section of the Beijing Metro Line 5, where the ladder track is installed, to analyze the ground vibration induced by urban rail transit. The vibration responses were measured at rails, sleepers, bridge deck as well as the surrounding ground surface when the train running on the ladder track and the common slab track. A prediction formula for ground vibration induced by trains running on elevated bridges is established, based on the statistical regression analysis of the calculation results and verified by measured data.

1. INTRODUCTION

With the rapid development of high-speed railway and large-scale construction of urban rail transit, elevated structures are used more and more in rail transit lines as a major form. According to statistics, in China, more than 70 percent of the support structures in the high-speed railways adopt elevated bridges [1, 2]. The emergence and development of elevated rail transit has provided a rapid and convenient traffic media, while at the same time, caused some new problems. The elevated rail transit systems run in urban regions, and often through city downtowns and resident areas. They may induce serious environmental vibrations and have become a problem that should be considered in planning and designing elevated rail transit

* Corresponding author; Tel: (86) 10-51683786; Fax: (86) 10-51683764; E-mail: wwguo@bjtu.edu.cn.

systems. There have been some lessons to learn from the construction of the Beijing metro line, Shanghai metro line, Japan Shinkansen and several other ones, where the vibration problem was not fully concerned during their construction, and has led to technical difficulty and higher cost for the reform engineering during the operation stage.

Researches have been done on the environmental vibration caused by trains on elevated bridges by field experiments and numerical simulations, and a lot of results and meaningful conclusions have been presented [3−11]. Takemiya [12] carried out a field measurement on the Shinkansen-train induced ground vibrations from a viaduct and made a comparison of measured and computed responses of the ground near the pier. Yang et al. [13] investigated the vibration of a simple beam subjected to the train passage, in which the train is modeled as two rows of moving concentrated load series, the bridge as an elastic support beam, and the ground as the semi-infinite body. Jiang and Zhang [14] analyzed the ground vibrations induced by the elevated rail transit by simplifying the whole system as a 2D train-bridge subsystem and a 3D frame-foundation-soil subsystem.

Most of the studies were carried out for bridges with common tracks, whereas little research has been done for bridges with vibration mitigation tracks. Many kinds of vibration reduction measures have been studied by designers and researchers in China and abroad. The floating ladder track, as a new reduction track that has been well used in Japan and America, is one of them. Theoretical analyses and engineering applications in Japan proved that the ladder track, with light weight, sufficient and effective elasticity, low maintenance and low cost properties, is an ideal track system that can effectively reduce vibration and noise of the track while keeping good train running safety and stability [15−19].

The propagation of environmental vibrations induced by rail transit along the distance has the tendency of gradual attenuation. With continuous efforts of scholars at home and abroad, considerable progress has been made in the prediction of environmental vibrations induced by the elevated rail transit [14, 20−24]. The conclusion was given by the Japanese Railway Technical Research Institute that the accelerations at different measurement points on the bridge or the nearby ground increase with the train speed. In addition, the Kyoto Institute for Environmental Pollutions and the Kawasaki Institute for Environmental Pollutions in Japan proposed, respectively, the prediction formulae for the environmental vibrations induced by elevated rail transit. In China, Xie et al. [25] and Zhang et al. [26] suggested the statistic regression prediction formulae for the ground vibration induced by the elevated rail transit based on the measurements from the Wuhan rail transit Line 1 and the Shanghai Mingzhu Line, respectively. In the above-mentioned prediction formulae, however, the factors such as the property of in-situ soil layers, the types of foundations, and the structural form of bridges were not considered. Therefore, these formulae can hardly be applied directly to predict the environmental vibrations induced by urban rail transit. In this chapter, a prediction formula which can fully reflect the influence factors of free field vibrations by the elevated rail transit is established through statistical regression analysis in combination with theoretical analysis, numerical simulation and field test results.

2. VIBRATION SOURCES OF ELEVATED RAIL TRANSIT SYSTEM

In the elevated rail transit system, trains travel on the bridge at a certain speed, producing dynamic impact on the bridge and induce structural vibration. Bridge vibration extends to the piers through bearings, and furthermore to the ground soil further via piers and foundations, which is thought to be one of sources of environmental vibration (see Figure 1).

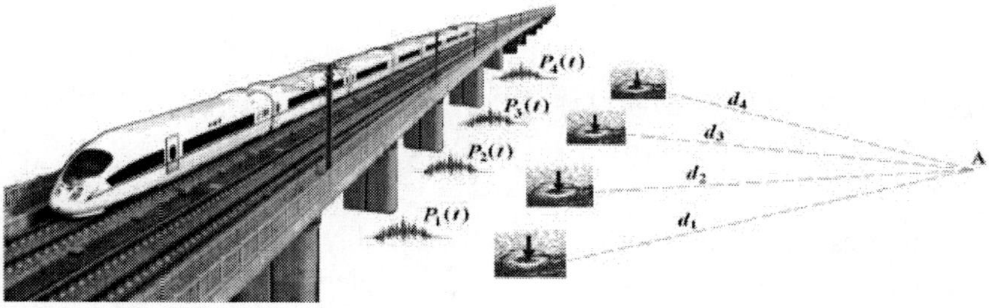

Figure 1. Illustration of environmental vibration source in elevated rail transit system.

When a train travels on a viaduct, the vibrations of the train and the viaduct inter-act on and inter-affect each other, forming a complex train-bridge dynamic interaction system. Figure 2 shows the self-excitation sources and external ones that cause vibrations of train-bridge system.

- Excitations of train-bridge system
  - External excitations
    - Wind loads
    - Seismic actions
    - Collision foces
    - Moving gravity of train
  - Self-excitations
    - Track irregularities
    - Wheel hunting movement
    - Eccentricity of wheel-set
    - Unevenness of wheel treads

Figure 2. Excitations of train-bridge dynamic interaction system.

The related influence factors of train-bridge system can be summarized as follows:

(1) Train speeds. Vehicles' running on the bridge at a certain speed forms the moving gravity loading on bridge structures. The dynamic effects are related to the composition and running speed of the train, the axle-load and axle distances of the vehicles.

(2) Track irregularities. Track geometry variations or irregularities form a random excitation source of self-excited forced vibration of train-bridge system when vehicles passing a bridge at a certain speed.

(3) Dynamic interaction between train vehicle and bridge structure. It consists of the dynamic effects of bridge and track deformation and vibration on running trains, the resonance appeared at a certain train speed, etc. The resonance of train-bridge system is associated with train speeds, natural frequencies of train and bridge, spring and damping characteristics of bogie and wheel hunting movement.

(4) Hunting movement of vehicle wheels. Due to the wheel with reversed conic wheel tread running on the straight track, and the gap between rim and the side surface of rail, a hunting movement appears when the wheel runs on the rail. It is one of the main factors inducing the lateral vibration of the train-bridge system.
(5) Centrifugal force of vehicles. When a train runs on a curved bridge, the centrifugal forces of moving vehicles form lateral moving loads, inducing lateral and torsional vibration of the system.

3. Analytical Model of Train-Bridge-Environment Vibration System

When a train runs on a railway bridge, it will inevitably cause the bridge to vibrate. The bridge vibrations will propagate to the base through the piers and foundations, and further to the ground soil, ending up with vibrations of ground and surrounding buildings. This process leads to a complicated system. Hence, the environment vibration caused by elevated rail transit is usually analyzed through establishing the systematic train-bridge-ground-building dynamic model.

It is difficult to simulate the behavior of the train traveling in a time-varying system with an integral spatial model. A commonly-used method is dividing the train-bridge-environment vibration system into two subsystems, namely, the train-track-bridge interaction system and the pier-foundation-soil-building model.

3.1. Analytical Model of Train-Track-Bridge Interaction System

A 2D model representing the train-track-bridge interactions is usually adopted in the environmental vibration research. The dynamic loads of running train vehicles on the pier-foundation-soil-building system can be obtained by simulation with the real parameters of the train and the bridge.

3.1.1. Analytical Model of Moving Vehicle Load-track-bridge System

Figure 3 gives a vibration source model for the elevated rail transit system, where the interactions of rail-ballast-beam-bearing-pier are taken into consideration.

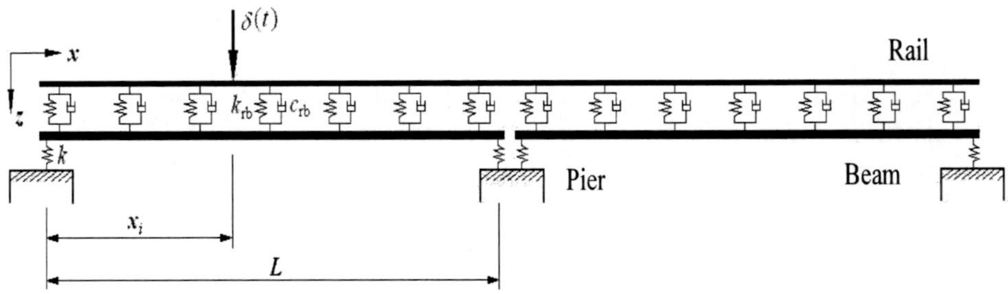

Figure 3. Analytical model of moving vehicle load-track-bridge system.

The following assumptions are adopted in the model:

(1) The rails are considered as elastically-supported Euler beams with indefinite length.
(2) The bridge is thought to be an Euler beam, elastically-supported at span-ends with the stiffness coefficient of k.
(3) The stiffness and damping between rails and track bed are modeled by spring-damper, with the stiffness and damping coefficients being k_{rb} and c_{rb}, respectively.
(4) The pier-tops are assumed as fixed supports, for the displacements of them are too small to affect the forces transferred from the superstructure.

The pier-top forces induced by a train passage are solved by the following procedures: first, the dynamic response of the beam subjected to a pulse load $\delta(t)$ at the location of x_i on the bridge is solved, and then the bearing reaction forces and the bridge responses under moving loads are obtained.

3.1.1.1. Dynamic Responses of Beam Subjected to a Pulse Load $\delta(t)$

The motion equation for the elastically-supported beam with elastic bearings subjected to a pulse load $\delta(t)$ at the location of x_i (Ref. Figure 3) can be expressed as:

$$E^* I \frac{\partial^4 w_b(x,t)}{\partial x^4} + \bar{m}_b \frac{\partial^2 w_b(x,t)}{\partial t^2} = \delta(t)\delta(x - x_i) \tag{1}$$

where, $w_b(x, t)$ denotes the vertical displacement of the bridge at position x and time t; $E^* = E(1+2\xi_b i)$ is the complex elastic modulus when the internal damping of the bridge is taken into account; ξ_b is the damping ratio of the bridge, I the inertia moment of beam cross section, $\bar{m}_b$ the mass density of the bridge in unit length, and x_i the initial position of load $\delta(t)$.

The following boundary conditions are introduced for the beam with elastic bearings:

$$E^* I \frac{\partial^2 w_b(x,t)}{\partial x^2}\Big|_{x=0} = 0 \; ; \; E^* I \frac{\partial^2 w_b(x,t)}{\partial x^2}\Big|_{x=L} = 0 \tag{2a}$$

$$E^* I \frac{\partial^3 w_b(x,t)}{\partial x^3}\Big|_{x=0} = -k w_b(0,t) \; ; \; E^* I \frac{\partial^3 w_b(x,t)}{\partial x^3}\Big|_{x=L} = -k w_b(L,t) \tag{2b}$$

and the initial conditions of the beam are:

$$w_b(x,0) = \dot{w}_b(x,0) = 0 \tag{3}$$

The partial differential Equation (1) can be solved by the modal deposition method, in which the geometry coordinate of the structure is transformed into the generalized modal coordinate. For a continuous structure in single dimension, the transform is expressed as:

$$w_b(x,t)=\sum_{i=1}^{n} q_i(t)\cdot\varphi_i(x) \tag{4}$$

where $q_i(t)$ is the generalized coordinate with respect to time t; $\varphi_i(x)$ is the mode-shape function.

When the first two modes are considered, namely, n=2, the expressions of $\varphi_1(x)$ and $\varphi_2(x)$ can be obtained by combining with the boundary and initial conditions as:

$$\varphi_1(x)=\frac{\sin(\pi x/L)+\kappa}{1+\kappa} \tag{5}$$

$$\varphi_2(x)=\frac{\sin(2\pi x/L)-2\kappa x/L+\kappa}{1+\kappa} \tag{6}$$

where $\kappa = E^* I\pi^3/kL^3$ indicates the stiffness ratio between the beam and the elastic bearing. It is noticeable that $\varphi_1(x)$ and $\varphi_2(x)$ are orthogonal in terms of $\int_0^L \varphi_1(x)\varphi_2(x)\mathrm{d}x=0$.

Substituting Equations (4) to (6) into (1) and using the orthogonal condition, the vibration equation of the beam can be simplified as the differential equations about $q_1(t)$ and $q_2(t)$:

$$g_{11}q_1(t)+g_{12}\ddot{q}_1(t)=\delta(t)\varphi_1(x_i) \tag{7}$$

$$g_{21}q_2(t)+g_{22}\ddot{q}_2(t)=\delta(t)\varphi_2(x_i) \tag{8}$$

where g_{11}, g_{12}, g_{21}, g_{22} are expressions of known parameters, representing the coefficient of the differential equation.

Further, the Fourier transform $\overline{F}(\omega)=\int F(t)e^{-i\omega t}\mathrm{d}t$ is conducted with respect to time t, which favors solving the problem in frequency domain. The expressions of $\overline{q}_1(\omega)$ and $\overline{q}_2(\omega)$ can be obtained through the Fourier transform as:

$$\overline{q}_1(\omega)=\frac{2L^3(1+\kappa)^2\cdot\varphi_1(x_i)}{mL^4[\pi(2\kappa^2+1)+8\kappa]\omega^2+E^* I\pi^4(\pi+4\kappa)} \tag{9}$$

$$\overline{q}_2(\omega)=\frac{6L^3(1+\kappa)^2\pi\cdot\varphi_2(x_i)}{L^4(3\pi+12\kappa+2\kappa^2\pi)\omega^2+48E^* I\pi^4(\pi+2\kappa)} \tag{10}$$

So far, the displacement of bridge in the frequency domain can be expressed as:

$$\overline{w}_b(x,\omega)=q_1(\omega)\varphi_1(x)+q_2(\omega)\varphi_2(x) \tag{11}$$

Substituting Equations (5), (6), (9) and (10) into (11), the dynamic response of the bridge in the frequency domain can be obtained when the elastically-supported beam with elastic bearings is subjected to a pulse load $\delta(t)$ at location x_i.

3.1.1.2. Displacements of Rails and Rail-Beam Interaction Forces Induced by Moving Load

The displacement of the rail and the rail-beam interaction forces are illustrated in Figure 4, which are induced by the moving axle-load passing the tracks on the bridge.

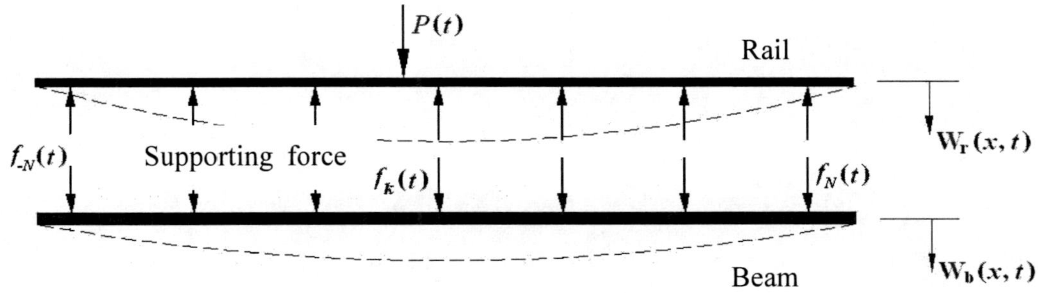

Figure 4. Displacement of the rail and the rail-beam interaction forces.

When a train travels on an Euler beam with indefinite length, the rail displacement satisfies:

$$\overline{\mathbf{W}}_{\mathrm{r}}^{\delta} = \overline{\mathbf{B}}_{\mathrm{rs}} - \overline{\mathbf{B}}_{\mathrm{r}} \cdot \bar{\mathbf{f}}^{\delta} \tag{12}$$

$$\overline{\mathbf{W}}_{\mathrm{r}} = A(k_x,\omega)\mathbf{E}_s - \overline{\mathbf{B}}_{\mathrm{r}} \cdot \bar{\mathbf{f}} \tag{13}$$

where $A(k_x,\omega) = \frac{1}{V}\tilde{\overline{B}}_{\mathrm{r}}(k_x,\omega)\overline{C}(\omega)$, and $\tilde{\overline{B}}_{\mathrm{r}}(k_x,\omega)$ is the Green's Function in the wavenumber-frequency domain; vector $\overline{\mathbf{B}}_{\mathrm{rs}} = \left\{\overline{B}_{\mathrm{r}}(x_{-N},\omega) \ \ldots \ \overline{B}_{\mathrm{r}}(x_0,\omega) \ \cdots \ \overline{B}_{\mathrm{r}}(x_N,\omega)\right\}^{\mathrm{T}}$ is the Green's Function of the first type of vertical displacement of rails in frequency domain; vector $\overline{\mathbf{B}}_{\mathrm{r}}$ is a $(2N+1)\times(2N+1)$ matrix and expressed as:

$$\overline{\mathbf{B}}_{\mathrm{r}} = \begin{bmatrix} \overline{B}_{\mathrm{r}}^{(-N)(-N)} & \cdots & \overline{B}_{\mathrm{r}}^{(-N)0} & \cdots & \overline{B}_{\mathrm{r}}^{(-N)N} \\ & \ddots & \vdots & \vdots & \vdots \\ & & \overline{B}_{\mathrm{r}}^{00} & \cdots & \overline{B}_{\mathrm{r}}^{0N} \\ \text{sym} & & & \ddots & \vdots \\ & & & & \overline{B}_{\mathrm{r}}^{NN} \end{bmatrix} \tag{14}$$

where $\overline{B}_{\mathrm{r}}^{qm}$ denotes the element $\overline{B}_{\mathrm{r}}(x_q - x_m,\omega)$.

The force vector between rails and bridge deck can be expressed as:

$$\bar{\mathbf{f}}=(k_{\text{rb}}+i\omega c_{\text{rb}})(\overline{\mathbf{W}}_{\text{r}}-\overline{\mathbf{W}}_{\text{b}}) \tag{15}$$

where $\bar{\mathbf{f}}$, $\overline{\mathbf{W}}_{\text{r}}$ and $\overline{\mathbf{W}}_{\text{b}}$ are, respectively, column vectors consisting of forces between rails and bridge deck, vertical displacement of rails and vertical displacement of bridge, at the location of 2*N*+1 sleepers under all axle-loads of the moving train, expressed as:

$$\bar{\mathbf{f}}=\{\bar{f}_{-N}^{\text{final}}(\omega) \quad \cdots \quad \bar{f}_{0}^{\text{final}}(\omega) \quad \cdots \quad \bar{f}_{N}^{\text{final}}(\omega)\}^{\text{T}} \tag{16a}$$

$$\overline{\mathbf{W}}_{\text{r}}=\{\bar{w}_{\text{r}}(x_{-N},\omega) \quad \cdots \quad \bar{w}_{\text{r}}(x_{0},\omega) \quad \cdots \quad \bar{w}_{\text{r}}(x_{N},\omega)\}^{\text{T}} \tag{16b}$$

$$\overline{\mathbf{W}}_{\text{b}}=\{\bar{w}_{\text{b}}(x_{-N},\omega) \quad \cdots \quad \bar{w}_{\text{b}}(x_{0},\omega) \quad \cdots \quad \bar{w}_{\text{b}}(x_{N},\omega)\}^{\text{T}} \tag{16c}$$

The vectors $\bar{\mathbf{f}}^{\delta}$ and $\overline{\mathbf{W}}_{\text{r}}^{\delta}$ represent, respectively, the force and vertical displacement vectors of rails at the location of 2*N*+1 sleepers under the pulse load $\delta(t)$.

3.1.1.3. Vertical Displacement of Bridge under All Axle-loads of Moving Train

According to the expression of bridge displacement caused by the impulse load acting on the beam, the bridge displacement under all axle-loads of the moving train can be obtained. When the *m*-th force acting on the bridge is expressed as $f_m(t)\delta(x-x_m)$, the bridge displacement under all *N* axle-loads can be written as

$$\bar{w}_{\text{b}}(x,\omega)=\sum_{m=-N}^{N}\bar{f}_{m}^{\text{final}}(\omega)w_{\text{b}}^{\delta}(x-x_{m},\omega) \tag{17}$$

When $x=x_q=q\cdot d$ (*d* is the interval between two adjacent sleepers), $\bar{w}_{\text{b}}(x_q,\omega)$ indicates the bridge displacement in the location of the *q*-th sleeper induced by all moving axle-loads. It can be expressed as:

$$\bar{w}_{\text{b}}(x_q,\omega)=\sum_{m=-N}^{N}\bar{f}_{m}^{\text{final}}(\omega)w_{\text{b}}^{\delta}(x_q-x_{m},\omega) \tag{18}$$

Since the bridge displacement at each sleeper location meets Equations (18) and (19), $\overline{\mathbf{W}}_{\text{b}}$ can be obtained by combining all the bridge displacements at 2*N*+1 sleeper locations as:

$$\overline{\mathbf{W}}_{\text{b}}=\overline{\mathbf{B}}_{\text{b}}\bar{\mathbf{f}} \tag{19}$$

where $\overline{\mathbf{B}}_{\text{b}}$ is a (2*N*+1)×(2*N*+1) matrix, $\bar{B}_{\text{b}}^{qm}$ represents its elements $w_{\text{b}}^{\delta}(x_q-x_m,\omega)$, which is shown in details as:

$$\overline{\mathbf{B}}_{\rm b}=\begin{bmatrix} \overline{B}_{\rm b}^{(-N)(-N)} & \cdots & \overline{B}_{\rm b}^{(-N)0} & \cdots & \overline{B}_{\rm b}^{(-N)N} \\ & \ddots & \vdots & \vdots & \vdots \\ & & \overline{B}_{\rm b}^{00} & \cdots & \overline{B}_{\rm b}^{0N} \\ \text{Sym} & & & \ddots & \vdots \\ & & & & \overline{B}_{\rm b}^{NN} \end{bmatrix} \tag{20}$$

3.1.1.4. Pier-Top Force under All Axle-Loads of Moving Train

Combining Equations (13), (15) and (19), the vector expression of the forces between rail and bridge deck can be derived as:

$$\bar{\mathbf{f}}=\left(\overline{\mathbf{B}}_{\rm b}-\frac{\mathbf{I}}{k_{\rm rb}+i\omega c_{\rm rb}}+\overline{\mathbf{B}}_{\rm r}\right)^{-1}\cdot A(k_x,\omega)\cdot\overline{\mathbf{E}}_s \tag{21}$$

The force on the pier-top under all axle-loads can be computed by multiplying the bridge displacement at the bearing position and the stiffness of the bearing. In practice, the $2N+1$ sleepers are in the two adjacent spans and the sleeper support in the bearing position is thought to be the intermediate one. Thus, according to Equation (18), the force on the pier-top during a train passage can be obtained as:

$$\begin{aligned} F(\omega) &= \overline{w}_{\rm b}(L,\omega)\cdot k\cdot\eta(x_m)+\overline{w}_{\rm b}(0,\omega)\cdot k\cdot\eta(x_m) \\ &= k\cdot\sum_{m=-N}^{0}\overline{f}_m^{\rm final}(\omega)w_{\rm b}^{\delta}(L-x_m,\omega)+k\cdot\sum_{m=0}^{N}\overline{f}_m^{\rm final}(\omega)w_{\rm b}^{\delta}(-x_m,\omega) \end{aligned} \tag{22}$$

where $\eta(x_m)$ is a sign function, $\eta(x_m)=1$ when x_m is located on both spans the pier supports and $\eta(x_m)=0$ when x_m on other spans.

The calculated forces on pier-tops of the bridge will spread further to the base via piers and foundations, which forms exactly the vibration sources in the analysis of environment vibration for elevated rail transit.

3.1.2. Analytical Model of Train-viaduct System

The dynamic interactions of train-bridge system have been studied systematically by the scholars worldwide, and various analysis models valid for different cases have been established. Different from the previous research, the numerical analysis model developed herein is to obtain the loads of vibration source acting on the pier-ground-building subsystem model.

In this train-bridge subsystem model, the train model is composed of various locomotives and vehicles, the bridge model consists of multi simply-supported spans or continuous spans, and the bridge piers lie on either pile or direct flat foundations (see Figure 5).

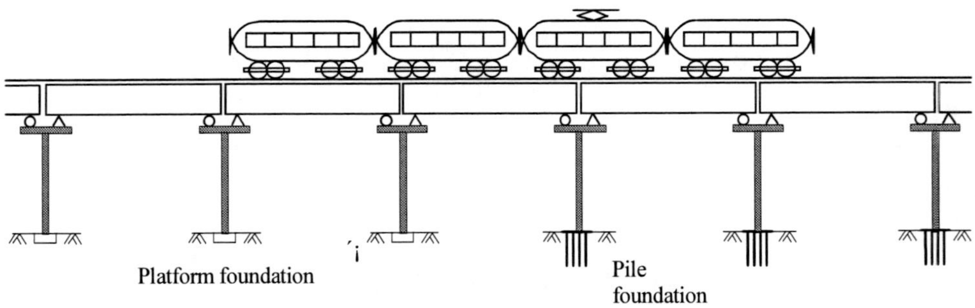

Figure 5. Illustration of analytical model of elevated rail transit system..

3.1.2.1. Train Model

The train model consists of locomotives and passenger cars, both of which are vibrating systems with multiple DOFs composed of car-bodies, bogies, wheel-sets, suspension springs and dashpots. Herein a train model composed of articulated vehicles coupled with each other by the bogies is introduced as follows (see Figure 6).

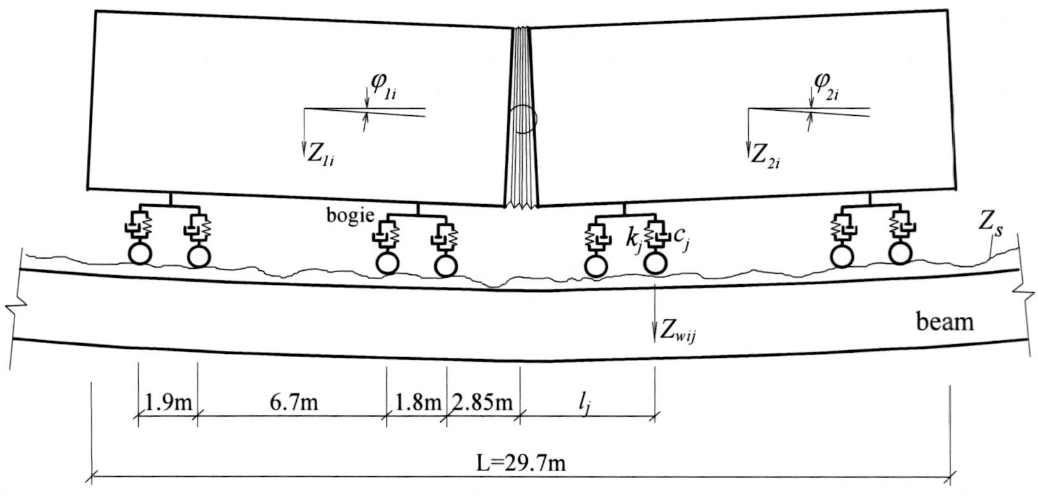

Figure 6. Illustration of a 8-axle articulated train model.

The following assumptions are adopted for the 8-axle articulated vehicle model:

(1) A 2D model is considered, namely, only the vertical vibrations of the system are taken into account.
(2) Each vehicle consists of two car-bodies with an articulated connection.
(3) All the car-bodies, bogies and wheel-sets in each vehicle are regarded as rigid components, namely, the elastic deformation during vibration is ignored.
(4) The suspension system of the vehicle is simplified as a single suspension system, where the car-body and wheel-sets are connected by equivalent springs and dampers, whose stiffness and damping coefficients are obtained by converting the corresponding primary and secondary spring stiffness and damping coefficients in terms of equivalent springs and dampers.

(5) A total of 4 DOFs are assigned to the car-body of the *i*-th vehicle, to account for the floating Z_{1i}, Z_{2i} and pitching φ_{1i}, φ_{2i}. While only 1 DOF is assigned to each wheel-set, i.e., the floating movement. There are only 3 independent DOFs because the two car-bodies are coupled with each other by the bogies. For example, the pitching of the 2nd car-body can be expressed with other 3 DOFs, namely, $\varphi_2=Z_2/L-Z_1/L-\varphi_1$. The idealized configuration for each 8-axle articulated vehicle in a train can therefore be modeled by an 11-DOF dynamic system.

The equation of motion of vehicles can be established by the structural dynamics method, and written as:

$$[M_v]\{\ddot{v}\}+[C_v]\{\dot{v}\}+[K_v]\{v\}=\{F_v\} \tag{23}$$

where $[M_v]$, $[K_v]$ and $[C_v]$ are the mass, stiffness and damping matrices, $\{v\}$, $\{\dot{v}\}$ and $\{\ddot{v}\}$ the displacement, velocity and acceleration vectors of vehicles, respectively; $\{F_v\}$ is the force vector of the vehicle.

3.1.2.2. Bridge Model

Nowadays, bridges are usually analyzed and designed with the help of finite element method. The equation of motion the bridge structures can be written as:

$$[M_b]\{\ddot{Z}\}+[C_b]\{\dot{Z}\}+[K_b]\{Z\}=\{F_b\} \tag{24}$$

where $[M_b]$, $[K_b]$ and $[C_b]$ are the mass, stiffness and damping matrices, $\{Z\}$, $\{\dot{Z}\}$ and $\{\ddot{Z}\}$ the displacement, velocity and acceleration vectors of the bridge, respectively; $\{F_b\}$ is the force vector of the bridge.

If the external forces (such as wind load, seismic action, collision forces) are ignored in the train-bridge dynamic analysis, the coupled motion equations for the train-bridge system can be derived by combining Equations (23) and (24), together with the wheel-bridge relationship.

Although the original finite element model can be directly used to constitute the basic equations for the coupled train-bridge system, the direct integration of these equations in the time domain is very cumbersome. Therefore, the modal superposition technique is adopted to solve the equations of motion of the bridge, which makes the coupled FE equations of motion related to the bridge nodes become the superposition of independent modal equations. Owing to the fact that the dynamic response of a structure is dominantly influenced by its several lowest modes, this approach has a great advantage that an adequate estimation of the dynamic response can be obtained by considering only a few modes of vibration, even for a spatial structural system with hundreds of DOFs.

In this way, the vehicle equations are combined with the bridge modal equations. The number of mode shapes of the bridge taken into account in the computation should be large enough to include the effects of both the global deformation of the bridge and the local deformation of the structural elements supporting the track. This decision may be made through a convergent study of the effects of the number of mode-shapes or through a comparison with the measurement data.

When the Modal Synthesis Method is adopted for the bridge model, i.e., the bridge is discreted with the generalized coordinates, the movement of bridge deck at any section can be represented by the superposition of the bridge mode shapes as:

$$Z_{\mathrm{b}}(x) = \sum_{n=1}^{N} q_n \phi_n(x) \tag{25}$$

where ϕ_n denotes the n-th mode-shape function of the bridge at the position of x along the bridge, N the number of mode shapes concerned in the calculation, and q_n the generalized coordinate of the n-th mode.

The modal analysis is conducted with respect to the bridge as a whole. It is assumed that the vibration mode of bridge nodes is consistent with that of track on bridge deck, and the mode shape between the deck nodes obtained from the eigenvalue analysis is determined by the Lagrange interpolation. If the modes of bridge vibration are normalized with $\phi_n^{\mathrm{T}} \mathbf{M} \phi_n = 1$, the n-th modal equation of the bridge can be expressed as

$$\ddot{q}_n + 2\xi_n \omega_n \dot{q}_n + \omega_n^2 q_n = F_n \tag{26}$$

where ξ_n and ω_n stand for the damping ratio and circular frequency, and F_n is generalized force, of the n-th mode, respectively.

3.1.2.3. Dynamic Analysis Model for Train-bridge System

Assuming the wheel/track relationship with close contact and no elastic deformation is considered, the wheel-set movement $Z_{\mathrm{w}ij}$ and the bridge movement $Z_{\mathrm{b}}(x_{ij})$ satisfy:

$$Z_{\mathrm{w}ij} = Z_{\mathrm{b}}(x_{ij}) + Z_{\mathrm{s}}(x) = \sum_{n=1}^{N} q_n \phi_{ij}^n + Z_{\mathrm{s}}(x) \tag{27}$$

where Z_{s} represents the relative displacement between wheel-set and beam, i.e., the track irregularities. It can be seen that the wheel-set displacement can be expressed as the linear combination of bridge modal displacement q_n (n=1,2,... N).

Clearly, the displacements of the wheel do not need to be included in the equations of motion for the train-bridge system. This can reduce the computational effort significantly. Substituting Equation (27) into the abovementioned vehicle and bridge equations, the coupled equations of motion for the train-bridge system can be derived and expressed as follows:

$$\begin{bmatrix} M_i + \frac{J_2}{L^2} & -\frac{J_2}{L^2} & \frac{J_2}{L} \\ -\frac{J_2}{L^2} & M_i + \frac{J_1}{L^2} & -\frac{J_2}{L} \\ \frac{J_2}{L} & -\frac{J_2}{L} & J_1 + J_2 \end{bmatrix} \begin{Bmatrix} \ddot{Z}_{1i} \\ \ddot{Z}_{2i} \\ \ddot{\varphi}_{1i} \end{Bmatrix}$$

$$+\begin{bmatrix} \sum_{j=1}^{4} c_j + \sum_{j=5}^{8} \frac{l_j^2}{L^2} c_j & -\sum_{j=5}^{8}\left(\eta \frac{l_j}{L} + \frac{l_j^2}{L^2}\right) c_j & -\sum_{j=1}^{4} \eta l_j c_j + \sum_{j=5}^{8} \frac{l_j^2}{L} c_j \\ -\sum_{j=5}^{8}\left(\eta \frac{l_j}{L} + \frac{l_j^2}{L^2}\right) c_j & \sum_{j=5}^{8}\left(1+\eta \frac{l_j}{L}\right)^2 c_j & \sum_{j=5}^{8}\left(\eta - \frac{l_j}{L}\right) c_j \\ -\sum_{j=1}^{4} \eta l_j c_j + \sum_{j=5}^{8} \frac{l_j^2}{L} c_j & \sum_{j=5}^{8}\left(\eta - \frac{l_j}{L}\right) c_j & \sum_{j=1}^{8} l_j^2 c_j \end{bmatrix} \begin{Bmatrix} \dot{Z}_{1i} \\ \dot{Z}_{2i} \\ \dot{\varphi}_{1i} \end{Bmatrix}$$

$$+\begin{bmatrix} \sum_{j=1}^{4} k_j + \sum_{j=5}^{8} \frac{l_j^2}{L^2} k_j & -\sum_{j=5}^{8}\left(\eta \frac{l_j}{L} + \frac{l_j^2}{L^2}\right) k_j & -\sum_{j=1}^{4} \eta l_j k_j + \sum_{j=5}^{8} \frac{l_j^2}{L} k_j \\ -\sum_{j=5}^{8}\left(\eta \frac{l_j}{L} + \frac{l_j^2}{L^2}\right) k_j & \sum_{j=5}^{8}\left(1+\eta \frac{l_j}{L}\right)^2 k_j & \sum_{j=5}^{8}\left(\eta - \frac{l_j}{L}\right) k_j \\ -\sum_{j=1}^{4} \eta l_j c_j + \sum_{j=5}^{8} \frac{l_j^2}{L} k_j & \sum_{j=5}^{8}\left(\eta - \frac{l_j}{L}\right) k_j & \sum_{j=1}^{8} l_j^2 k_j \end{bmatrix} \begin{Bmatrix} Z_{1i} \\ Z_{2i} \\ \varphi_{1i} \end{Bmatrix}$$

$$= \sum_{n=1}^{N} \begin{Bmatrix} \sum_{j=1}^{4} \phi_{ij}^n (k_j q_n + c_j \dot{q}_n) - \sum_{j=5}^{8} \eta \frac{l_j}{L} \phi_{ij}^n (k_j q_n + c_j \dot{q}_n) \\ \sum_{j=5}^{8}\left(1-\eta \frac{l_j}{L}\right) \phi_{ij}^n (k_j q_n + c_j \dot{q}_n) \\ -\sum_{j=1}^{8} \eta l_j \phi_{ij}^n (k_j q_n + c_j \dot{q}_n) \end{Bmatrix}$$

$$+\begin{Bmatrix} \sum_{j=1}^{4} [k_j Z_s(x_{ij}) + c_j \dot{Z}_s(x_{ij})] - \sum_{j=5}^{8} \eta \frac{l_j}{L} [k_j Z_s(x_{ij}) + c_j \dot{Z}_s(x_{ij})] \\ \sum_{j=5}^{8}\left(1-\eta \frac{l_j}{L}\right) [k_j Z_s(x_{ij}) + c_j \dot{Z}_s(x_{ij})] \\ -\sum_{j=1}^{8} \eta l_j [k_j Z_s(x_{ij}) + c_j \dot{Z}_s(x_{ij})] \end{Bmatrix}, \quad (i=1,2,\cdots N_v) \tag{28a}$$

$$\ddot{q}_n + 2\xi_n \omega_n \dot{q}_n + \omega_n^2 q_n = \sum_{i=1}^{N_v} \phi_{ij}^n \Bigg\{ -\sum_{j=1}^{8}\sum_{m=1}^{N} \phi_{ij}^m [m_{ij}\ddot{q}_m + c_i \dot{q}_m + k_i q_m] + \sum_{j=1}^{4} (k_j Z_{1i} + c_j \dot{Z}_{1i})$$

$$-\sum_{j=5}^{8} \eta \frac{l_j}{L} (k_j Z_{1i} + c_j \dot{Z}_{1i}) + \sum_{j=5}^{8}\left(1+\eta \frac{l_j}{L}\right)(k_j Z_{2i} + c_j \dot{Z}_{2i}) - \sum_{j=1}^{8} \eta l_j (k_j \varphi_{1i} + c_j \dot{\varphi}_{1i})$$

$$-\sum_{j=1}^{8} [m_{ij} \ddot{Z}_s(x_{ij}) + c_j \dot{Z}_s(x_{ij}) + k_j Z_s(x_{ij})] + \frac{1}{8} g(M_i + m_{ij}) \Bigg\}, \quad (n=1,2,\cdots\cdots N_q) \tag{28b}$$

Since the coefficients ϕ in the equations above are the functions of train speed V and time t, they are always changing along with the change of wheel-set positions when the train runs on the bridge. Equation (28) is actually the second-order linear non-homogeneous differential equations with time-varying coefficients.

The dynamic reaction time history of bearings can be obtained through the analysis of train-bridge subsystem. When rubber bearings are adopted, the dynamic reaction $R_r(t)$ can be computed via the following equation:

$$R_r(t) = \sum_{n=1}^{N} q_n(t)\phi_r^n \cdot 4S_r \cdot E_r / H_r \quad (29)$$

where ϕ_r^n is the value of the n-th mode-shape function $q_n(t)$ at the location of bearing; S_r, E_r and H_r are, respectively, the cross section area, elastic modulus and thickness of the rubber bearing.

The dynamic reaction of a pier is the sum of the dynamic reactions of two bearings it supports. Shown in Figure 7 is a dynamic reaction history of a 24m-span simply-supported bridge, where can be clearly observed the dynamic actions of the train during different stages, namely, entering bridge, on the bridge, and leaving the bridge.

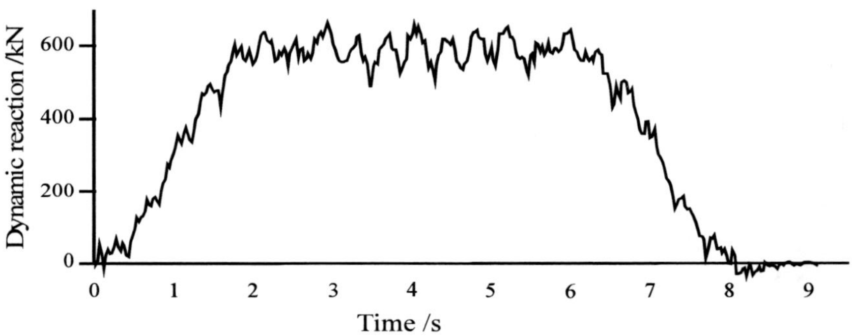

Figure 7. A dynamic reaction history of a 24m-span simply-supported bridge.

It is demonstrated in the dynamic reaction curve that the main frequency is about 2.6 Hz, which just coincides with the loading frequency when the train (the average distance between bogies is 7.425 m) travels at the speed of 70 km/h, namely, $f = V / L_a = 19.44/7.425 \approx 2.6$ Hz, indicating the loading frequency plays a key role herein.

Compared in Table 1 are the static and dynamic bearing and pier reactions and the corresponding dynamic magnification factors of the bridge. It is observed that the dynamic reaction force on the pier-top induced by train-load (reactions 1+2) is much smaller than the double of a single reaction, due to the phase difference between the reactions of two bearings on a pier. Note that the distance between two bearings is 1.1 m, and the average distance between bogies is 7.425 m, thus, the loading phase difference is about 170°, almost opposite. In conclusion, it is necessary to use a 2-span bridge model for calculating pier reaction.

Table 1. Bearing reaction forces and dynamic magnification factors

Reaction force	Bearing 1	Bearing 2	Resultant of bearing reactions (1+2)
Static loading /kN	363.9	363.9	591.5
Dynamic loading /kN/	435.0	435.0	705.0
Dynamic Magnification factor	1.195	1.195	1.192

3.2. Analytical Model of Pier-Foundation-Soil-Building Interaction System

In dynamic analysis of the pier-foundation-soil-building interaction system, both the 2D plane strain model and the 3D spatial FE model can be used. Under certain conditions, the 3D semi-analytical model is also adopted. The vibration responses of any point at ground or in the building can be obtained by inputting the dynamic loads calculated by train-track-bridge system model into the pier-foundation-soil-building system model.

3.2.1. 2D Plane Model

A 2D plane strain model of pier-foundation-soil-building is shown in Figure 8, where the dynamic action of the bridge superstructure and train-load is simplified as the dynamic force acting on the pier-top calculated by the train-track-bridge system model. The pier is simulated with beam elements, and the masses of the girder and the train are regarded as lumped masses on the top of the pier. The structural characteristics of foundation are considered in the finite element mesh. The simulation of the base and boundaries of the soil and the foundation-soil interaction meets the principles of general plane finite element model.

The vibration induced by the train is transmitted to the soil through pier and foundation. This can be regarded as a point vibration source. When the distance from the ground observation point to the center line of the bridge is far enough (more than 2 times of the bridge span length), a 2D plane model can be adopted for its advantage of convenient modeling and easy calculation. While when the distance is shorter than 2 times of the bridge span length, the superposition of impacts from several point vibration sources and the phase difference of vibrations at the observation point from two or more adjacent piers should be considered.

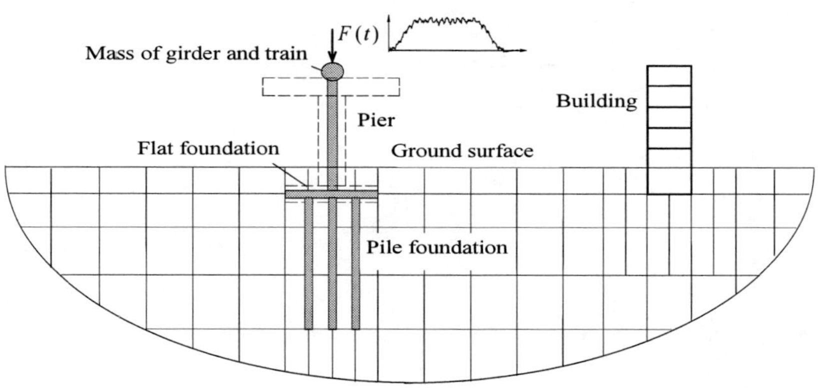

Figure 8. A 2D model for pier-foundation-soil-building interaction system.

3.2.2. 3D Spatial Model

For the environmental vibration analysis induced by the elevated rail transit, it is usually concerned about the vibration of the ground or nearby buildings at a distance to the bridge longitudinal axis in the case of multi-span bridge. During the train passage, the dynamic train-load is transmitted to the soil by a series of piers to form array point vibration source. Shown in Figure 9 is the 3D pier-foundation-soil-building spatial model established by the actual array point vibration source, which directly reflects the superposition of dynamic effect at the observation point from each point vibration source and the influence of phase difference of vibrations from adjacent piers reaching the observation point.

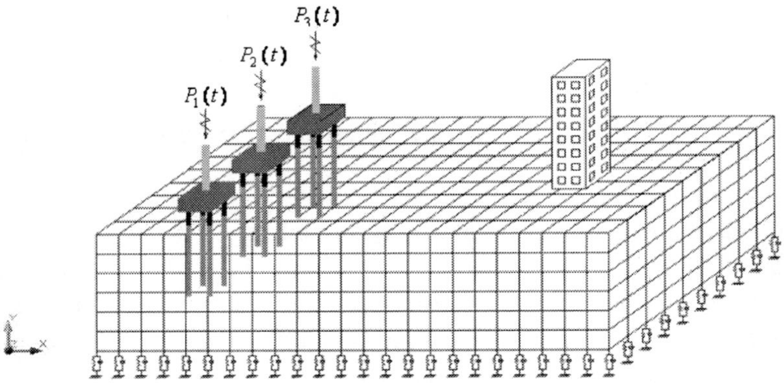

Figure 9. A 3D model for pier-foundation-soil-building interaction system.

3.2.3. Semi-analytical Model Based on Green's Function

To analyze the ground vibrations with analytic analysis, the dynamic loads calculated by the train-track-bridge system model can be input to an elastic half-space or visco-elastic half-space strata. This is the semi-analytical model of the pier-foundation-soil system, as shown in Figure 10.

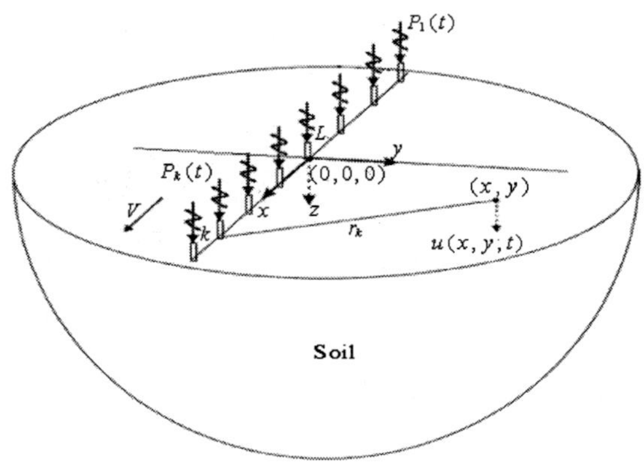

Figure 10. Semi-analytical model based on Green's function for pier-foundation-soil system.

Assuming the bridge has identical spans with the length L, when the train travels through the bridge at a constant speed V, the forces acting on the pier-foundation-soil system via all piers are the same but with a time interval. Suppose the force acting on the first pier is P_1(t), the expression of the force on the k-th pier can be expressed as:

$$P_k(t) = P_1(t - kL/V) \tag{30}$$

Then the total force by all piers acting on the system can be expressed as:

$$F(x,y,t)=\sum_{k=1}^{n}P_1(t-kL/V)\delta(x-kL)\delta(y) \tag{31}$$

3.2.3.1. Vibration Response of Ground Soil under Dynamic Load of Single Pier

As shown in Figure 11, the interaction force between the bridge foundation and ground soil can be regarded as the external load acting on the half-space soil layer. The frequency spectrum of displacement at the ground observation points induced by the dynamic load of the k-th pier can be written as:

$$w_k^*(x,y,\omega)=\bar{w}_k^*(r,\omega)=P_k^*(\omega)\mathrm{G}_{zz}(r_k,\omega) \tag{32}$$

where $P_k^*(\omega)$ is the frequency spectrum of the external excitation on the k-th pier obtained by using Fourier transform to the force on the k-th pier $P_k(t)$; $r_k=\sqrt{(x-x_k)^2+y^2}>0$, where ($x_k$, 0) and ($x$, y) are, respectively, the coordinates of the k-th pier and the ground observation point. G_{zz} is the Green's function of elastic half-space stratum under harmonic point load [18, 19]:

$$G_{zz}(r,\omega)=\frac{1}{2\pi G}\begin{cases}\dfrac{1-\nu}{(r-2r_{eq}/\pi)}\exp\left[-i\left(\dfrac{\omega(r-2r_{eq}/\pi)(1-i\eta)}{\gamma C_R}\right)\right], & (r-\dfrac{2r_{eq}}{\pi}\le\beta\lambda_R)\\ \dfrac{1-\nu}{\sqrt{\beta\lambda_R(r-2r_{eq}/\pi)}}\exp\left[-i\left(\dfrac{\omega(r-2r_{eq}/\pi)(1-i\eta)}{\gamma C_R}\right)\right], & (\beta\lambda_R<r-\dfrac{2r_{eq}}{\pi}\le 1.5\lambda_R)\\ \dfrac{1-\nu}{\sqrt{\beta\lambda_R(r-2r_{eq}/\pi)}}\exp\left[-i\left(\dfrac{\pi}{4}+\dfrac{\omega(r-2r_{eq}/\pi)(1-i\eta)}{\gamma C_R}\right)\right], & (1.5\lambda_R<r-\dfrac{2r_{eq}}{\pi})\end{cases} \tag{33}$$

where C_R is the Rayleigh wave velocity of the foundation soil; G and ν are, respectively, the shear modulus and Poisson's ratio of the soil; $\lambda_R=2\pi C_R/\omega$ is the wavelength of Rayleigh wave; γ is the correction factor of the Rayleigh wave velocity in the near field; r_{eq} is the equivalent radius of the foundation; $(1-i\eta)$ is the material damping related to Rayleigh wave velocity, and $\eta=0$ if the half-space stratum is elastic, which means the vibration energy is completely dissipated in the form of radiation damping during energy transmitting to infinity boundary; $\beta=(2/3)\beta^*$ is the coefficient to describe the boundary between the near field and the far field, and β^* can be deduced as [19, 20]:

$$\beta^*=\left(\frac{1-\nu}{2\pi}\right)^2\frac{\left[8q_R-(48-32q_p^2)q_R^3+48(1-q_p^2)q_R^5\right]^2}{(2q_R-1)^4(q_R^2-q_p^3)} \tag{34}$$

where $q_R=C_s/C_R$.

In the analysis above, the contribution of shear wave and dilatational wave to the ground response are ignored because the Rayleigh wave occupies most of the energy of ground surface vibration.

The time history response of ground vibration can be obtained by the inverse Fourier transform to Equation (32):

$$w_k^*(x,y,t) = \int_{-\infty}^{\infty} w_k^*(x,y,\omega)\mathrm{e}^{i\omega t}\,\mathrm{d}\,\omega \tag{35}$$

3.2.3.2. Vibration Response of Ground Soil under Dynamic Loads of Multi-piers

When the train runs on a bridge with multi-spans, the ground vibration response at any point can be deduced as the superposition of vibration response at this point due to dynamic loads of all piers.

In addition, when the train runs on the bridge from -∞ end to +∞ end at a constant speed V, the frequency spectrum of all pier excitations can be acquired using Fourier transform to Equation (31) as:

$$F^*(x,y,\omega) = \frac{1}{2\pi}\int_{-\infty}^{\infty} F(x,y,t)e^{-i\omega t}\,\mathrm{d}\,t = \sum_{k=-\infty}^{\infty} P_k^*(\omega)e^{-i\omega(kL/V)}\delta(x-kL)\delta(y) \tag{36}$$

Then the frequency spectrum of ground vibration under dynamic loads of multi-piers can be obtained by using the Green's function (33) and integrating along the x-axis:

$$w^*(x,y,\omega) = \int_{-\infty}^{\infty} F^*(x',0,\omega)\mathrm{G}_{zz}(r',\omega)\,\mathrm{d}\,x' \tag{37}$$

where $r' = \sqrt{(x-x')^2 + y^2} > 0$.

Substituting Equations (36) and (33) into (37), the following expression can be derived:

$$w^*(x,y,\omega) = P_k^*(\omega)\sum_{k=-\infty}^{\infty} e^{-i\omega(kL/V)}\,\mathrm{G}_{zz}(r_k,\omega) \tag{38}$$

where $r_k = \sqrt{(x-kL)^2 + y^2} > 0$.

The time history response of ground vibration under dynamic loads of all piers can be solved by the inverse Fourier transform to Equation (38):

$$w(x,y,t) = \int_{-\infty}^{\infty} w^*(x,y,\omega)e^{i\omega t}\,\mathrm{d}\,\omega \tag{39}$$

In the derivation above, it is assumed that the composition of each bridge span is identical with the same span length L. However, this method can also be used for those cases where the bridge span lengths are different if the force on k-th pier is expressed in $P_k(t)$ and $(x-kL)$ is replaced with $(x-x_k)$ directly.

4. Experimental Study of Environmental Vibrations Induced by Elevated Rail Transit System

A field measurement was carried out at the elevated trial section of the Beijing Metro Line 5. The trial section crosses two continuous box-girders on straight line, with a 3×30m continuous span and a 3×27m continuous span, as shown in Figure 11. There are two types of tracks adopted in the trial section. The ladder track is set on the down direction to the Taipingzhuang Station, and the common slab track on the up direction to the Taipingzhuang North Station (see Figure 12).

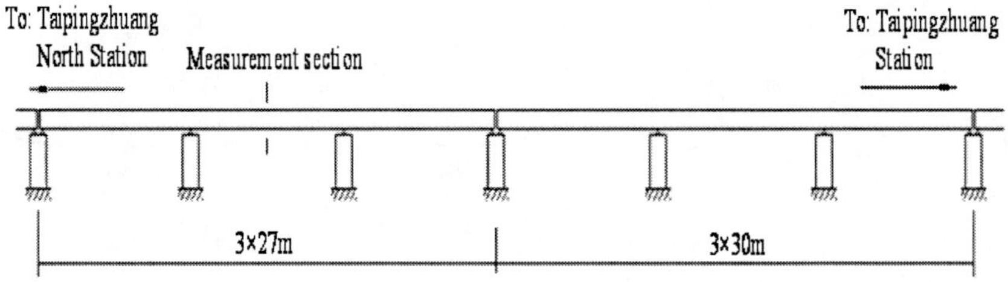

Figure 11. The trail section of elevated bridge on Beijing Metro Line 5.

Figure 12. The ladder track on the elevated bridge (the other line is common slab track).

4.1. Track Irregularity Measurement

The primary dynamic inputs to train-bridge system come from track geometry variations or irregularities. Track irregularities refer to the deviations of the rails that support and guide the wheels from the ideal rail, which includes static irregularity at unloaded status and dynamic irregularity under loading. Track irregularities may occur because of initial installation errors, degradation of support materials, and dislocation of track joints, etc. The four commonly-used geometrical parameters to define tangent track irregularities are illustrated in Figure 13. They are the vertical profile z_v, the cross-level z_c, the alignment y_a, and gauge y_g. The cross-level irregularity can also be expressed in terms of the angle formed by the level difference between the left rail and the right rail, namely, $\theta_c = z_c/2b$.

A measurement was performed on the track irregularities of the ladder track and the slab track at the trial section. The lateral and vertical track irregularities of the inner rail and outer rail of the ladder track were measured with theodolite and level instruments, from which the track central line orientation, rotational and gauge irregularities were obtained. The statistical standard variation of track irregularities and the power spectrum were calculated by the following formulae [21]:

$$\sigma_\eta^2 = \frac{1}{X}\int_0^X \left[\eta(x) - \mu_\eta\right]^2 \mathrm{d}x = \frac{1}{X}\int_0^X \eta^2(x)\,\mathrm{d}x - \mu_\eta^2 = \phi_\eta^2 - \mu_\eta^2 \tag{40}$$

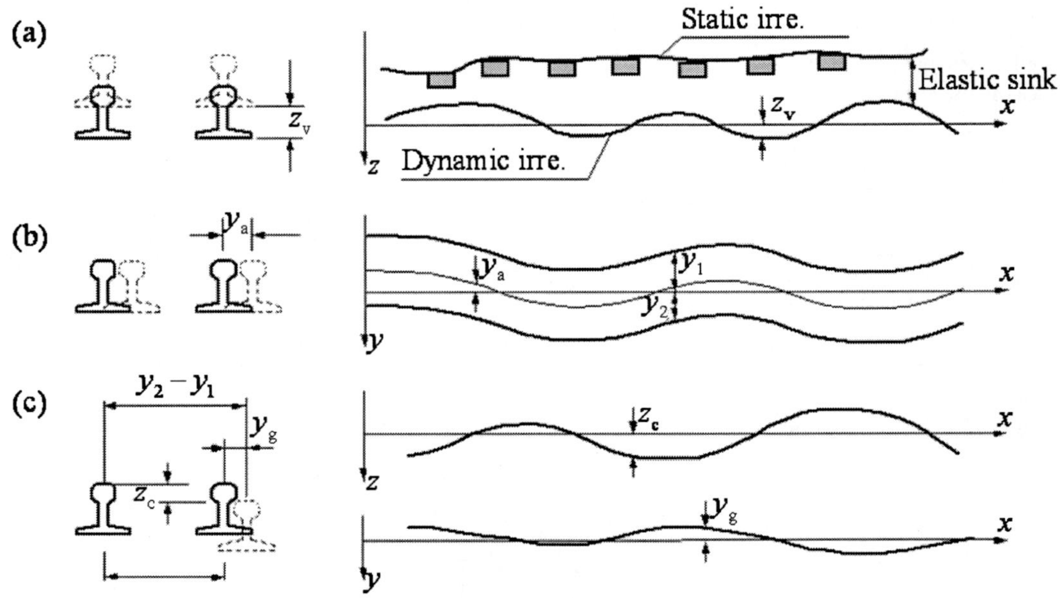

Figure 13. Track irregularities: (a) vertical profile z_v; (b) alignment y_a; (c) cross-level z_c and gauge y_g.

$$S_\eta(f) = \lim_{\Delta f \to \infty} \frac{1}{\Delta f}\left[\lim_{X \to \infty} \frac{1}{X}\int_0^X \eta^2(x, f, \Delta f)\,\mathrm{d}x\right] \tag{41}$$

where $\eta(x)$ is the profile, μ_η is the mean, X is the total sampling length, f is the space frequency, and Δf is the frequency interval of track irregularity.

Table 2. Maximum and standard deviation of track irregularities (unit: mm)

Item			Maximum	Standard deviation
Ladder track	Outer rail	Vertical	4.78	2.01
		Lateral	3.10	1.14
	Inner rail	Vertical	4.62	1.86
		Lateral	4.37	1.58
	Track center line	Vertical	4.73	1.79
		Lateral	2.87	1.09
		Cross level	6.42	1.42
		Gauge	0.53	0.18
Common slab track	Vertical		3.30	1.39
	Lateral		4.86	1.32

Shown in Table 2 are the statistical parameters of the measured track irregularities. It is seen that the maximum and standard deviation of the measured track irregularities are reasonable compared with the railway track code in China [22].

Shown in Figures 14 and 15 are, respectively, the power spectra of the vertical and lateral rail irregularities of the ladder track and common slab track. One can see that the irregularity wavelength focuses on 0.5-80 m, in which the amplitude of long wave is big and that of middle wave is small, and the shorter the wavelength is, the smaller the amplitude will be. Besides, the peaks in the power spectra show that there are many periodic wave components in the random track irregularity in this section. This corresponds with the irregularity state of new rail where wave wearing was formed in rail rolling process.

According to the measurement data, the irregularity spectra can be fitted as:

$$S(f) = \frac{A(f^2 + Bf^3 + C)}{f^4 + Df^3 + Ef^2 + Ff + G} \tag{42}$$

where the unit of S(f) is mm^2/(1/m); f is the spatial frequency of the track irregularities with the unit of 1/m; A, B, C, D, E, F, and G are the parameters to be fitted. The fitted parameters are listed in Table 3.

Table 3. Fitted parameters for the irregularity spectra

Direction	A	B	C	D	E	F	G
Vertical	2.2882	-1.0088	-0.7884	1.5073	0.4538	-1.9475	1.8125
Rotation	2.5437	-0.9791	-0.7533	1.6389	0.9103	-2.6442	2.2103
Gauge	-7.1553	0.7810	1.6202	-2.2860	4.0987	-0.1224	4.8230
Lateral	-2.3943	1.6510	0.0000	2.9867	4.2552	-0.0041	0.0000

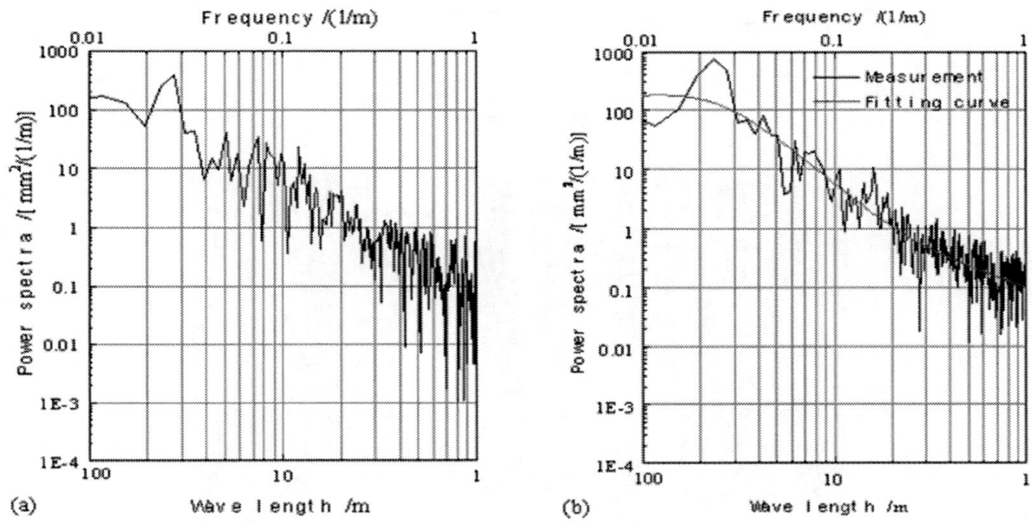

Figure 14. Power spectra of vertical rail irregularity: (a) common slab track, (b) ladder track.

4.2. Outline of Experiment

The field measurement test was carried out at the viaduct on the Beijing Metro Line 5 in November 2008. The acceleration responses of rails, sleepers, bridge deck and ground surface were measured during train passage, and there were 30 groups of data obtained.

The accelerometers LC0123ICP were mounted on the rail bottoms and sleepers of the ladder track and the common slab track, and the concrete beam deck, as shown in Figure 16.

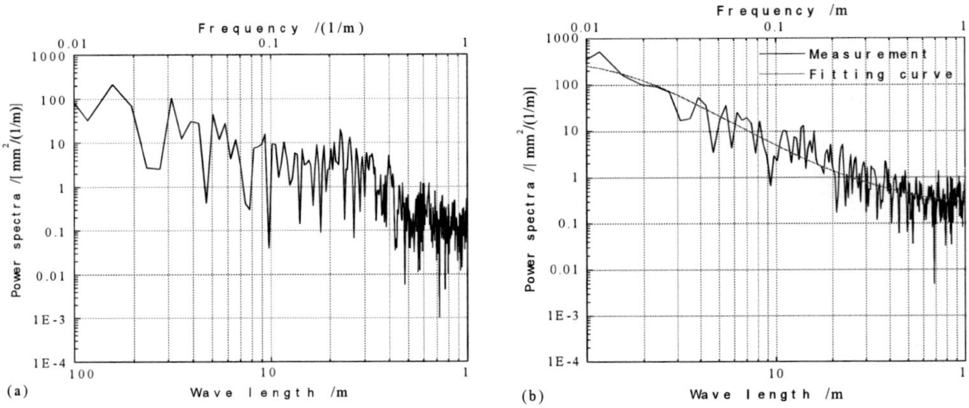

Figure 15. Power spectra of lateral rail irregularity: (a) common slab track, (b) ladder track.

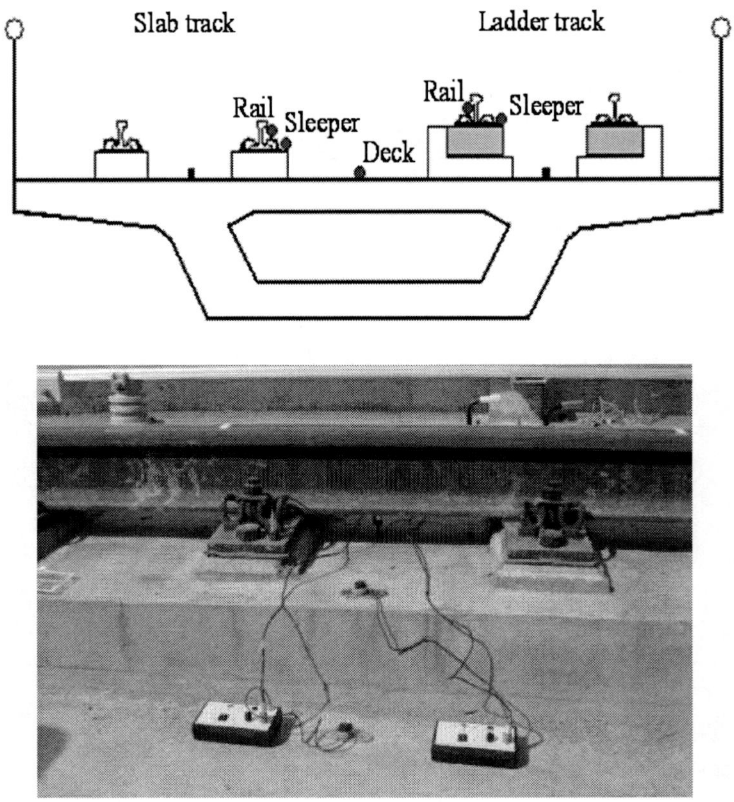

Figure 16. Accelerometers and locations of measuring points on the bridge.

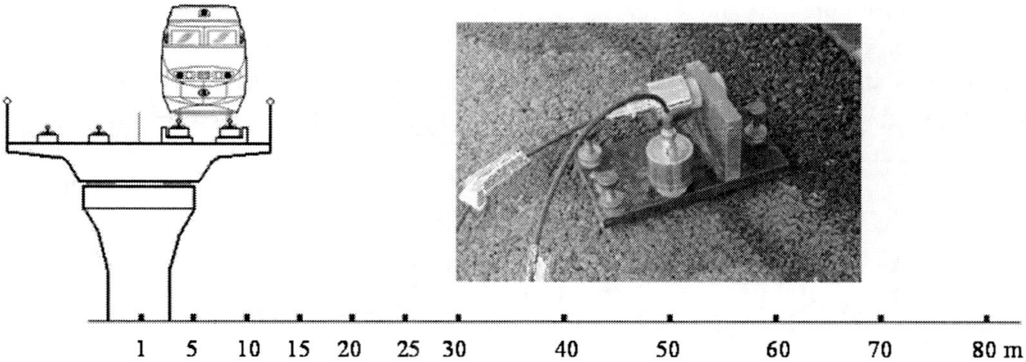

Figure 17. Accelerometer and locations of measuring points on the ground.

Figure 18. The train running through the bridge during the experiment.

There were 20 measurement points for the ground vibration, which were located at the distances of 0−80 m to the central line of the viaduct, respectively, as shown in Figure 17. Both the vertical (Z direction) and horizontal (perpendicular to the bridge central line) vibrations were measured.

The Intelligence Data Acquisition System Wavebook-516E/WBK18 with 32 CHs was adopted for signal processing. The sampling frequency was 5120 Hz.

The experimental load was the train running on Metro Line 5, which is composed of six vehicles (see Figure 18). The train speed was 55 km/h. The train consists of six carriages. Each carriage has two bogies and four axles, with the axle-weight of 140 kN. The total length of the train is 120 m. Since the measurement site is very close to the Taipingzhuang North Station, the train speeds on the up-line and the down-line are different: The train ran on the down line ladder track at the normal speed 70 km/h, while the train ran on the up line common slab track at 40−50 km/h. In the experiment, the controlled train speed 50 km/h was realized on both the up and down lines.

4.2.1. Vibration of Track and Bridge

The typical acceleration histories of the rail, the sleeper and the beam deck measured at the ladder track and the common slab track are shown in Figure 19, respectively.

Shown in Figures 20 and 21 are, respectively, the measured acceleration histories and their different frequency components when train running on ladder track at speed of V=69.6 km/h. In this study, the vibration level of structures is based on the reference value of 10^{-5}m/s^2. From the figures one can see clearly the influence of wheel axle-loads on the responses of the bridge deck.

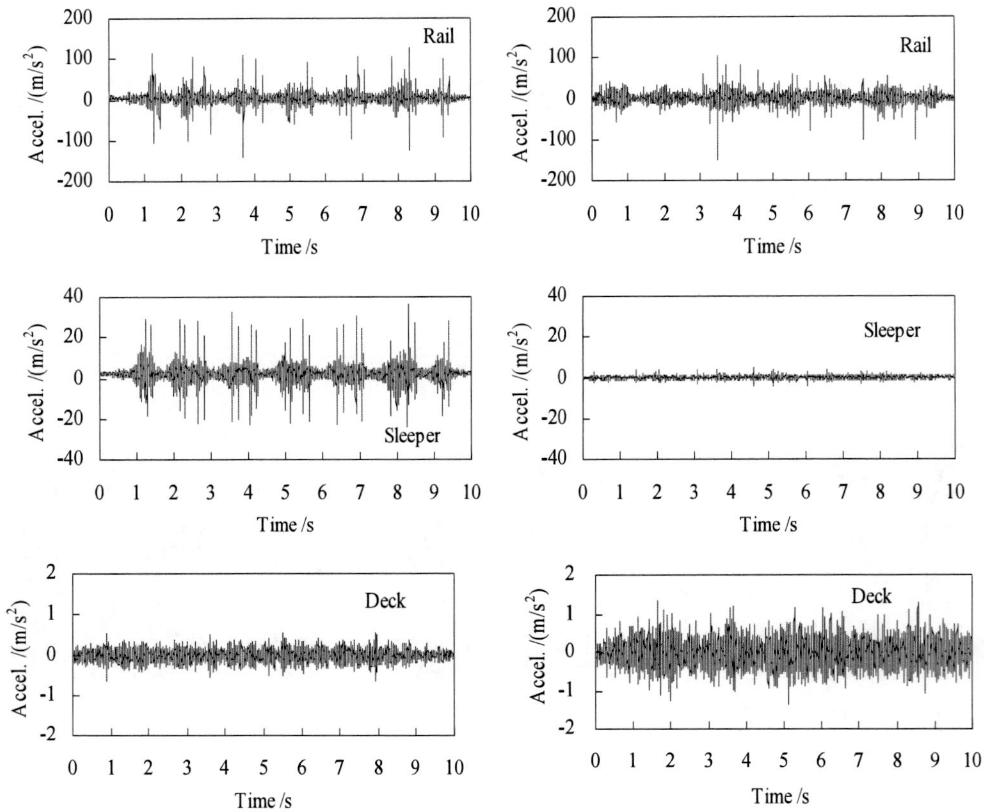

Figure 19. Acceleration histories when train traversed the bridge on ladder track (left: V=48.1 km/h) and common slab track (right: V=47.6 km/h).

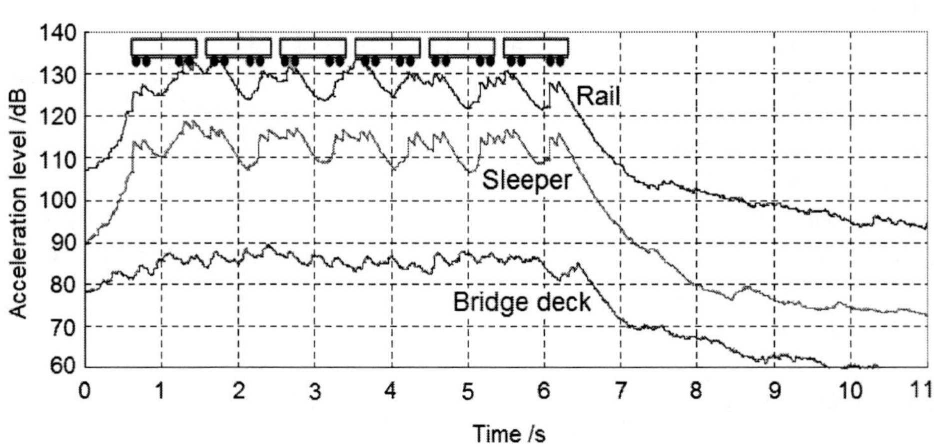

Figure 20. Acceleration level histories of rail, sleeper and deck (V=69.6 km/h).

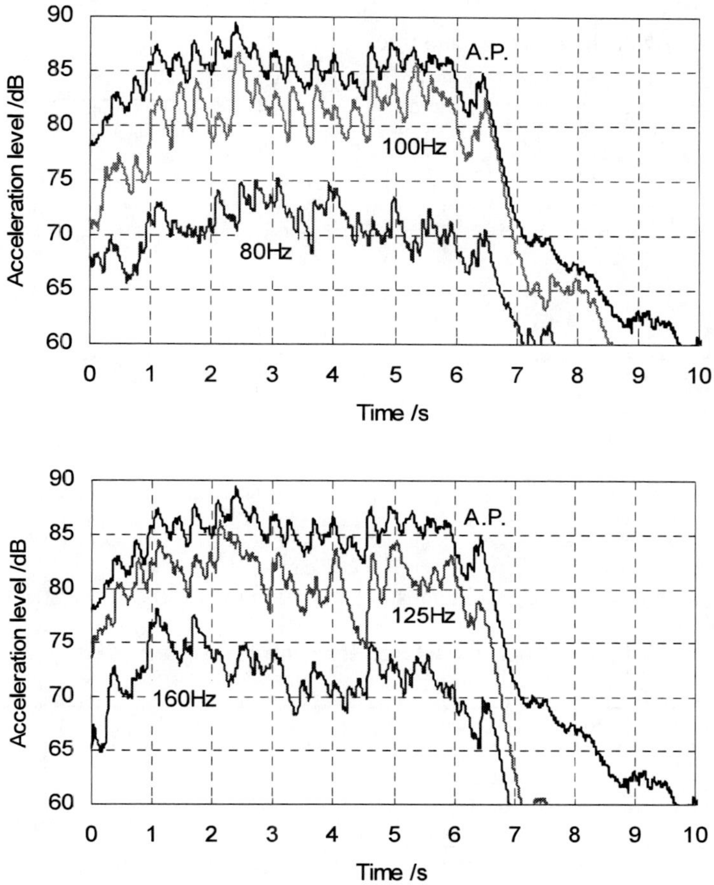

Figure 21. Acceleration level histories of bridge deck in different frequency components (V=69.6 km/h).

Figure 22 shows the distributions of the measured acceleration levels of rail and bridge deck versus train speed, and Table 4 lists the maximum and average acceleration levels of the rails, sleepers and deck, when the train ran on the ladder track and the slab track, respectively.

Based on the data from the figure and the table one can find that:

(1) When the train ran on the ladder track and the slab track, the average accelerations of rails were 129 dB and 122 dB, those of sleepers were 116.75 dB and 95.5 dB, and those of bridge deck were 85.75 dB and 90.0 dB, respectively. It should be noticed that the average speeds of the train on the common slab track and the ladder track were 46.5km/h and 69.35km/h respectively. Even in such a condition that the train speed on the ladder track was 50% higher than that on the slab track, the ladder track reduced the vibration by 4.5dB in maximum and 2.25 dB in average, respectively.

(2) Under the controlled speed 50km/h of the train on both two types of tracks, the maximum accelerations of rails on the two types of tracks were 124 dB, and those of sleepers were 113 dB and 97 dB, respectively, thus the acceleration of sleeper on ladder track was higher than that of slab track by 16 dB. While the accelerations of

bridge deck are 85 dB and 91 dB respectively, thus the acceleration of train on ladder track was lower than that on slab track by 6 dB.

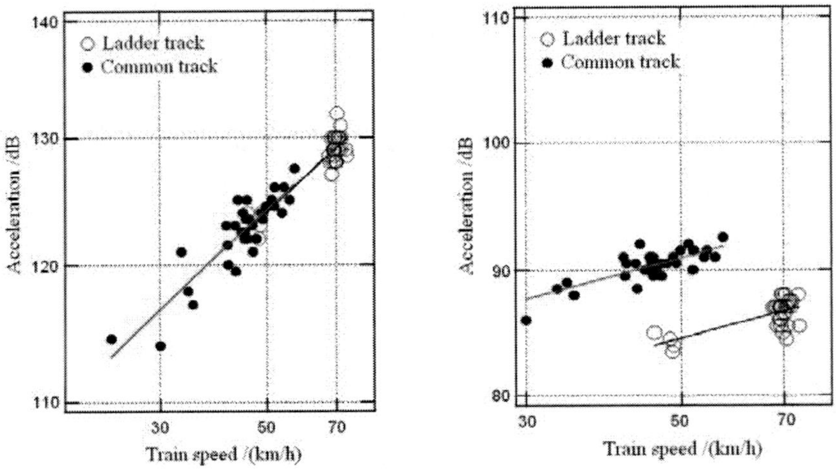

Figure 22. Distribution of measured maximum accelerations at rail (left) and deck (right) vs train speed.

Table 4. Measured acceleration level (unit: dB)

Position	Ladder track (Down line)			Common track (Up line)			Reduction (50 km/h)
	Maximum	Average	50 km/h	Maximum	Average	50 km/h	
Rail	132	129	124	127.5	122.0	124	0
Sleeper	118	116.75	113	98.5	95.5	97	+16
Deck	88	85.75	85	92.5	90	91	-6
Train speed /(km/h)	73.7	69.35	50	57.3	46.5	50	--

Shown in Figure 23 are the one-third octave frequency properties of the rails, sleepers and bridge deck when the train runs on two types of tracks, under the same train speed of 50 km/h.

Shown in Figure 24 are the transfer relations of the acceleration levels from rail to bridge deck of the two types of tracks, obtained by spectrum analysis on the measured data.

The insertion loss index in one-third octave frequency is adopted to compare the vibration reduction property at bridge deck when the train passing on the ladder track and on the common slab track (see Figure 25). The insert loss index is expressed by:

$$IL = 20 \times \log\left(\frac{a_{slab}}{a_{ladder}}\right) \tag{43}$$

where a_{slab} and a_{ladder} are the accelerations of bridge deck when the train passing on the ladder track and the common slab track, respectively.

From the sample distribution of insert loss one can see that the transfer ratios of the two types of tracks are different in different frequency bands. In most frequency bands of 1-5 Hz and above 10 Hz, the transfer ratio of ladder track is higher than that of the slab track, with the maximum difference 22.0 dB. In the frequency band of 5-10 Hz, the transfer ratio of ladder track is lower than that of slab track, with the maximum difference −5.8 dB. In all-pass frequency analysis, the average vibration reduction effect of ladder track is 8.6 dB better than the slab track.

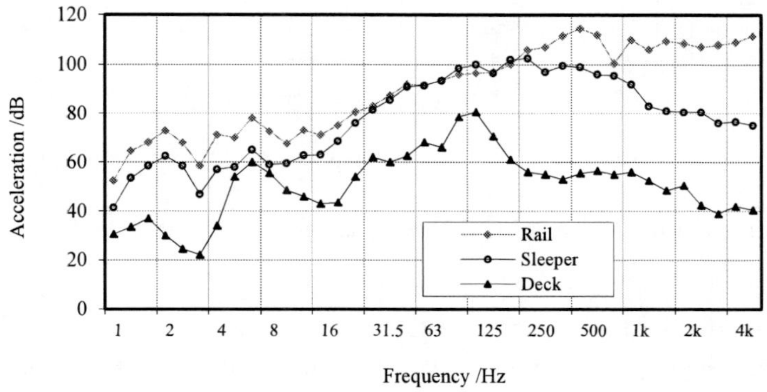

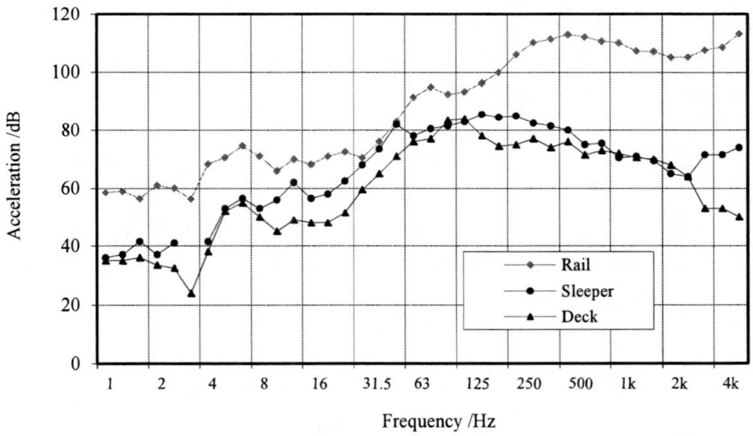

Figure 23. Accelerations measured at rails, sleepers and bridge deck of two types of tracks: ladder track (top) and common slab track (bottom).

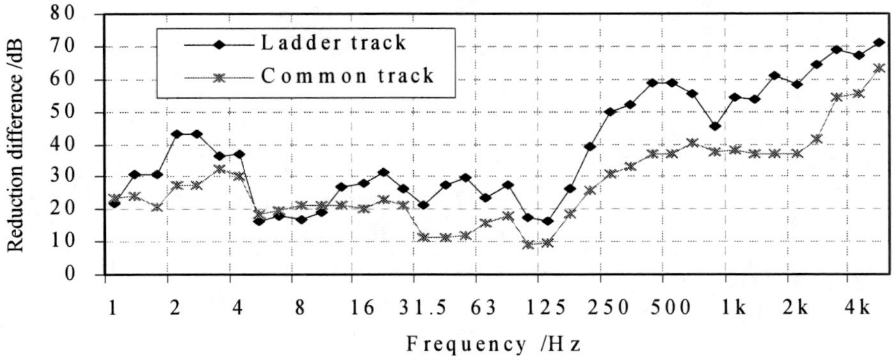

Figure 24. Comparison of the transfer ratios between two types of tracks.

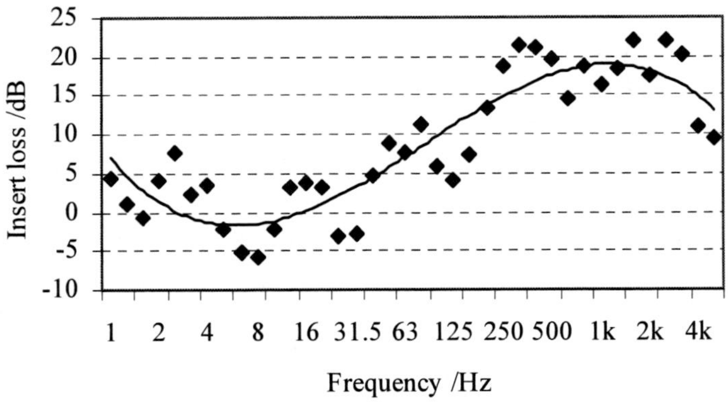

Figure 25. Insert loss of bridge vibrations under train passage on ladder track with respect to common slab track.

From the trend distribution one can see that the whole insert loss trend curve is within the positive value area, with the minimum value −2.0 dB and the maximum value 18.6 dB, and the average vibration reduction of the ladder track in all-pass frequency is 8.6 dB.

Figure 26 shows the one-third frequency distributions of bridge deck when train running on the bridge at different speeds. It can be seen that for a 20 km/h train speed difference, the acceleration difference is about 2.2 dB, and the maximum influence at certain frequency band is more than 10 dB.

4.2.2. Vibration of Ground

The ground accelerations were measured when the train running through the bridge on the two types of tracks. The vertical acceleration histories of ground under the train passage at 50 km/h are shown in Figure 27.

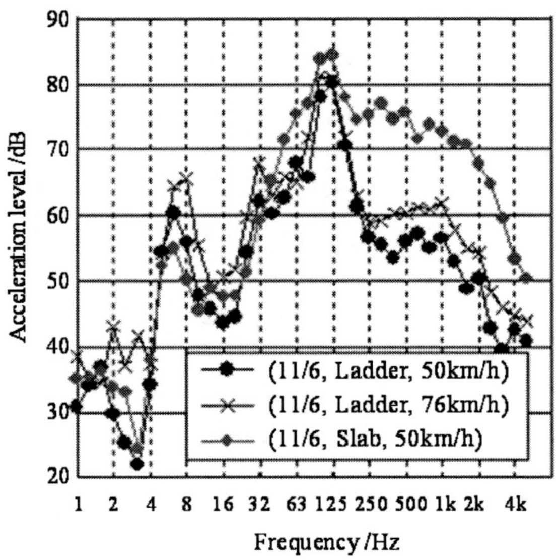

Figure 26. Influence of train speed on the responses of bridge deck.

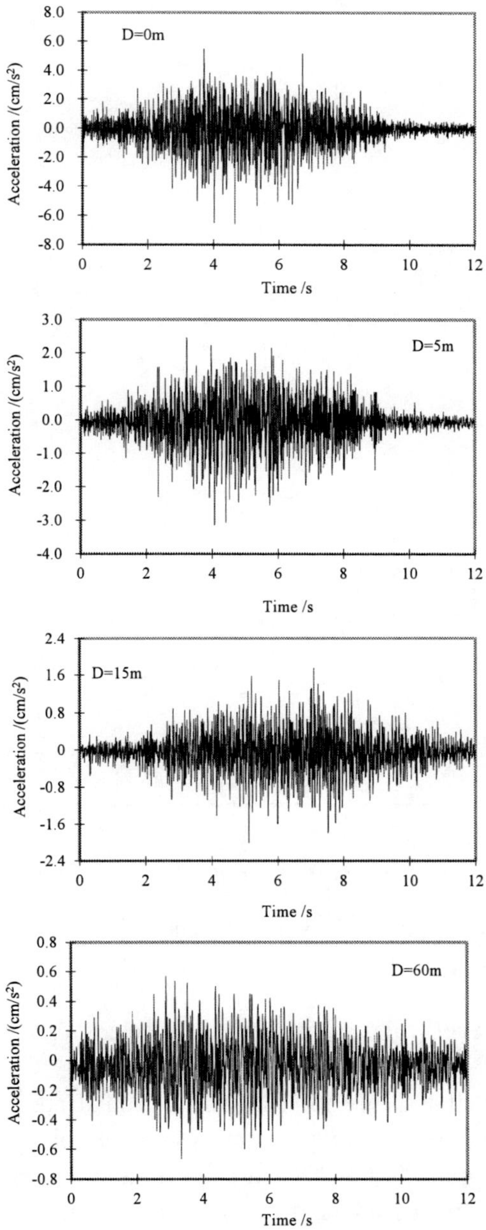

Figure 27. Vertical ground acceleration curves when the train running on the bridge.

Illustrated in Figure 28 are the distributions of ground acceleration vibration levels versus the distance to the bridge pier, where the plotted symbols are the measured ground acceleration levels under each train passage, while the solid lines and the dashed lines are the averages of them under train passage on the common slab track and the ladder track, respectively. The acceleration level is calculated by acceleration RMS when the train traverses the bridge, using the formula

$$L_a = 20\log\left(\frac{a}{a_{ref}}\right) \tag{44}$$

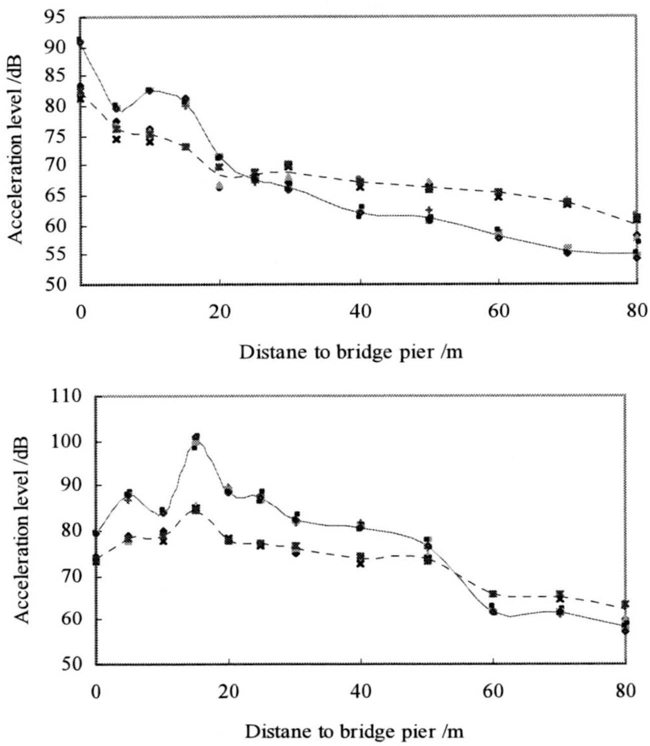

Figure 28. Distribution of vertical (top) and horizontal (bottom) ground acceleration levels when the train runs through the bridge.

where L_a is acceleration level (dB), a_{ref} is the base acceleration 1×10^{-6} m/s^2, a is the RMS acceleration (m/s^2).

For vertical vibration, the acceleration levels of ground are 55–91 dB when the train running on the common slab track, while the ones when the train running on the ladder track are 60–83 dB. The main trend is that the vertical vibration of ground attenuates with the distance increase, but an obvious rebound occurs at the distance of 10–20 m to the bridge pier. The vertical ground vibration is bigger when the train running on the common slab track than that when the train running on the ladder track at the distance less than 25 m, and it becomes smaller when the distance exceeds 25 m.

For horizontal (perpendicular to the longitudinal axis of the bridge) vibration, the acceleration levels of ground are 58–100 dB when the train running on the common slab track, while those when the train running on the ladder track are 62–85 dB. The main trend is that the horizontal vibration of ground also attenuates with the distance increase, but several obvious rebounds occur at the distances of 5 m, 15 m, 50 m and 70 m to the bridge pier. The horizontal ground vibration under the train passage on the common slab track is bigger than that under the train passage on the ladder track at the distance less than 53 m, and it becomes smaller when the distance exceeds 53 m.

The Fast Fourier Transform is adopted to analyze the data in frequency domain, which are the ground accelerations measured at 0m, 5m, 10m, 15m and 60m, corresponding to the characteristic points in Figure 28. The obtained one-third-octave frequency spectra of ground in vertical and horizontal directions are shown in Figures 29 and 30, respectively, from which the frequency characteristics of ground vibrations versus the distance to vibration source when the train running on the common slab track and the ladder track can be observed.

It can be observed from the figures that when the train runs on the ladder track, the frequency band is widened compared with that when the train runs on common slab track, and the peak vibration components in 50−80 Hz are greatly depressed. Therefore, the ladder track has certain reduction effect on ground vibration for the environmental vibrations induced by elevated rail transit bridges.

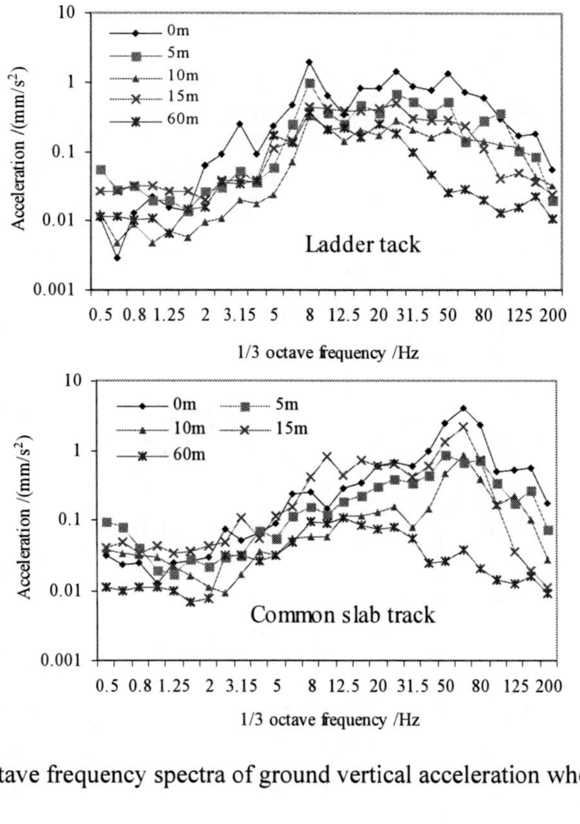

Figure 29. One-third-octave frequency spectra of ground vertical acceleration when the train runs through the bridge.

5. Influence Analysis and Prediction on Environmental Vibration Induced by Elevated Rail Transit

In this section, a 3D train load-pier-foundation-soil dynamic analysis model (Figure 31) is established to investigate the influences of different factors (train, bridge and soil) on ground vibration. On this basis, a prediction model for environmental vibration induced by trains running on elevated bridges is proposed according to the statistical regression analysis of the calculation results, and the model is verified by the measurement data.

The size of the model: 840 m in longitudinal direction of the bridge, 300 m in transverse direction, and 84 m in depth. The model includes a series of bridge piers (at least 10 piers). The soil is simulated by the 8-node isoparametric element. The FE grid size is 2.5 m near the pier, and 5 m in the far-field from the pier. The viscoelastic boundary condition is adopted. The damping ratio of soil is taken as 0.1, and the integration time step is 0.001s.

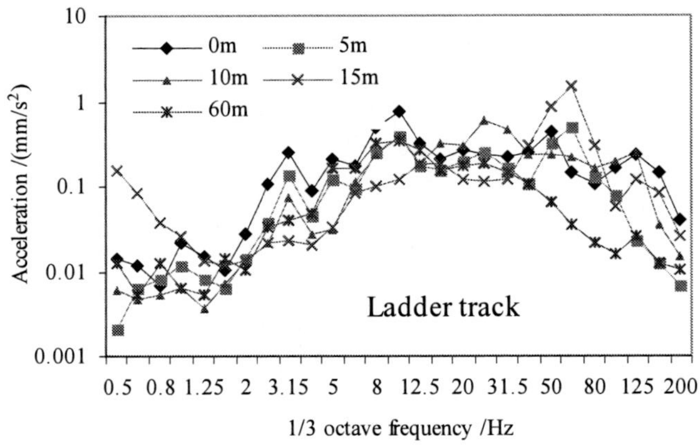

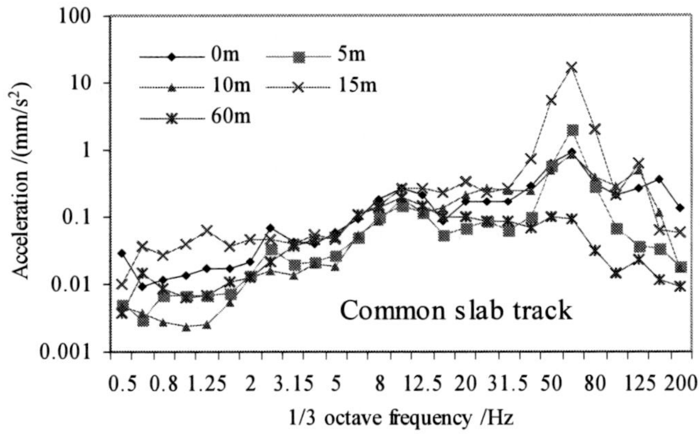

Figure 30. One-third-octave frequency spectra of ground horizontal acceleration when the train runs through the bridge.

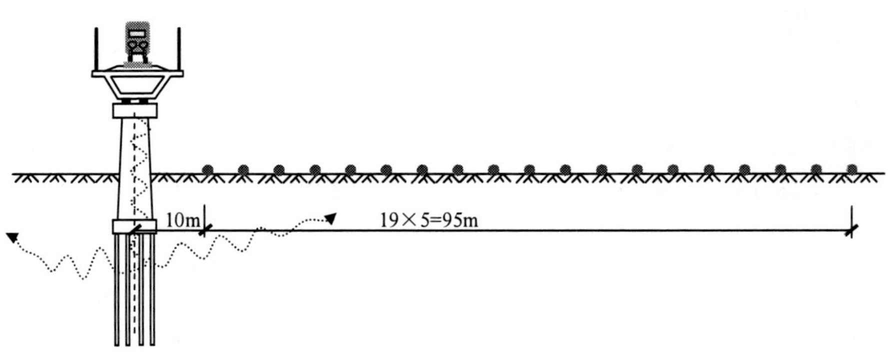

Figure 31. The 3D train-pier-foundation-soil dynamic analysis model.

5.1. Analysis of Environmental Vibration Induced by Elevated Rail Transit

5.1.1. Basic Input Data

The analysis is carried out by changing the main influence factors, based on a group of basic input parameters. The basic input parameters of the bridge, track, foundation soil and train vehicles are listed in Tables 5 to 8, respectively.

Table 5. Standard parameters of bridge

Superstructure and bearings				Pier foundation	
Span /m	30	Elastic modulus /(MPa)	325	Pile type	Round section
Cross-sectional area /m^2	4.132	Girder type	Box-shape	Section radius /m	0.5
Mass per unit length /(kg/m)	10330	Damping ratio of the structure	0.05	Pile length /m	40
Concrete grade	C40	Section moment of inertia /m^4	0.74	Number of piles	4
Support stiffness /(kN/m)	4.5×10^6	Support damping /(kN·s/m)	2000	Pier height /m	6.0

Table 6. Standard parameters of track

Rail	Mass per unit length /(kg/m)	51.514	Sleeper	Mass /(kg)	240
	Elastic modulus /(MPa)	2.06×10^5		Spacing /cm	62.5
	Section moment of inertia /(m^4)	2.04×10^4	Stiffness and damping between rail and sleeper	Stiffness /(MN/m)	300
	Damping ratio	0.01		Damping /(kN·s/m)	100

Table 7. Standard parameters of foundation soil

Density /(kg/m^3)	Compressive modulus / MPa	Poisson's ratio	Damping ratio
1960	200	0.3	0.1

5.1.2. Numerical Results

The vibration responses of ground surface at different points are analyzed, using the above basic parameters for train and bridge. Figure 32 shows the calculated vertical acceleration time history curves of the point 15 m from the bridge when the train runs at 60 km/h through the bridge, as well as the measured result on the Beijing Metro Line 5 for comparison.

To better compare the calculated results and the measured ones, Figure 33 depicts the 3D distributions of the vertical ground accelerations versus the distance and the time. It can be seen both the vibration amplitudes and the attenuation trends of the numerical results are in good accordance with the measured ones.

Table 8. Vehicle parameters

Item	Motor-car	Trailer-car
Length of coach L /m	22.1	19.0
Distance between two bogies $2s$ /m	15.7	12.6
Distance between two wheel-sets $2d$ /m	2.5	2.2
Mass of car body m_c /kg	26400	22800
Mass moment of inertia of car body around y-axis J_c /(kg-m^2)	1.07×10^6	6.62×10^5
Mass of bogie m_t /kg	2550	2550
Mass moment of inertia of bogie around y-axis J_t /(kg-m^2)	2410	2410
Mass of wheel-set m_w /kg	1760	1760
Stiffness of the primary suspension system k_w /(kN/m)	1500	1500
Damping of the primary suspension system c_w /(kNs/m)	5	5
Stiffness of the secondary suspension system k_t /(kN/m)	275	275
Damping of the secondary suspension system c_t /(kNs/m)	30	30

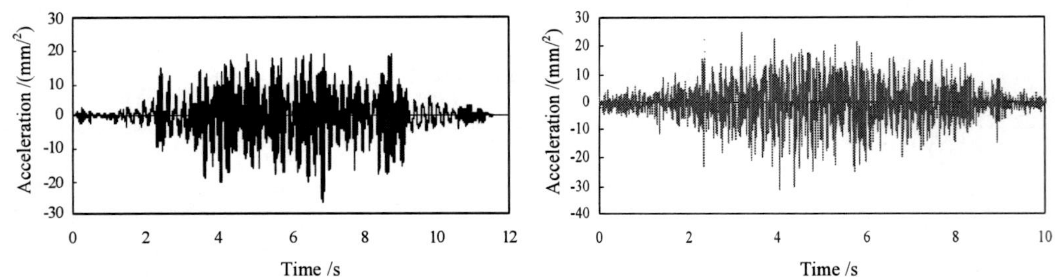

Figure 32. Calculated (left) and measured (right) vertical acceleration time history curves.

5.1.3. Influential Factor Analysis of Free-field Vibration Induced by Elevated Transit

The effects of several main parameters on the free-field ground vibration are analyzed by changing the input data to the train load-pier-foundation-soil dynamic system model.

5.1.3.1. Bridge Span

To analyze the influence of bridge span on the ground vibration, different spans ranging from 7.5 m to 60 m are considered in the analysis, while the other parameters remain the basic input. The result shows that the ground vibration increases with the span length, for it mainly depends on the pier-top force of the bridge, and the longer the bridge span, the heavier the train load, and the much bigger the force acting on the pier-top. By statistical regression analysis, the relationship between the adjustment for ground vibration level and the bridge span length is established as follows:

$$L_{bs} = 0.165 \times l - 5 \tag{45}$$

in which L_{bs} is the adjustment of ground vibration level (dB); l is the bridge span length (m).

One can see from the formula that when the bridge span changes from 7.5 m to 60 m, the ground vibration level increases about 8.5 dB.

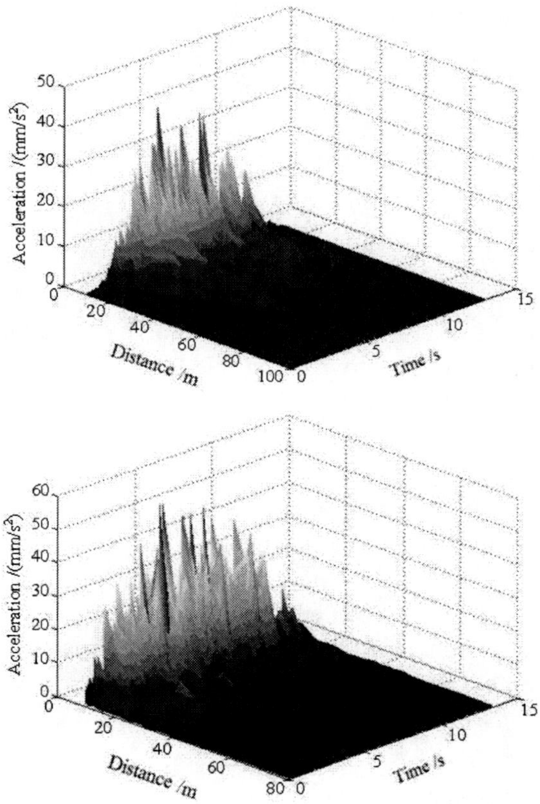

Figure 33. The 3D distributions of computed (top) and measured (bottom) vertical ground accelerations versus the distance and the time.

5.1.3.2. Bridge Girder Mass

To analyze the influence of bridge girder mass on the ground vibration, different mass densities per unit span length ranging from 6 t/m to 14 t/m are considered for the girders in the analysis, while the other parameters remain the basic input. The result shows that the ground vibration increases with the girder mass. By statistical regression analysis, the relationship between the adjustment for ground vibration level and the girder mass density is established as follows:

$$L_{\mathrm{bm}} = 0.62 \times (8 - \bar{m}) \tag{46}$$

in which L_{bm} is the adjustment of ground vibration level (dB); $\bar{m}$ is the girder mass density per unit span length (t/m).

One can see from the formula that when the mass density changes from 6 t/m to 14 t/m, the ground vibration level drops about 5 dB.

5.1.3.3. Bridge Foundation

In this case, different lengths of pile foundation of the bridge ranging from 0 m to 60 m are considered, while the other parameters remain the basic input. The result shows that the ground vibration increases with the pile length. By statistical regression analysis, the relationship between the adjustment for ground vibration level and the pile length is established as follows:

$$L_{\mathrm{pl}} = \begin{cases} 6.6 + 0.22(30 - l) & (l < 30\mathrm{m}) \\ 6.6 & (30 \le l \le 40\mathrm{m}) \\ 6.6 + 0.20(40 - l) & (l > 40\mathrm{m}) \end{cases} \tag{47}$$

in which L_{pl} is the adjustment of ground vibration level (dB); l is the pile length (m).

One can see from the formula that when the pile length changes from 0 m to 60 m, the ground vibration level drops about 10 dB.

5.1.3.4. Bearing Stiffness

There are three kinds of bridge bearings, rigid bearing, common rubber bearing and shock absorption bearing, included in the study, with the corresponding stiffness values of 2×10^8 kN/m, 1.9×10^6 kN/m and 3×10^5 kN/m, respectively, while the other parameters remain the basic input. The result shows that the ground vibration increases with the stiffness of the bearing. By statistical regression analysis, the adjustments of ground vibration levels for the three types of bearings are obtained as follows:

$$L_{\mathrm{bt}} = \begin{cases} +6 & \text{(rigid bearing)} \\ 0 & \text{(common robber bearing)} \\ -5 & \text{(shock absorption bearing)} \end{cases} \tag{48}$$

in which L_{bt} is the adjustment of ground vibration level (dB) for different bearings.

5.1.3.5. Train Speed

Different train speeds ranging from 40 km/h to 100 km/h are considered in the analysis, while the other parameters remain the basic input. The result shows that the ground vibration increases with the train speed. By statistical regression analysis, the relationship between the adjustment for ground vibration level and the train speed is established as follows:

$$L_{\mathrm{tv}} = 0.075 \times V - 4.5 \tag{49}$$

in which L_{tv} is the adjustment of ground vibration level (dB); V is the train speed (km/h).

One can see from the formula that a 10 km/h increase of train speed induces about 0.75 dB increase in the ground vibration level.

5.1.3.6. Axle-load of Vehicle

The ground vibrations are calculated under different axle-loads of the vehicle ranging from 95 kN to 160 kN, while all the bridge parameters remain the basic input. The result shows that the ground vibration increases with the increase of axle-load of the vehicle. By statistical regression analysis, the relationship between the adjustment for ground vibration level and the axle-load of vehicle is established as follows:

$$L_{tp} = 0.08 \times P - 8.0 \tag{50}$$

in which L_{tp} is the adjustment of ground vibration level (dB); P is the axle-load of vehicle (kN).

One can see from the formula that when the axle-load changes from 95 kN to 160 kN, the ground vibration level increases about 5.2 dB.

5.1.3.7. Foundation Soil Property

Generally speaking, the soil property depends on the shear-wave velocity c_s and the Poisson's ratio ν. Herein only the shear-wave velocity is used to indicate the soil property, since the Poisson's ratio has little effect on the ground vibration. In the analysis, however, the shear-wave velocity of the soil cannot be input as an independent parameter, which is determined by the formula $c_s = \sqrt{E_s / 2(1+\nu)\rho}$, where E_s is the elastic modulus of soil and ρ the density. The shear-wave velocity ranges from 140 m/s to 450 m/s according to the general scope of the elastic modulus and density of medium soft and medium hard soil. The result shows that the ground vibration increases with the shear-wave velocity of soil. By statistical regression analysis, the relationship between the adjustment for ground vibration level and the shear-wave velocity is established as follows:

$$L_{sv} = 0.0234 c_s - 8.591 \tag{51}$$

in which L_{sv} is the adjustment of ground vibration level (dB); c_s is the shear-wave velocity of soil (m/s).

5.2. Prediction of Environmental Vibration Induced by Elevated Rail Transit

The prediction formula for free-field vibration induced by elevated rail transit system is established, which considers the span length and mass of beam, type of bridge foundation, type of bridge bearing, speed of train, axle-load of vehicle, and shearing velocity of subsoil. The prediction formula is expressed as follows:

$$L_z = A + B\lg(r) + L_{bs} + L_{bm} + L_{pl} + L_{bt} + L_{tv} + L_{tp} + L_{sv} \tag{52}$$

where: $A+B\times\lg(r)$ shows the attenuation of ground vibration versus distance r between the receiver and the source, with constants $A=100$ and $B=-23$. The other terms are, respectively, the adjustments of ground vibration levels for the span-length and mass of beam, type of bridge foundation, type of bridge bearing, speed of train, axle-load of vehicle, and shearing velocity of subsoil. Details of them can be found in Table 9.

Table 9. Adjustments for the prediction formula

Parameters	Adjustment of vibration level /dB		
Length of beam L_{bs}	L_{bs}=0.165×l-5.0		
Mass of beam L_{zm}	L_{zm}=0.62×(8-$\bar{m}$), where $\bar{m}$ is mass density per unit length (t/m).		
Type of bridge foundation (L_{pl})	Platform foundation		6.6
	Pile	l≤30m	0.22 (30-l)
		30<l≤40m	0
		l>40m	0.2 (40-l)
Type of bearing L_{bt}	Rigid bearing		6
	Common bearing		0
	Isolation bearing		-5
Train speed L_{tv}	L_{tv}=0.075×V-4.5		
Axle-load L_{tp}	L_{tp}=0.08×P-8.0		
Soil shear-wave velocity L_{sv}	L_{sv}=0.0234×C_s-8.6		

Using the prediction formula (52), with the real parameters of train vehicles, bridge structures and soils, the ground vibration levels at the site close to an elevated bridge on the Beijing Metro Line 5 are predicted, as shown in Figure 34, where the train speed is 60 km/h. The experimentally measured data are also plotted in the figure for comparison.

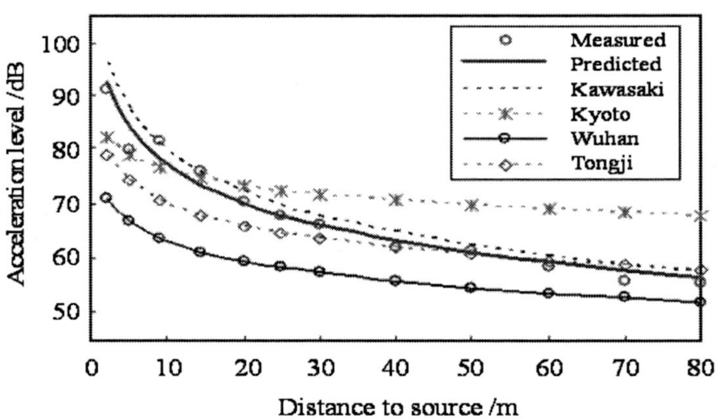

Figure 34. Prediction of ground vibration induced by elevated rail transit system.

There are some other prediction formulae for ground vibrations induced by elevated rail transit system proposed by the researchers in China and abroad [23, 24], which are based on the analytical and numerical results, or on the in situ measured data. Some typical models are as follows:

(a) The prediction formula proposed by the Kyoto Institute for Environmental Pollutions in Japan:

$$VL = 85 - 9\lg(d) \tag{53}$$

where d is the distance (m) from the observation point to the source (longitudinal center).

(b) The prediction formula proposed by the Kawasaki Institute for Environmental Pollutions in Japan:

$$VL = 93.92 - 25\lg(d) + 0.18V \tag{54}$$

where d is the distance (m) to the bridge longitudinal center, V is the train speed (km/h).

(c) The prediction formula proposed by the Wuhan University of Technology:

$$VL = 63 - 12.3 \times \lg(\frac{r}{10}) \tag{55}$$

where r is the distance (m) to the bridge longitudinal center.

(d) The prediction formula proposed by the Tongji University:

$$VL = 70 - 13.6 \times \lg(\frac{r}{10}) \tag{56}$$

where r is the distance (m) to the bridge longitudinal center.

The predicted results of these models are also plotted in Figure 34. From the comparison among them and the measured data, one can conclude that:

The prediction formula (52) proposed in this chapter considers more parameters than the other models, and thus has better applicability.

Compared with formulae (53), (55) and (56), formula (52) gives better prediction precision in both magnitudes and distribution trends.

Compared with formula (54), formula (52) shows a similar distribution trend, while it gives better prediction values at the locations near the source.

CONCLUSION

In this chapter, the previous and current development of research on the environmental vibration caused by the elevated rail transit are investigated and summarized. Based on the studies of predecessors, a 3D interaction model of coupled train-bridge-ground-building vibration system is established, which is composed of the train-track-bridge submodel and the pier-foundation-soil-building submodel. The two submodels interact with each other through the contact forces on the pier tops of the viaduct.

A field measurement was carried out at the elevated trial section of the Beijing Metro Line 5, where both the ladder track and the common non-ballasted slab track are installed to analyze the ground vibrations induced by urban rail transit. The vibration responses were measured of rails, sleepers, bridge deck as well as the surrounding ground surface when the train running on the ladder track and the common non-ballasted slab track. The data are analyzed in both the time domain and the frequency domain. It can be concluded from the experimental analysis that the ladder track installed on the elevated bridge has special vibration reduction property.

The influences of the main factors on ground vibration are investigated. According to the statistical regression analysis of the calculation results, a prediction formula for environmental vibration induced by the elevated rail transit is established, in which are considered the bridge span lengths, girder masses, foundation types, bearing types, train speeds, axle-loads of vehicle, and shear-wave velocities of subsoil. The prediction formula is then applied to the site near an elevated bridge on the Beijing Metro Line 5, using the real parameters of train vehicles, bridge structures and subsoil. The numerical results demonstrate that the proposed formula can efficiently predict the ground vibration level.

Acknowledgments

This study is sponsored by the National Basic Research Program of China ("973" Program, 2013CB036203), the Fundamental Research Funds for the Central Universities (2012JBM075), and the Natural Science Foundation of China (51078029).

References

[1] Zheng J. (2008). *The High-speed Railway Bridges in China*. Beijing: Higher Education Press.

[2] Xia H., Zhang N., Guo W.W. (2011). Application of train-bridge-interaction analysis to bridge design of high-speed railways in China. *Proc. IWHIR'2011, Hong Kong*: p.355–372.

[3] Bian X.C. (2004). Analysis of viaduct-ground vibrations due to high-speed train moving loads. *Journal of Vibration Engineering*, 19: 438–445.

[4] Chang L., Spanos P.D., Yan W.M., et al. (2009). Ground vibrations induced by vehicles moving on elevated roads: random vibration model and experimental data. *Proc. of ISEV'2009*, Beijing: p.169–177.

[5] Chen J.G., Xia H., Yao J.B. (2009). Experimental study of environmental vibration induced by viaduct rail transit. *Proc. of ISEV'2009*, Beijing: p.358–362.

[6] Chua K.H., Lo K.W., Koh C.G. (1997). Performance of urban rail transit system: vibration and noise study. *Journal of Performance of Constructed Facilities*, 5: 67–75.

[7] Lombaert G, Degrande G. (2009). Ground-borne vibration due to static and dynamic axle loads of InterCity and high-speed trains [J]. *Journal of Sound and Vibration*, 319: 1036–1066.

[8] Degrande G. and Lombaert G. (2000). High-speed train induced free field vibrations: in situ measurements and numerical modeling. *Proc. Wave2000*, Okayama: p.29–42.

[9] He X.W., Hayashikawa T., Kawatani M. and Matsumoto T. (2009). Analytical evaluation on train-induced ground vibration around shinkansen viaducts. *Proc. of ISEV'2009*, Beijing: p.87–92.

[10] Jiang T., Zhang X. (2004). In Situ Experimental and Numerical predication of Environmental Vibration by Urban Viaduct rail transit. *Journal of Tongji University (Natural Science)*. 32: 565–569.

[11] Li Z.Y., *et al.* (2007). Analysis of ground vibration induced by high-speed train. *Journal of TongJi University (Natural Science)*, 35: 909–914.

[12] Takemiya H. (2008). Analyses of wave field from high-speed train on viaduct at shallow/deep soft grounds. *Journal of Sound and Vibration*, 310: 631–649.

[13] Yang Y.B., Yau J.D. and Hsu L.C. (1997). Vibration of simple beams due to trains moving at high speed. *Journal of Engineering Structures*, 19: 936–944.

[14] Jiang T., Zhang X. (2004). In situ experimental and numerical predictions of environmental vibration induced by urban viaduct rail transit. *Journal of Tongji University*, 32(5): 565–569.

[15] Wakui H., Matsumoto N., Okuda H. and Asanuma K. (2002). Performance and application of ballast ladder track. *New Railway structure*, 52(4): 32–34.

[16] Wakui H., Matsumoto N., Okuda H. and Asanuma K. (2006). Structure and design of ladder sleeper. *New Railway Structure*, 56(3): 26–28.

[17] Inoue H., Oyado M., Tottori S., et al. (2006). Mechanical properties and load-bearing performances of Ladder sleeper. *RTRI Report*, 10(9): 33–38.

[18] Okuda H., Sogabe M., Matsumoto N., Wakui H. (2006). Environmental performance improvement of railway structural system using ladder track. *Research Report of Railway Technical Research Institute*. Japan.

[19] Zou Y.W. (2006). Analysis of ladder track's effect on reducing vibration of elevated bridge in rail transit. *Master Dissertation of Beijing Jiaotong University*, Beijing, China.

[20] Sheng X., Jones C.J.C., D. Thompson J. (2006). Prediction of ground vibration from trains using the wavenumber finite and boundary element methods. *Journal of Sound and Vibration*, 293(3–5): 575–586.

[21] Takemiya H., Goda K. (1997). Prediction of ground vibration induced by high-speed train operation. *The 5th International Conference on Sound and Vibration. Adelaide*, 5: p.2681–2688.

[22] Tian C.L. (1993). Prediction method for vibration of elevated rail transit. *Noise and Vibration Control*, 6: 35–36.

[23] Wang B.S., Xu Z.J., Yang Y.W. (2008). Predicting train-induced vibration of building in planning. *Journal of Zhejiang University (Engineering Science)*, 42(9): 1502–1505.

[24] Lei X.Y., Liu Q.J., Zhu C.J. (2008). Prediction and Assessment of Environment Vibration Induced by Elevated Railway. *Noise and Vibration Control*, 12(6): 108–112.

[25] Xie W.P., Zhao W., Chen W.P. (2007). Discussion of Some Questions In Technical Guidelines for Environment Impact Assessment of Urban Rail Transit. *Journal of Huazhong University of Science and Tchnology*, 4: 1–4.

[26] Zhang X. (2002). Practical and Theoretical Research on the Environmental Vibration Caused by Elevated Railroad Traffic. *Doctoral Dissertation of Tongji University*, Shanghai, China.

[27] Qiu J.J. (2005). *Practical and evaluation research on the environmental vibration caused by elevated railroad traffic*. Doctoral dissertation, Wuhan University of Technology.

[28] Miller G. F. and Pursey H. (1954). The field and radiation.

[29] Wolf J.P. (1994). *Foundation vibration analysis using simple physical models*. Englewood Cliffs, NJ: PTR Prentice Hall.

[30] Rücker R. (1982). Dynamic behavior of rigid foundations of arbitrary shape on a half-space. *Earthquake Engineering and Structural Dynamic*s, 10: 675–690.

[31] Xia H. (2010). *Traffic induced environmental vibrations and controls*. Beijing: Science Press.

[32] Ministry of railways of China. (1996). *Administration standard for track irregularities in trunk railways*. Beijing: China Railway Publishing House.

[33] Chen J.G. (2009). Prediction and Peremeter research on Viaduct Train-induced Environmental Vibrations. *Doctoral Dissertation of Beijing Jiaotng University*, Beijing, China.

[34] Fujikake T. (1986). A prediction method for the propagation of ground vibration from railway trains. *Journal of Sound and Vibration*, 111(2): 357–360.

In: Traffic Induced Environmental Vibrations and Controls ISBN: 978-1-62618-240-0
Editors: H. Xia and R. Calçada

Chapter 12

TRAIN-INDUCED VIBRATIONS OF ELEVATED RAILWAY STATIONS

Nan Zhang * ***and Qian Cheng***
School of Civil Engineering, Beijing Jiaotong University,
Beijing, China

ABSTRACT

In this chapter, the design and functional information of some typical elevated railway stations are introduced, and the structural characteristics and vibration problems of elevated railway stations are summarized. The dynamic interaction system of elevated railway stations and high-speed trains under the loads of train running, braking or starting are analyzed based on the vehicle model and the wheel-rail interaction assumptions considering the longitudinal motion. As a case study, the Wuhan Railway Station is analyzed with the cases of the high-speed train and ordinary speed train passing or braking in the station building, the effect of train speed, types and loading positions are analyzed and concluded.

INTRODUCTION

In recent years, there occurred some large-scale multi-function railway stations composed of underground works, elevated bridges and elevated large-span station halls in China [1, 2], such as the Shanghai South Station, the Beijing South Station, the Guangzhou New Railway Station and the Wuhan High-speed Railway Station, and some others. These stations unify the functions of metro station, urban railway station, conventional railway station and high-speed railway station as a whole, and thus are very convenient for the transfer of passengers between different transit media. They are also appreciated for their novel appearance and high efficiency. While on the other hand, trains passing, braking or starting on the elevated structure of the station will cause the dynamic interaction between the train and the elevated

---

* E-mail address: nanzhang@263.net.

structure. The vibration may cause the fatigue failure of the structure and shorten the service life of the station, and affect the comfort of passengers and station staff. The problem of running train induced vibrations and noises of elevated railway structures have been complained by passengers in the vehicle and the station hall. This problem has been concerned and is under study by researchers and designers in China.

TYPICAL ELEVATED RAILWAY STATIONS

In the past decades, with the rapid development of high-speed railways, and due to the demands on convenience and efficiency of traffic transfer, some old stations were reformed and some new large-scale elevated stations reflecting modern technological levels have been built in many European countries.

(1) Berlin Central Station

The Berlin Central Station (Berlin Hauptbahnhof), cost about 13 billion Euros and more than 10 years of construction, is the largest engineering project in Germany after the War [3]. It began full operation on 28 May 2006. The main structure of the station consumed 8.5 million tons of steel and 500,000 cubic meters of concrete, and the roofs and the walls are with more than 9,000 pieces of special glass.

The Central Station is a comprehensive large-scale transfer center, covering an area of 15,000 square meters. The main part of the station is a 5-storey transfer hall. The top layer is the east-west platform 12m elevated from the ground, and the bottom layer is the north-south platform 15m underground. The three middle layers are for transfer, including the entrance hall on the ground floor, distributed of shops, restaurants and other ancillary facilities. A sectional perspective view of the station is shown in Figure 1.

Figure 1. Sectional perspective view of the Berlin Central Station.

Once it opened, the Berlin Central Station immediately became the center of the European railway system, which connects all transports to a station in the city center, and provides large-scale vertical transfer in a hall. In addition to high-speed trains and other long-distance trains, in this station stop suburban railways, metros, trams, buses, taxis, or even bicycles and tourist tricycles.

From the Berlin Central Station, we saw the new design concept for the metropolitan railway stations: making full use of traffic advantage of city central area, integrating multi-transports for efficient transfer, achieving "zero distance" vertical transfer, and minimizing conflicts of ground transportation. This is the future trend in railway station construction.

(2) Lisbon Orient Station

The Orient Station in Lisbon was originally a transportation hub station to greet the 1998 World Expo [4]. For adapting to the effective convergence of high-speed trains and other transportations, the station was completed on May 19, 2008 after continuous expansion. The Portuguese government called this station as a modern palace hall and one of the marvelous landmarks of Lisbon.

The Oriental station is a large transport hub integrating railway station, bus terminal, car parking and underground subway platforms. The railway platforms locate on the second floor elevated 11m above the ground and covered with gorgeous translucent glass and white steel roof frames, and so it is called a "hillside birch forest", as shown in Figure 2.

Figure 2. Scenery graph of the Lisbon Orient Station.

As a new gate of Lisbon, the Orient Station becomes the most influential building in the New Orient District, not only because it is a perfect transportation hub, but also because it reflects the architectural aesthetic as well as the best coordination with the surrounding environment.

In the last decades, with the rapid development of high-speed railways in China, the construction of railway stations also reached a climax [1]. Since the beginning of this century, the Ministry of Railways has organized extensive investigation on the railway stations in the developed countries, and proposed five requirements of "functional, systematic, advanced, cultural and economical" on planning and construction of railway stations. More and more elevated railway stations have been built and opened to operations, with the function developed from a sole passenger transport to the city transport hub and integrated services.

(1) Nanjing South Station

The Nanjing South Station, linking five railway lines, is an important railway terminal station, as shown in Figure 3. As one of the five originating stations of Beijing-Shanghai high-speed railway, it opened on June 28, 2010. A total construction area of the station is about 458,000 m^2, and the area of main station building is about 281,500 m^2.

The Nanjing South Station is a transport hub integrating railway, roadway, metro and bus systems. The railway platforms are on the second floor, with the length of 660m and the width of 450m, elevated 12.5m above the ground. There are 15 platforms in the station; platforms 1 to 5 are for the Beijing-Shanghai High-Speed Railway, 6 to 11 for the Shanghai-Nanjing-Wuhan-Chengdu Railway and the Nanjing-Hangzhou Railway, 12 to 15 for the Nanjing-Anqing Railway. The waiting room is on the third floor elevated 24.5m above the ground.

As a representative of the modern high-speed railway stations in China, the Nanjing South Station integrates in design the up-to-date technologies, environmental features, urban culture and traffic building space, presenting a distinctive characteristic with the combination of significance, functionality, systematization, progressiveness and culture.

Figure 3. Aerial view of the Nanjing South Station.

(2) Beijing Southern Station

The New Beijing Southern Railway Station is the largest railway station in Asia, with a total area of 94 hectares and total budget of 12.5 billion RMB. The construction area of the station is more than 220,000 m^2, which may contain 10,500 passengers in the waiting hall. There are 24 tracks and 13 platforms in the station, including 5 tracks and 3 platforms for conventional trains, 12 tracks and 6 platforms for high-speed trains, and 7 tracks and 4 platforms for inter-city trains.

Shown in Figure 4 are the outline and the main structure of the New Beijing Southern Railway Station, which is a multifunctional railway station combining high-speed railway, common railway, subway lines, urban railway, buses and taxis as a whole. The station has five floors. With the elevated floor for taxis and social cars, the ground floor for buses and passengers entrance. The minus 1 floor for vehicle parking and traffic exchange and passengers' exit, and also reserved the connections to the urban railway, the minus 2 floor for Subway Line 4, and the minus 3 floor for Subway Line 14. The highest roof of the hall is 40 m above the ground.

The new Beijing Southern Railway Station will be the railway terminal for high-speed railways, inter-city railways and urban railways. The Beijing-Tianjin inter-city railway, with its total length of 115 km and the train speed of 200 km/h, had opened before the Beijing Olympic Games, 2008. The Beijing-Shanghai high-speed railway, with the total length of 1300 km and the train speed of 350 km/h, has been in service since 2011.

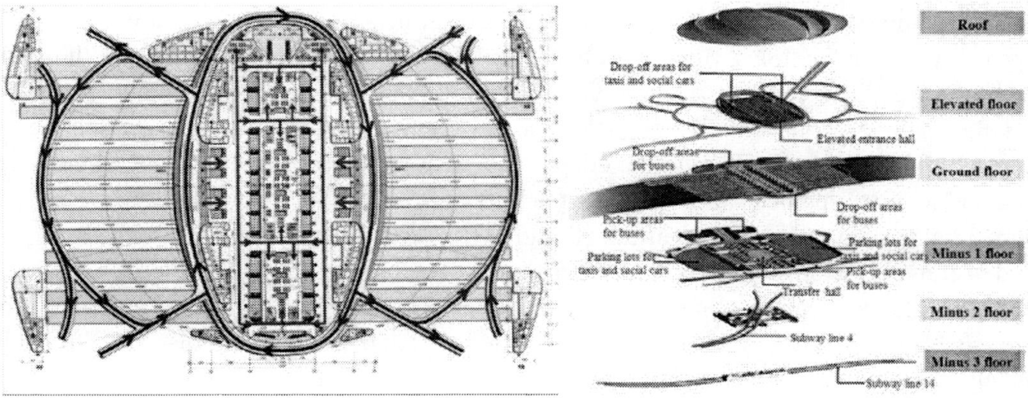

Figure 4. The New Beijing Southern Station.

(3) Guangzhou South Station

The Guangzhou South Railway Station, opened in January 2010, is one of the four major railway passenger centers planned by the Ministry of Railways of China. It is not only one of the three originating stations on the Wuhan-Guangzhou high-speed railway, but also a junction station connecting the Guangzhou-Shenzhen-Hong Kong high-speed railway, the Guangzhou-Zhuhai intercity rail transit and the Guangzhou-Maoming railway.

The New Guangzhou South Station has a total construction area of 565,800 m^2, with a 580 m length in north-south direction and a 475m width in east-west direction. The height of the station exceeds 50 m and there is a main hall with a large-span arch roof. The whole building includes the main station building, platform-pillar-free awning, elevated platforms, parking lots under platform and service rooms. The main structure includes four layers, with three layers on the ground and one underground. The underground layer is the entry and exit hall for metros, connecting Metros Line 2 and Line 7. The railway platforms are on the ground floor, the border ports and VIP room, and the waiting hall for passengers on the third floor. Its sectional schematic drawing is shown in Figure 5.

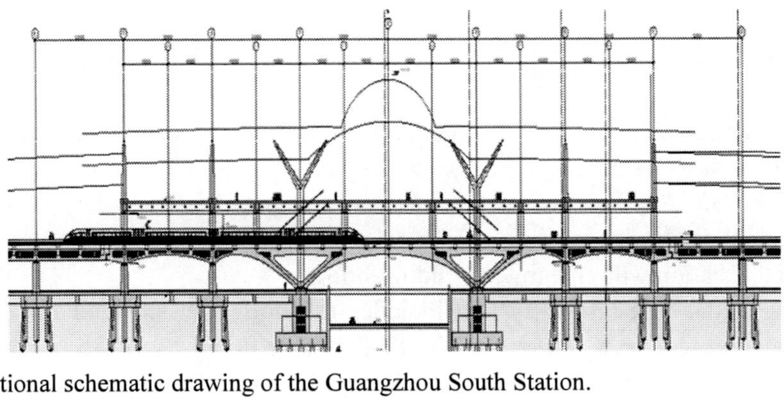

Figure 5. Sectional schematic drawing of the Guangzhou South Station.

The new Guangzhou Railway Station is a 3-storey building. The first floor is for entrance exit with a height of 12 m, and the second floor for platforms with a height of 9 m. The third floor is the waiting hall with a height of 27 m. The structure of the station adopts the frame system. The frame columns of the first floor are the bridge piers with the section of 3m×6m, and the main column grid of 32m×21.5m. The roof girders of the first layer are the railway beams supporting the trains. Both the columns and the girders are much thicker than those for civil buildings. The columns of the second floor are relative slender, which are 1.5m×2m, and the main grid is 32m×21.5m. Most girders adopt PC structure with the section of 1.2m×2.6m, while the 44.75 m girder adopts steel truss structure. The third floor adopts steel columns with the diameter of 2.0 m and thickness of 6 cm, and the column grids are from 68m×32m to 68m×64m. The roof adopts light steel structure. The overall stiffness of this kind of frame structure system is relative small and the vibration-resistant ability is weak.

In the station, there are in total 28 tracks, with 24 for arrival and departure and 4 for through. The train can pass through the station on the main tracks with the speed of 200 km/h, namely, through the frame structure. The mass and stiffness of the second floor with waiting rooms are relatively small, which is not help to control vibration due to the train. Structural joints are set between the main tracks and the arrival and departure tracks, which can effectively reduce the direct vibration impact to the waiting room,

In addition, some other large-scale elevated stations for high-speed railways, such as the Shanghai Hongqiao Station, Wuhan Station, Shijiazhuang South Station, Zhengzhou station, Shenzhen Station, have been opened to public or are under construction.

STRUCTURAL SYSTEMS AND VIBRATION FEATURES OF ELEVATED STATION

The elevated railway stations must meet such basic requirements as ease passenger flows, safe boarding and landing, comfortable environment, compact layout and convenient management. The selection of structure type for elevated stations is decided by the synthetic factors like the spatial locations in the environment, city planning, architectural forms, construction methods and passengers flow organizations, etc; considering the overall balance and harmonization.

3.1. Structural Systems of Elevated Stations

Elevated station is neither a simple station construction nor a single bridge structure, but a combined structure of station and bridge [5, 6]. According to the structural system, elevated stations can be divided into two categories: the integrated house-bridge system and the separated station-and-bridge system.

(1) Elevated station with integrated building-bridge system

The integrated house-bridge system includes the bridge system and the spatial frame system, which are steel or PC structures with three or four floors, as shown in Figure 6.

The bridge structure system is constructed of girders, piers and foundations, and then the platforms are built on the bridge, see Figure 6 (left). For this system, the structural performance is similar to that of common bridge, but higher stiffness and better stability are required. Since the mass center is mainly concentrated at the top part of the structure, it is sensitive to earthquake and structural vibrations induced by the running trains.

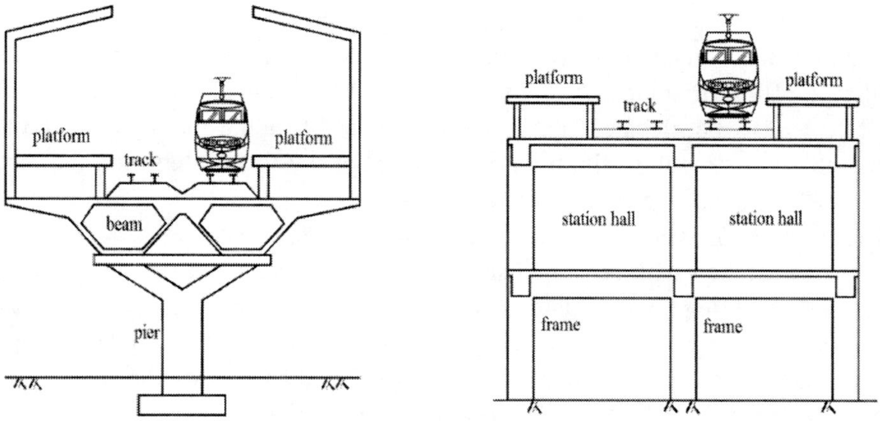

Figure 6. Elevated stations with integrated house-bridge system: bridge system (left), and frame system (right).

The frame system is made up of spatial frame structures and continuous girders supported by the frame, see Figure 6 (right). In this system, the spatial frame structure works as both building structure and bridge piers, by which the girders of the bridge are supported. This system has the advantages of good integrality, high stiffness and uniform mass distribution. Therefore, its earthquake-resistant ability is relatively better. However, its foundation is easy to subside unevenly under irregular loads, easy to induce structural vibrations by the running trains, and it is difficult to design because two types of codes are required.

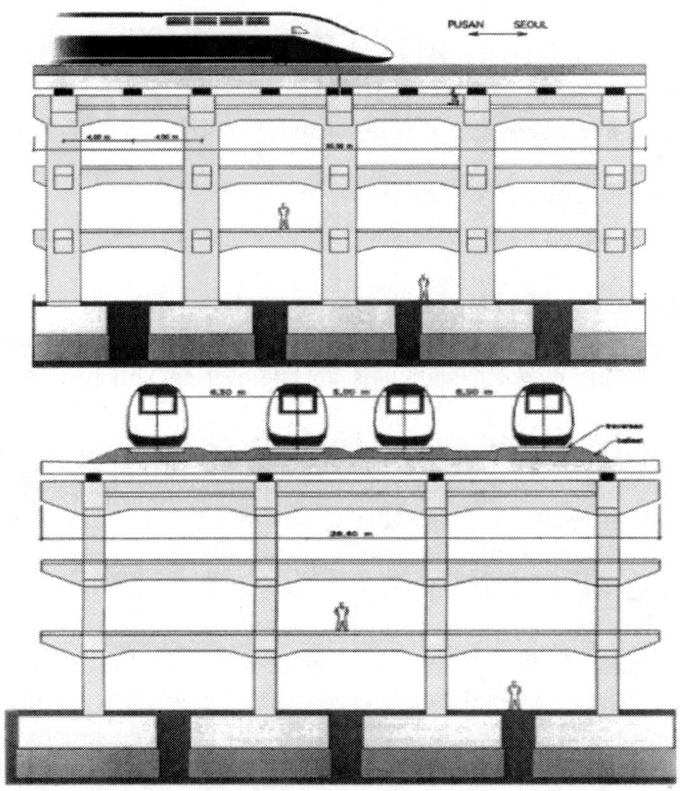

Figure 7. The frame structure system of Cheonan Station.

The Cheonan Station on the Seoul-Busan high-speed railway in Korea is a typical elevated station with frame structure system, as shown in Figure 7. It is located in a 10 km wide valley at 100 km southeast to Seoul. Due to the complex terrain of the valley, the railway line had to be elevated up. The elevated line and the main station building are reinforced concrete structures, the train passes through the fourth floor and the three floors below are used for the office and commerce.

(2) Elevated station with separated building-and-bridge system

The elevated station with separated station-and-bridge system is composed of two parts: the frame and the bridge. The frame is for the station building, and the bridge is for the railway track. The two parts are completely separated in structure, as shown in Figure 8. For this system, the vibration induced by traffic is weaker than that of the integrated station-bridge system since the traffic loads are only supported by the bridge.

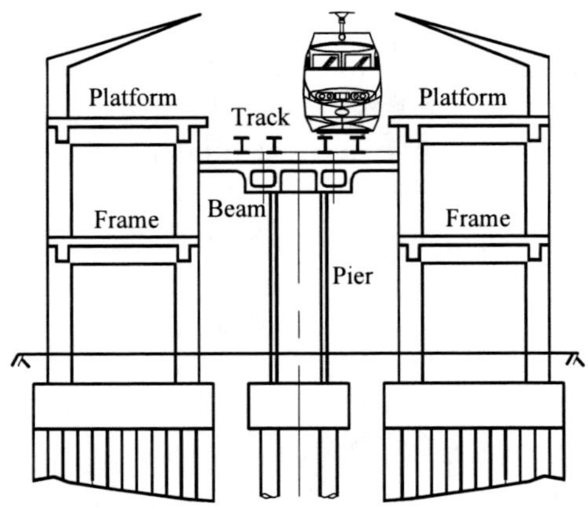

Figure 8. Elevated station with separated building-and-bridge system.

(3) Structural features of elevated stations

The elevated station structure, compared with general buildings and bridges, has the following features:

For the elevated station, the train runs on the upper layer of the structure. The load of train occupies a large proportion of dynamic component and the loading point is continuously changing. Therefore, the elevated station is an engineering structure mainly withstanding dynamic loads.

The elevated station is a special structure integrating station building and bridge together. Owing to the difference of destruction mechanism between the bridge and building, the design methods are different. The allowable stress method is used in bridge design and the limit state method in building design. There is no uniform method in the design of the elevated station structure now, and different methods of calculation are used for different types of components.

The elevated station is the directly receiver of vibration and noise, and becomes the secondary vibration and noise radiation source. At this point, it is similar with the viaduct bridge, but there is a big difference: the bridge belongs to an open structure, while the station is a closed structure, thus the components radiating the secondary structure sound are the floors of the station instead of bridge girders. Therefore, for the elevated station, vibration proof and noise reduction is one of the main tasks in structural design and building material selection, and this is the significance for studying dynamic interaction of the train-elevated station system.

3.2. Vibration Problems of Elevated Stations

Dynamic responses may appear in the vehicles and bridge system when the train passes through the elevated railway station, and large vibration may happen in the station building, which will have an impact on the comfort of passengers and the surrounding environment.

The vibration of elevated stations mainly includes the following problems:

(1) Dynamic effects of running train

When the train runs in the elevated station, the station structure will suffer the dynamic impact from the train and vibrate. This vibration directly affects the working performance and service life of the station [7-12]. (1) The loading effect of train vehicles at certain speed, namely the vertical and lateral forces formed by a series of gravity loads, which is related to the composition of the train, the wheel load arrangement and train speed. (2) The stochastic and periodical factors such as the geometric and dynamic irregularities of the track, the hunting movement of the bogies and wheels, and the faults of rails and wheels, form the self-excitations of the system. (3) The dynamic interactions of vehicles and bridges and the corresponding resonant vibrations, including the bridge resonance induced by the periodical action of moving load series of the weights and centrifugal forces of vehicles, the loading rate of moving loads, and the periodical loading of train vehicles excited by rail irregularities and hunting movements, and the vehicle resonance induced by periodical action of regular arrangement of bridge spans and their deflections.

(2) The impact of passengers

There are several vibration problems to be considered for the passengers on the elevated railway structures [13]. (1) The activities of passengers on the platform may induce structural vibrations of elevated station, especially the large roofs of huge station halls supported by high truss members. (2) Passengers in the waiting hall and on the platform may feel discomfort of the vibrations induced by moving trains or by other machineries equipments. (3) The riding comfort of passengers in the vehicle may be affected by the vibration of vehicles when the train moves through the elevated station.

(3) Impact of train braking and traction

Vibrations may be induced by the braking force and traction force when a train stops or starts on the elevated station structures. For elevated railway stations, especially for those with high steel member supported roofs, the vibration induced by braking and traction may become considerable, since the forces behave as a sudden load with a sharp change in magnitude and may induce a strong impact on the structure [14-19].

(4) Impact of the train traction motors and other equipments

The running of traction motors and locomotives, vehicles and other equipments in station, such as ventilation and air conditioning equipment, will cause the elevated platform or waiting room floor to vibrate, which affect the comfort of passengers, especially those sensitive to vibrations. The frequencies of these machineries are within the range of 20 Hz to several hundred Hertz.

The vibrations induced by the above factors including their combinations should be carefully studied, estimated in the design stage of elevated stations, and necessary measures be taken to prevent and reduce the vibrations.

Dynamic interaction analysis of train and elevated station system is a complicated problem, affected by lots of random factors.

Up to now, there lack systematic research on the vibration problem of elevated stations. The elevated station is neither a simple house building nor a single bridge structure, but a special structure integrating both of them. For this integrated building-bridge system, it needs necessary study on its structural styles, dynamic performances, train-induced responses, vibration and noise control measures, structural fatigues, and detailed structures. With the development of elevated rail transit, the vibration and noise research on the elevated stations excited by running trains has become an important branch in the environmental vibration engineering.

VIRRATION ANALYSIS METHOD FOR TRAIN-ELEVATED STATION SYSTEM

When a train passes through an elevated station at a certain speed, or brake, starts in the station, the station structure will be excited to vibrate [20–24]. At the same time, the structural vibration is in turn an excitation for the train. Therefore, the vibration of the train and the station structure is a process of mutual influence and coupling, which is generally studied by establishing the analysis model of train-station system and via numerical simulation method.

4.1. Train Braking and Starting Force

One of the dynamic analysis characteristics on elevated stations is to consider the deceleration condition when the train arrives in the station and the acceleration condition when the train departs from the station. Therefore, the determination of braking force and starting force is the key issue in dynamic analysis of train-station system.

The braking force is related to the type and the braking mode of the train. For different braking systems, the courses of the braking force are different.

The air braking system is generally used in freight trains in China [25], which takes the pressed air as the motive force and controls the brake shoe pressure by changing the pressure of the brake cylinder.

The China Academy of Railway Sciences measured the car-body deceleration curve and the pier-top displacement curve when the train braked on the Shahe Bridge (train composition DF_{11}+YW_{22}+YW_{25K}+YZ_{25K}+YZ_{25K}) and the Luojiatun Bridge (train composition DF_{11}+ YZ_{25K}+YZ_{25K}), respectively, in the field experiment on the Qinhuangdao-Shenyang high-speed Railway, as shown in Figure 9 [26]. The initial speed of the train braking was 50 km/h.

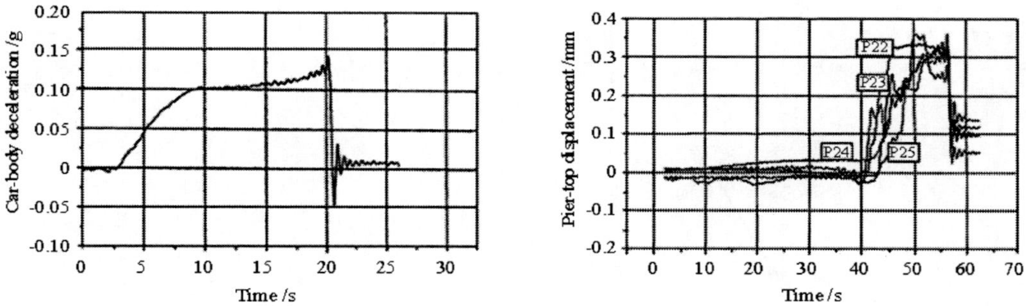

Figure 9. Time history curves of typical car-body deceleration (left) and pier-top displacement (right).

The left figure shows that the car-body deceleration curve consists of three stages of linear-ascending, flat steady and descending.

Stage I: the linear-ascending, during which the braking force increased linearly from zero to the flat segment.

Stage II: the flat steady stage, during which the braking force is relatively steady and gradually increases to the peak until the train stops.

Stage III: the sudden descending stage, during which the braking force decreased to zero.

In the analysis, the train deceleration curve can be determined by the following simplified methods:

(1) According to the design code for line, bridge, tunnel and station with train speed of 200 km/h, the maximum braking force ratio is taken as 0.164.
(2) According to the measured results shown in Figure 9, the braking force increases linearly in an 8.2 s period after the train brakes.
(3) In addition to the measured results in the Qinhuangdao-Shenyang high-speed Railway, there are many other studies pointed out the above-mentioned stages of car-body deceleration during the train brake. So 0.164 is taken as the braking force ratio at the flat steady stage, namely, the braking force ratio is maintained at 0.164 during the flat steady stage until the train stops.

The car-body deceleration curve is obtained by the above method, as shown in Figure 10. The train braking force ratio increases linearly from 0 to t_1 and maintain constant 0.164 at flat steady stage after t_1 until the train stops at t_2.

The trains braking or starting in elevated stations are mainly high-speed EMU, and their braking mode is different from that of freight trains and common passenger trains. Li et al. (1999) mainly studied the emergency braking condition of freight trains, and found the braking cylinder pressure increased to the maximum after braking and then remained unchanged, and the braking force mainly depended on the friction coefficient of braking shoe which is inversely proportional to the instantaneous speed. This explains why the braking force rate changed sharply at the train speed close to zero, as shown in figure. For the high-speed EMU, however, the braking system is controlled with computer, which can achieve accurate control of the braking cylinder pressure, so the train braking force remains stable in the braking course.

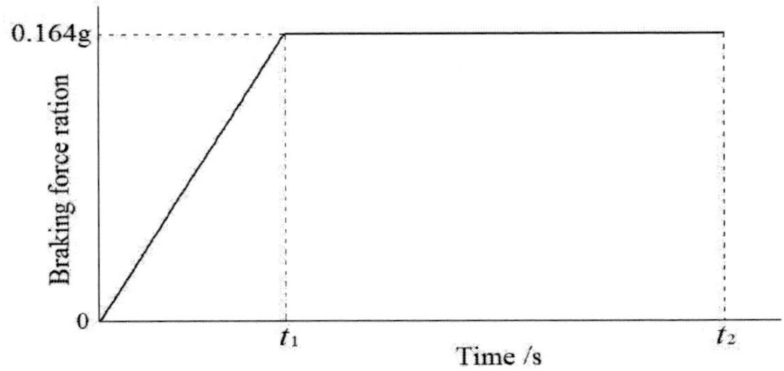

Figure 10. The simplified course of train braking force ratio.

The braking deceleration curve in Figure 10 simplified from the measured results is simple and clear, and easy to calculate, but it is based on the maximum braking force ratio, generally corresponding to the emergency braking condition. Since the braking of the train in the station is usually a normal braking, the emergent braking and the classified conventional braking are considered separately in this analysis, based on which the high-speed EMU braking force calculation method is established.

As the high-speed EMU achieves accurate control on the braking cylinder pressure by constantly revised braking command, the actual braking decelerations in all levels of braking can be in accordance with the target theoretical deceleration.

For example, the CRH-2 EMU in China adopts the computer controlled electrical command-style electro-pneumatic braking mode. The braking force can be adjusted automatically according to the speed-adhesion mode, which includes 1–7 grade common braking modes and a quick braking mode, and the related target deceleration curves are shown in Figure 11.

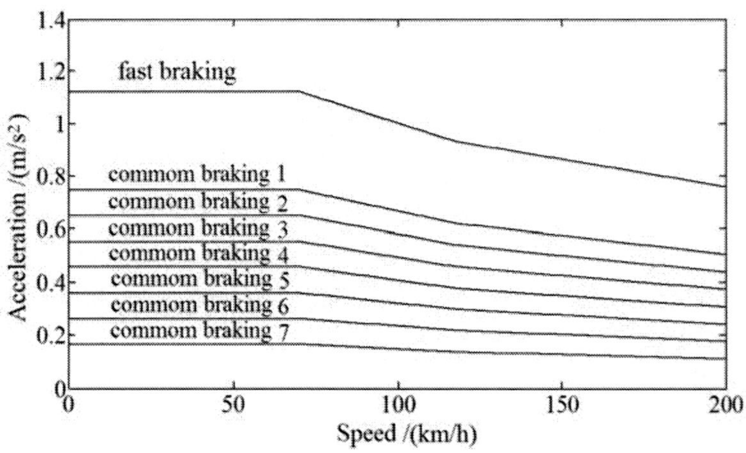

Figure 11. Braking deceleration curves of CRH-2 EMU.

In the entire braking course, the theoretical deceleration keeps rising at the high-speed stage, and maintains the constant theoretical deceleration when the train slows down to a lower speed stage until the train speed becomes zero.

The braking deceleration curves, as shown in Figure 11, are composed of three sections: the decelerations are constant at the speed of 0-70 km/h, and linear variations at 70-118 km/h and 118-200 km/h, respectively.

For braking calculation of high-speed EMU trains, the actual braking courses of acceleration, velocity and displacement can be obtained from these target deceleration curves, and then the longitudinal interaction force between train and station structure can be calculated, which will be described in Section 8.3.2.

For starting force, the China Academy of Railway Sciences proposed the time history curve of the longitudinal force of CRH-2 EMU based on theoretical and experimental study, as shown in Figure 12.

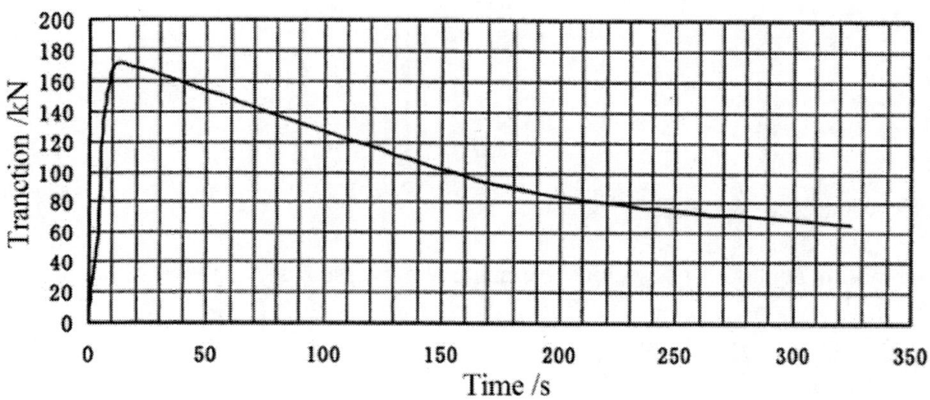

Figure 12. Time history curve of the longitudinal force of CRH-2 EMU.

It can be seen that the starting force increases rapidly with the increase of the traction force, and the maximum force occurs at the beginning of starting course. Some research proposed that the average starting acceleration of CRH-2 EMU is 0.406 m/s^2. For simple calculation, the train is considered to start with uniformly acceleration, and the traction force is calculated according to the average acceleration.

4.2. Analysis Model of Train-Station System

The dynamic analysis model of train-station system consists of the vehicle model, the station structure model and the vehicles-structure coupling relationship model.

4.2.1 Vehicle Model

The train is composed of several independent vehicle elements, where the interaction between any two adjacent vehicle elements is ignored. Each vehicle is a complicated MDOF vibration system consisting of a car-body, bogies, wheel-sets, suspension springs and dashpots. The car-body, bogies and wheel-sets are regarded as rigid elements. Each car-body or bogie has six DOFs in directions of x, y, z, θ, φ and ψ, and each wheel-set has five DOFs in direction of x, y, z, θ and ψ. Therefore, the idealized configuration for each vehicle with 2-bogies and 4-axles in a train can therefore be modeled as a 38-DOFs dynamic system, as shown in Figure 13.

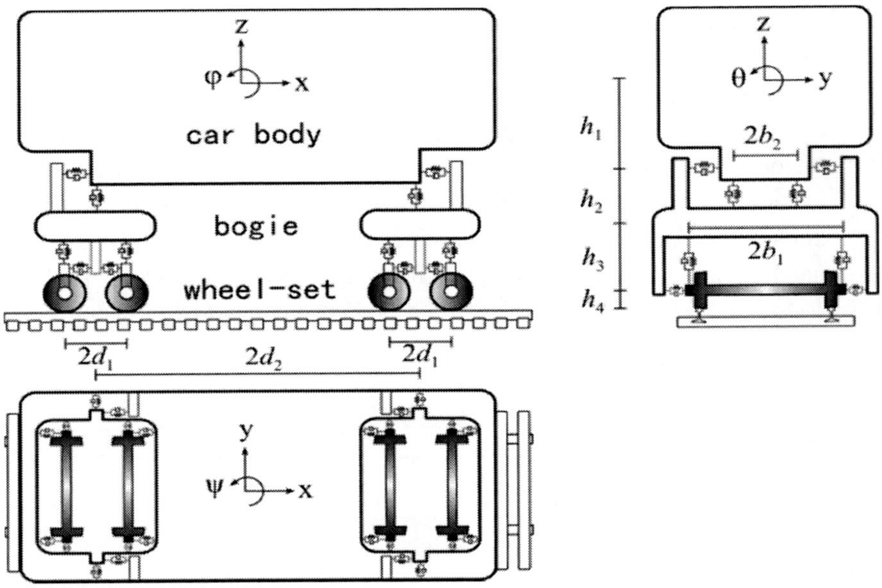

Figure 13. Vehicle element model.

The motion equations of a vehicle element are expressed as:

$$\mathbf{M}_v\ddot{\mathbf{X}}_v + \mathbf{C}_v\dot{\mathbf{X}}_v + \mathbf{K}_v\mathbf{X}_v = \mathbf{F}_v \tag{1}$$

where $\mathbf{M}_v$, $\mathbf{C}_v$, $\mathbf{K}_v$ and $\mathbf{P}_v$ are, respectively, the mass, damping and stiffness matrices and force vector of the vehicle. $\mathbf{X}_v$ is the displacement vector. For a train with n vehicle element, they can be expressed as

$$\begin{aligned}
\mathbf{M}_v &= \text{diag}[\mathbf{M}_{v1} \quad \mathbf{M}_{v2} \quad \cdots \quad \mathbf{M}_{vi} \quad \cdots \quad \mathbf{M}_{vn}] \\
\mathbf{C}_v &= \text{diag}[\mathbf{C}_{v1} \quad \mathbf{C}_{v2} \quad \cdots \quad \mathbf{C}_{vi} \quad \cdots \quad \mathbf{C}_{vn}] \\
\mathbf{K}_v &= \text{diag}[\mathbf{K}_{v1} \quad \mathbf{K}_{v2} \quad \cdots \quad \mathbf{K}_{vi} \quad \cdots \quad \mathbf{K}_{vn}] \\
\mathbf{X}_v &= [\mathbf{X}_{v1} \quad \mathbf{X}_{v2} \quad \cdots \quad \mathbf{X}_{vi} \quad \cdots \quad \mathbf{X}_{vn}]^T \\
\mathbf{F}_v &= [\mathbf{F}_{v1} \quad \mathbf{F}_{v2} \quad \cdots \quad \mathbf{F}_{vi} \quad \cdots \quad \mathbf{F}_{vn}]^T
\end{aligned} \tag{2}$$

For the ith vehicle element with 4-axles, the sub-displacement matrix $\mathbf{X}_{vi}$, sub-mass matrix $\mathbf{M}_{vi}$ and sub-stiffness matrix $\mathbf{K}_{Vi}$ can be respectively expressed as:

$$\mathbf{X}_{vi} = \left[\mathbf{X}_c \ \mathbf{X}_{b1} \ \mathbf{X}_{b2} \ \mathbf{X}_{w1} \ \mathbf{X}_{w2} \ \mathbf{X}_{w3} \ \mathbf{X}_{w4}\right]^T \tag{3}$$

$$\mathbf{M}_{vi} = \text{diag}\left[\mathbf{M}_c \ \mathbf{M}_{b1} \ \mathbf{M}_{b2} \ \mathbf{M}_{w1} \ \mathbf{M}_{w2} \ \mathbf{M}_{w3} \ \mathbf{M}_{w4}\right] \tag{4}$$

$$
\mathbf{K}_{vi} = \begin{bmatrix}
\mathbf{K}_{c} & & & & & & \\
\mathbf{K}_{b1,c} & \mathbf{K}_{b1} & & & \text{sym} & & \\
\mathbf{K}_{b2,c} & \mathbf{0} & \mathbf{K}_{b2} & & & & \\
\mathbf{0} & \mathbf{K}_{w1,b1} & \mathbf{0} & \mathbf{K}_{w1} & & & \\
\mathbf{0} & \mathbf{K}_{w2,b1} & \mathbf{0} & \mathbf{0} & \mathbf{K}_{w2} & & \\
\mathbf{0} & \mathbf{0} & \mathbf{K}_{W3,b2} & \mathbf{0} & \mathbf{0} & \mathbf{K}_{w3} & \\
\mathbf{0} & \mathbf{0} & \mathbf{K}_{W4,b2} & \mathbf{0} & \mathbf{0} & \mathbf{0} & \mathbf{K}_{w4}
\end{bmatrix} \tag{5}
$$

where $\mathbf{X}_{c} = \left[x_{c}\ y_{c}\ z_{c}\ \theta_{c}\ \varphi_{c}\ \psi_{c}\right]^{T}$

$$\mathbf{X}_{bj} = \left[x_{bj}\ y_{bj}\ z_{bj}\ \theta_{bj}\ \varphi_{bj}\ \psi_{bj}\right]^{T},\quad (j=1,2)$$

$$\mathbf{X}_{wl} = \left[x_{wl}\ y_{wl}\ z_{wl}\ \theta_{w}\ \psi_{wl}\right]^{T},\quad (l=1,2,3,4)$$

$$\mathbf{M}_{c} = \mathrm{diag}\left[m_{c}\ m_{c}\ m_{c}\ I_{xc}\ I_{yc}\ I_{zc}\right]$$

$$\mathbf{M}_{bj} = \mathrm{diag}\left[m_{bj}\ m_{bj}\ m_{bj}\ I_{xbj}\ I_{ybj}\ I_{zbj}\right],\quad (j=1,2)$$

$$\mathbf{M}_{wl} = \mathrm{diag}\left[m_{wl}\ m_{wl}\ m_{wl}\ I_{xwl}\ I_{zwl}\right],\quad (l=1,2,3,4)$$

$$
\mathbf{K}_{c} = 4\begin{bmatrix}
k_{x2} & & & & & \\
0 & k_{y2} & & & \text{sym} & \\
0 & 0 & k_{z2} & & & \\
0 & h_1 k_{y2} & 0 & h_1^2 k_{y2} + b_2^2 k_{y2} & & \\
-h_1 k_{y2} & 0 & 0 & 0 & h_1^2 k_{y2} + d_2^2 k_{y2} & \\
0 & 0 & 0 & 0 & 0 & k_{x2} b_2^2 + k_{y2} d_2^2
\end{bmatrix}
$$

$$
\mathbf{K}_{bj,c} = 2\begin{bmatrix}
-k_{x2} & 0 & 0 & 0 & h_1 k_{x2} & 0 \\
0 & -k_{y2} & 0 & -h_1 k_{y2} & 0 & (-1)^j d_2 k_{y2} \\
0 & 0 & -k_{z2} & 0 & d_2 k_{z2} & 0 \\
0 & h_2 k_{y2} & 0 & h_1 h_2 k_{y2} - b_2^2 k_{z2} & 0 & -(-1)^j d_2 h_2 k_{y2} \\
-h_2 k_{x2} & 0 & 0 & 0 & h_1 h_2 k_{x2} & 0 \\
0 & 0 & 0 & 0 & 0 & -b_2^2 k_{x2}
\end{bmatrix},\quad (j=1,2)
$$

$$
\mathbf{K}_{bj} = 2\begin{bmatrix}
2k_{x1} + k_{y2} & & & & & \\
0 & 2k_{y1} + k_{y2} & & & \text{sym} & \\
0 & 0 & 2k_{z1} + k_{z2} & & & \\
0 & 2h_3 k_{y1} - h_2 k_{y2} & 0 & K_{44} & & \\
h_2 k_{x2} - 2h_3 k_{x1} & 0 & 0 & 0 & K_{55} & \\
0 & 0 & 0 & 0 & 0 & K_{66}
\end{bmatrix},\quad (j=1,2)
$$

$$K_{44} = h_2^2 k_{y2} + 2h_3^2 k_{y1} + b_2^2 k_{z2} + 2b_1^2 k_{z1}$$

$$K_{55} = h_2^2 k_{x2} + 2h_3^2 k_{x1} + 2d_1^2 k_{z1}$$

$$K_{66} = 2d_1^2 k_{y1} + b_2^2 k_{x2} + 2b_1^2 k_{x1}$$

$$\mathbf{K}_{wl,bj} = 2\begin{bmatrix} -k_{x1} & 0 & 0 & 0 & h_3 k_{x1} & 0 \\ 0 & -k_{y1} & 0 & -h_3 k_{y1} & 0 & (-1)^{l-2j} d_1 k_{y1} \\ 0 & 0 & -k_{z1} & 0 & -(-1)^{l-2j} d_1 k_{z1} & 0 \\ 0 & 0 & 0 & -b_1^2 k_{z1} & 0 & 0 \\ 0 & 0 & 0 & 0 & 0 & 0 \\ 0 & 0 & 0 & 0 & 0 & -b_1^2 k_{x1} \end{bmatrix}, \quad \begin{pmatrix} j=1, l=1,2 \\ j=2, l=3,4 \end{pmatrix}$$

$$\mathbf{K}_{wj} = \begin{bmatrix} 2k_{x1} & & & & \\ 0 & 2k_{y1} & & \text{sym} & \\ 0 & 0 & 2k_{z1} & & \\ 0 & 0 & 0 & 2b_1^2 k_{z1} & \\ 0 & 0 & 0 & 0 & 2b_1^2 k_{z1} \end{bmatrix}, \quad (l = 1,2,3,4)$$

where the subscript c, b and w stand for the car-body, bogie and wheel-set; j=1 or 2 stands for the front or rear bogie; when j=1, l=1 or 2, the wheel-set is linked to the front bogie; when j=2, l=3 or 4, the wheel-set is linked to the near bogie. Respectively, k_{x1}, k_{y1}, k_{z1} are the spring coefficients of the first suspension system in three directions, and k_{x2}, k_{y2}, k_{z2} are those of the second suspension system. The other dimensions are illustrated in Figure 12.

The damping matrix can be obtained by simply replacing "k" in the corresponding stiffness matrix by "c".

4.2.2 Elevated Station Model

The following assumptions are adopted for the elevated station model:

(1) The structure subsystem is a linear system, which means the global mass, damping and stiffness matrices of the structure subsystem are constants.
(2) The structure node spacing is sufficiently small, and the structural displacement, velocity and acceleration at non-node locations can be calculated by linear interpolation; any non-node force of the structure subsystem is assigned to the two adjacent nodes according to the inverse distance relationship.
(3) The forces acting on the structure subsystem from the vehicle subsystem in y, z and θ directions are considered, and in x direction, only the starting and braking forces of the vehicle on the bridge are considered.
(4) The track irregularities remain unchanged during the passage of trains.
(5) The intervals of track irregularity sampling points are sufficiently small, and the track irregularities between the sampling points can be calculated by linear interpolation.

The equations of motion for the station structure subsystem can be expressed as:

$$\mathbf{M}_s\ddot{\mathbf{X}}_s + \mathbf{C}_s\dot{\mathbf{X}}_s + \mathbf{K}_s\mathbf{X}_s = \mathbf{F}_s \tag{6}$$

where $\mathbf{M}_s$, $\mathbf{C}_s$, $\mathbf{K}_s$ and $\mathbf{F}_s$ are, respectively, the global mass, damping and stiffness matrices and the force vector of station subsystem. The matrices can be obtained by the finite element method or the superposition method.

4.2.3. Train-station Interaction Model

The train-station interaction model is a key issue in dynamic response analysis of train-station system. The longitudinal, lateral and vertical displacement relationships and interacted forces should be defined in the model.

The train-station interacted force is applied to the left and right wheel-rail contact points, as shown in Figure 14.

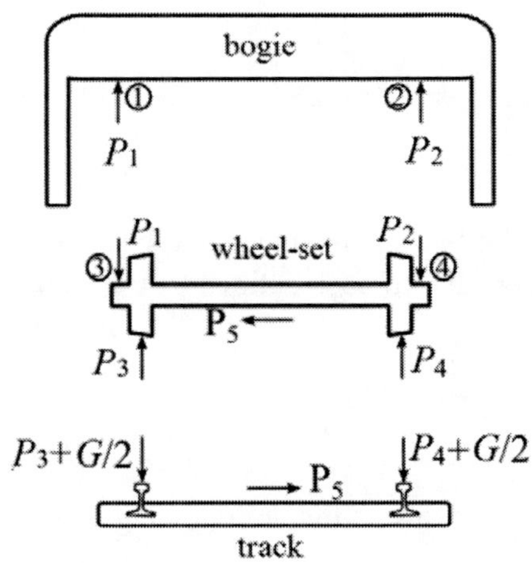

Figure 14. The wheel-rail interacted force.

The vertical wheel-rail force is determined by the wheel-rail close contact assumption: the wheel-set motion is the sum of the station structure motion and additional motion caused by irregularity, so the vertical wheel-rail force is sum of the first suspension force, wheel-set inertia force and the static wheel weight. The lateral wheel-rail force is determined by the Kalker creep theory, on which some simplifications are made to ensure the linear relationship between the vehicle and the station subsystems.

For an individual wheel-set, P_1 and P_2 are the vertical interacted force between the bogie and the wheel-set; P_3 and P_4 are the vertical wheel-rail interacted force; P_5 is the lateral wheel-rail interacted force; G is the static axle load.

The vertical displacements of the upper suspension points 1, 2 and the lower suspension points 3 and 4 are:

$$\begin{cases} z_1 = z_b + \eta d_1 \varphi_b - b_1 \theta_b \\ z_2 = z_b + \eta d_1 \varphi_b + b_1 \theta_b \\ z_3 = z_I - b_1 \theta_I \\ z_4 = z_I + b_1 \theta_I \end{cases} \tag{7}$$

where z_b, θ_b and φ_b are the bogie displacements in z、θ and φ directions, respectively; z_I and θ_I are the vertical and torsional track irregularities, respectively; η=is the sign function, which equals −1 for the front wheel-set and 1 for the rear one. Then the first suspension forces P_1, P_2 and the vertical wheel-rail forces P_3, P_4 can be expressed as:

$$\begin{cases} P_1 = k_{z1}(z_1 - z_3) + c_{z1}(\dot{z}_1 - \dot{z}_3) \\ P_2 = k_{z1}(z_2 - z_4) + c_{z1}(z_2 - z_4) \\ P_3 = \left(\dfrac{1}{2} + \dfrac{b_1}{g_0}\right)P_1 + \left(\dfrac{1}{2} - \dfrac{b_1}{g_0}\right)P_2 + \dfrac{G}{2} + \dfrac{m_w \ddot{z}_I}{2} - \dfrac{I_{xw}\ddot{\theta}_I}{g_0} \\ P_4 = \left(\dfrac{1}{2} - \dfrac{b_1}{g_0}\right)P_1 + \left(\dfrac{1}{2} + \dfrac{b_1}{g_0}\right)P_2 + \dfrac{G}{2} + \dfrac{m_w \ddot{z}_I}{2} + \dfrac{I_{xw}\ddot{\theta}_I}{g_0} \end{cases} \tag{8}$$

where g_0 is the rail gauge; m_w and I_{xw} are the mass and the x-inertia of the wheel-set.

The lateral wheel-rail interacted force is defined by the Kalker creep theory:

$$P_5 = f_{22}\frac{\dot{y}_w - \dot{y}_{IL}}{V} + f_{22}\frac{\dot{y}_w - \dot{y}_{IR}}{V} = 2f_{22}\frac{\dot{y}_w - \dot{y}_I}{V} \tag{9}$$

The creep coefficient f_{22} is a constant by the following assumptions:

(a) The wheel-rail normal contact force is taken as the static wheel load, $G/2$.

(b) The TB-wheel and the 60 kg/m rail are considered, which are commonly adopted in Chinese railway system. The contact point is at a 1:20 cone surface for the TB-wheel and at a cylindrical surface with 300 mm radius for the 60 kg/m rail.

Thus, the lateral wheel-rail force P_5 is in proportion to the wheel-rail relative velocity, and the vehicle velocity terms can be moved from the right-hand side to the left-hand side in Equation (1) as:

$$\mathbf{M}_v\ddot{\mathbf{X}}_v + (\mathbf{C}_v + \mathbf{C}_c)\dot{\mathbf{X}}_v + \mathbf{K}_v\mathbf{X}_v = \mathbf{F}_v \tag{10}$$

where $\mathbf{C}_c$ is the additional damping matrix due to the wheel-rail creep force.

In the longitudinal dynamic analysis of the elevated station, the target control acceleration, velocity and displacement of the wheel-set longitudinal motion are given in the train braking or starting courses, and the longitudinal displacement and deformation of wheel-rail contact condition meet the following assumption:

$$\begin{cases} x_o = x_w + x_s \\ \dot{x}_o = \dot{x}_w + \dot{x}_s \\ \ddot{x}_o = \ddot{x}_w + \ddot{x}_s \end{cases} \tag{11}$$

where x_o and x_w are, respectively, the target control and actual longitudinal displacement of the wheel-set, and x_s the longitudinal displacement of the station structure, at the wheel-rail contacting position. The longitudinal wheel-rail force is also applied to the left and right wheel-rail contact point.

For the vehicle equations, the degrees-of-freedom in directions of x, z and θ are not independent, so the nonzero terms in $\mathbf{F}_v$ can be expressed as:

$$\begin{cases} F_{vb,jl,x} = 2k_{x1}\sum_l x_{wl} + 2c_{x1}\sum_l \dot{x}_{wl} \\ F_{vb,jl,z} = 2k_{z1}\sum_l z_{Il} + 2c_{z1}\sum_l \dot{z}_{Il} \\ F_{vb,jl,\theta} = 2b_1^2(k_{z1}\sum_l \theta_{Il} + c_{z1}\sum_l \dot{\theta}_{Il}) \\ F_{vb,jl,\varphi} = -2h_3(k_{x1}\sum_l x_{wl} + c_{x1}\sum_l \dot{x}_{wl}) + 2d_1(-1)^l(k_{z1}\sum_l z_{Il} + c_{z1}\sum_l \dot{z}_{Il}) \\ F_{vw,jl,y} = 2f_{22}\dot{y}_{Il}/V \end{cases} \tag{12}$$

where the subscript wl stands for the longitudinal motion of the lth wheel-set, and the subscript Il stands for the track irregularities at the position of the lth wheel-set.

The five equations in Equation (12) reflect the x, z, θ and φ forces acting on bogies and the y force acting on wheel-sets. The force vector $\mathbf{F}_b$ in Equation (6) can also be obtained by summing the effect of all wheel-sets in all vehicle elements, and the contributions of the mth wheel-set linked to the jth bogie in x, z, θ and φ directions are expressed as:

$$\begin{cases} F_{s,jl,x} = -m_{wl}\ddot{x}_{wl} + 2k_{x1}(x_{bj} - x_{wl} - h_3\varphi_{bj}) + 2c_{x1}(\dot{x}_{bj} - \dot{x}_{wl} - h_3\dot{\varphi}_{bj}) \\ F_{s,jl,y} = 2f_{22}\left(\dot{y}_{wl} - \dot{y}_{Il}\right)/V \\ F_{s,jl,z} = 2k_{z1}\left[z_{bj} + (-1)^l\varphi_{bj} - z_{Il}\right] + 2c_{z1}[\dot{z}_{bj} + (-1)^l\dot{\varphi}_{bj} - \dot{z}_{Il}] - G - m_w\ddot{z}_{Il} \\ F_{s,jl,\theta} = 2k_{z1}b_1^2\left(\theta_{bj} - \theta_{Il}\right) + 2c_{z1}b_1^2\left(\dot{\theta}_{bj} - \dot{\theta}_{Il}\right) - I_{xw}\ddot{\theta}_{Il} \end{cases}$$

4.2.4. Interaction Equations and Inter-system Iteration

The dynamic equilibrium equations for the train-station interacted system can be expressed by the simultaneous equations of the train subsystem and the station structure subsystem. The interaction equations are:

$$\begin{cases} \mathbf{M}_{v1}\ddot{\mathbf{X}}_{v1} + (\mathbf{C}_{v1} + \mathbf{C}_{c1})\dot{\mathbf{X}}_{v1} + \mathbf{K}_{v1}\mathbf{X}_{v1} = \mathbf{F}_{v1} \\ \mathbf{M}_{v2}\ddot{\mathbf{X}}_{v2} + (\mathbf{C}_{v2} + \mathbf{C}_{c2})\dot{\mathbf{X}}_{v2} + \mathbf{K}_{v2}\mathbf{X}_{v2} = \mathbf{F}_{v2} \\ \cdots\cdots\cdots\cdots\cdots\cdots\cdots\cdots\cdots \\ \mathbf{M}_{vn}\ddot{\mathbf{X}}_{vn} + (\mathbf{C}_{vn} + \mathbf{C}_{cn})\dot{\mathbf{X}}_{vn} + \mathbf{K}_{vn}\mathbf{X}_{vn} = \mathbf{F}_{vn} \\ \mathbf{M}_{s}\ddot{\mathbf{X}}_{s} + \mathbf{C}_{s}\dot{\mathbf{X}}_{s} + \mathbf{K}_{s}\mathbf{X}_{s} = \mathbf{F}_{s} \end{cases} \tag{13}$$

where n is the number of vehicle elements of the train. The first n equations in Equation (13) are for the vehicle subsystem, the last equation is for the station structure subsystem.

The force vector $\mathbf{F}_{vi}$ are the functions of the exciter y_{Il}, z_{Il}, θ_{Il} and x_{wl} in Equation (12), which is the sum of the track irregularity and the station structure motion. The force vector $\mathbf{F}_s$ is the function of the exciter and the vehicle motion. Thus, the equations are coupled and can be solved by an iteration procedure, as shown in Figure 15.

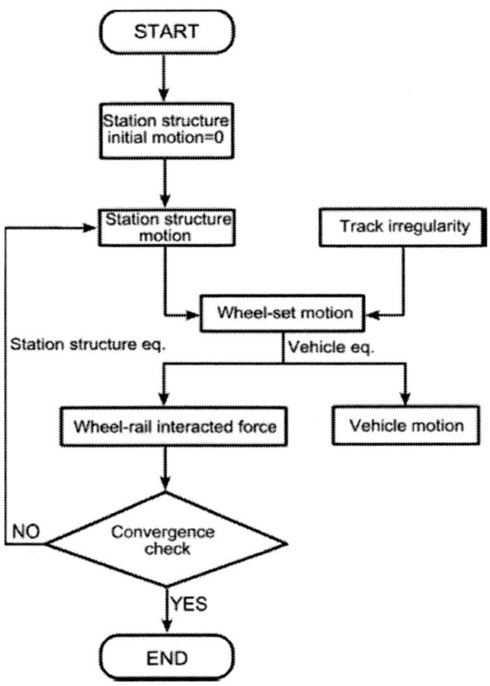

Figure 15. Iteration procedure for solving train-station interaction system.

5. Case Study of Train-Station Interaction System

The vibration influence of running train on buildings is a complex problem. The vibration energy is transmitted from the moving trains to the bridge through the track, then to the pier and foundation through the bridge bearing, and finally to the nearby buildings through the soils. Such vibration transmissions are a complex process, any part of which can play an important role. Since the vibration of elevated stations under moving trains is related to the structural system and the function deployment of the station building, its transmission is more complicated than ordinary buildings.

In the research on elevated stations, the authors conducted dynamic interaction analyses of train-station systems, such as the Sihui Station and Zhichunlu Station in Beijing, the Guangzhou South Railway Station, the Wuhan Railway Station, the Shenzhen North Railway Station and the Nanjing South Railway Station. The vibration responses of the stations under moving trains, and the running safety and stability indices such as derailment factors, offload factors, lateral wheel/rail forces, car-body accelerations and riding comfort indices have been obtained, which provided a basis for the design of these stations.

In this section, by taking the Wuhan Railway Station with integrated house-bridge system as an illustrative case study, the analytical method and the vibration characteristics of elevated stations are presented.

Table 1. Calculation cases for the Wuhan Railway Station

Case number		Axial line	The first line	The second line
Trains running on a single bound	1	Axis K	SS8-G, 120 km/h	N/A
	2	Axis E	CRH-2, 200 km/h	N/A
	3	Axis H	ICE-3, 80 km/h	N/A
	4	Axis L	SS8-G, 45km/h	N/A
	5	Axis B	CRH-2, 80 km/h	N/A
	6	Axis F	CRH-2, 80 km/h	N/A
Trains running on double lines (with the same speed in opposite bounds)	7	Axis K	SS8-G, 120 km/h	SS8-G, 120 km/h
	8	Axis E	CRH-2, 200 km/h	CRH-2, 200 km/h
	9	Axis H	ICE-3, 80 km/h	ICE-3, 80 km/h
	10	Axis L	SS8-G, 45 km/h	SS8-G, 45 km/h
	11	Axis B	CRH-2, 80 km/h	CRH-2, 80 km/h
	12	Axis F	CRH-2, 80 km/h	CRH-2, 80 km/h
Trains running on double bounds (with the same speed and arrived time in opposite bounds)	13	Axis K	SS8-G, 120 km/h	SS8-G, 60 km/h
	14	Axis E	CRH-2, 200km/h	CRH-2, 100km/h
	15	Axis H	ICE-3, 80km/h	ICE-3, 40km/h
	16	Axis L	SS8-G, 45km/h	SS8-G, 22.5 km/h
	17	Axis B	CRH-2, 80km/h	CRH-2, 40km/h
	18	Axis F	CRH-2, 80km/h	CRH-2, 40km/h
Train braking on a single bound	19	Axis L	SS8-G, 45→0km/h	N/A
	20	Axis B	CRH-2, 80→0km/h	N/A
	21	Axis F	CRH-2, 80→0km/h	N/A
	22	Axis H	ICE-3, 80→0km/h	N/A
Trains braking on double lines (with the same arrived time in opposite bounds)	23	Axis L	SS8-G, 45→0 km/h	SS8-G, 45→0km/h
	24	Axis B	CRH-2, 80→0 km/h	CRH-2, 80→0 km/h
	25	Axis F	CRH-2, 80→0 km/h	CRH-2, 80→0 km/h
	26	Axis H	ICE-3, 80→0 km/h	ICE-3, 80→0 km/h

The Wuhan Railway Station is one of the most important passenger stations in the Wuhan railway hub, the originating station of the Wuhan-Guangzhou high-speed railway. It is the control project of the Beijing-Guangzhou high-speed railway in the high-speed railway network of China, and was put into operation at the end of 2009. The whole project consists of station building canopies without platform pillars, elevated pedestrian platforms, ground pedestrian squares and parking lots. It is 550 m long from north to south and 343.5 m wide from east to west. with the total construction area 114602 m^2.

The station consists of two elevated layers (18.8m and 25m), a platform layer (10.250m), a ground layer (±0.000 m) and an underground layer. The metro station locates on the underground layer, which is the terminal for urban rail transit. The railway platform is a bridge structure and the upper layer is a passenger station building, which is a large-span spatial steel structure with its roof 58 m above the ground. There are two railway yards in the station, one for high-speed railways and the other for conventional railways. There are 11 platforms and 20 tracks, including 4 main lines and 16 arrival and departure lines. The main lines allow the train to pass through at 200 km/h or higher speed.

The new architectural technique of Wuhan Railway Station is reflected by its structural style, namely the huge-span structure and the integrated house-bridge system. To meet the requirement on service function and architectural form, both the span and the story height are designed very large. The main structure consists of five main arches and forty semi-arches, with the largest arch span at the central hall 116 m and the smallest 80 m. As an integrated house-bridge system, the interaction between the superstructure and the bridge was fully considered to ensure the design consistency of the two types of structures.

The vibration state of the Wuhan Railway Station under running trains and the interaction forces between the train and the structure are analyzed herein, including 26 calculation cases, as shown in Table 1, in which the K to F axes are shown in Fig. 16.

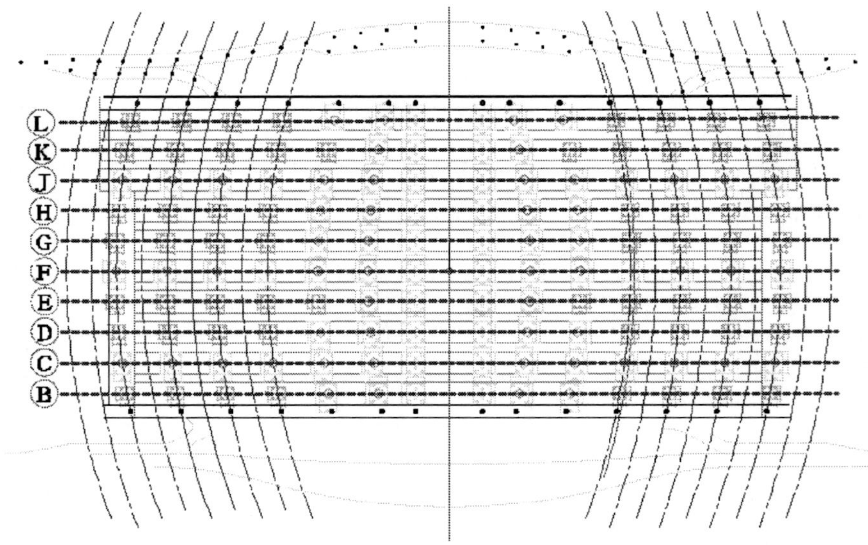

Figure 16. The plane of the station structure.

In the Table, SS8-G represents the conventional train composed of 1 SS8 locomotive hauling 12 YZ25G passenger cars, CRH-2 represents the high-speed EMU train made in China, and ICE-3 the high-speed EMU train made in German. The basic parameters of the trains are listed in Table 2.

Table 2. Basic parameters of the trains

Train	Car type	Total length /m	Distance between bogies /m	Wheel-base /m	Axle load /kN
SS8-G	Locomotive	17.516	9.000	2.900	228.00
	Car	26.570	18.500	2.400	152.68
CRH-2	Motor-car	25.000	17.500	2.500	135.00
	Trailer-car	25.000	17.500	2.500	120.00
ICE-3	Motor-car	24.775	17.375	2.500	160.00
	Trailer-car	24.775	17.375	2.500	146.00

The finite element model of the station structure is shown in Figure 17. In the model, the track beams from the B axis to L axis are reformed partly, the piers and girders are modeled with actual section, and the secondary dead-load 260 kN/m of double track is added into the self-weight of the girders. The supports are connected with the nodes at the pier-tops using master-and-slave nodes. For simple-supported beams, the nodes at the supports or the master-and-slave nodes at the station center side are restrained in x, y, z and θ directions, while at the other side are restrained in x, z and θ directions. For continuous beams, the nodes at the second support are restrained in x, y, z and θ directions, while at the other supports are restrained in x, z and θ directions. All the piers are fixed at bottoms.

The girders and piers are simulated with spatial beam elements. There are 11 nodes for each simple-supported beam and 35 nodes for the continuous beam, and in total 145 nodes ($10\times11+35=145$) for the whole model.

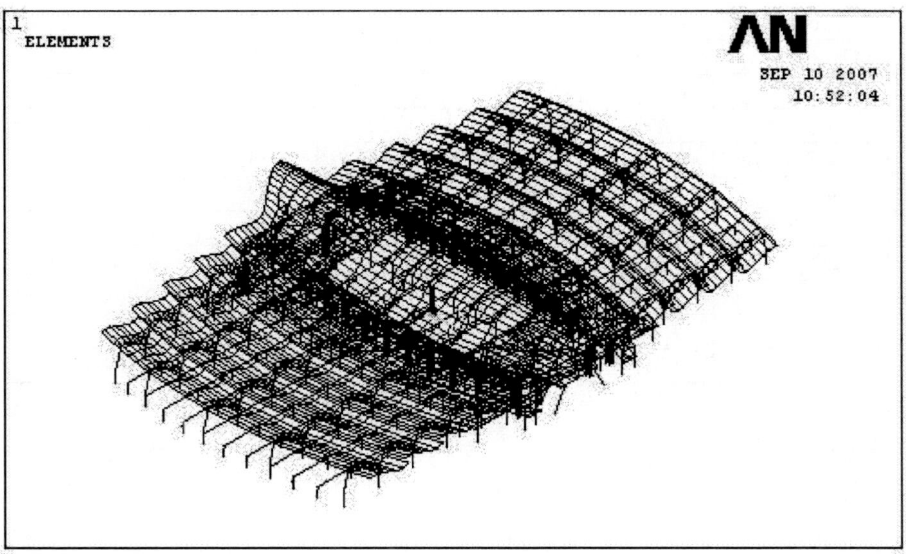

Figure 17. The finite element model of the station structure.

Listed in Table 3 are the calculated control indices of vehicle responses. The results shows that under 26 different cases, all the indices for vehicle running safety, stability and riding comfort meet the relevant criteria, in which the riding comfort indices of the vehicle are evaluated as good for Cases 8 and 14, and as excellent for other cases.

Table 3. Control indices of vehicle responses

Case number	Vehicle type	Derailment factor	Offload factor	Car body acc. /(m/s^2)		Sperling index	
				Vertical	Lateral	Vertical	Lateral
1	Locomotive	0.325	0.373	0.266	0.252	2.232	2.288
	Car	0.307	0.265	0.199	0.160	2.195	2.094
2	Car	0.512	0.198	0.186	0.565	2.176	2.587
3	Car	0.237	0.232	0.196	0.236	2.188	2.268
4	Locomotive	0.314	0.347	0.259	0.150	2.277	2.055
	Car	0.288	0.252	0.185	0.083	2.166	1.858
5	Car	0.402	0.170	0.125	0.304	2.052	2.323
6	Car	0.404	0.167	0.125	0.304	2.058	2.333
7	Locomotive	0.335	0.392	0.283	0.323	2.221	2.378
	Car	0.319	0.278	0.206	0.169	2.189	2.074
8	Car	0.552	0.254	0.210	0.661	2.220	2.714
9	Car	0.258	0.234	0.194	0.214	2.147	2.166
10	Locomotive	0.334	0.450	0.411	0.191	2.530	2.175
	Car	0.310	0.202	0.177	0.106	2.133	2.021
11	Car	0.474	0.230	0.186	0.432	2.172	2.418
12	Car	0.470	0.229	0.186	0.432	2.169	2.404
13	Locomotive	0.372	0.454	0.411	0.262	2.545	2.300
	Car	0.320	0.299	0.244	0.174	2.210	2.240
14	Car	0.563	0.250	0.237	0.667	2.228	2.705
15	Car	0.278	0.251	0.205	0.235	2.195	2.255
16	Locomotive	0.368	0.408	0.357	0.173	2.446	2.210
	Car	0.273	0.213	0.185	0.097	2.157	1.949
17	Car	0.410	0.182	0.132	0.309	2.044	2.334
18	Car	0.402	0.183	0.132	0.309	2.050	2.337
19	Locomotive	0.307	0.339	0.243	0.143	2.270	2.048
	Car	0.287	0.250	0.185	0.083	2.162	1.857
20	Car	0.400	0.173	0.130	0.304	2.071	2.319
21	Car	0.401	0.172	0.130	0.304	2.077	2.317
22	Car	0.241	0.262	0.246	0.236	2.204	2.267
23	Locomotive	0.325	0.388	0.343	0.173	2.341	2.070
	Car	0.310	0.252	0.185	0.102	2.164	2.029
24	Car	0.404	0.174	0.135	0.304	2.076	2.332
25	Car	0.403	0.174	0.135	0.304	2.071	2.330
26	Car	0.240	0.263	0.246	0.236	2.203	2.259

The maximum responses at several key places of the station (see Figure 18) are calculated under Case 25, when two CRH-2 trains start to brake at 80 km/h until stop on the F axis track beam in opposite directions, and the first wheels of the two trains reach the midpoint of the track beam at the same time. The location sketch of key points is also shown in Figure 18. Shown in Figures 19 to 21 are, respectively, the time history curves of accelerations at the floor center of the elevated waiting hall, the top of the central reticulated shell, and the overhanging point of the canopy.

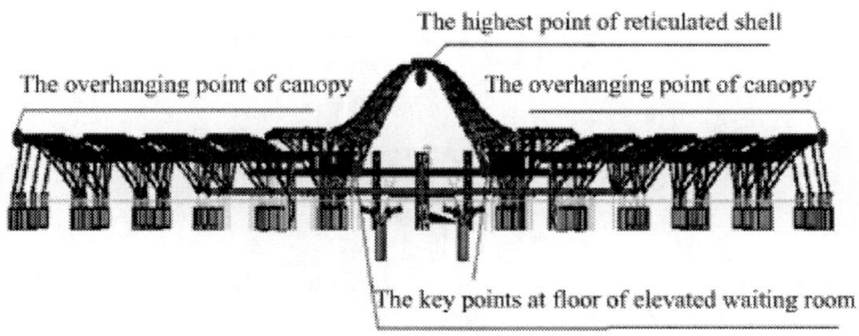

Figure 18. The location sketch of the key points.

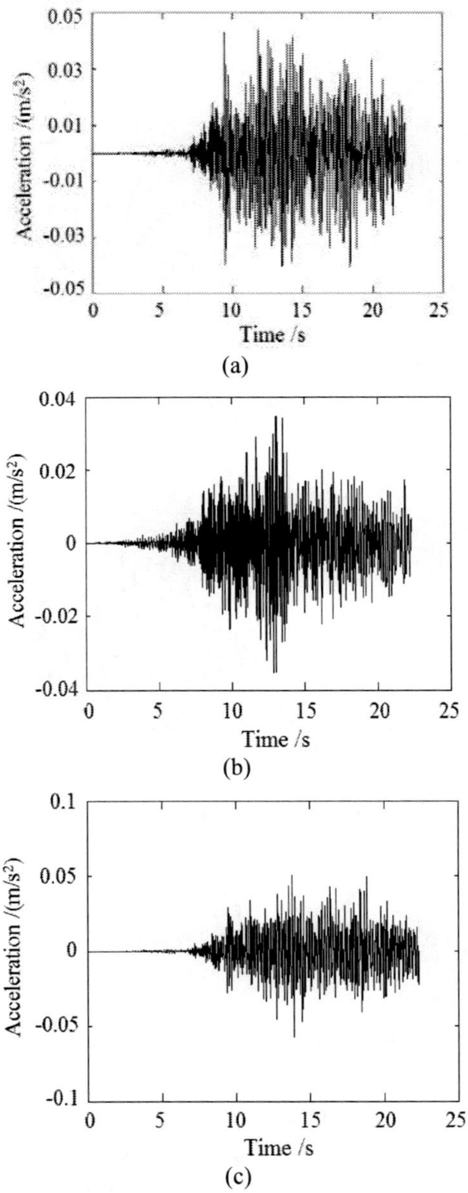

Figure 19. Time history curves of accelerations at floor center of elevated waiting room: (a) longitudinal, (b) horizontal and (c) vertical.

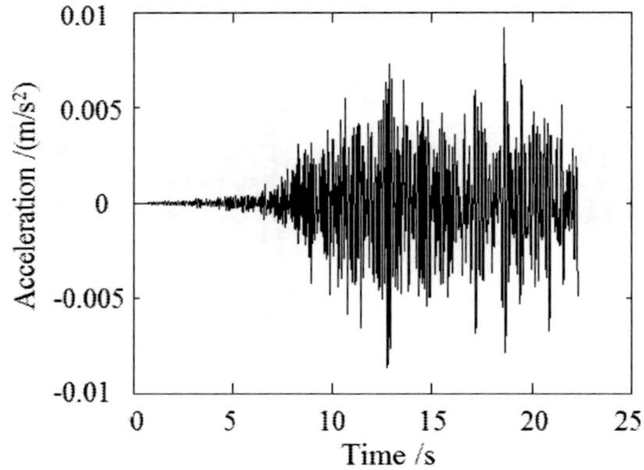

(a)

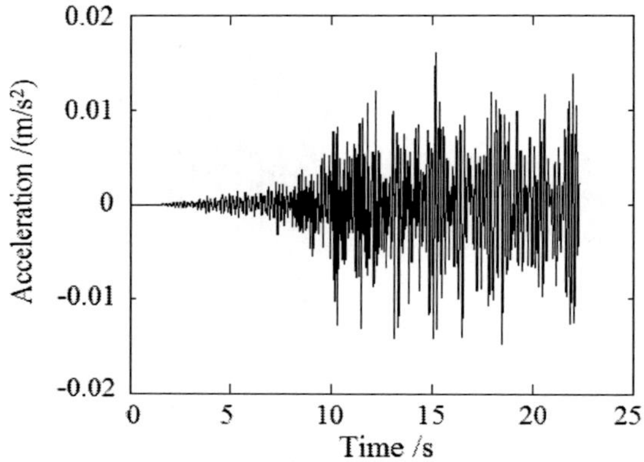

(b)

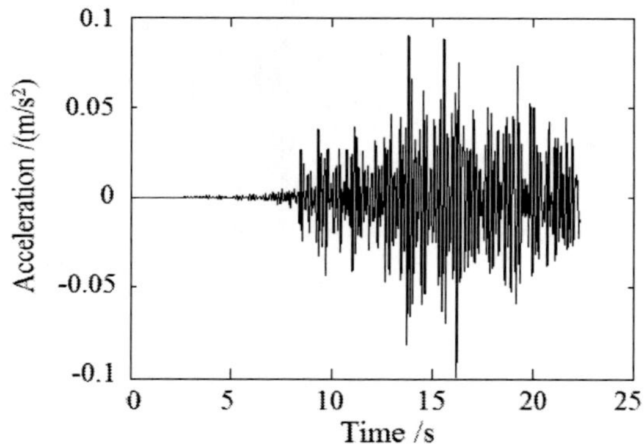

(c)

Figure 20. Time history curves of accelerations at the top of the reticulated shell: (a) longitudinal, (b) horizontal and (c) vertical.

Table 4. Maximum accelerations at several key places of the station

Position	Acceleration /(mm/s^2)		
	Longitudinal	Horizontal	Vertical
Floor of the elevated waiting hall	44	35	61
Top of central reticulated shell	9	12	100
Overhanging point of the canopy	86	59	220

The maximum accelerations in longitudinal, horizontal and vertical directions at the floor center of the elevated waiting hall, the top of central reticulated shell, and the overhanging point of the canopy are summarized in Table 4.

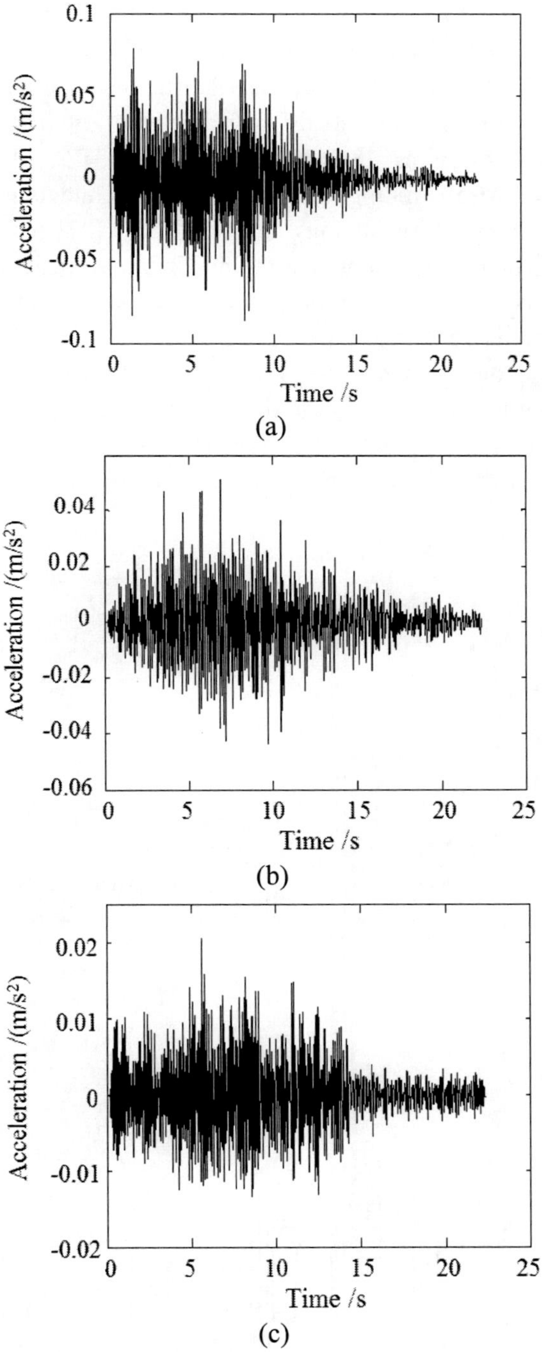

Figure 21. Time history curves of accelerations at the overhanging point of station canopy: (a) longitudinal, (b) horizontal and (c) vertical.

When the external excitation frequency or striding frequency of people on the floor slab is close to the natural frequency of the floor slab, a resonance of the floor slab will be induced, which may affect the comfort and produce irritable and unsafe feeling of the people. Large resonant response also affects safety and service life of the structure. In evaluating the vibration of waiting halls, the resultant value of horizontal and vertical accelerations is adopted as an allowable index. The comfort of human being inside a building under vibration is a complicated issue. Different persons will have different sensitivities and endurances to vibration caused by different dynamic loads under different environments. Thus, there are several comfort criteria, such as ISO6897-1984, ISO2631-2, BS6611, the Technical Guidelines for Steel High-rise Buildings in China, and the Guideline of Japan Architecture Association. The ATC (Applied Technology Council, USA) [27] recommends in its Design Guidelines: Minimizing Floor Vibrations that the maximum vertical acceleration of building floor should be used as a control parameter, and the limit values for operation rooms, residences/offices, shopping centers and pedestrian bridges are 0.0025g, 0.005g, 0.015g and 0.05g, respectively within the frequency range from 4 to 8 Hz. There is no direct reference for the vibration standard for the waiting room in the railway station. Since the waiting room and platforms of station are noisy public cases, the limit may be adopted between those of shopping centers and pedestrian bridges, namely between 0.015 and 0.05g, by referencing the ATC's suggestion.

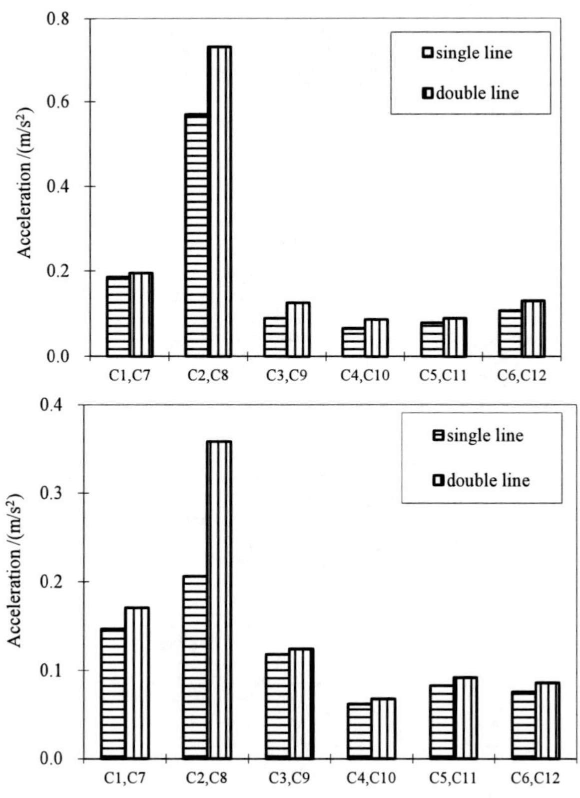

Figure 22. Maximum mid-span acceleration of simply-supported beams under single line and double line running cases: vertical (uper) and lateral (lower).

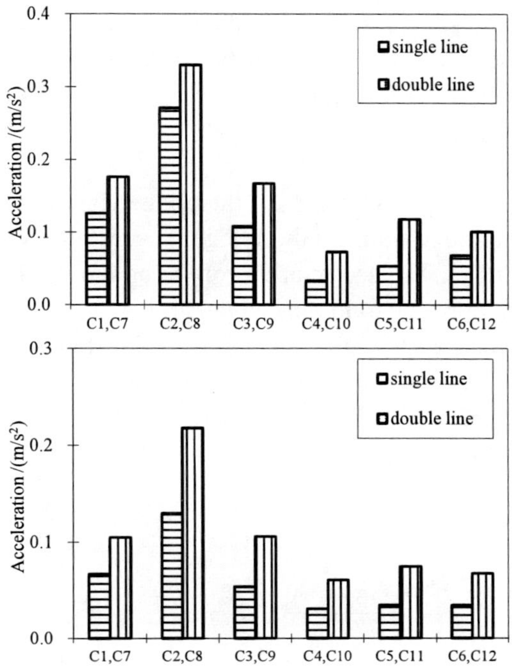

Figure 23. Maximum mid-span acceleration of continuous beams under single line and double line running cases: vertical (upper) and lateral (lower).

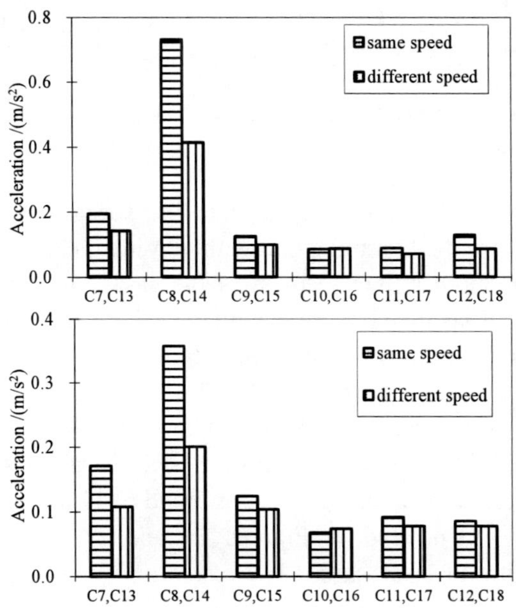

Figure 24. Maximum mid-span acceleration of simply-supported beams when two trains run at the same speed and different speeds: vertical (upper) and lateral (lower).

From the calculated results, the resultant value of longitudinal, horizontal and vertical peak accelerations of the floor slab of elevated waiting hall is 0.0085 g, which meets the limit. The resultant acceleration at the overhanging canopy is 0.025 g, exceeding the limit value of

0.015g, but it still meets the peak limit for outside pedestrian bridges. Since the overhanging canopy is away from passengers, larger limit can be considered.

For the simply-supported and continuous track beams, the peak accelerations under different cases are calculated.

(1) Comparison of single line and double line cases

As can be seen from the Figures 22 and 23, the bridge vibration is larger when two trains run on the bridge than only one train on the bridge, because the vibration energy is higher when two trains on the bridge. The accelerations of the bridges under two trains are 1.048 to 2.226 times of those when only one train runs on the bridge.

(2) Comparison of two trains with the same speed and different speeds

When the two trains run on the bridge at the same speed for Cases 7 to 12, and run at different speeds for Cases 13 to 18, the acceleration responses are calculated, and the results are shown in Figures 24 and 25.

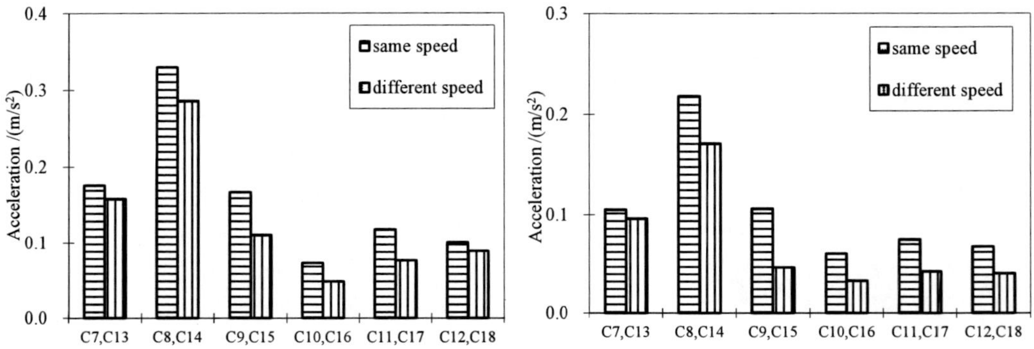

Figure 25. Maximum mid-span acceleration of continuous beams when two trains run at the same speed and different speeds: vertical (left) and lateral (right).

For the cases where the two trains run with identical speed, both the two train speeds are *V*, while for the cases where the two trains run with non-identical speeds, the train speed is *V* on one line, and *V*/2 on another line. In most of the cases, the trains with identical speed induce larger bridge vibrations than the trains with non-identical speeds. The accelerations of the bridges under two trains with identical speed are 0.919~2.255 times of those when two trains run non-identical speeds.

(3) Comparison of train running on different track beams

The vibrations of the bridge are calculated when CRH-2 train passes through the B and F axes at the identical speed, corresponding to Cases 5 and 6, 11 and 12, 17 and 18, 20 and 21, 24 and 25, respectively. Compared in Figures 26 and 27 are the maximum mid-span accelerations of the bridge in these cases. The results show no significant change of track beam vibrations when the train runs on the B axis and the F axis.

(4) Comparison of trains entering the station on single line and on double lines

The vibrations of the bridge are calculated when the trains enter the station on single line for Cases 19 to 22, and on double lines for the Cases 23 to 26. Compared in Figures 28 and 29 are the maximum mid-span accelerations of the bridge in these cases.

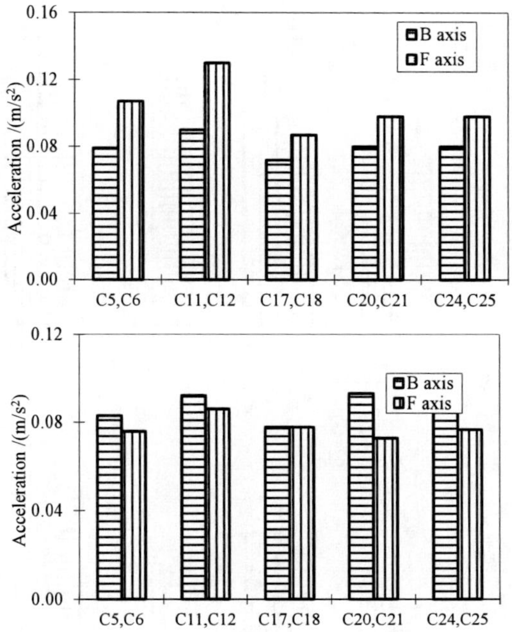

Figure 26. Maximum mid-span acceleration of simply-supported beams when two trains run at different track beams: vertical (upper) and lateral (lower).

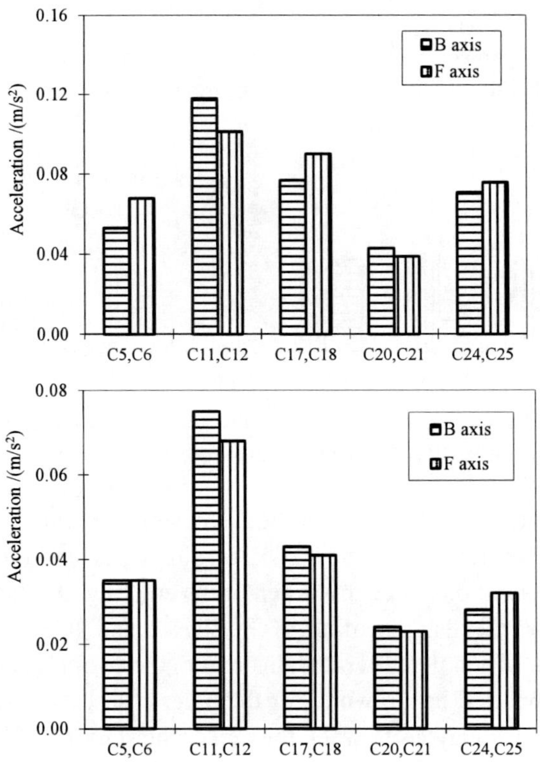

Figure 27. Maximum mid-span acceleration of continuous beams when two trains run at different track beams: vertical (upper) and lateral (lower).

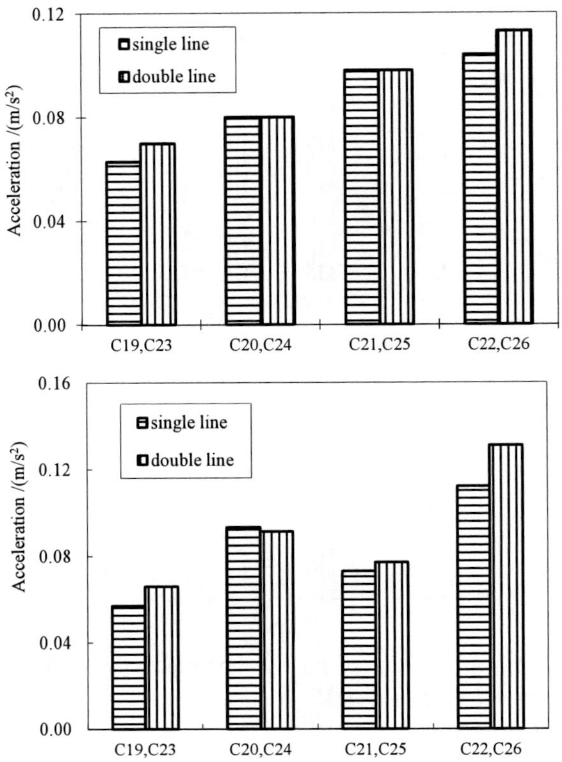

Figure 28. Maximum mid-span acceleration of simply-supported beams when train entering the station in single line and in double lines: vertical (upper) and lateral (lower).

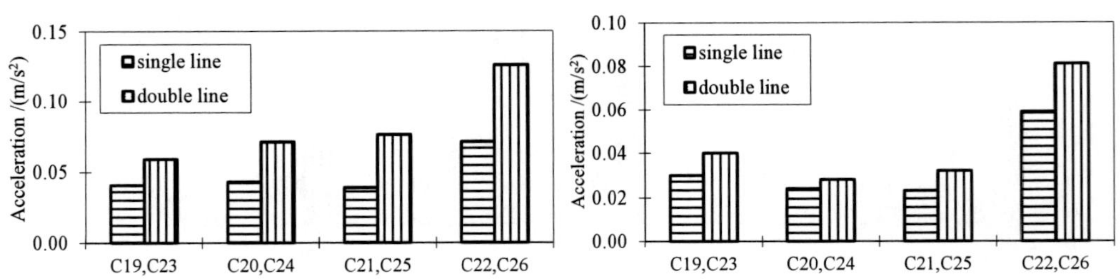

Figure 29. Maximum mid-span acceleration of continuous beams when train entering the station in single line and in double lines: vertical (left) and lateral (right).

The results show the mid-span accelerations of simply-supported beam are not much different when the CRH-2 trains enter the station on single line and on double lines, while for the other cases, the accelerations of beams when trains entering the station on double lines are greater. This is because the lengths of SS8+G, CRH-2 and ICE-3 train are, respectively, 331m, 195m and 392m. When the CRH-2 trains brake, the two trains do not cross each other on the same simply-supported beam, while for the other cases; the responses of the beam are greater, since the two trains may cross when they enter the station on double lines.

(5) Comparison of trains braking and running through the bridges

The vibrations of the bridge are calculated when trains brake in the station for Cases 19 to 26, and trains run through the station for Cases 4−6, 3, 10−12 and 9, respectively. Compared in Figures 30 to 33 are the maximum mid-span accelerations of the bridge in these cases.

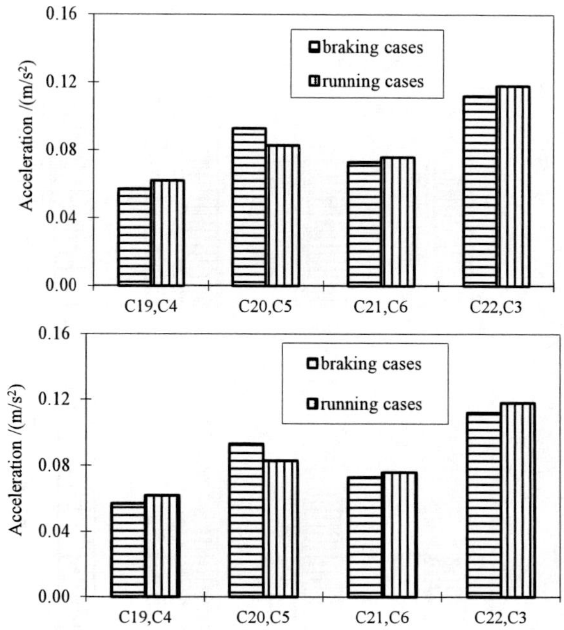

Figure 30. Maximum mid-span acceleration of simply-supported beams when train brakes and runs in single line: vertical (upper) and lateral (lower).

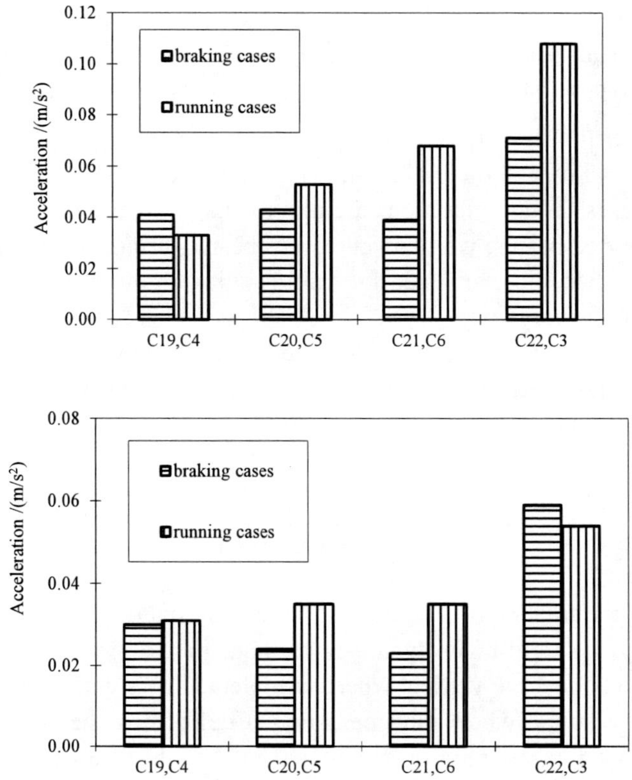

Figure 31. Maximum mid-span acceleration of continuous beams when train brakes and runs in single line: vertical (upper) and lateral (lower).

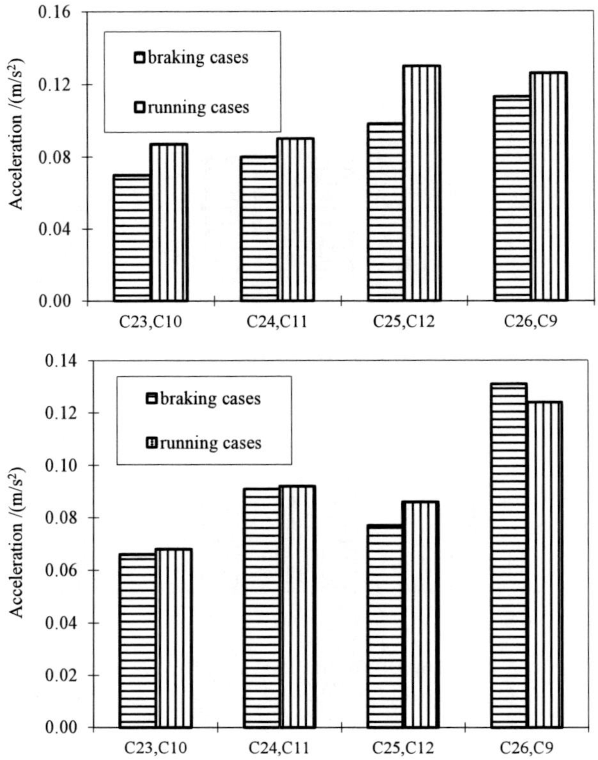

Figure 32. Maximum mid-span acceleration of simply-supported beams when train brakes and runs in double lines: vertical (upper) and lateral (lower).

When trains enter the station, the braking distances of conventional trains and high-speed trains are 95 m and 240 m, and the initial braking acceleration is small.

In single line cases, bridge vibrations are approximately the same when trains brake and run on simply supported beams, while on continues beams, bridge vibrations induced by train braking are smaller than those induced by train running through, because the train speeds have been reduced to be very low.

In double line cases, bridge vibrations induced by train running through are greater than those by train braking, because the speeds for running trains are much faster than the corresponding braking trains.

CONCLUSION

In this chapter, the train-station interaction model is established and solved, in which the train is modeled by the rigid-body dynamics method. In the model, the station building is modeled by the FE method, the vertical wheel-rail interaction is defined by the corresponding assumptions, and the lateral wheel-rail interaction is defined by the simplified Kalker creep theory. The corresponding assumption is adopted in the wheel-rail longitudinal interaction, which should be updated by more detail and accurate calculations, considering the relative displacement and the nonlinear contact forces between the wheel and the rail. By the updated wheel-rail relation, the 3D interaction can be simulated and the effect of the braking to the running safety of the train can be studied.

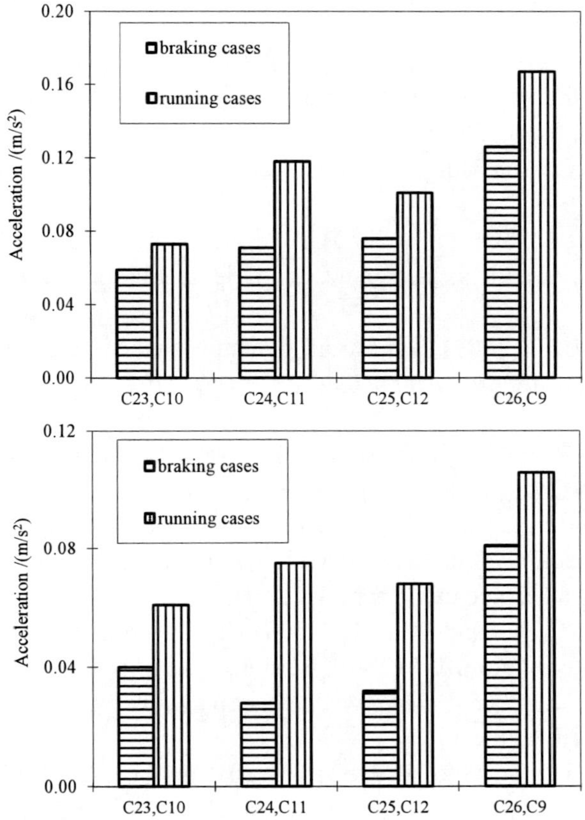

Figure 33. Maximum mid-span acceleration of continuous beams when train brakes and runs in double lines: vertical (upper) and lateral (lower).

The train-station coupling dynamic equilibrium equations are solved by the inter-system iteration procedure. As a case study, the dynamic response of the Wuhan Railway Station is analyzed when the trains running through, braking and starting on the platforms. The results show that the maximum accelerations at the waiting room floors are between the criteria for the markets and the outside pedestrian bridges in the ACT code.

ACKNOWLEDGMENTS

This study is sponsored by the Major State Basic Research Development Program of China ("973" Program: 2013CB036203), the 111 project (Grant No B13002), the National Natural Science Foundation of China (Grant No. 51078029) and the New Century Excellent Talents Plan.

REFERENCES

[1] Zheng J., et al. (2008). *High-speed railway bridges in China*, Beijing: Higher Education Press.

[2] Xia H., Zhang N. Guo W.W., Cao Y.M. (2007). Running train induced vibrations and noises of elevated railway structures and their influences on the environment. Proc. ISEV'2007, Taipei, p. 27–36.

[3] Meinhard von Gerkan, Jürgen Hillmer (2009). Berlin Hauptbahnhof. *Architectural Journal*, 4: 46–51.

[4] Yuan F., Verena M. Schindler (2000). A resplendent revival of civilization-review on santiago calatrava and his design work "Oriente Station". *Time + Architecture*, 4: 52–55.

[5] Gao R., Jiang L.C. (2001). Types and dynamic analysis of elevated railway station structures. *International symposium on traffic induced vibration and controls*, Beijing, p. 295–302.

[6] Gao R., Xia H. (2005). Traffic-induced vibrations and noises in elevated railway structures. *ISEV 2005*, Okayama, p. 569–573.

[7] Frýba L. (1999). *Dynamics of Railway Bridges*. Thomas Telford: London.

[8] Xia H., De Roeck G. and Jose M. Goicolea (2012). *Bridge vibration and control.* New York: Nova science publishers, Inc.

[9] Cheng Q., Zhang N. and Xia H. (2009). Train-induced Structural Vibrations of Elevated Railway Station. *ISEV2009*, Beijing, p. 178–183.

[10] Xia H., Zhang N. (2005). *Dynamic Interaction of Vehicles and Structures*. Science Press: Beijing. (in Chinese).

[11] Zhang N., Xia H. and Guo W.W., et al. (2010). A vehicle-bridge linear interaction model and its validation. *International Journal of Structural Stability and Dynamics*, 10(2): 335–361.

[12] Zhai W.M. (2007). Vehicle-track Coupling Dynamics. Science Press: Beijing. (in Chinese)

[13] Xia H., et al. (2010). *Traffic Induced Environmental Vibrations and Controls*. Science Press: Beijing. (in Chinese).

[14] Cheng H.T. (1999). *Research on dynamics of long and heavy haul train during braking and flexible multibody of freight car system.* Beijing: China Academy of Railway Sciences.

[15] Li H.N. (2001). *A study on the braking load of trains and its action mechanism on bridges* [D]. Beijing: Northern Jiaotong University.

[16] Lei J.Q., Li H.N. and Feng D. (2006). Experimental study and computational analysis on braking force between running train and railway bridge. *Engineering Mechanics*, 23(3): 134–140.

[17] Yin C.X. (2005). *Study on nonlinear static and dynamic analysis and emulation of additional longitudinal forces in railway bridges*. Beijing: China Academy of Railway Sciences.

[18] Wu M.L., Cheng G.H. and Wang X.Y. (2009). Discussion of braking deceleration control of railway vehicles. *Journal of the China Railway Society*, 31(1): 94–97.

[19] Zhang N., Xia H., Cheng Q., et al. (2011). Analysis method for a vehicle structure

coupled system under braking force. *Journal of Vibration & Shock*, 30(2): 138−143.
[20] Flýba L. (1974). Quasi-static distribution of braking and starting forces in rails and bridge. *Rail International*, 5 (11): 698−716.
[21] Gupta R.K., Traill-Nash R.W. (1980). Bridge dynamic loading due to road surface irregularities and braking of vehicle. *Earthquake Engineering and Structural Dynamics*, 8 (1): 83–96.
[22] Krylov V.V. (1996). Generation of ground vibrations by accelerating and braking road vehicles. *Acoustics*, 82 (4): 642–649.
[23] Law S.S., Zhu X.Q. (2005). Bridge dynamic responses due to road surface roughness and braking of vehicle. *Journal of Sound and Vibration*, 282(3–5):805–830.
[24] Ju S.H., Lin H.T. (2007). A finite element model of vehicle–bridge interaction considering braking and acceleration. *Journal of Sound & Vibration*, 303: 46−57.
[25] Rao Z. (2011). *Train Braking*. China Railway Publishing House: Beijing. (in Chinese).
[26] Ye Y.S., et al. (2004). *Experimental research on bridges, embankments and tracks on Qinhuangdao-Shenyang high-speed railway.* Technical Report, Beijing, China Academy of Railway Science.
[27] Applied Technology Council. (1996). *Minimizing floor vibration.* ATC Report, USA.

INDEX

#

A

B

C

D

E

J

K

L

M

N

O

P

Q

R

S

T